EXPERIMENTAL COSMOLOGY AT MILLIMETRE WAVELENGTHS

Related Titles from the AIP Conference Proceedings Subseries on Astronomy and Astrophysics

609   Astrophysical Polarized Backgrounds: Workshop on Astrophysical Polarized Backgrounds
Edited by Stefano Cecchini, Stefano Cortiglioni, Robert Sault, and Carla Sbarra, March 2002, 0-7354-0055-5

599   X-Ray Astronomy: Stellar Endpoints, AGN, and the Diffuse X-Ray Background
Edited by Nicholas E. White, December 2001, 0-7354-0043-1

598   Solar and Galactic Composition: A Joint SOHO/ACE Workshop
Edited by Robert F. Wimmer-Schweingruber, December 2001,CD-ROM included, 0-7354-0042-3

587   Gamma 2001: Gamma-Ray Astrophysics 2001
Edited by Steven Ritz, Neil Gehrels, and Chris R. Shrader, October 2001, 0-7354-0027-X; CD-ROM: 0-7354-0030-X

586   Relativistic Astrophysics: 20th Texas Symposium
Edited by J. Craig Wheeler and Hugo Martel, October 2001, 0-7354-0026-1

575   Astrophysical Sources for Ground-Based Gravitational Wave Detectors
Edited by Joan M. Centrella, July 2001, 0-7354-0014-8

566   Observing Ultrahigh Energy Cosmic Rays from Space and Earth: International Workshop
Edited by Humberto Salazar, Luis Villaseñor, and Arnulfo Zepeda, May 2001, 0-7354-0002-4

558   High Energy Gamma-Ray Astronomy: International Symposium
Edited by Felix A. Aharonian and Heinz J. Völk, April 2001, 1-56396-990-4

476   3 K Cosmology: EC-TMR Conference
Edited by Luciano Maiani, Francesco Melchiorri, and Nicola Vittorio, May 1999, 1-56396-847-9

470   After the Dark Ages: When Galaxies Were Young (The Universe at 2<z<5)
Edited by Stephen S. Holt and Eric P. Smith, April 1999, 1-56396-855-X

To learn more about these titles, or the AIP Conference Proceedings Series, please visit the webpage **http://proceedings.aip.org**

EXPERIMENTAL COSMOLOGY AT MILLIMETRE WAVELENGTHS

2K1BC Workshop

Breuil-Cervinia (AO), Valle d'Aosta, Italy   9-13 July 2001

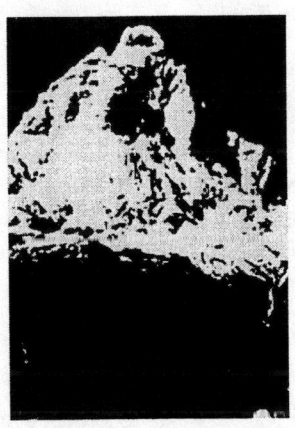

EDITORS
Marco De Petris
University of Rome "La Sapienza", Italy
Massimo Gervasi
University of Milan Bicocca, Italy

Melville, New York, 2002
AIP CONFERENCE PROCEEDINGS ■ VOLUME 616

Editors:

Marco De Petris
Dipartimento di Fisica
Università degli Studi di Roma "La Sapienza"
Piazzale A. Moro, 2
00185 Rome
ITALY

E-mail: marco.depetris@roma1.infn.it

Massimo Gervasi
Dipartimento di Fisica "G. Occhialini"
Università degli Studi di Milano Bicocca
Piazza della Scienza, 3
20126 Milan
ITALY

E-mail: massimo.gervasi@mib.infn.it

Authorization to photocopy items for internal or personal use, beyond the free copying permitted under the 1978 U.S. Copyright Law (see statement below), is granted by the American Institute of Physics for users registered with the Copyright Clearance Center (CCC) Transactional Reporting Service, provided that the base fee of $19.00 per copy is paid directly to CCC, 222 Rosewood Drive, Danvers, MA 01923. For those organizations that have been granted a photocopy license by CCC, a separate system of payment has been arranged. The fee code for users of the Transactional Reporting Service is: 0-7354-0062-8/02/$19.00.

© 2002 American Institute of Physics

Individual readers of this volume and nonprofit libraries, acting for them, are permitted to make fair use of the material in it, such as copying an article for use in teaching or research. Permission is granted to quote from this volume in scientific work with the customary acknowledgment of the source. To reprint a figure, table, or other excerpt requires the consent of one of the original authors and notification to AIP. Republication or systematic or multiple reproduction of any material in this volume is permitted only under license from AIP. Address inquiries to Office of Rights and Permissions, Suite 1NO1, 2 Huntington Quadrangle, Melville, NY 11747-4502; phone: 516-576-2268; fax: 516-576-2450; e-mail: rights@aip.org.

L.C. Catalog Card No. 2002103820
ISBN 0-7354-0062-8
ISSN 0094-243X

Printed in the United States of America

CONTENTS

Preface ix
Group Photo xi

BALLOON BORNE EXPERIMENTS

ℓ-Space Spectroscopy of the Cosmic Microwave Background with
the BOOMERanG Experiment 3
    P. de Bernardis, P. A. R. Ade, J. J. Bock, J. R. Bond, J. Borrill, A. Boscaleri, K. Coble, C. R. Contaldi,
    B. P. Crill, G. De Gasperis, G. De Troia, P. Farese, K. Ganga, M. Giacometti, E. Hivon, V. V. Hristov,
    A. Iacoangeli, A. H. Jaffe, W. C. Jones, A. E. Lange, L. Martinis, S. Masi, P. Mason, P. D. Mauskopf,
    A. Melchiorri, P. Natoli, T. Montroy, C. B. Netterfield, E. Pascale, F. Piacentini, D. Pogosyan, G. Polenta,
    F. Pongetti, S. Prunet, G. Romeo, J. E. Ruhl, F. Scaramuzzi, and N. Vittorio

The MAXIMA and MAXIPOL Experiments 12
    P. L. Richards, M. Abroe, P. Ade, A. Balbi, J. Bock, J. Borrill, A. Boscaleri, P. de Bernardis, J. Collins,
    P. G. Ferreira, S. Hanany, V. V. Hristov, A. H. Jaffe, B. Johnson, A. T. Lee, T. Matsumu, P. D. Mauskopf,
    C. B. Netterfield, E. Pascale, B. Rabii, G. F. Smoot, R. Stompor, C. D. Winant, and J. H. P. Wu

Interstellar Dust in the BOOMERanG Maps 18
    S. Masi, P. A. R. Ade, J. J. Bock, A. Boscaleri, B. P. Crill, P. de Bernardis, M. Giacometti, E. Hivon,
    V. V. Hristov, A. E. Lange, P. D. Mauskopf, T. Montroy, C. B. Netterfield, E. Pascale, F. Piacentini,
    S. Prunet, and J. Ruhl

An Update on the TopHat Experiment 23
    G. W. Wilson, J. Aguirre, J. Bezaire, E. S. Cheng, P. R. Christensen, S. Cordone, D. A. Cottingham,
    T. Crawford, D. J. Fixsen, L. Knox, R. Kristensen, S. S. Meyer, H. U. Nogaard-Nelsen, R. F. Silverberg,
    and P. T. Timbie

Archeops: A Balloon Experiment to Measure CMB Anisotropies with a Broad Range of
Angular Sizes 31
    A. Benoit and the Archeops Collaboration

Noise Properties of the BOOMERANG Instrument 39
    F. Piacentini, P. A. R. Ade, J. J. Bock, J. R. Bond, J. Borrill, A. Boscaleri, K. Coble, C. R. Contaldi,
    B. P. Crill, P. de Bernardis, G. De Troia, P. Farese, K. Ganga, M. Giacometti, E. Hivon, V. V. Hristov,
    A. Iacoangeli, A. H. Jaffe, W. C. Jones, A. E. Lange, L. Martinis, S. Masi, P. Mason, P. D. Mauskopf,
    A. Melchiorri, T. Montroy, C. B. Netterfield, E. Pascale, D. Pogosyan, G. Polenta, F. Pongetti, S. Prunet,
    G. Romeo, J. E. Ruhl, and F. Scaramuzzi

Millimeter and Submillimeter Observations from the Atacama Plateau and
High Altitude Balloons 44
    M. Devlin

A Simple and Reliable Star Sensor for Spinning Payloads 52
    F. Nati, P. de Bernardis, S. Masi, D. Yvon, and the Archeops Team

Attitude Control System for Balloon-Borne Experiments 56
    E. Pascale and A. Boscaleri

Three Sun Sensors for Stratospheric Balloon Payloads 59
    G. Romeo, P. de Bernardis, G. Di Stefano, S. Masi, F. Piacentini, F. Pongetti, and S. Rao

GROUND BASED EXPERIMENTS

First Season Observations with the Degree Angular Scale Interferometer (DASI) 65
    E. M. Leitch, J. E. Carlstrom, N. W. Halverson, J. Kovac, C. Pryke, W. L. Holzapfel, and M. Dragovan

The Very Small Array 72
    A. C. Taylor

Spurious Signals in VSA Data 79
    M. E. Jones

Millimeter and Submillimeter Observations from the South Pole 83
    A. A. Stark

MAD-4-MITO, a Multi Array of Detectors for Ground-based mm/submm SZ Observations 92
    L. Lamagna, M. De Petris, F. Melchiorri, E. Battistelli, M. De Grazia, G. Luzzi, A. Orlando,
    and G. Savini

SZ Surveys with the Arcminute MicroKelvin Imager 97
    C. Holler

Cross-talk in Close-packed Interferometer Arrays 102
    R. Subrahmanyan

Results from the First Engineering Run of BOLOCAM and Plans for the Future 107
    P. Mauskopf, P. Ade, J. Bock, S. Edgington, S. Golwala, A. Goldin, J. Glenn, D. Haig, V. Hristov,
    B. Knowles, A. Lange, H. Nguyen, and B. Rownd

The Diabolo Photometer and the Future of Ground-based Millimetric Bolometer Devices 116
    F.-X. Désert, A. Benoît, P. Camus, M. Giard, E. Pointecouteau, N. Aghanim, J.-P. Bernard, N. Coron,
    J.-M. Lamarre, P. Marty, J. Delabrouille, and V. Soglasnova

MASTER: Millimetre and Sub-millimetre Triple hEterodyne Receiver 123
    E. S. Battistelli, G. Boella, F. Cavaliere, M. Gervasi, A. Passerini, G. Sironi, M. Zannoni, D. Andreone,
    L. Brunetti, V. Lacquaniti, S. Maggi, R. Steni, E. Natale, G. Tofani, E. Bava, U. Pisani, J. R. Thorpe,
    and M. De Petris

MBI: Millimetre-Wave Bolometric Interferometer 126
    S. Ali, P. Rossinot, L. Piccirillo, W. K. Gear, P. Mauskopf, P. Ade, V. Haynes, and P. Timbie

Fastscanning: A New Observing Technique for Bolometer Arrays on Ground-Based Telescopes 129
    L. A. Reichertz, B. Weferling, W. Esch, and E. Kreysa

POLARIZATION EXPERIMENTS

Polarization Observations with the Cosmic Background Imager 135
    J. K. Cartwright, A. C. S. Readhead, M. C. Shepherd, S. Padin, T. J. Pearson, and G. B. Taylor

The BaR-SPOrt Experiment: The Science 140
    E. Carretti, G. Bernardi, S. Cecchini, S. Cortiglioni, C. Macculi, E. Morelli, C. Sbarra, G. Ventura,
    J. Monari, S. Poppi, G. Boella, S. Bonometto, M. Gervasi, G. Sironi, M. Tucci, M. Zannoni, M. Baralis,
    O. Peverini, R. Tascone, R. Fabbri, V. Natale, M. Bruscoli, A. Boscaleri, E. Pascale, and L. Nicastro

BaR-SPOrt: A Technical Overview 145
    C. Macculi, G. Bernardi, E. Carretti, S. Cecchini, S. Cortiglioni, E. Morelli, C. Sbarra, G. Ventura,
    J. Monari, S. Poppi, G. Boella, S. Bonometto, M. Gervasi, G. Sironi, M. Tucci, M. Zannoni, M. Baralis,
    O. Peverini, R. Tascone, R. Fabbri, V. Natale, M. Bruscoli, A. Boscaleri, E. Pascale, and L. Nicastro

Millimeter Wave Passive Devices for Measurements of the Polarized Sky Emission 150
    R. Tascone, D. Trinchero, M. Baralis, O. A. Peverini, A. Olivieri, E. Carretti, and S. Cortiglioni

MITO-Pol, a Polarimeter for the Testa Grigia Observatory 157
    R. Maoli, G. Savini, and F. Melchiorri

The Milano Polarimeter: An Instrument to Search for Large Scale Polarization of the Cosmic Microwave Background 164
    M. Zannoni, E. Battistelli, G. Boella, F. Cavaliere, M. Gervasi, A. Passerini, and G. Sironi

Scanning Polarimeters for Measurements of CMB Polarization 168
    S. Masi, P. de Bernardis, G. De Troia, P. Natoli, F. Piacentini, and G. Pisano

POLAR: Instrument and Results 175
    B. G. Keating, A. de Oliveira-Costa, C. W. O'Dell, L. Piccirillo, N. C. Stebor, M. Tegmark,
    and P. T. Timbie

COMPASS: A 2.6m Telescope for CMBR Polarization Studies 183
    L. Piccirillo, G. Dall'Oglio, P. Farese, J. Gundersen, B. Keating, S. Klawikowski, L. Knox, A. Levy,
    P. Lubin, C. O'Dell, P. Timbie, and J. Ruhl

PolKa: A Tunable Polarimeter for mm/submm Wavelengths 187
    G. Siringo, L. A. Reichertz, and E. Kreysa

SATELLITE EXPERIMENTS

Planck Low Frequency Instrument 193
    N. Mandolesi, M. Bersanelli, R. C. Butler, C. Burigana, D. Maino, A. Mennella, G. Morgante,
    L. Valenziano, and F. Villa

ODIN Preliminary Results, HERSHEL and Other Stories 202
    P. Encrenaz

Submillimeter and Millimeter Wave Sky Mapping in the Space Project Submillimetron 205
    V. D. Gromov, N. S. Kardashev, and L. S. Kuzmin

Herschel Space Observatory............210
    V. Natale

The High Frequency Instrument of Planck: Requirements and Design 213
    J. M. Lamarre, B. Maffei, P. A. R. Ade, M. Piat, J. Bock, J. L. Puget, P. de Bernardis, M. Giard,
    A. Lange, A. Murphy, J. P. Torre, A. Benoit, R. Bhatia, F. R. Bouchet, R. Sudiwala, and V. Yourchenko

The 4K Reference Load for the Planck Low Frequency Instrument............219
    L. Valenziano, M. Bersanelli, R. C. Butler, F. Cuttaia, N. Mandolesi, A. Mennella, G. Morigi,
    G. Morgante, M. Sandri, L. Terenzi, and F. Villa

The Planck Telescope............224
    F. Villa, M. Bersanelli, C. Burigana, R. C. Butler, N. Mandolesi, A. Mennella, G. Morgante, M. Sandri,
    L. Terenzi, and L. Valenziano

Analysis of Thermally-induced Effects in Planck Low Frequency Instrument............229
    A. Mennella, M. Bersanelli, C. Burigana, D. Maino, R. Ferretti, G. Morgante, M. Prina, N. Mandolesi,
    C. Butler, L. Valenziano, and F. Villa

Measuring CMB Polarization with ESA PLANCK Submm-Wave Telescope............234
    V. Yurchenko

The Infrared Telescope for Submillimetron Mission............239
    V. I. Bujakas, V. N. Leonov, and V. F. Troitsky

Planck Low Frequency Instrument: Beam Patterns............242
    M. Sandri, M. Bersanelli, C. Burigana, R. C. Butler, M. Malaspina, N. Mandolesi, A. Mennella,
    G. Morgante, L. Terenzi, L. Valenziano, and F. Villa

Sources Variability with Planck LFI............245
    L. Terenzi, M. Bersanelli, C. Burigana, R. C. Butler, G. De Zotti, N. Mandolesi, A. Mennella,
    G. Morgante, M. Sandri, L. Valenziano, and F. Villa

DETECTORS, OPTICS AND CRYOGENICS

Bolometers for Millimeter-wave Cosmology............251
    J. J. Bock

New Technologies for the Detection of Millimeter and Submillimeter Waves............259
    P. L. Richards, J. Clarke, J. M. Gildemeister, T. Lanting, A. T. Lee, M. J. Myers, D. Schwan,
    J. T. Skidmore, H. G. Spieler, and J. Yoon

Bolometer Arrays for mm/submm Astronomy............262
    E. Kreysa, H.-P. Gemünd, A. Raccanelli, L. A. Reichertz, and G. Siringo

A Filled Bolometer Array Camera for Ground-Based Observations............270
    V. Reveret, P. Agnese, P. Andre, E. Doumayrou, R. Gastaud, J. Le Pennec, and L. Rodriguez

Mesh Filters for the mm/submm Atmospheric Windows............273
    V. A. Soglasnova and I. A. Maslov

Partially-Coherent Long-Wavelength Optical Simulation Techniques for Microwave Background Astronomy............274
    S. Withington, C. Y. Tham, and G. Yassin

Corrugated Horn Design for HFI on PLANCK............282
    J. A. Murphy, R. Colgan, E. Gleeson, B. Maffei, C. O'Sullivan, and P. A. R. Ade

Millimetre-Wave Optics Design and Verification............290
    C. O'Sullivan, J. A. Murphy, S. Withington, G. Yassin, E. Atad-Ettedgui, W. Duncan, D. Henry,
    W. Jellema, and H. van de Stadt

Electromagnetic Modelling of Few-Moded Winston Cones in the Far-Infrared............295
    E. Gleeson, J. A. Murphy, S. E. Church, R. Colgan, and C. O'Sullivan

Two Hydrogen Sorption Cryocoolers for the Planck Mission............298
    G. Morgante, D. Barber, P. Bhandari, R. C. Bowman, P. Cowgill, D. Crumb, T. Loc, A. Nash, D. Pearson,
    M. Prina, A. Sirbi, M. Schemlzel, R. Sugimura, and L. A. Wade

Excess Noise in Cryogenic Detectors due to Vibrating 1 K Pots............303
    A. Raccanelli, L. A. Reichertz, and E. Kreysa

SZ EFFECT, MM SOURCES, AND SIMULATIONS

The Sunyaev-Zeldovich Effect: Recent Progress and Future Prospects.................309
    Y. Rephaeli

Non-thermal vs. Thermal SZ Effect in Galaxy Clusters.................316
    S. Colafrancesco, P. Marchegiani, and E. Palladino

Balloon-borne and Ground-based Sub-millimetre Cosmological Surveys: Breaking the "Redshift Deadlock".................322
    D. H. Hughes, I. Aretxaga, E. Chapin, and E. Gaztañaga

The Microwave Spectra of Planets.................330
    T. Encrenaz and R. Moreno

Big Bang Nucleosynthesis, Cosmic Microwave Background Anisotropies and Dark Energy.................338
    M. Signore and D. Puy

Primordial Molecules at Millimeter Wavelengths.................346
    D. Puy and M. Signore

Sunyaev-Zel'dovich Effect and Morphology of Galaxy Clusters.................351
    R. Piffaretti, P. Jetzer, D. Puy, and S. Schindler

Constraints on the Accuracy of Photometric Redshifts Derived from BLAST and Herschel/SPIRE Sub-mm Surveys.................354
    I. Aretxaga, D. H. Hughes, E. Chapin, and E. Gaztañaga

Simulating the Performance of Large-format Sub-mm Focal-plane Arrays.................357
    E. Chapin, D. H. Hughes, B. D. Kelly, and W. S. Holland

Iterative Map-making Methods for Cosmic Microwave Background Data Analysis.................360
    X. Dupac

Spatial Features of Non-thermal SZ Effect in Galaxy Clusters.................363
    E. Palladino, S. Colafrancesco, and P. Marchegiani

Utilization of a Center for Cosmic Structure to Stimulate Undergraduate Education-YCOOP.................367
    M. S. Spergel

Concluding Remarks.................370
    F. Melchiorri

List of Participants.................373

Author Index.................379

Preface

The 2K1BC[§] Workshop brought together people from all over the world involved in cosmological observations at the same wavelength range with different platforms. The setting for this International meeting was Breuil-Cervinia: a beautiful town at the foot of the Matterhorn mountain ("*the most famous rock in Europe*" with an altitude of 4.478 meters, known in Italy as the Cervino), in the Aosta Valley in Northern Italy. The place fully satisfies the requirements of a warm welcome to an International event with wonderful surroundings. Furthermore on the top of the Testa Grigia mountain (widely known as Plateau Rosa), at an altitude of 3480 meters, there is a scientific laboratory which was built more than 50 years ago managed by ICG/CNR in Turin. Here there is a 2.6 meter in diameter telescope in operation to observe the sky at millimeter wavelengths: the MITO project (Millimetre and Infrared Testagrigia Observatory).

During a five day conference of talks and a poster session, current and upcoming experiments were reported with exciting results and future forecasts. Balloon borne projects demonstrated the high accuracy reached to detect the tiny temperature fluctuations in CMBR (Cosmic Microwave Background Radiation). The ambitious satellite programs are promising to give a definitely clearer picture of observational cosmology. At the same time ground based observations are the complement to these projects necessary to extend the investigation of CMBR and related millimetric foregrounds. The study of CMBR polarization seems to be the main challenge in the near future and the major projects point in this direction. Particular attention was devoted, during the meeting, to the development of new detectors and observational techniques. In fact the fruitful interaction among theoreticians and experimentalists was highlighted in order to optimize the observations and get better results.

The people who attended the workshop really represented the scientific community in the field. In fact all the major projects were presented and discussed. There was only one exception: unfortunately we were not able to have any contribution on MAP (Microwave Anisotropy Probe), because this satellite experiment was launched just a few days before the 2K1BC Workshop.

The papers in this volume are organized according to topics following roughly the same order of the sessions of the meeting, including the poster contributions. In particular we have isolated, in a separate section, contributions regarding detectors, optics and cryogenics or techniques not directly related to any of the several experiments described elsewhere. On the other hand it is clear to the Editors, as well as to all the authors, that the present organization of the papers is just one of many possibilities. In fact most of the contributions could be correctly inserted into some of the other sections. Regarding this point we would like to thank all the authors for facilitating a fast publishing of this volume by respecting the paper submission deadlines.

The Workshop was made possible thanks to the several institutions which believed in this first exciting experience: University of Rome "La Sapienza", University of Milano–Bicocca, Department of Physics of the University of Rome "La Sapienza", Department of Physics "G. Occhialini" of the University of Milano-Bicocca, Gruppo Italiano di Fisica Cosmica (GIFCO), Consorzio Interuniversitario di Fisica Spaziale (CIFS), Agenzia Spaziale Italiana (ASI), European Space Agency (ESA), Istituto Nazionale di Fisica Nucleare - Sezione di Roma, Regione Valle d'Aosta - Assessorato Istruzione e Cultura, Comune di Valtournenche and Consorzio per lo Sviluppo Turistico del Cervino.

We are also grateful to the following sponsors: Rivoira, Agilent Technologies, Tektronix, QMC Instruments Ltd, Superelectric and Istituto San Paolo di Torino. Their funding largely contributed to the smooth running of the conference.

[§] 2K1BC = 2001 at Breuil-Cervinia

We have pleasure to acknowledge everyone who gave a valid contribution in organizing the event: Fernanda Lupinacci with the help of Sandra Capaldi and Stefano Galimberti. Stefano Petrocchi guaranteed all the projections without last minute hitches, Roberto Miglio and Antonio De Lucia linked Breuil-Cervinia to the rest of the world (*i.e.* Milan and Turin airports). Maurizio Perciballi took photos during the event and *quickly* impressed all of us in the Group Photo. The Experimental Cosmology Group of Rome and the Milan Radio Group members, and in particular: Angiola Orlando, Luca Lamagna, Federico Nati, Francesco Piacentini, Giorgio Savini, Elia Battistelli and Andrea Tartari gave continuous help behind the scenes. Our thanks also go to Giuseppe Maquignaz, Federica Pession, Cristina Uvire, Paola Magliozzi and Claudio Coriasco for their help on site and to Giorgio Pession, the Valtournenche mayor, for his welcome speech.

We also thank the Cervino S.p.A. cable car society and the friends of the "Rifugio delle Guide del Cervino" for helping with the organization of the Plateau Rosà excursion. Their help is continuous also during all the MITO observational campaigns.

The 2K1BC Local Organising Committee was made up of the Editors, Paolo de Bernardis, Roberto Maoli, Silvia Masi, Francesco Melchiorri and Giorgio Sironi while in the Scientific Organising Committee we had Pierre Encrenaz, Paul L. Richards, Ernst Kreysa, Phil Lubin and Carlo Castagnoli.

Finally the success of the 2K1BC has to be shared with all the LOC and SOC members, all the chairmen, all the contributors, all the participants and all the people in Breuil-Cervinia for putting up with us throughout the Workshop!

2K1BC Workshop Editors
Marco De Petris and Massimo Gervasi

Participants of the 2K1BC Workshop enjoying a sunny coffee break at the base of the Matterhorn mountain (Cervino 4.478 m a.s.l.). The characteristic cloud above the summit of the mountain gives the photo the final touch.

2K1BC Workshop
Experimental Cosmology at millimetre wavelengths

BALLOON BORNE EXPERIMENTS

ℓ-space spectroscopy of the Cosmic Microwave Background with the BOOMERanG experiment

P. de Bernardis[1], P.A.R. Ade[2], J.J. Bock[3], J.R. Bond[4], J. Borrill[5], A. Boscaleri[6], K. Coble[7], C.R. Contaldi[4], B.P. Crill[8], G. De Gasperis[9], G. De Troia[1], P. Farese[7], K. Ganga[10], M. Giacometti[1], E. Hivon[10], V.V. Hristov[8], A. Iacoangeli[1], A.H. Jaffe[11], W.C. Jones[8], A.E. Lange[8], L. Martinis[12], S. Masi[1], P. Mason[8], P.D. Mauskopf[13], A. Melchiorri[14], P. Natoli[9], T. Montroy[7], C.B. Netterfield[15], E. Pascale[6], F. Piacentini[1], D. Pogosyan[4], G. Polenta[1], F. Pongetti[16], S. Prunet[4], G. Romeo[16], J.E. Ruhl[7], F. Scaramuzzi[12], N. Vittorio[14]

[1] *Dipartimento di Fisica, Universitá La Sapienza, Roma, P.le A. Moro, 2, 00185, Italy.* [2] *Queen Mary and Westfield College, London, UK.* [3] *Jet Propulsion Laboratory, Pasadena, CA, USA.* [4] *C.I.T.A., University of Toronto, Canada.* [5] *N.E.R.S.C., LBNL, Berkeley, CA, USA.* [6] *IROE-CNR, Firenze, Italy.* [7] *Dept. of Physics, Univ. of California, Santa Barbara, CA, USA.* [8] *California Institute of Technology, Pasadena, CA, USA.* [9] *Department of Physics, Second University of Rome, Italy.* [10] *IPAC, Caltech, Pasadena, CA, USA.* [11] *Department of Astronomy, Space Sciences Lab and Center for Particle Astrophysics, University of CA, Berkeley, CA 94720 USA.* [12] *ENEA, Frascati, Italy.* [13] *Dept. of Physics and Astronomy, Cardiff University, Cardiff CF24 3YB, Wales, UK.* [14] *Nuclear and Astrophysics Laboratory, University of Oxford, Keble Road, Oxford, OX 3RH, UK.* [15] *Depts. of Physics and Astronomy, University of Toronto, Canada.* [16] *Istituto Nazionale di Geofisica, Roma, Italy.*

Abstract. The BOOMERanG experiment has recently produced detailed maps of the Cosmic Microwave Background, where sub-horizon structures are resolved with good signal to noise ratio. A power spectrum (spherical harmonics) analysis of the maps detects three peaks, at multipoles $\ell = (213^{+10}_{-13}), (541^{+20}_{-32}), (845^{+12}_{-25})$. In this paper we discuss the data analysis and the implications of these results for cosmology.

INTRODUCTION

The Cosmic Microwave Background (CMB) is a window on the Early Universe. It comes from an epoch when the age of the Universe was a few hundred thousand years (50000 times younger than today), the temperature was about 3000 K (1000 times more than today) and the density was one billion times larger than today [1],[2]. The very presence of the CMB is a compelling proof of the existence of a hot and dense initial phase in the Universe [3], [4]. The physics of the photons-baryons plasma present at that epoch, its interaction with the underlying dark matter distribution, and the resulting observable effects on the CMB have been studied in great detail [5].

If the inflationary scenario is true, photons coming from that epoch carry information which has been encoded at much earlier epochs, thus enabling us to investigate the history of the Universe as early as 10^{-36}s after the Big-Bang. An image of the CMB is a (processed) image of quantum fluctuations present in the Universe before the inflation phase [6], at energies of the order of the GUT energies.

In this framework, the statistical properties of the image of the CMB $\Delta T(\alpha, \delta)$ can be derived from first principles, given a small set of cosmological parameters. $\Delta T(\alpha, \delta)$, is expected to be a 2-D random gaussian field, with statistical properties fully described by its power spectrum c_ℓ. Here $c_\ell = \langle |a_{\ell,m}|^2 \rangle$, where $\Delta T(\alpha, \delta) = \sum_{\ell,m} a_{\ell,m} Y^\ell_m(\alpha, \delta)$. Analyzing the image and its power spectrum, it is then possible to estimate the cosmological parameters [7].

Many experimental teams have actively worked on the measurement of the spectrum, of the anisotropy power spectrum and of the polarization of the CMB. The purely Planckian nature of the spectrum has been established by the FIRAS spectrometer on board of the COBE satellite [8]. It is the proof of the cosmological nature of the CMB and of the Hot Big Bang theory proposed by Gamow in the 50s. The intrinsic, faint large scale anisotropy has been first detected by the DMR instrument on board of the COBE satellite [9]. Its low level and its power spectrum supported the inflationary hypothesis [6]. The degree and sub-degree-scale anisotropy has been detected by several ground based and balloon-borne experiments. Only recently, however, it has been possible to first detect the presence of peaks in the power spectrum [10], [11], [12], and to produce images where the sub-degree anisotropy is clearly visible [13], [14], [15], [16]. The presence of multiple peaks is the confirmation of the presence of acoustic oscillations in the plasma before recombination [5] and allows the detection of several important cosmological parameters [17], [18], [33].

In this paper we report results from the BOOMERanG experiment, a balloon-borne microwave telescope with cryogenic bolometric detectors. Several aspects of the instrument and other results are described in companion papers in this same conference proceedings [19], [20], [21], [22], [23], [24]: here we focus on the analysis techniques and on the cosmological significance of the results.

ℓ-SPACE SPECTROSCOPY

The detection of structures in the CMB is a difficult experimental problem. The size of the observable temperature fluctuations is of the order of a few tens μK, while instrumental, local and astrophysical backgrounds can be as large as few K. The differences in spectral and angular distributions allow the experimentalists to separate the cosmological component from the contaminations, but elaborate modulation techniques are needed [25], [19], [26]. Interferometers directly sample the correlation function of the temperature fluctuations [27], while total power receivers sample the temperature map: the power spectrum is derived by first organizing the time-ordered observations in a map and then performing the harmonic analysis.

In the case of BOOMERanG modulation is achieved by scanning the sky at constant elevation and constant azimuth speed. The signal from the detector is AC coupled and high pass filtered. Since most of the contaminating signals are either constant or smooth in the sky, they are efficiently rejected by this modulation. The disadvantage of this technique is that it results in an anisotropic filtering of the sky maps, which has to be taken properly in account in the data analysis (see below).

A further level of modulation comes from sky rotation. In fact, the central azimuth of the azimuth scans tracks the azimuth of the best sky region, while elevation is not changed during the scan, and is only changed in steps every several hours. The result of this strategy is that, due to sky rotation, the scans are gradually tilted in the sky, and the same pixel will be re-observed during the same day in differently tilted scans. This produces significant cross-linking in the sky coverage, which is important for the map-making algorithm used to create the image of the sky. The same process is repeated for several days. The comparison of maps obtained in different days, when the payload has drifted by thousands of km and the ground configuration is completely different, is a very effective tool to exclude contamination of the sky maps coming from the telescope sidelobes.

The peak to peak length of our azimuth scans is $\Delta A \sim 60^o$, so that the scans in the sky have a length $\Delta\theta \sim \Delta A \cos e \sim 42^o$. This length has been selected as the best compromise between several factors: sky coverage, avoidance of the sun, repetition frequency of scans, detector's speed, 1/f knee in detectors noise, etc. As a result, our ℓ-space resolution is limited to $\Delta\ell \sim \pi/\Delta\theta \sim 4$. This is more than enough to resolve the acoustic peaks present in the power spectrum of the CMB anisotropy, which have a width $\Delta\ell_p \sim 100$ [5]. In practice, we degrade the instrumental resolution to $\sim \Delta\ell_p/2$ by binning in ℓ, in order to improve the signal to noise ratio in each ℓ bin. The estimates of the power spectrum averaged over wide ℓ bins are called bandpowers. The finite length of the scans also limits the lowest multipole detectable $\ell_{min} \sim \Delta\ell$, but in practice a much higher $\ell_{min} \sim 25$ is set by the presence of drifts and $1/f$ noise. The maximum multipole observable in constant speed scans depends on the angular resolution of the telescope, on the time response of the detector [28] and on its noise; in general the sensitivity of the instrument to different multipoles is described by a suitable window function taking into account these effects (see below). In the case of BOOMERanG $\ell_{max} \lesssim 1200$.

THE 1998/99 LDB DATA

The BOOMERanG payload was flown by NASA-NSBF on Dec.29, 1998, from McMurdo (Antarctica). It remained at float for 10.6 days, circumnavigating Antarctica at an average altitude of 37 km. About 57 million 16-bit samples of the signal were collected for each of the 16 detectors. The data were edited for known instrument glitches, temperature fluctuations, and cosmic rays events. Less than 5% of the bolometer data has been found to be contaminated. Constrained realizations of noise were substituted to the contaminated signals.

The pointing has been reconstructed from the signals of the laser gyroscopes, of the differential GPS, and of the sun sensors. In the most recent pointing solution, repeated observations of compact sources show that the accuracy of the reconstruction is $\lesssim 2.5$ arcmin *rms*. Random errors in the pointing have the effect to smear-out the signals from small sources. This adds in quadrature to the intrinsic angular resolution of the telescope (9.5' FWHM at 150 GHz). The finite size of the pixelization has a similar effect. In ℓ space these three effects are modelled by the low pass filter $W(\ell)$ shown in fig.1 (called window function), which has to be taken into account in the reconstruction of the angular power spectrum of the sky. It is evident how these effects limit the sensitivity of our observations at high multipoles.

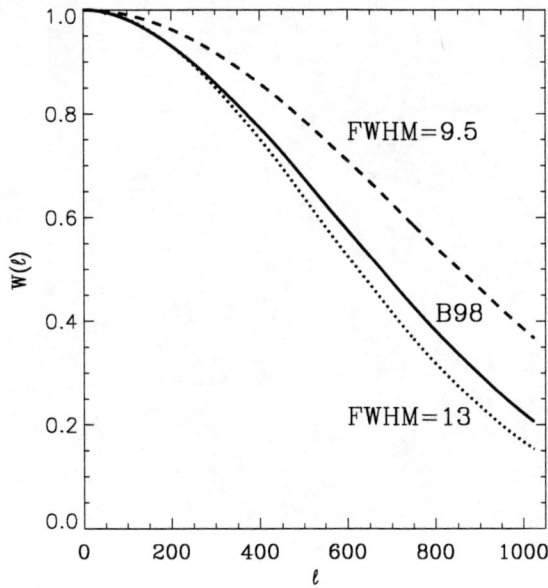

FIGURE 1. Window function $W(\ell)$ for the BOOMERanG instrument (B98) and for gaussian beams with FWHM=9.5' and 13'.

The time-domain high-pass filter is not included in the window function of fig.1 since its effect is, in general, anisotropic. It can be shown, however, that for our particular scan strategy the time-domain high pass filter acts as a high pass filter in the multipoles domain [29].

Sky maps have been constructed from the time ordered data and pointing using four independent methods: naive maps (just coadding data on the same pixel); maximum likelihood maps obtained using the MADCAP package ([30]); maximum likelihood maps obtained using the iterative method of [31]; suboptimal maps obtained using the fast map making method of [29].

All methods produce very similar maps. In fig. 2 we show the central region of the 150GHz map from the B150A channel obtained with the iterative method [31]. The color code used in the BOOMERanG maps correspond to temperature fluctuations of a 2.73K blackbody.

Degree-scale structures with amplitude of the order of $100 \mu K$ are evident in the map at 150 GHz. Consistent structure is also evident in the maps at 90 and 240 GHz. The similarity of the temperature maps obtained at different frequencies [13] is the best evidence for the CMB origin of the detected fluctuations. Foregrounds contamination can be constrained significantly in the center of the observed sky region [13], [20].

THE POWER SPECTRUM

Netterfield et al. [15] have computed the power spectrum of the central region of the BOOMERanG 150GHz maps (1.8% of the sky). The result has been obtained with a monte-carlo technique, which allows to estimate effectively the effects of sky coverage, anisotropic filtering, system noise and beam on the measured power spectrum [29]. Using about 10 times more data and about 80% more sky than the original data release, with an improved pointing solution

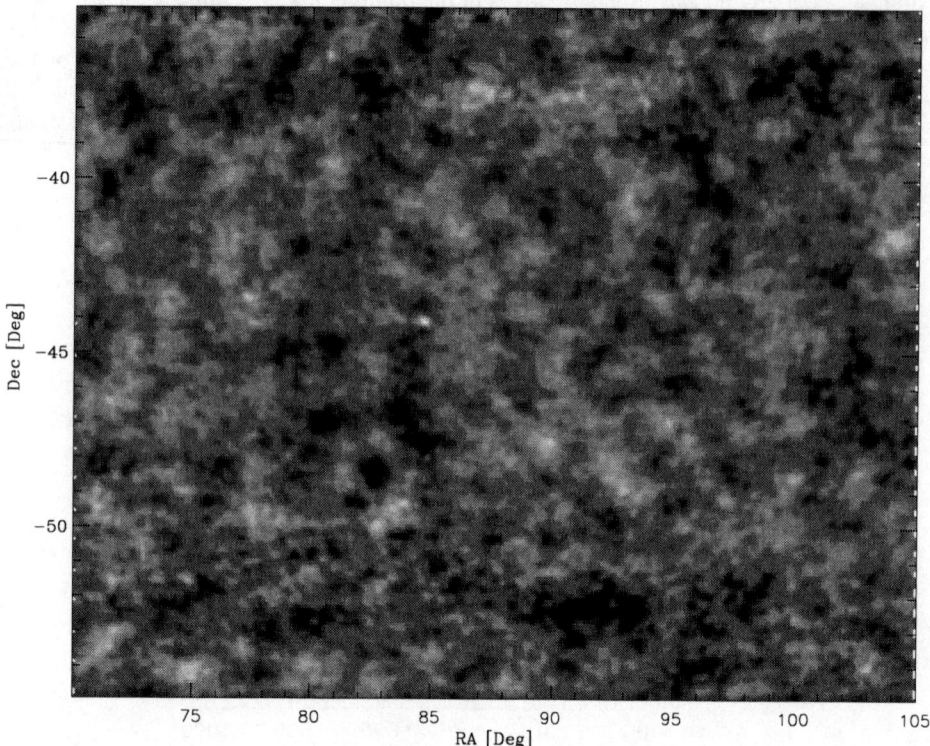

FIGURE 2. Map of the microwave sky measured by BOOMERanG at 150 GHz. The resolution of the observations is 12 arcmin. Healpix pixelization with 3.5' pixel side has been used. The units are for thermodynamic temperature fluctuations of a 2.73K blackbody. The same structures are visible in the 90 GHz and in the 240 GHz maps of BOOMERanG.

and a better measurement of the effective beam, it was possible to detect three peaks in the power spectrum of the CMB (see fig.3). Bandpowers have been computed following the recipe of [50]. These data are compared to the simultaneous data releases of the DASI [32] and MAXIMA [33] teams in fig.4.

Given the orthogonality of the experimental and analysis methods, the agreement of the three results is very good, at least visually. The existing anti-correlations in the bandpowers, and the presence of some overlap in the sky coverage of the BOOMERanG and DASI data should be taken into account for a more quantitative comparison. Such a comparison will be the best argument to exclude significant systematic effects in the three spectra.

COSMOLOGY-INDEPENDENT DATA ANALYSIS

Let's consider first the problem of assessing the statistical properties of the image of the sky we have obtained at 150 GHz. We are interested to see if the temperature of the CMB in the sky is distributed as a Gaussian. If this is true, then the power spectrum measures all the information encoded in the image. The gaussianity of the image is not a trivial result of the central limit theorem. As a matter of facts, images of the sky with the same resolution, but at higher frequencies, are highly non-gaussian. A visual exam of the IRAS maps of interstellar dust emission at 100 μm is very convincing in this regard [34]. In fig.5 we compare the 1-P distribution of the 150GHz map (dominated

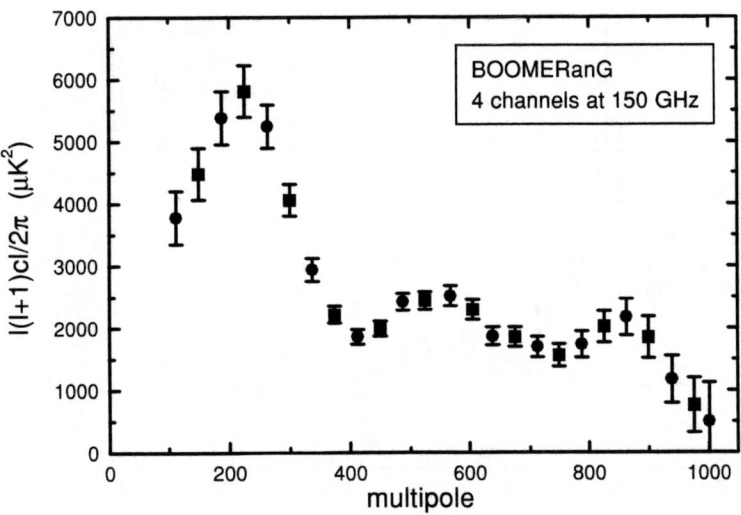

FIGURE 3. CMB anisotropy power spectrum detected by BOOMERanG. The dataset plotted as circles and the dataset plotted as squares are significantly correlated, and are both plotted to show the effect of different binning in ℓ-space. The different bandpowers of the same dataset are instead effectively uncorrelated. The gain (10% at 1σ) and beam (1.4' at 1σ) calibration errors are not plotted, since they are totally correlated for all the bandpowers.

by CMB anisotropy) to the 1-point distribution of the same patch of the sky measured by BOOMERanG at 410 GHz (dominated by thermal emission of interstellar dust). Even this naive test shows that the CMB is something special, being very accurately gaussian distributed. Of course, for a more quantitative test of the 1-P distribution it is necessary to take into account the correlation properties of the signal and of the noise in the data.

The simplest non-gaussianity estimators in the pixel space are Skewness and Kurtosis of the 1-point distribution, and the Minkowski functionals. [35] have analyzed the 150GHz maps of BOOMERanG using these five estimators, and a Monte-Carlo approach to account for the correlations in the data, to compute the statistical significance of the results and assess the effect of systematics. All the tests are consistent with the gaussian hypothesis, as reported in table 1. Small non gaussian signals added to the gaussian CMB fluctuations can be excluded with different levels of significance, depending on the nature of the contaminants. For example the rms of fluctuations distributed as a 1 DOF χ^2 must be less than 3% of the rms of the CMB. Fluctuations due to instrumental effects, as well as fluctuations due to the dust foreground and to extragalactic point sources are found to be irrelevant. Spectral methods to study gaussianity are also being exploited. There are many kinds of non-gaussianity [36], and a through analysis will require the use of many different methods.

Given the results above, however, it is reasonable to assume that the power spectrum is the only tool we need to study the statistical properties of the image. [15] have discussed how the measured power spectrum of the sky is robust against variations of the ℓ-binning, channel selection, data subset selection, effects of uncertainties in the beam and effects of the noise. [17] have shown that the three peaks and two dips present in the power spectrum are statistically significant. The first peak is at $\ell_1 = (213^{+10}_{-13})$ (the errors correspond to a 1σ confidence interval in the location if the peak). Its amplitude is detected at $\gtrsim 5\sigma$, while for the second peak (at $\ell_2 = (541^{+20}_{-32})$) and third peak (at $\ell_3 = (845^{+12}_{-25})$) the amplitudes are detected at basically 2σ. Several methods to measure the location and amplitude of the peaks have been compared, all producing very consistent results. In particular, the results of fits using empirical functions are consistent with the results of fits using a database of adiabatic inflationary spectra of the CMB [17].

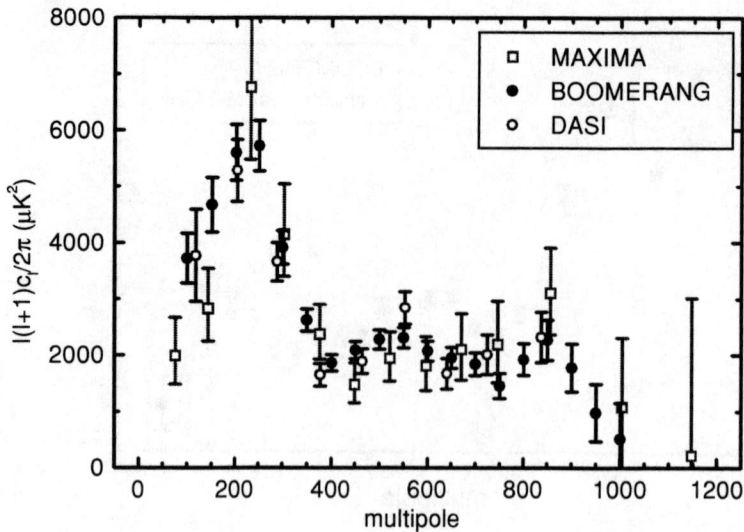

FIGURE 4. CMB anisotropy power spectrum detected by BOOMERanG, MAXIMA and DASI. Approximately uncorrelated bandpowers are plotted for each of the experiments. The error bars represent statistical errors only.

COSMOLOGICAL SIGNIFICANCE AND PARAMETERS ESTIMATION

The presence of "acoustic" features in the power spectrum of the CMB has been forecast long time ago [37],[38]. In the hot big bang model, it is the result of acoustic waves present in the pre-recombination Universe, and of their behaviour inside the acoustic horizon, in the presence of fluctuations in the density of the dominating form of matter. The same density fluctuations which are sources of the acoustic oscillations in the pre-recombination plasma are also responsible for the gravitational collapse starting after recombination, and leading to the formation of the large scale structures we see in the nearby Universe. Wiggles seem to be present in the spectrum of the large scale distribution of galaxies detected by the recent 2dF survey [39]; if this will be confirmed and if the wiggles will be found to be consistent with the modulation of the transfer function produced by "acoustic" features, it will be a wonderful success of this theory.

The alternative model of formation of galaxies from topological defects should lead to non gaussian CMB anisotropies [40], and fails to reproduce the observed power spectrum [41]: it can only be subdominant [42].

A large amount of work has been spent to accurately predict the power spectrum of the CMB in the adiabatic inflationary scenario. As of today, detailed fast codes are available [43], [44], and can be used to setup large databases of power spectra, to which the measured spectrum can be compared. More work is still to be done to properly include in the analysis the possible existence of isocurvature modes [45].

In this framework, the location of the first peak in the power spectrum mainly depends on the curvature of the Universe. The first peak is due to those fluctuations that enter the horizon shortly before recombination, and have just enough time to fully compress before recombination happens. The size of these perturbations is thus very similar to the size of the acoustic horizon at recombination. We thus have a "standard ruler", a few hundred thousand light years long, placed at a distance of about 14 billion light years. If we take into account the fact that the Universe has expanded by a factor ~ 1000 since recombination, we conclude that these fluctuations should appear as $\sim 1^o$ spots in the CMB. This is correct only if the geometry of the Universe is Euclidean, not curved, i.e. if the average mass-energy density of the Universe is the critical one ($\Omega = 1$). If, instead, the mass-energy density is higher than critical ($\Omega > 1$), the geometry of space will have a positive curvature and the photons will travel along curved geodesics. The excess density will act as a magnifying glass, and the same fluctuations in the CMB will appear as spots larger than 1°. The opposite will happen if the density is lower than critical, acting as a de-magnifying glass and producing a typical angular size of the

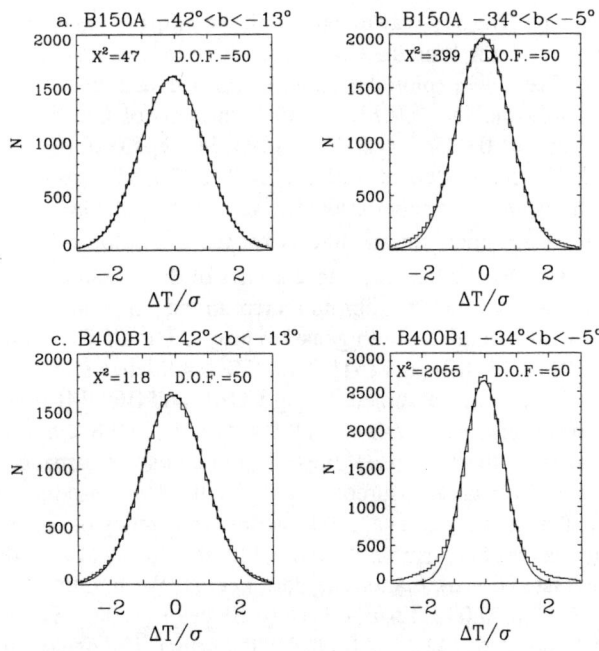

FIGURE 5. In the top row, we plot the 1-P distribution of the ratio $\Delta T_i / \sqrt{\sigma_i^2 + \sigma_{sky}^2}$ for the 150 GHz map (dominated by CMB fluctuations) in a box at high galactic latitude (a) and in a box at intermediate galactic latitudes (b). In the lower row the same distributions are plotted for the map at 410 GHz, which is dominated by interstellar dust

fluctuations smaller than 1°. In any case the presence of a typical size of the fluctuations θ will produce a peak in the power spectrum of the CMB at $\ell_1 = \pi/\theta$. By measuring the location of the peak it will thus be possible to measure Ω. The quantitative treatment of this angular-size vs distance test can be found in [46], [47]. In general, ℓ_1 decreases when Ω increases, but the location of the first peak is also controlled by Ω_Λ. Only if $\Omega_\Lambda = 0$ the simple relationship $\ell_p \sim \Omega^{-\frac{1}{2}}$ holds. The measurement of ℓ_1 from the BOOMERanG power spectrum is $\ell_1 = (213^{+10}_{-13})$, which is consistent with a flat geometry of the Universe. Rigorous confidence intervals for the parameter Ω can be found with the Bayesian analysis of the full power spectrum as described below; frequentist methods are also being developed [48].

The ratio between the amplitude of the second peak and the amplitude of the first one depends mainly on the physical density of baryons $\Omega_b h^2$ and on the tilt of the density fluctuations spectrum. A high density of baryons favors compressions against rarefactions: the odd-order peaks (compression) are enhanced while the even-order peaks (rarefaction) are depleted. From the BOOMERanG power spectrum the ratio is $(5450 \pm 350)/(2220 \pm 330) = (2.45 \pm 0.52)$. Assuming a scale invariant power spectrum of the density fluctuations ($n = 1$), this corresponds to a physical density $\Omega_b h^2 \sim 0.02$. Again, better constraints are found by means of the Bayesian analysis of the full power spectrum as described below. In fact, a tilt of the density fluctuations spectrum ($n < 1$) has the same effect of a high baryons density in depleting the second peak with respect to the first one, so there is a degeneracy between the two parameters. But the effects of the two quantities on the amplitude of the third peak are different. The amplitude of the third peak is increased by a high baryon density, while it is decreased by a red ($n < 1$) primordial density fluctuation spectrum. The result is that extending the observations to $\ell \sim 1000$ breaks the degeneracy between n and $\Omega_b h^2$, thus allowing a determination of both the parameters.

It is important to stress the fact that our result for $\Omega_b h^2$ agrees with the constraint on $\Omega_b h^2$ from the Big Bang Nucleosynthesis. In fact, the physical density of baryons affects the yield of the nuclear reactions happening in the first few minutes after the big bang. The resulting primordial abundances of light elements are measured by the optical absorption spectra of primordial clouds of matter [49]. It is evident that both the physics and the experimental methods involved in these two measurements of $\Omega_b h^2$ are completely orthogonal to the CMB ones. The fact that the two estimates of $\Omega_b h^2$ agree so well should be considered a great success of the Hot Big Bang model.

The multiple peaks and dips are a strong prediction of the simplest adiabatic inflationary models, and more generally of models with passive, coherent perturbations. Although the main effect giving rise to them is regular sound compression and rarefaction of the photon-baryon plasma at photon decoupling, there are a number of influences that make the regularity only roughly true. The best way to extract all the information encoded in the data is by comparison to a large database of C_ℓ spectra. In order to limit the size of the database, we considered for the first approach the class of adiabatic inflationary models. We have explored a parameter space with 6 discrete parameters and a continuous one. The parameters ranged as follows: $\Omega_m = 0.11,...,1.085$, in steps of 0.025; $\Omega_b = 0.015,...,0.20$, in steps of 0.015; $\Omega_\Lambda = 0.0,...,0.975$, in steps of 0.025; $h = 0.25,...,0.95$, in steps of 0.05; spectral index of the primordial density perturbations $n_s = 0.50,...,1.50$, in steps of 0.02, $\tau_C = 0.,..,0.5$, in steps of 0.1. The overall amplitude C_{10}, expressed in units of C_{10}^{COBE}, is allowed to vary continuously. We used the BOOMERANG power spectrum expressed as 18 bandpowers C_b [15] and we computed the likelihood for the cosmological model C_b^T as $\exp(-\chi^2/2)$, where $\chi^2 = (C_b - C_b^T) M_{bb'}^{-1} (C_{b'} - C_{b'}^T)$. Here $M_{bb'}$ is the covariance matrix of the measured bandpowers; C_b^T is an appropriate band average of C_ℓ. A 10% Gaussian-distributed calibration error in the gain and a 1.4' (13%) beam uncertainty were included in the analysis as additional parameters with gaussian priors. The COBE-DMR bandpowers used were those of [50], obtained from the RADPACK distribution [51]. The 95% confidence intervals for the parameters we find in this way depend to some extent on the priors assumed. Using COBE and BOOMERanG data only, with a weak prior $0.45 < h < 0.90$, significantly constrains three parameters: $0.9 < \Omega < 1.15, 0.8 < n < 1.1$ and $0.015 < \Omega_b h^2 < 0.029$. Using more restrictive priors, deriving from the properties of the large scale distribution of Galaxies (σ_8 and Γ), or the data of high-redshift supernovae, or the measurement of h by the HST, produces narrower, consistent intervals for these parameters [15], [17]. This fact suggests a good overall consistency of the present cosmological paradigm. Including these priors, it is also possible to constrain the two additional forms of mass-energy contributing to the total mass-energy density in the Universe, i.e. dark matter and dark energy. We find that the 95% confidence intervals for $\Omega_b h^2$ and Ω_Λ are $0.36 < \Omega_\Lambda < 0.72$ and $0.09 < \Omega_b h^2 < 0.18$ (LSS prior); $0.52 < \Omega_\Lambda < 0.88$ and $0.01 < \Omega_b h^2 < 0.17$ (SN1a prior); $0.40 < \Omega_\Lambda < 0.84$ and $0.06 < \Omega_b h^2 < 0.26$ (HST h prior). The detection of a non-zero Ω_Λ comes thus from independent paths and sets a formidable challenge to our understanding of fundamental physics [52].

CONCLUSIONS

The BOOMERanG experiment has produced multi-frequency maps of the microwave sky, where the structure of the CMB has been resolved with high signal to noise ratio. The structures in the CMB are gaussian, and their power spectrum features three peaks. This is consistent with the presence of acoustic oscillations in the primeval plasma. It also fits the predictions of the adiabatic inflationary scenario. The values of the cosmological parameters inferred in this scenario point to a flat universe with nearly scale-invariant initial adiabatic perturbations and a significant contribution of dark energy to the total density of the Universe.

Significant work remains to be done with the BOOMERanG data. We are working to analyze all the remaining channels sensitive to the CMB for a more accurate determination of the power spectrum. In parallel, we are currently assessing the gaussianity of the maps with different methods. Also, we are using the observations of Galactic sources in order to improve the gain and beam calibration accuracy. A search for Sunyaev-Zeldovich effect in the direction of the rich clusters of Galaxies present in the observed region is also being carried out. Methods for components separation and the data of the higher frequency channel are being used to investigate the properties of interstellar cirrus clouds.

ACKNOWLEDGMENTS

This activity has been supported by the University of Rome La Sapienza, Programma Nazionale di Ricerche in Antartide and Agenzia Spaziale Italiana in Italy, by NASA and NSF in USA, by PPARC in UK and by Univ. of Toronto in Canada.

REFERENCES

1. Weinberg S., 1977, The first three minutes, Basic Books, ISBN-0465024378
2. Peebles P.J.E, 1994, Principles of Physical Cosmology, Princeton Series in Physics
3. Gamow G., 1946, Phys.Rev., 70, 572
4. Dicke R.H., et al., 1965, Ap.J., 142, 414
5. Hu W., Sugiyama N. & Silk J., 1997, Nature, 386, 37
6. Kolb E.W. and Turner M.S, 1990, The Early Universe, Addison-Wesley

7. see e.g. Efstathiou G., and Bond, J. R., 1999, MNRAS, 304, 75
8. Mather J., et al., 1990, Ap.J., 354, L37
9. Smoot G., et al., 1992, Ap.J.,1992, 396, L1
10. Miller, A. et al. 1999, Ap.J., 524, L1
11. Torbet E., et al., 1999, Ap.J., 521, L79-L82
12. Mauskopf P., et al., 2000, Ap.J., 536, L59-L62
13. de Bernardis, P., et al. 2000, Nature, 404, 955-959
14. Hanany, S. et al., 2000, Ap.J., 545, L5-L9
15. Netterfield B., et al., 2001, submitted to Ap.J., astro-ph/0104460
16. Leitch E.M., et al., 2001, astro-ph/0104488
17. de Bernardis, P., et al. 2002, Ap.J. in press, astro-ph/0105296
18. Pryke C., et al., astro-ph/0104490
19. Piacentini F., et al., these proceedings; see also astro-ph/0105148
20. Masi S., et al., these proceedings; see also Ap.J, 553, L93-L96, 2001, (astro-ph/0101539)
21. Masi, S., et al., 'Scanning polarimeters for measurements of CMB polarization', these proceedings.
22. Hristov V.V., et al., these proceedings.
23. Bock., J., et al., these proceedings.
24. Pascale E. and Boscaleri A., these proceedings.
25. Miller, A. et al. 2001, astro-ph/0108030
26. Lee A., et al., 1999, in "3K cosmology", AIP Conf. Proc. 476; astro-ph/9903249
27. White M., et al., astro-ph/9912422
28. Hanany S., et al., 1999, MNRAS, astro-ph/9801291
29. Hivon E. et al., 2001, astro-ph/0105302
30. Borrill, J., in 3K Cosmology astro-ph/9911389
31. Natoli P., et al. 2001, submitted to Ap.J., astro-ph/0101252
32. Halverson N.W. et al. 2001, astro-ph/0104489
33. Lee A.T. et al. 2001, astro-ph/0104459
34. Schlegel D.J. et al. 1999, Ap.J. 500, 525.
35. Polenta G., et al., in preparation.
36. Belen Barreiro R., 1999, astro-ph/9907094.
37. Sunyaev, R.A. & Zeldovich , Ya.B., 1970, Astrophysics and Space Science 7, 3-19
38. Peebles, P.J.E, and Yu J.T., 1970, Ap.J., 162, 815
39. Percival W.J., 2001, astro-ph/0105252
40. see e.g. Avelino P.P., Shellard E.P.S., Wu J.H.P., and Allen B. 1998, Ap.J., 507, L101.
41. see e.g. Durrer R., Kunz M., and Melchiorri A. 1998, Phys. Rev. D V59, 123005)
42. Bouchet F.R. et al. 2001, astro-ph/0005022
43. Lewis A., Stewart E., Lasenby A., astro-ph/9911176
44. Seljak, U. & Zaldarriaga, M. 1996, Ap.J. 469, 437
45. Turok N., et al., 2001, astro-ph/0012141
46. Weinberg S., 2000, astro-ph/0006276
47. Melchiorri A. and Griffiths L.M., astro-ph/0011147
48. Gewiser E., astro-ph/0105010
49. Burles, S., Nollett, K.M. & Turner, M.S. 2000, astro-ph/0010171
50. Bond, J.R., Jaffe, A.H. & Knox, L. 1998, Phys. Rev. D57, 2117
51. Knox L. 2000, http://flight.uchicago.edu/knox/radpack.html
52. Weinberg S., 1989, Rev. Mod. Phys., 61, 1

The MAXIMA and MAXIPOL Experiments

P.L. Richards,[1,3,16] M. Abroe,[13] P. Ade,[4] A. Balbi,[2,3,5] J. Bock,[6,7] J. Borrill,[3,8] A. Boscaleri,[9] P. de Bernardis,[10] J. Collins,[1] P.G. Ferreira,[11,12] S. Hanany,[3,13] V.V. Hristov,[7] A.H. Jaffe,[3] B. Johnson,[13] A.T. Lee,[1,2,3] T. Matsumu,[13] P.D. Mauskopf,[14] C.B. Netterfield,[15] E. Pascale,[9] B. Rabii,[1,2,3,16] G.F. Smoot,[1,2,3,16] R. Stompor,[3,16,17] C.D. Winant,[1,3,16] J.H.P. Wu[18]

[1] *Department of Physics, University of California, Berkeley, CA 94720-7300, U.S.A.*
[2] *Division of Physics, Lawrence Berkeley National Laboratory, Berkeley, CA 94720*
[3] *Center for Particle Astrophysics, University of California, Berkeley, CA 94720-7300, U.S.A.*
[4] *Queen Mary and Westfield College, London, UK*
[5] *Dipartimento di Fisica, Università Tor Vergata, Roma, Italy*
[6] *Jet Propulsion Laboratory, Pasadena, CA, U.S.A.*
[7] *California Institute of Technology, Pasadena, CA, U.S.A.*
[8] *National Energy Research Scientific Computing Center, Lawrence Berkeley National Laboratory, Berkeley, CA 94720, U.S.A.*
[9] *IROE-CNR, Florence, Italy*
[10] *Dipartimento di Fisica, Università La Sapienza, Rome, Italy*
[11] *Astrophysics, University of Oxford, Oxford, UK*
[12] *CENTRA, Instituto Superior Technico, Lisbon, Portugal*
[13] *School of Physics and Astronomy, University of Minnesota/Twin Cities, Minneapolis, MN, U.S.A.*
[14] *Department of Physics and Astronomy, University of Wales, Cardiff, UK*
[15] *Department of Physics and Astronomy, University of Toronto, Canada*
[16] *Space Sciences Laboratory, University of California, Berkeley, CA 94720-7300, U.S.A.*
[17] *Copernicus Astronomical Center, Warsaw, Poland*
[18] *Department of Astronomy, University of California, Berkeley, CA 94720-3411 U.S.A.*

Abstract. MAXIMA is a balloon-based bolometric experiment to measure the temperature anisotropy of the CMB over spatial frequency range $36 \leq \ell \leq 1235$. The MAXIMA-1 flight produced a 124 square degree temperature anisotropy map with a beam diameter of 10 arcmin. These data have been used to produce a power spectrum which is in excellent agreement with data from BOOMERANG and DASI, but covers a wider range of angular scales. The MAXIMA power spectrum is consistent with the prediction of Λ CDM models and has been used to constrain cosmological models. The MAXIMA experiment is described and an outline is given of the contents and significance of papers written by the MAXIMA team. MAXIMA is being modified to measure the polarization anisotropy of the CMB. A brief description of this MAXIPOL experiment is also given.

THE MAXIMA-1 EXPERIMENT

Measurements of the anisotropy of the cosmic microwave background (CMB) can discriminate between cosmological models and determine cosmological parameters with high accuracy (Kamionkowski and Koswoski, 1999) [1]. MAXIMA is a balloon-borne experiment optimized to map the CMB anisotropy over hundreds of square degrees with an angular resolution of 10 arcmin and to produce measurements of the CMB power over a wide range of angular scales corresponding to $36 \leq \ell \leq 1235$.

Instrumentation

Lee, et. al. [2] gives a detailed description of the MAXIMA system. It is an off-axis Gregorian telescope mounted on an attitude-controlled balloon platform. The primary mirror rotates so as to scan the telescope beams in azimuth. Re-imaging optics in the receiver cryostat are cooled to ~3 K. The receiver consists of 16 photometers: eight operating at 150 GHz, four at 240 GHz and four at 410 GHz. All have 10' FWHM beams. The CMB radiation is detected with spider web bolometers (Bock, et. al. [3]) operated at 0.1K. The bolometers are AC biased to avoid low frequency amplifier noise, and the detector time constants, determined in flight, range from 5 to 13 ms. The combined sensitivity of the eight 150 GHz photometers is 41 microK × rt(sec), which is the best of any CMB experiment to date.

The gondola azimuth is driven by a reaction wheel using information from a two-axis magnetometer and three rate gyroscopes. Pointing is reconstructed using a CCD camera bore-sighted to the center of the primary mirror scan and an offset CCD camera that views Polaris. An on-board computer locates the two brightest stars in each CCD field at a 5 Hz rate.

Observations

The instrument was launched from the National Scientific Balloon Facility in Palestine, Texas at 1.6 UT 1998 August 2. The CMB anisotropy was measured in two overlapping observations. Each was conducted at a constant elevation angle and consisted of two independent modulations in azimuth: a slow modulation of the entire instrument +/- 5.6 deg amplitude, 68 sec period and 45 deg elevation for the 1st scan; (+/- 3.3 deg, 47 sec, 31 deg elevation for the 2nd scan) and a fast, saw tooth modulation using the primary mirror (+/- 2 deg, 0.5 Hz for both scans). The first observation began at 4.3 UT and lasted 1.7 hours. The second began at 6.0 UT and lasted 1.4 hours. Due to the rotation of the sky, the two scans were cross-linked at an angle of 22 deg. The gondola altitude varied from 38.0 to 38.8 km during the CMB observations.

Calibration and Beams

A full beam calibration of the 150 and 240 GHz photometers was obtained from the CMB dipole. The instrument was rotated 100 times with a period of 20 seconds from 3.6 to 4.2 UT. The data from each rotation was fitted to a model including the CMB dipole and Galactic dust emission (Finkbeiner, et. al. [4]). An additional component was found which correlated with the 410 GHz channel signal. This component is consistent with a large angular scale atmospheric contribution and was subtracted. Beam contour maps and were obtained from the Jupiter observation from 7.5 to 8.1 UT. The beam profiles were integrated and used with the angular diameter and brightness temperature of Jupiter (Goldin, et. al. [5]) to calibrate the photometers. A very tight 4% temperature calibration uncertainty was obtained at 150 and 240 GHz. A detailed analysis has been published by Wu et al. [6] of the beam profiles and the procedures required for map making with slightly asymmetric beams.

Data Analysis

The raw data for each photometer consisted of 1.2 million samples for the first CMB scan and 1 million samples for the second. The data were deglitched, the electronic and bolometric transfer functions deconvolved, and estimates were made of the noise power spectrum of the data stream. It was assumed that the time domain data were dominated by noise and the procedure of Ferreira and Jaffe [6] were used to confirm the validity of this assumption. High and low pass filters were used to eliminate frequencies where there are no appreciable optical signals.

The calibrated time stream data were combined with the attitude pointing solution to produce a maximum likelihood pixelized map of temperature anisotropy and a pixel-pixel noise correlation matrix using the techniques described in Wright [8], Tegmark [9], Bond et. al. [10], and Ferreira and Jaffe [7]. Individual 100 square degree maps were generated from the outputs of four individual photometers. After checking for consistency and verifying that the noise in different maps was uncorrelated, a noise weighted average temperature anisotropy map was produced. The maximum likelihood angular power spectrum C_ℓ of the map was generated using the MADCAP (Borrill [11]) implementation of the Newton-Raphson iterative maximization of the likelihood function following Bond, Jaffe & Knox [12]. A complete description of the map making procedure has been presented in Stompor et al. [13].

Analysis of data from the MAXIMA-2 flight, launched in June 1999, is in progress. This flight measured over 200 square degrees of the sky. Approximately 25% of this observing region overlaps with the MAXIMA-1 observations.

Foregrounds

Foreground sources of confusion may include emission from the atmosphere, interstellar dust emission, synchrotron radiation, free-free emission, and point sources. The spectrum of the observed signal is not consistent with that of the atmosphere which has little power at the angular scales measured. The temporal stability of the observed sky structure is inconsistent with an atmospheric or ground-based origin. The ground-based telescope sidelobe measurements (Lee, et. al. [2]) also provides evidence that sidelobe response from the ground or the moon is not significant.

The 100 micron IRAS/DIRBE map (Schlegel et. al. [14]) was extrapolated to the MAXIMA wavebands and cross-correlated with the CMB maps. No correlations large enough to effect CMB power spectrum estimation were found. Using the galactic emission model of Bouchet and Gispert [15] it was found that both Bremstrahlung and Synchrotron emission are expected to contribute fluctuations with a magnitude less than 1 microK at 150 GHz. The scan region was chosen to avoid known bright radio point sources and none were detected in our CMB maps. No corrections to the data were required for any source of foreground contamination.

Power Spectrum and Cosmological Parameters

The MAXIMA map and power spectrum out to $\ell = 785$ computed with 5 arcmin pixels and an analysis of the resulting constraints on cosmological parameters were released in Hanany et al [16] and Balbi et al. [17] a few weeks after the first announcement from BOOMERANG [18]. The two data sets are essentially independent, consistent and complementary [16]. BOOMERANG has more statistical weight but MAXIMA extends to higher angular frequencies and has a more accurate calibration. In a joint effort by the two teams, the two data sets were combined by marginalizing over differences in calibration and beam uncertainties and then used to test cosmological models and determine values of cosmological constants in Jaffe et al. [19]. The combined power spectrum clearly defines the first acoustic peak and is consistent with a range of flat Λ CDM models with a relatively large baryon fraction. The data are consistent with the second and third acoustic peaks predicted by the models. The best fit is for a second peak that is somewhat smaller than previously expected [19]. This implies a somewhat larger baryon fraction than that predicted from big bang nucleosynthesis. The data do not demonstrate the existence of a second or third peak because they are also consistent with a horizontal straight line for values of ℓ above the first acoustic peak. A seven parameter Bayesian analysis was used to deduce cosmological parameters. The results support the picture of a flat universe, a flat primordial power spectrum, and a baryon fraction of ~5%. When combined with supernova data to break the degeneracy in the Ω_Λ Ω_m plane, the picture emerges of a universe that is ~70% dark energy, and ~25% dark matter. If the supernova data are not considered, and data for large scale structure are used to break the degeneracy, the conclusion remains essentially the same. The general agreement obtained for the values of total density and the fractions of dark energy, dark matter and baryons should be considered a triumph for experimental cosmology.

High Resolution Analysis

The power spectral analysis of the MAXIMA-1 data presented [16,17] out to $\ell=785$ was limited by the 5 arcmin pixels used in the computations. These computations were redone using 23,000 3 arcmin pixels from the fully cross-linked central 60 square degrees of the MAXIMA-1 scans. The resulting power spectrum presented in Lee et al. [20] extends to $\ell=1235$. It is consistent with the previous work and includes the ℓ-range of the anticipated 3rd acoustic peak. Stompor et al. [21] includes a statistical analysis to show that the point at $\ell=825$ where a third acoustic peak is expected is higher than the surrounding points with 95% confidence. The existence of multiple acoustic peaks is strong confirmation of inflationary models. Constraints on cosmological parameters based on the high resolution analysis are presented in Stompor et al., [21] where a degeneracy between the primordial slope n_S and the optical depth τ_c at the surface of last scattering is discussed. The value of $\Omega_b h^2$ from MAXIMA alone is consistent with the baryon fraction from Big Bang nucleosynthesis. The discrepancy noted in Jaffe et al. [19] from the combined MAXIMA BOOMERANG analysis arose because the statistical weight of the BOOMERANG data dominated in the neighborhood of the second acoustic peak.

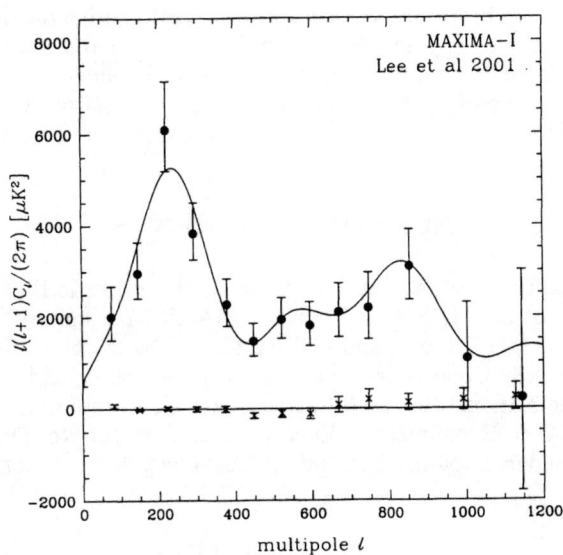

FIGURE 1. Composite angular power spectrum of the CMB anisotropy from the MAXIMA-1 map (filled circles). The points for $l < 335$ are from the power spectrum of the full 5' pixelized map from Hanany et al. [16], and the points for $l > 335$ are from the power spectrum of the central region of the 3' pixelized map. The error bars are 68% confidence intervals calculated using the offset lognormal likelihood functions of Bond, Jaffe, & Knox [12]. The solid curve is the best-fit (ΛCDM) inflationary adiabatic cosmology to the Stompor et al. [21] MAXIMA-1 and COBE/DMR power spectrum. The model has $\Omega_{cdm}, \Omega_\Lambda, \tau_c, n_s, h) = (0.07, 0.78, 0.0, 0.0, 1.0, 0.53)$ Balbi et al. [17]. The crosses are the estimated power spectrum of the difference between two independent maps, the first given by one of the three photometers and the other from the sum of the other two. The data in this figure are the same as those from Lee et al. [20], except for the expanded scale. The theoretical fit has been changed to that of Stompor et al. [21].

Gaussianity

Gaussianity of cosmological perturbations in the universe is one of the key predictions of standard inflation, but not of other models of structure formation such as cosmic defects. Tests for Gaussianity have been carried out using many methods for several CMB data sets, especially the COBE DMR map. All tests of the DMR data indicate Gaussianity, except for two that may have non-cosmological origins. The DMR data are not ideal for these tests since they probe scales larger than the causal horizon and thus combine uncorrelated perturbations. The central-limit theorem shows that non-Gaussianity will be erased on such large angular scales.

Wu et al. [22] have used the methods of moments, the Kolmogorov test, the X^2 test and Minkowski functionals in eigen, real, Wiener-filtered and signal-whitened spaces to test the MAXIMA-1 data. The data are consistent with Gaussianity. This result is consistent with standard cosmological inflation and place limits on the existence of cosmic defects. It also justifies the methods of data analysis used to produce maps, power spectra and cosmological parameters, which are valid only for Gaussian fluctuations. In a related work, Santos, et al. [23] investigate the bispectrum of the MAXIMA-1 data and conclude that the results are consistent with a Gaussian sky.

MAXIPOL

The MAXIMA photometer is being converted to a polarimeter to measure the polarization anisotropy of the CMB. The conversion is simple to describe, but difficult to carry out in practice. Grid polarizers are being placed in front of the photometer horns in the focal plane. A cold anti-reflection coated half-wave plate is located at the Lyot stop and rotated at a few Hz, which is fast compared with the gondola scan rate. The direction and strength of a polarized signal from the sky will then be obtained by synchronous demodulation of the data. Relatively small regions of the sky with low dust contamination will be mapped deeply so as to detect the E-mode polarization of the CMB. A turnaround flight is planned for Ft. Sumner, N.M. in spring 2002.

ACKNOWLEDGMENTS

We thank Danny Ball and the other staff at NASA's National Scientific Balloon Facility in Palestine, TX for their outstanding support of the MAXIMA program. MAXIMA is supported by NASA Grants NAG5-3941, NAG5-6552, NAG5-4454, GSRP-031, and GSRP-032, and by the NSF through the Center for Particle Astrophysics at UC Berkeley, NSF Cooperative Agreement AST-9120005, and KDI Grant 9872979. The data analysis used resources of the National Energy Research Scientific Computing Center which is supported by the Office of Science of the U.S. Department of Energy under Contract No. DE-AC03-76SF00098, and the resources of the Minnesota Supercomputing Institute. PA acknowledges support from PPARC rolling grant, UK.

REFERENCES

1. Kamionkowski, M., and Kosowsky, A., The Cosmic Microwave Background and Particle Physics, *Ann. Rev. Nucl. Part. Sci.* **49**, 77-123 (1999).
2. Lee, A. T., et. al., MAXIMA: An Experiment to Measure Temperature Anisotropy in the Cosmic Microwave Background, in L. Maiani, F. Melchiorri and N. Vittorio (eds.), *AIP Conf. Proc.* **476**, pp 224-236. Proceedings of 3K Cosmology Conference, Rome, Italy, astro-ph/9903249 (1998).
3. Bock, J. J., et al., A Novel Bolometer for Infrared and Millimeter-Wave Astrophysics, *Space Sciences Reviews* **74**, 229-235 (1995).
4. Finkbeiner, D. P., et. al., Extrapolation of Galactic Dust Emission at 100 Microns to CMBR Frequencies Using FIRAS, *Ap. J.* **524**, 867 (1999)
5. Goldin, A. B., et. al., Whole Disk Observations of Jupiter, Saturn and Mars in Millimeter-Submillimeter Bands, *Ap.J.Lett.* **488**, L161, (1997).
6. Wu, J.H.P., et al., Asymmetric Beams in Cosmic Microwave Background Anisotropy Experiments, Astro-ph/0007212, *Ap. J. Suppl.* **132**, 1 (2001).
7. Ferreira, P. G., and Jaffe, A. H., Simultaneous Estimation of Noise and Signal in Cosmic Microwave Background Experiments, *Monthly Notices of the Royal Astron. Soc.* **312**, 89 (2000).
8. Wright, E., Scanning and Mapping Strategies for CMB Experiments, Report UCLA-ASTRO-ELW-96-03, (1996) presented at the IAS CMB Data Analysis Workshop in Princeton, astro-ph/9612006.
9. Tegmark M., How to Make Maps from CMB Data Without Losing Information, astro-ph/9611130, *Ap. J.* **480**, 87 (1997)
10. Bond, J. R., et. al., Computing Challenges of the Cosmic Microwave Background, *Computing in Science and Engineering* **1**, 21 (1999).
11. Borrill, J, MADCAP-The Microwave Anisotropy Dataset Computational Package, *Proceedings of the Fifth European SGI/Cray MPP Workshop*, astro-ph/9911389 (1999).
12. Bond, J. R., Jaffe, A. H., and Knox, L., Estimating the Power Spectrum of the Cosmic Microwave Background, *Phys. Rev.* **D57**, 2117 (1998).
13. Stompor, R., et al., Making Maps of the Cosmic Microwave Background: The MAXIMA Example, astro-ph/0106451, Phys. Rev. D (to be published).
14. Schlegel D. J., et. al., Maps of Dust Infrared Emission for Use in Estimation of Reddening and Cosmic Microwave Background Radiation Foregrounds, *Ap. J.* **500**, 525 (1998), and Application of SFD Dust Maps to Galaxy Counts and CMB Experiments, in S. Colombi and Y. Mellier (eds.) *Wide Field Surveys in Cosmology*, (in press) astro-ph/9809230.
15. Bouchet, R. B., and Gispert, R., Foregrounds and CMB Experiments: I. Semi-Analytical Estimates of Contamination, astro-ph/9903176, *New Astron.* **4**, 443-479 (1999).
16. Hanany, S., et al., MAXIMA-1: A Measurement of the Cosmic Microwave Background Anisotropy on Angular Scales of 10 Arcminutes to 5 Degrees, *Ap. J. Lett.* **545**, L5 (2000).

17. Balbi, A. et al., Constraints on Cosmological Parameters from MAXIMA-1, astro-ph/0005124, Ap. J. **545**, L1 (2000).
18. de Bernardis, P., et al., A Flat Universe from High-Resolution Maps of the Cosmic Microwave Background Radiation, *Nature* **404**, 955-959 (2000).
19. Jaffe, A.H., et al. Cosmology from MAXIMA-1, BOOMERANG, and COBE/DMR Cosmic Microwave Background Observations, *Phys. Rev. Lett.* **86**, 3475 (2001).
20. Lee, A.T., et al., A High Spatial Resolution Analysis of the MAXIMA-1 Cosmic Microwave Background Anisotropy Data, astro-ph/0104459, Ap. J. Letters (to be published).
21. Stompor, R., et al., Cosmological Implications of the MAXIMA-1 High Resolution Cosmic Microwave Background Anisotropy Measurement, astro-ph/0105062, Ap. J. Lett. (to be published).
22. Wu, J.H.P., et al., Tests for Gaussianity of the MAXIMA-1 CMB Map, astro-ph/0104248, Phys. Rev. Lett. (submitted).
23. Santos, M.G., et al., An Estimate of the Cosmological Bispectrum from the MAXIMA-1 CMB Map, astro-ph/0107588, Phys. Rev. Lett. (submitted).

Interstellar dust in the BOOMERanG maps

S. Masi[1], P.A.R. Ade[2], J.J Bock[3,4], A. Boscaleri[5], B.P. Crill[3], P. de Bernardis[1], M. Giacometti[1], E. Hivon[3], V.V. Hristov[3], A.E. Lange[3], P.D. Mauskopf[2], T. Montroy[6], C.B. Netterfield[7], E. Pascale[5], F. Piacentini[1], S. Prunet[8], J. Ruhl[6]

[1] *Dipartimento di Fisica, Universitá La Sapienza, Roma, P.le A. Moro, 2, 00185, Italy.* [2] *Department of Physics and Astronomy, Cardiff CF24 3YB, Wales, UK.* [3] *California Institute of Technology, Pasadena, CA, USA.* [4] *Jet Propulsion Laboratory, Pasadena, CA, USA.* [5] *IROE-CNR, Firenze, Italy.* [6] *Dept. of Physics, Univ. of California, Santa Barbara, CA, USA.* [7] *Depts. of Physics and Astronomy, University of Toronto, Canada.* [8] *C.I.T.A., University of Toronto, Canada.*

Abstract. Interstellar dust (ISD) emission is present in the mm-wave maps obtained by the BOOMERanG experiment at intermediate and high Galactic latitudes. We find that, while being sub-dominant at the lower frequencies (90, 150, 240 GHz), thermal emission from ISD is dominant at 410 GHz, and is well correlated with the IRAS map at 100 μm. We find also that the angular power spectrum of ISD fluctuations at 410 GHz is a power law, and its level is negligible with respect to the angular power spectrum of the Cosmic Microwave Background (CMB) at 90 and 150 GHz.

INTRODUCTION

Interstellar dust, in the form of "cirrus" clouds, produce low level, patchy emission in the far-IR / mm-waves, even at high Galactic latitudes [1]. This has been recognized as a possible contaminant for sensitive CMB measurements at high frequencies (see e.g. [2], [3]) and also at low frequencies (see e.g. [4], [5]). Power spectra of this clumpy emission have been estimated from the IRAS survey at 3000 GHz and from the DIRBE survey at 1250 GHz [6] [7] [8] [9] [10]. All get $c_\ell \sim \ell^{-\beta}$ with $\beta \sim 2.5 \div 3$. In this paper we obtain a similar result for the data of the BOOMERanG map at 410 GHz, and we use it to estimate the level of Galactic contamination in the BOOMERanG maps at 90, 150 and 240 GHz. BOOMERanG [11], [12], [13], [14] has a dust monitor built-in, the 410 GHz channel. This samples dust emission at a frequency much closer to the CMB than IRAS and DIRBE. In fact, the 410 GHz map is dust dominated. We also detect dust emission at 90, 150 and 240 GHz. In the following we focus on the dust as a foreground, as in [15], and give a hint of new results relative to the temperature distribution of high latitude clouds.

BOOMERANG DATA

The BOOMERanG survey has been carried out at high and intermediate Galactic latitudes, in a southern sky region which contains the sky patch where interstellar dust emission is minimum. This region is approximately a rectangle in RA and dec, with $240^o \lesssim \ell \lesssim 290^o$ and $-50^o \lesssim b \lesssim -10^o$. In order to compare the BOOMERanG maps to other maps of the sky it is necessary to take into account the filtering procedure which has been applied to the time ordered data obtained during the scans. In fact, the signals from the BOOMERanG detectors have been high-pass filtered in the time domain to get rid of 1/f noise and drifts. A 16mHz hardware high pass was present in the amplification chain, and a 0.1Hz high pass filter has been applied in the data reduction software. This means that the structures in the sky aligned to the scans and larger than $\sim 10^o$ have been effectively suppressed. Any comparison to other maps should be done by filtering them in the same way, and using the same map making procedure. We have applied such a procedure to the FDS maps [16] derived from the IRAS and DIRBE maps. The procedure consisted in convolving the FDS map with the BOOMERanG beam, then sample the convolved map in the same 57×10^6 directions observed by BOOMERanG, thus obtaining a fake time ordered dataset. This has been high pass filtered before applying the same map making procedures used for the BOOMERanG maps [17].

THE 410 GHZ MAP

The map at 410 GHz and the FDS maps at the same frequency have been published in [15]. If one computes the difference of the two maps, only the cores of cold dust clouds, relatively brighter at 410 GHz than at 3000 GHz, remain prominent in the difference map [15]. This means that the two components model of FDS [16] works remarkably well in general, but deviations are present in selected regions of the sky, pointing to a more complex dust distribution. We have computed the power spectrum of the brightness fluctuations at 410 GHz in the three different regions of the 410GHz map. These sample three representative areas of the sky at low, intermediate and high Galactic latitudes. At low and intermediate latititudes the measured power spectra follow approximately power laws $c_\ell \sim \ell^{-2.3}$ and $c_\ell \sim \ell^{-3.1}$ respectively. At the high galactic latitudes the data are dominated by detector noise, and we can only give upper limits for the power spectrum of the sky. As expected because of the similarity of the maps, the spectra measured at 410 GHz are consistent with corresponding spectra computed from the FDS maps [15].

LONGER WAVELENGTHS

At 90, 150 and 240 GHz the ISD emission is subdominant with respect to anisotropy of the CMB, at least at intermediate and high Galactic latitudes. We can search for a subdominant dust signal by correlating with the IRAS 3000 GHz signal (or its FDS filtered extrapolation at the considered frequency). If we scatter plot the BOOMERanG signals (in the pixels of the map) against the comparison filtered FDS map, we expect that only ISD signals will produce a positive correlation. In fact, thermal emission from ISD has a broad spectrum peaking around 3000GHz, but extending all the way to 90GHz; on the contrary, the detector noise contributions to the signals in BOOMERanG and IRAS are uncorrelated, and the CMB anisotropy is present only in the bands measured by BOOMERanG. Its effect can only reduce the statistical significance of the positive correlation produced by ISD, but not change its estimate. We have produced scatter plots of this kind in several subregions of the sky observed by BOOMERanG. In fig.1,2,3 we plot some of these scatter plots.

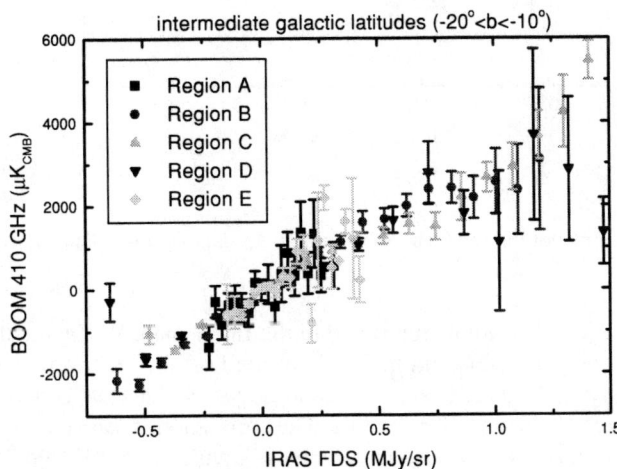

FIGURE 1. Scatter plot of the 410 GHz data vs the IRAS data - intermediate Galactic latitudes.

The data have been averaged in equal IRAS brightness bins to help a visual evaluation of the correlation. The analysis, however, has been carried out on the original pixels. We use Pearson's linear correlation coefficient R to estimate the statistical significance of the correlations, and the slope S of the best fit line as an estimate of the average

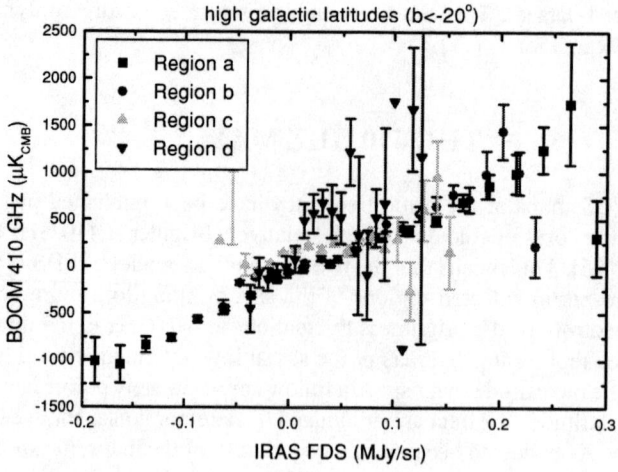

FIGURE 2. Scatter plot of the 410 GHz data vs the IRAS data - high Galactic latitudes.

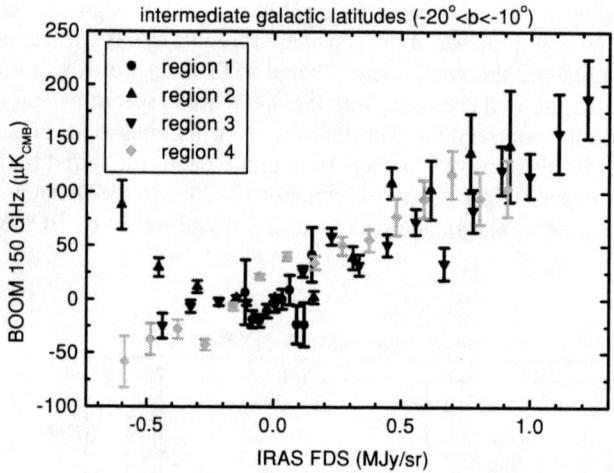

FIGURE 3. Scatter plot of the 150 GHz data vs the IRAS data - intermediate Galactic latitudes.

ratio between ISD emission in the BOOMERanG maps and in the IRAS map. We divide the map in several regions, and for each region we compute the regression and the slope of the best fit. The average of the best fits is our best estimate of the general slope $S(\nu)$; the standard error on the average accounts for variations of dust properties and measurement errors. The results are visible in tables 1 and 2. The correlation is statistically significant at intermediate Galactic latitudes in all channels, while at high Galactic latitudes is significant at 410 and 240 GHz.

Once they are converted in brightness units, the slopes $S(\nu)$ give the spectrum of dust brightness fluctuations normalized at 410 GHz. The resulting spectrum is consistent with a single temperature Rayleigh-Jeans thermal spectrum times an emissivity $\varepsilon = \varepsilon_o \nu^\gamma$, with $\gamma = (1.2 \pm 0.3)$ at low latitudes, and $\gamma = (2.3 \pm 1.0)$ at high latitudes.

TABLE 1. Correlations at intermediate Galactic latitudes (22843 7' pixels with $-20^o < b < -10^o$)

frequency (GHz)	R	$S(\mu K_{CMB}/(MJy/sr))$
90	0.032	58 ± 49
150	0.085	93 ± 23
240	0.156	254 ± 46
410	0.298	3200 ± 190

TABLE 2. Correlations at high Galactic latitudes (68987 7' pixels with $b < -20^o$)

frequency (GHz)	R	$S(\mu K_{CMB}/(MJy/sr))$
90	-0.028	-20 ± 110
150	0.003	46 ± 29
240	0.041	258 ± 52
410	0.138	4700 ± 1500

THE DUST AS A FOREGROUND

Now we can combine the results obtained in the two previous paragraphs. The IRAS correlated dust fluctuations detected at 410 GHz by BOOMERanG can be extrapolated to 240, 150, 90 GHz using the measured slopes: $c_\ell(\nu) = [S(\nu)/S(410)]^2 c_\ell(410)$. In this way we find that the contamination from IRAS-correlated dust at 150 GHz is two orders of magnitude below the measured power spectrum of the sky [15], thus confirming that the region of sky observed by BOOMERanG is dominated at 150 GHz by CMB anisotropy. The contamination from the uncorrelated component depends on its spectrum. For reasonable spectra, it is smaller than the correlated part [15].

Our analysis shows that the FDS extrapolation works very well at intermediate latitudes, and within a factor 2 even at high Galactic latitudes. In the disk centered at $b = -27^o$ the rms fluctuation of the SFD IRAS map is 0.15 MJy/sr (extrapolated at 410 GHz). About 20% of the sky is as good or better than such disk, i.e. in about 20% of the sky the dust power spectrum is < 1% of the CMB power spectrum at 150 GHz.

Using multiband instruments, good foreground subtraction will be possible using components separation techniques, even in regions more contaminated than in our reference disk.

COLD DUST CLOUDS

We are studying several selected regions in the dust map of BOOMERanG. Particularily interesting are cold clouds, especially at high latitudes. In fig.4 we show the same high latitude region in the IRAS map, in the BOOMERanG 410 GHz map, and in the map obtained by subtracting the 240 GHz and the 150 GHz maps. If the subtraction is carried out with the maps expressed in CMB temperature units, it removes efficiently the CMB anisotropy, leaving only the noise components and the dust, which in these units is much brighter at 240 GHz than at 150 GHz. General agreement is evident from the comparison. A few clouds colder than the average are also evident in the comparison. We are currently investigating the physical nature of these clouds.

CONCLUSIONS

BOOMERanG has detected thermal emission from ISD cirrus at intermediate and high Galactic latitudes. The 410 GHz map is morphologically very similar to extrapolation of the IRAS(3000GHz) and DIRBE(1250GHz) maps. The angular power spectrum of the dust dominated 410 GHz map is a power law $c_\ell \sim \ell^{-\beta}$ with $2 \lesssim \beta \lesssim 3$. We have detected a component correlated with the IRAS/DIRBE map in all the BOOMERanG bands at $-10^o > b > -20^o$, and in the 150, 240 and 410 GHz bands at higher Galactic latitudes. This dust contamination is negligible with respect to the CMB anisotropy at high Galactic latitudes, accounting for less than 1% of the total angular power spectrum for multipoles $\ell > 100$ at $\nu < 180 GHz$.

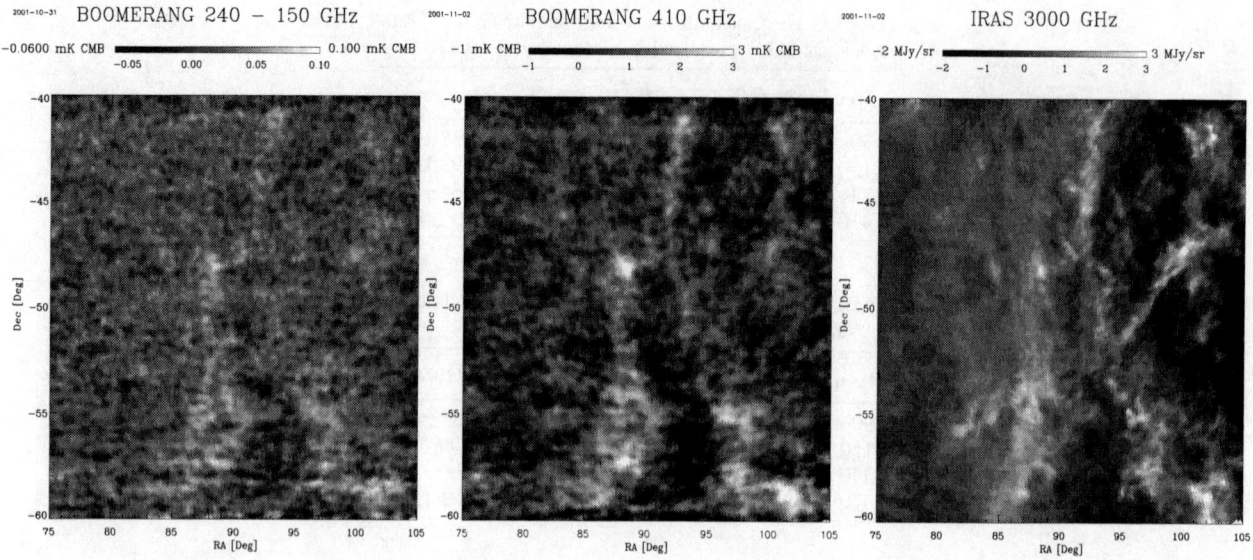

FIGURE 4. A high galactic latitude region with cirrus clouds as seen from BOOMERanG in the 240-150 GHz map (left panel), BOOMERanG (in the 410GHz map, center panel) and IRAS (in the 3000 GHz map filtered in the same way as BOOMERanG - right panel).

ACKNOWLEDGMENTS

This activity has been supported by the University of Rome La Sapienza, Programma Nazionale di Ricerche in Antartide and Agenzia Spaziale Italiana in Italy, by NASA and NSF in USA, by PPARC in UK and by Univ. of Toronto in Canada.

REFERENCES

1. Low F.J. et al. 1984, Ap.J. 278, L19.
2. Masi S. et al. 1995, Ap.J., 452, 253.
3. Masi S. et al. 1996, Ap.J., L47, 463.
4. Leitch E., et al. 1997, Ap.J., 486, L23.
5. de Oliveira-Costa 1997, Ap.J., 482, L17.
6. Gautier T.N. III, et al., 1992, A.J., 103, 4.
7. Low F.J. and Cutri R.M., 1994, Infrared Phys. Techn., 35, 291.
8. Guarini G. et al., 1995, Ap.J. 442, 23.
9. Wright E.L., 1998, Ap.J. 496, 1.
10. Schlegel D.J. et al. 1999, Ap.J. 500, 525.
11. Masi S., et al., in "3K cosmology", AIP Conf. Proc. 476, 237, (1999); astro-ph/9911520
12. de Bernardis, P., et al. 2000, Nature, 404, 955-959
13. Crill B. P., 2002, in preparation
14. Piacentini F., et al. 2001, "The BOOMERANG North America Instrument: a balloon-borne bolometric radiometer optimized for measurements of cosmic background radiation anisotropies from 0.3 to 4 degrees", astro-ph/0105148 (Ap.J. in press); see also his paper in this book.
15. Masi S., et al., 2001, Ap.J, 553, L93-L96, (astro-ph/0101539)
16. Finkbeiner D.P. et al. 1999, Ap.J., 524, 867.
17. Hivon E. et al., 2001, astro-ph/0105302

An Update on the TopHat Experiment

Wilson, G.W.[*], Aguirre, J.[†], Bezaire, J.[†], Cheng, E.S.[**], Christensen, P.R.[‡],
Cordone, S.[§], Cottingham, D.A.[¶], Crawford, T.[∥], Fixsen, D.J.[††], Knox, L.[‡‡],
Kristensen, R.[‡], Meyer, S.S.[†], Norgaard-Nelsen, H.U.[‡], Silverberg, R.F.[**] and
Timbie, P.T.[§§]

[*]*Dept. of Astronomy, University of Massachusetts, Amherst, MA, 01003*
[†]*Dept. of Physics, University of Chicago, Chicago, IL, 60637*
[**]*NASA/GSFC Laboratory for Astronomy and Solar Physics, Code 685.0, Greenbelt, MD 20771*
[‡]*Danish Space Research Institute, Copenhagen, Denmark*
[§]*ISCO International, Mt. Prospect, IL, 60056*
[¶]*Global Science and Technology, Inc., NASA/GSFC Laboratory for Astronomy and Solar Physics, Code 685.0, Greenbelt, MD 20771*
[∥]*Dept. of Astronomy, University of Chicago, Chicago, IL, 60637*
[††]*SSAI, NASA/GSFC Laboratory for Astronomy and Solar Physics, Code 685.0, Greenbelt, MD 20771*
[‡‡]*Dept. of Physics, University of California at Davis, Davis, CA 95616*
[§§]*Dept. of Physics, University of Wisconsin, Madison, WI, 53706*

Abstract. The TopHat Long Duration Balloon (LDB) experiment, which launched on January 4, 2001, mapped 6% of the sky in a region centered about the South Celestial Pole. The five spectral bands of the instrument span from 150 to 660 GHz and are sensitive to CMBR anisotropy and thermal galactic dust emission. Analysis is in progress. The parameters of the experiment, the observing scheme, and some preliminary results are discussed.

INTRODUCTION

Anisotropy in the Cosmic Microwave Background Radiation (CMB) provides a unique observational tool for probing the physical conditions of the very early Universe, determining precise values for fundamental cosmological constants, and providing key quantitative information on the evolution of large scale structure (see *e.g.*, [1], [2], [3], and [4]). Its spectrum, very close to an ideal blackbody at 2.725 K [5], contains a record of the energy release processes from the earliest of times. The small anisotropies in its spatial distribution ($\sim 50\mu K$), are imprints of the distribution of matter at the surface of last scattering at a redshift of $z \sim 1000$. These anisotropies therefore specify the initial conditions for all large-scale structure formation models.

A particularly interesting angular scale for observing CMB anisotropy is near 1°, where the first acoustic peak (or "Doppler peak") enhancement of the fluctuation power spectrum is expected to be observable. This is the angular scale that corresponds to the size of the horizon at the redshift of the effective source of CMB anisotropies. A number of very exciting detections of anisotropy at angular scales near and smaller than 1° have been reported recently (see *e.g.*, [6], [7], [8]). If these results stand the test of time, then they would be among the first generation of precision measurements of CMB anisotropy, providing values for key cosmological parameters at roughly 10% accuracy.

In this paper we describe the flight of one of the most recent precision measurements of the CMB anisotropy - the TopHat experiment. This unique balloon-borne payload, launched from McMurdo Station, Antarctica in January of 2001, is mounted on the *top* of a balloon in order to provide a large-area map with highly redundant sky sampling and a minimal risk of contamination from unknown systematic error sources.

THE TOPHAT INSTRUMENT

TopHat is a compact, light-weight far-infrared telescope designed to be mounted on top of a high-altitude balloon. The top of a balloon is a nearly ideal observing platform for the CMB; there are no hindrances over nearly 2π steradians above the telescope, the platform is above 99.7% of the atmosphere (and a larger fraction of the water vapor), and there is minimal wind shear since the balloon moves along with the air. Long duration balloon flights from Antarctica now allow up to two weeks of continuous observations - only space flight provides a better viewing opportunity.

Top-mounted packages, however, must adhere to stringent mass and size constraints imposed by the launch procedure and stability of the balloon platform. TopHat, for example, is limited to a total mass of 120 kg with its center of gravity no higher than 61 cm above the balloon's top. This includes the mirrors, electronics, mechanisms, shielding, and, of course, the cryostat.

The TopHat telescope is a compact on-axis Cassegrain system designed to rotate continuously, about the local vertical, at a rate of 4 rotations per minute. The detectors are five ion-implanted, monolithic Silicon bolometers cooled to 270 mK by a ^{3}He refrigerator. A sixth "dark" detector is also present to monitor potential sources of systematics. These detectors view the sky with a beam size of approximately 20' in five frequency bands spanning 150 GHz to 660 GHz. In this section, we describe the general features of the TopHat experiment. A future instrument paper will describe each of the subsystems in detail.

Instrument Design

The TopHat experiment is actually composed of two subsystems: the top payload - which flies on top of the balloon, and a second gondola which provides electrical power and a communications link to the top while hanging underneath the balloon.

The top payload, shown in Figure 1, is composed of the telescope, cryogenics, detectors, and readout electronics. The telescope is a 1 m diameter on-axis Cassegrain which produces a 20' FWHM tophat-shaped beam on the sky. The cryogenic system and detectors (both described further below) sit behind the primary mirror in a small volume as shown in Figure 1. Attitude determination sensors and a low-noise data acquisition system (not shown in the figure) take up most of the remaining volume underneath the primary mirror.

The entire telescope is enclosed in a light-weight, double-walled aluminum shield which protects the instrument from Sun and Earth illumination. The shield, telescope, attitude determination sensors, and data acquisition system all rotate at a constant rate of 4 RPM. By rotating the majority of items on top of the balloon, local contamination of the radiometer signals is kept constant while the sky is modulated overhead. This observing strategy is particularly powerful in its ability to mitigate systematic errors and is described in detail below.

An unpointed support gondola hangs below the balloon to provide the top system with power and a telemetry link. An electrical connection is made between the two systems via several redundant sets of wires that are sewn into the gores of the balloon during the balloon's manufacture. Power is generated via eight sets of solar panels which are hung from the four sides of the gondola. Data from the top system is written to two on-board disks and compressed versions are sent to the ground via a line-of-sight radio link or a (much slower) TDRSS satellite link in real time.

Detector System

The Tophat detector system is a dichroic photometer which separates incoming radiation into five spectral bands from 150 to 660 GHz. The infrared detectors are monolithic silicon bolometers built in the Device Development Lab at NASA Goddard Space Flight Center. The central disk is 2.4 mm in diameter and 5 μm thick, and is supported by four legs, each 30 μm wide by 5 μm thick, and 2.3 mm long. A thin coating of bismuth acts as the infrared absorber. The thermistor is formed by implantation doping of the silicon.

These bolometers are not "spiderwebbed" in any way, and the rate of cosmic ray strikes we observed in the Tophat flight is quite low (approximately one per 3 minutes).

The entire photometer is kept at 270 mK by the cryogenic system, described in the next section.

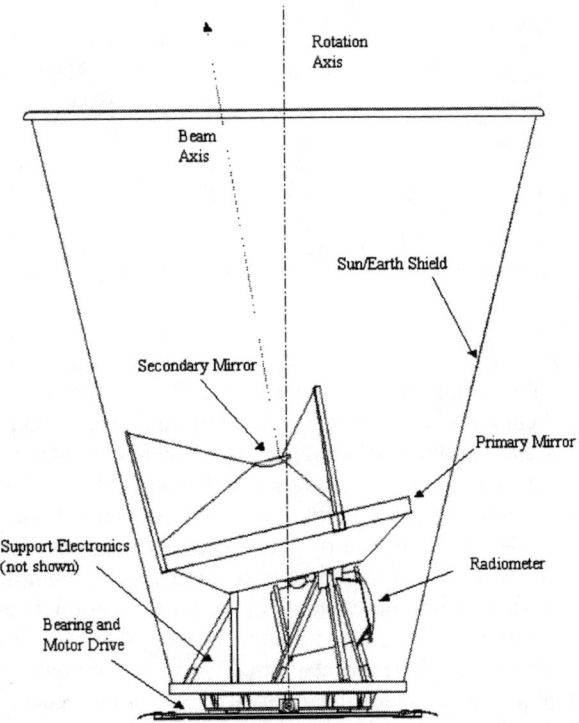

FIGURE 1. This figure shows the configuration of the TopHat telescope. The entire structure shown here is located on top of the balloon, and secured onto the balloon top "apex" plate. The main beam points 12° from the local vertical, defined by the balloon top plate, and spins on the main bearing at the bottom of the telescope. The bottom payload, which hangs below the balloon and contains the support electronics, hard disk archives, batteries, and solar panels, is not shown.

Cryogenics

The stringent weight and size requirements for top-mounted packages necessitated the development of a light-weight, compact, and long hold-time cryostat for the TopHat mission. Weighing more than 100 kg, cryostats which have previously flown on Antarctic LDB missions would consume the entire mass budget for the top of the balloon. By careful design, with attention to weight and heat flow, most of the internal supports can be reduced to tension struts that allow a small lightweight cryostat with a long holdtime.

The TopHat cryostat, the result of a 6 year development effort, is an internally pumped ^{3}He refrigerator with supporting ^{4}He and liquid nitrogen reservoirs. The cryostat has a mass of 10 kg when full of cryogens and is able to maintain an operating temperature of .27 K for a typical week of observing time accessible to a long-duration balloon mission. The cryostat is described in detail in [9].

In preparation for the TopHat flight, four copies (with minor variations) of the cryostat were produced. Over 40 cycles have demonstrated repeatable performance, both over time and with the different dewars. Hold times were measured (or extrapolated) to be consistent for the different cryostats with the limiting cryogen being the 1.8 L of ^{4}He. Tip-tests and the insertion of flingles to change the flow impedance of the vent tube showed the dominant heat transfer to the ^{4}He reservoir to be due to convective flow in the ^{4}He fill/vent tube. Table 1 lists the hold-time results of an in-lab test of the TopHat cryostat performed before launch.

Flight Plan and Observing Scheme

Antarctic LDB payloads are required to function nearly autonomously due to the low bandwidth of communications once the balloon is beyond the range of the line-of-sight telemetry link. A suitable telescope observing scheme must be both simple and repetitive to ensure continuous observations during the flight. The key scientific consideration

TABLE 1. Pre-Flight TopHat Cryostat Characteristics

Cryo. Stage	Volume L	Temp. K	Hold Time days
Pressure Jacket	20	300	–
Nitrogen	4.2	77	10
Isothermal Shield	–	20	–
^{4}He	1.8	4.2	10
^{3}He	0.04	0.270	32

that drives the design of the observing strategy is the ability to remove contamination, instrumental drifts, and local foregrounds while minimizing the loss of information about the CMB. The TopHat observing strategy is designed to uniquely use the top mounted configuration as well as a roughly constant-latitude flight from McMurdo, Antarctica to accomplish all of these goals. It produces multiple observations of a point on the sky at different times and at a number of different telescope orientations to measure and remove such systematic effects.

For TopHat, the beam is offset from the local vertical by 12°, and spins about the local vertical at 4 RPM. When flying from McMurdo (latitude 78° South), this allows each rotation to scan the SCP as a part of a circle that extends 24° (nominal) from the SCP. Over the course of a sidereal day a complete map of a 48°(nominal) cap centered on the SCP is made. Each subsequent sidereal day of observations produces another map. By spinning while observing at a constant elevation, TopHat builds upon the method pioneered by the FIRS experiment [10] and [11], for producing a sky map relatively free of inter-pixel correlations. This configuration extends the capability of balloon-borne measurements on 20' angular scales to include the ability to create large-area maps of the sky intensity. Figure 2 shows the nominal TopHat observing strategy superimposed on the DIRBE dust emission at the SCP.

THE FLIGHT

TopHat launched from McMurdo Station, Antarctica on January 4, 2001. A carefully choreographed launch involving two balloons, two gondolas, several large pieces of machinery, and approximately 15 people was executed smoothly and expertly by the National Scientific Balloon Facility (NSBF) crew. The erection dynamics of the balloon, known from test launches to be highly variable, were smooth and docile. Ironically, the most typical aspect of the launch - releasing the lower gondola - provided the greatest challenge to the NSBF crew who were required to break the ice runway's speed limit to catch up to the fleeing balloon. The overall launch, however, was nearly perfectly executed despite its complexity.

Flight Operations

Once launched, TopHat quickly climbed to its float altitude of 37.5 ± 0.14 km where it remained for the duration of the flight. After a short set of detector characterization tests the initial observations of CMB anisotropy began. On-board data acquisition and CMB observations occurred continuously over the following four sidereal days. In order to further modulate the sky signal with respect to local contamination, the telescope rotation direction was reversed once during the flight and the sign of the bolometer bias was changed twice. Finally, a load-curve was performed on each detector at the beginning of each sidereal day to ensure proper bias settings. Data acquisition was terminated after four sidereal days when it was discovered that the cryostat had run out of cryogens (see below).

Overall, the telescope performed nearly perfectly. All six bolometers functioned to expectation and all TopHat sensors and electronics worked as designed and without problems. Despite the telescope's nominal performance however, there were two uncontrollable and unexpected developments during the flight: an anomalous tilt of 3–5 degrees of the top plate of the balloon and a loss of the TDRSS telemetry link due to a broken NSBF transmitter. Fortunately, the loss of the TDRSS link to the package was mitigated by the retrieval of the on-board data disks. The autonomous operation of the payload allowed TopHat to continue CMB observations despite a lack of commanding and telemetry. The tilt of the top plate, however, has had several effects on the flight. Below we discuss some of these effects as well as our efforts to eliminate any resulting systematic contamination.

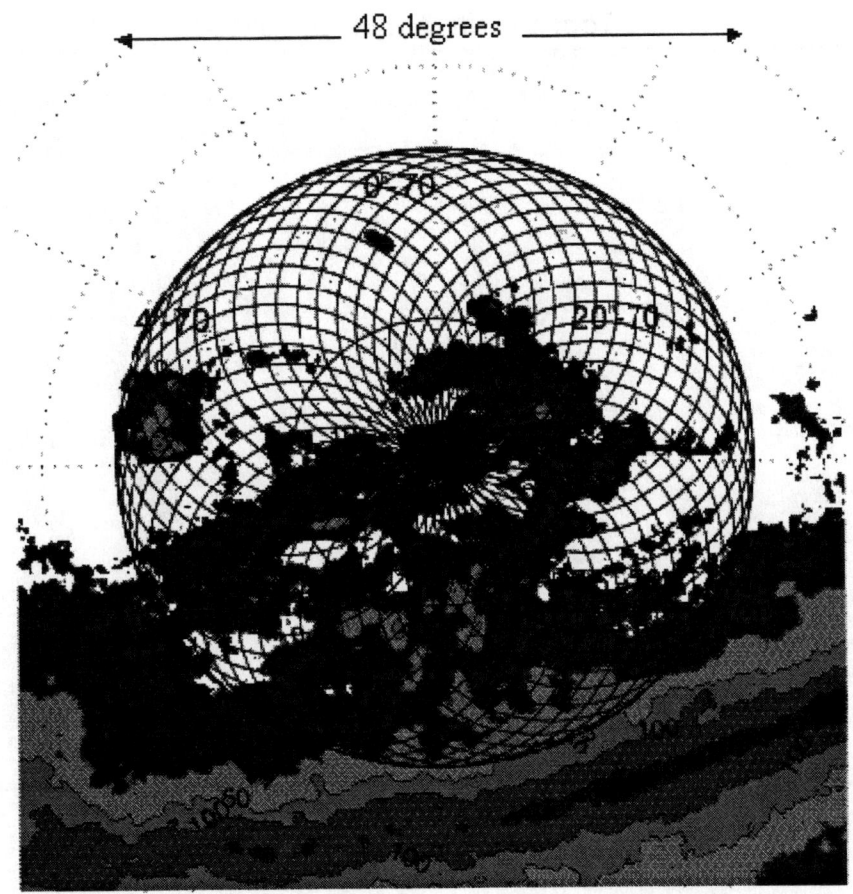

FIGURE 2. The TopHat observing strategy superimposed on the DIRBE dust emission near the south celestial pole.

Tilt Characteristics

Though as of this writing the cause of the top plate tilt remains unknown, we are able to characterize the tilt by using the telescope's housekeeping and attitude determination sensors. Figure 3 shows the amplitude of the tilt of the payload as a function of balloon altitude. The tilt clearly shows a positive correlation with altitude. In addition, we have found the tilt amplitude to depend to a small degree on the side of the balloon facing the Sun.

Fortunately, the tilt of the top plate is stable and changes only on time scales that are long compared to the telescope rotation - i.e., hours. This feature allows us to cleanly separate effects associated with the tilt from true sky signals. The stability of the tilt suggests that it is a global feature of the balloon. As expected, the top of the balloon proved to be a very stable platform. So far in our analyses we detect no instabilities or oscillations in the balloon platform which are driven by the telescope motion.

Effects on the Data

The top plate tilt leads to induced radiometric signals via the following mechanisms:

1. signals induced optically via increased atmospheric pickup
2. signals induced optically from variations in the temperature of the telescope optics due to solar heating
3. signals induced optically from gravitationally induced motions of elements in the optical chain
4. signals induced microphonically

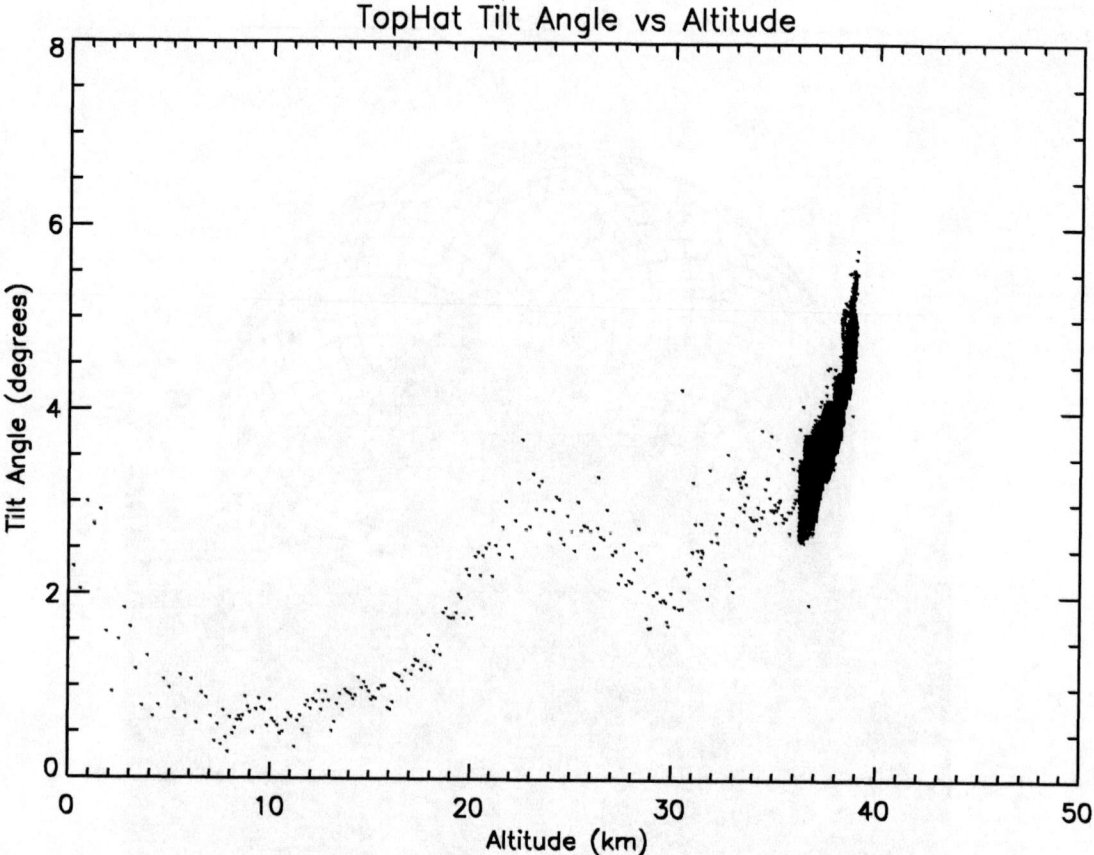

FIGURE 3. Tilt of the balloon top plate as a function of balloon altitude. Shown is a scatter plot of the tilt amplitude as a function of balloon altitude. All of the CMB data is taken at an altitude above 36.5 km (120 kft).

There is evidence in the raw data of all four classes of signals. Fortunately, due to the cross-linking of the observations on the sky and the various levels of modulation of the sky signal, tilt-induced contamination is not locked to the sky and can be separated from the desired signals which are modulated by the sky, balloon, and telescope rotation. The dominant signal from the tilt occurs at the frequency of rotation of the telescope. While this is far from the frequency where the most interesting CMB signals reside, this contamination does affect our measure of the CMB dipole - the primary calibration source for TopHat. Removal of this contamination will be required to accurately calibrate the experiment.

The greatest impact the tilt of the top plate has on the data comes in the reduction of available observing time. After four days at float altitude, the TopHat cryostat ran out of cryogens - between two and four days earlier than expected. This premature warming is likely due to an increased heat load on the ^{4}He tank resulting from a sloshing of the cryogens as the payload executed its observing strategy. Pre-flight lab measurements show the ^{4}He boiloff rate to be sensitive to rapid tilting of the cryostat.

Attitude Reconstruction

While the presence of the tilt complicates the determination of the telescope pointing, the highly redundant nature of the observations and the simplicity of the observing scheme make it possible to reconstruct the pointing to high accuracy. The telescope's attitude is determined by two methods: 1) by direct measurement using on-board attitude sensors (an inclinometer, sun sensors, a magnetometer, and a gyro) and 2) by minimizing the per-pixel variance in the resulting radiometric maps of the sky. The approach is iterative. First, a crude pointing reconstruction is made with method 1. Method 2 is then used to determine unknown (but stable) offsets between the axes of the various pointing

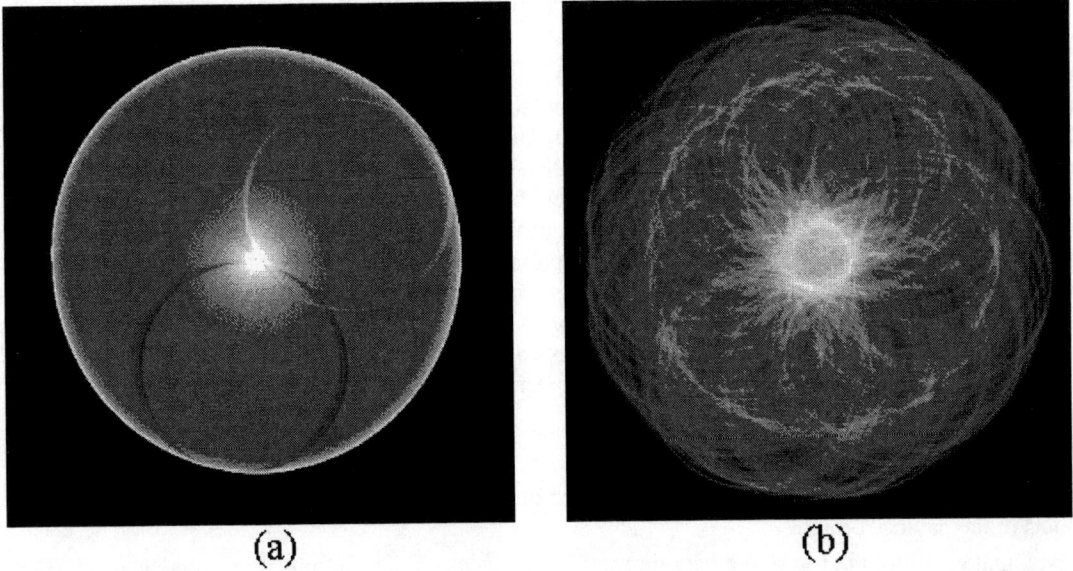

FIGURE 4. Two versions of the TopHat sky coverage: (a) in the absence of the anomolous tilt of the balloon top plate and (b) actual 4-day sky coverage. Note that despite the tilt, uniform sky coverage is achieved. The area of sky covered in (b) is approximately 1.36 times that of (a).

sensors and the IR beam. The process is then iterated until sky maps are created with structures that are self-consistent and agree with independent high-resolution maps such as IRAS and DIRBE.

Figure 4 shows the preliminary coverage map of the TopHat field after four days of observations. As can be seen from the figure, fairly uniform coverage has been achieved despite the tilt of the top plate of the balloon. With further iterations, we estimate that our final pointing uncertainty will be on-order 1 arcminute - better than the level required for precision maps of the CMB at the angular scales that can be addressed by TopHat.

Finally, as is evident from Figure 4, rather than observing the nominal 48° diameter cap centered on the SCP, TopHat has observed a cap with a diameter of approximately 56°- or an area of sky approximately 36% greater than originally planned. While the resulting per-pixel signal-to-noise will be lower than planned, this should only degrade the CMB power spectrum estimations at the smallest angular scales. Measurements at angular scales around the first acoustical peak and larger, on the other hand, should improve slightly due to the reduced sample variance.

DATA ANALYSIS PROGRESS

As of the time of this writing, data analysis of the time stream is in progress. A large amount of the initial effort has been spent in characterizing the tilt of the package, the instrument as a whole, and any potential systematic contamination of the data. Sky maps have been produced in all five frequency bands and rough comparisons have been made with dust maps derived from [12]. Galactic structure is evident in the maps of the high frequency channels and an estimate of the CMB dipole has been recovered from the lower frequency channels. TopHat results will be announced once we have completed the analyses and ensured that any residual contaminations from the tilt of the telescope have been removed.

Once the final set of sky maps for the five channels are produced, spectral decomposition will be used to generate maps of the CMB, the dust, and possibly the Far-Infrared background. These maps will represent the largest and deepest survey of the sky at these frequencies. Band power estimates of the CMB power spectrum will be made from the CMB maps and cosmological parameters will be estimated using these results.

SOME THOUGHTS ON TOP-MOUNTED PACKAGES

The development effort required to successfully fly TopHat from the top of a balloon has been atypical in both its extent and the level of collaboration with the National Scientific Balloon Facility. The TopHat scientists were required to develop a complex cryogenic instrument with a mass 10 times lower than that of a typical balloon-borne telescope. Simultaneously, the National Scientific Balloon Facility developed and tested a launch scenario designed to maximize the opportunities for launch while also ensuring the safety of the ground personnel.

The non-scientific payoff for this difficult work is the enabling of a new observing platform for astronomical and atmospheric studies by complex payloads. With the exception of the unanticipated tilt mentioned above, the top of the balloon proved to be a stable and docile environment with a full view of sky over more than 2π steradians. Only a full analysis of the TopHat data will reveal the scientific payoff of this exercise - however, it can be said that this is the first balloon-borne CMB mission where the balloon itself does not obscure the vast fraction of available sky.

In summary, we offer this list of observations on our experience on the top of the balloon:

1. The top of the balloon is stable and fairly rigid. No balloon motions or oscillations driven by our telescope motion have been detected in our data.
2. At 270lbs, TopHat is the heaviest payload launched on top of the balloon. This weight constraint drove many aspects of the payload design.
3. The anomalous tilt of the top of the balloon remains a mystery. It has an altitude dependence and (to a smaller degree) a dependence on the side of the balloon facing the sun.
4. The complexity of launching a massive top-mounted package severely limits flight opportunities due to the nearly ideal weather conditions required for launch.
5. With the exception of satellites, there is no better platform for low-atmosphere, low-background observations of the firmament.

CONCLUSIONS

The TopHat Long Duration Balloon experiment, which launched on January 4, 2001, mapped 6% of the sky in a region centered about the South Celestial Pole. TopHat represents the latest in the first series of experiments capable of making precision measurements of large fractions of the sky at millimeter and sub-millimeter experiments and will produce sky intensity maps in five distinct frequency bands ranging from 150–660 GHz. Analysis of the TopHat data set is ongoing.

REFERENCES

1. Hu, W., and White, M., *Astrophys. J.*, **471**, 30 (1996).
2. Knox, L., *Phys. Rev. D*, **52**, 4307 (1995).
3. Jungman, G., Kamionkowski, M., Kosowsky, A., and Spergel, D. N., *Phys. Rev. D*, **54**, 1332 (1995).
4. White, M., Scott, D., and Silk, J., *AR&AA*, **32**, 319 (1994).
5. Mather, J. C., Fixsen, D. J., Shafer, R. A., Mosier, C., and Wilkinson, D. T., *Astrophys. J.*, **512**, 511 (1999).
6. Halverson, N. W., et al. (2001), preprint `astro-ph/0104489`.
7. Netterfield, C. B., et al. (2001), preprint `astro-ph/0104460`.
8. Hanany, S., et al., *Astrophys. J.*, **545**, L5 (2000).
9. Fixsen, D. J., Cheng, E. S., Crawford, T. M., Meyer, S. S., Wilson, G. W., Oh, E. S., and Sharp, E. H., *RSI*, **72** (2001), in press.
10. Boughn, S. P., Cheng, E. S., Cottingham, D. A., and Fixsen, D. J., *Astrophys. J. Letters*, **391**, L49 (1992).
11. Page, L. A., Cheng, E. S., and Meyer, S. S., *Astrophys. J. Letters*, **355**, L1 (1990).
12. Finkbeiner, D. P., Davis, M., and Schlegel, D. J., *Astrophys. J.*, **524**, 867 (1999).

Archeops: A balloon experiment to measure CMB anisotropies with a broad range of angular sizes

A. Benoît* and the Archeops Collaboration[†]

*Centre de Recherche sur les Très Basses Températures, 25 Avenue des Martyrs BP166, F-38042 Grenoble Cedex 9, France
[†]from the following institutes:
California Institut of Technology, Pasadena USA
Centre d'Etude Spatiale des Rayonnements, Toulouse France
Centre de spectrométrie nucléaire et de spectrométrie de masse, Orsay France
Colllège de France, Paris France
DAPNIA CEA, Saclay France
Institut d'Astrophysique de Paris, Paris France
Institut d'Astrophysique Spatiale, Orsay France
Institut des Sciences Nucléaires de Grenoble, Grenoble France
IROE CNR, Firenze Italy
Jet Propulsion Laboratory, Pasadena USA
Laboratoire d'Astrophysique de l'Observatoire de Grenoble, Grenoble France
Laboratoire de l'accélérateur linéaire, Orsay France
Landau Institute of Theoretical Physics, Moscow Russia
Observatoire Midi-Pyrénées, Toulouse France
Queen Mary and Westfield College, London UK
Universita di Roma La Sapienza, Roma Italy
University of Minnesota at Minneapolis USA

Abstract. The Cosmic Microwave Background Radiation is the oldest photon radiation that can be observed, having been emitted when the Universe was about 300,000 year old. It is a blackbody at 2.73 K, and is almost perfectly isotropic, the anisotropies being about one part to 100,000. However, these anisotropies, detected by the COBE satellite in 1992, constrain the cosmological parameters such as the curvature of the Universe.
Archeops is a balloon-borne experiment designed to map these anisotropies. The instrument is composed of a 1.5 m telescope and bolometers cooled at 85 mK to detect radiation between 150 and 550 GHz. To lower atmosphere parasitic signal, the instrument is lifted at 32 km altitude with a stratospheric balloon during the arctic night. This instrument is also a preparation for the Planck satellite mission, as its design is similar.
We discuss here the results of the first scientific flight from Esrange (near Kiruna, Sweden) to Russia on January 29th 2001, which led to a 22% (sub)millimetre sky coverage unprecedented at this resolution.

THE SCIENTIFIC OBJECTIVE

The Cosmic Microwave Background

The Cosmic Microwave Background Radiation (CMBR) was emitted by the Universe when it was 300,000 years old just after the Big Bang. Its spectrum is known as a blackbody with a temperature of only 2.725 degrees above absolute zero. In various directions in the sky, we observe small temperature differences of the order of one part in 100000, that were measured for the first time by the COBE satellite [9]. These so-called anisotropies trace the fluctuations of the density of matter that occured before the decoupling of the CMBR. These fluctuations are thought to be the origin, by gravitational collapse, of the large-scale structure of the Universe (galaxies, clusters,...) that we observe today. Its pattern can also yield an indirect measurement of the density, age and curvature of the Universe (see *e.g.* [5]).

There have been many experiments that have already measured these anisotropies with various techniques, angular resolution, noise and scanning strategy. Most recent ones (e.g. TOCO, Boomerang [2, 7], and Maxima [4, 6]) have improved on COBE results by the wavelength coverage, the sensitivity and the angular resolution.

The observation strategy

Balloon experiments are either limited by integration time due to small duration flights (in USA or Europe) or Sun disturbance (in Antarctic Summer). This in turn forbids mapping large portions of the sky. An alternative is to use a flight during the polar night in the more accessible Arctic region.

The Archeops experiment[1] aims at mapping the anisotropies of the cosmic microwave background from small to large scales at the same time. For this purpose, a beam of about 8 arcminutes is swept through the sky by spinning a 1.5 m telescope pointing at 41 degree elevation around its vertical axis. A large fraction of the sky is covered when the rotation of the Earth makes the swept circle drift across the celestial sphere. This is only possible if the observations are done during the Arctic night and on a balloon where neither the Sun nor the atmosphere disturb the measurements. Ozone cloud emission and residual winds can be avoided with a high altitude strastospheric balloon.

From the Swedish balloon and rocket base in Esrange near Kiruna, in cooperation with Russian scientists, the CNES balloon team can launch balloons in the polar night, with a typical trajectory ending just before the Ural mountains in Russia. Integration times can be up to 24 hours in the December-January campaigns.

THE INSTRUMENT

A general description of the first Archeops instrument can be found in [1] where the first gondola used during the test flight (that happened in Trapani in July 1999) is described. The present experiment mainly uses the same concept.

The telescope, optics and detectors

The Archeops telescope is a two mirror, off-axis, tilted Gregorian telescope consisting of a parabolic primary (main diameter of 1.5 m diameter) and an elliptical secondary (this design is similar to the one proposed for Planck during phase-A). The telescope was designed to provide diffraction-limited performance when coupled to single mode horns producing beams with FWHM of 8 arcminutes or less at frequencies higher than 140 GHz. Both mirrors were milled from 8 inch thick billets of aluminum 6061-T6 and were thermally cycled twice during machining to relieve internal stresses. The primary and secondary mirrors weigh 45 kg and 10 kg respectively.

For CMB anisotropy measurements, control of spectral leaks and beam sidelobe response is critical. Archeops channels have been specifically designed to maximize the sensitivity to the desired signal, while rejecting out-of-band or out-of-beam radiation. We have chosen to use the configuration developed for Planck HFI, using a triple horn configuration for each photometric pixel, as shown schematically in Fig. 1.

In this scheme, radiation from the telescope is focussed into the entrance of a back-to-back horn pair. With no optical components in the path, control of the beam is close to ideal. Proper single mode corrugated feeds associated with a new profiled-flared design will be used to obtain 30dB telescope edge taper with the telescope/horns combination. The new profiled-flared horns avoid the use of a lens at the exit aperture of the second horn, creating a beam-waist where wavelength selective filters can be placed. Finally, the third horn maintains beam control and focuses the radiation onto the spider web bolometer placed at the exit aperture.

A convenient aspect of this arrangement is that the various components can be placed on different temperature stages in order to create thermal breaks and to reduce the level of background power falling onto the bolometer and fridge. In Archeops, the back-to-back horn pair is located on a cold plate cooled by Helium vapor at 8 K. Sidelobe response, beamwidth on the sky and spillover are accurately controlled by the design of the front horn. More low pass

---

[1] More details on the experiment can be found at http://www.archeops.org

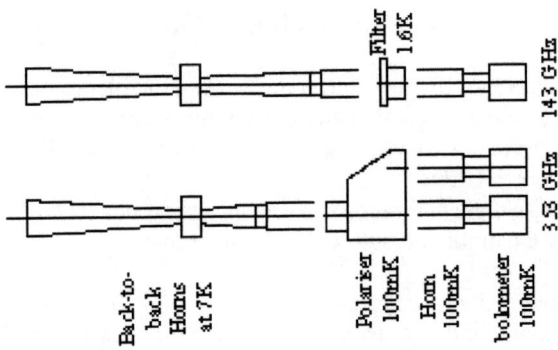

FIGURE 1. The cold optics (of which the 143 GHz channel is shown on the top) is made of a 10 K back-to-back horn, a 1.6 K filter stage and the 100 mK horn and bolometer stage. Light enters from the left. A typical 353 GHz channel has one 10 K stage for two bolometers. A polariser splitter is inserted at the 1.6 K stage (bottom configuration).

filters are placed on a screen at 1.6 K to further reject unwanted radiation from the inner sanctum where the 100 mK detectors are located.

Twenty two bolometers are placed on the 100 mK low temperature plate. There are 9 bolometers at 143 GHz, 7 at 217 GHz, 6 polarised bolometers at 353 GHz and two at 545 GHz. The higher frequency horns (545 GHz) are multimoded, as this increases the signal at this frequency and the side lobes rejection is less critical. One blind bolometer is placed on the same copper plate to study the electronic noise of the bolometers at this stage. These are placed at different points in the focal plane and observe the same sky pixel at a different time. Bolometers on the same line observe the sky with typicaly 100 msec time difference as bolometers on different lines observe the same pixel with a time difference of the order of a few minutes. The six 353 GHz channels are devoted to the measurement of galactic polarized emission. The bolometers are assembled in three pairs, with one single back-to-back horn and a polarizer splitter. The two bolometers of each pair measure the polarized intensity of the incoming signal in two orthogonal directions. Each pair makes a different angle with respect to the scan axis to enable the full determination of the Stokes parameters. Archeops will provide the first measurement of polarization in this range of frequencies with a sensitivity adequate for measuring galactic dust polarised emission, as well as a validation of the technical configuration for PLANCK-HFI.

The gondola and the pivot

The gondola is made with welded aluminium square tubes 30*30*2 and a careful design prevents from important deformations of the optical design in the presence of strength. Typical change in the relative mirror position stays below 0.2 mm when the gondola is lifted or tilted. Total mass of the main frame is 70 kg. The two mirrors and the cryostat are fixed to the frame and the elevation of the beam direction is fixed to the value of 41 degrees.

The pivot connects the flight chain of the balloon to the payload through a thrust bearing, providing the necessary degree of freedom for payload spin. Two deep groove bearings provide stiffness against transverse loads to the rotating steel shaft inside the pivot. The pivot includes a torque motor that acts against the flight chain to spin the payload. After initial acceleration, the motor provides just enough torque to compensate the friction in the thrust bearing and the small residual air friction.

The rotation of the payload is monitored by a vibrating structure rate gyroscopes that can detect angular speeds as low as 0.1 deg/s. These are sampled at 150 Hz by a 16 bit ADC and a PID feedback loop control is implemented in software, to drive the torque motor in the pivot.

The fast stellar sensor

A custom star sensor has been developed for pointing reconstruction in order to be fast enough to work on a payload rotating at 2-3 rpm. At this spin rate, the use of a pointed platform for the star sensor is impractical. Each independent beam (8 arcmin. wide) is scanned by the mm-wave telescope in about 10 ms, establishing a detector response time that excludes the use of present large-format CCDs.

We decided therefore to develop a simple night sensor, based on a telescope with photodiodes along the boresight of the mm-wave telescope. Thus, like the millimeter telescope, the star sensor scans the sky along a circle at an elevation of 41 deg.

A linear array of 46 sensitive photodiodes (Hamamatsu S-4111-46Q) were placed in the focal plane of a 40 cm diameter, 1.8 m focal length parabolic optical mirror. Each photodiode has a sensitive area of 4 mm (in the scan direction) by 1 mm (pitch in the cross-scan direction). The line of photodiodes is perpendicular to the scan and covers 1.4 degrees in elevation on the sky, with about 7.6 arcminute (along the scan) by 1.9 arcminutes (cross-scan) per photodiode. A top baffle, painted black inside and located above all nearby payload structures, prevents stray radiation.

The sensitivity of the photodiodes defines the average number of stars we can observe during one rotation of the payload. With a sensitivity limited to stars of magnitude 7, we can count between 50 and 100 stars per turn during night time. In order to control the pointing during the day, we use an optical filter in front of the diodes to minimize the perturbation due to stray light. A test flight in Kiruna (April 1999) shows that, even with the Sun at low elevation (< 5 deg. elevation) we can observe a few stars for each rotation of the gondola.

The star sensor software extracts from the time-sampled photodiode signals candidates star with detection time, measured flux, coordinate along diode array, and quality criteria. This software produces from raw data the list of time-ordered star candidates and makes it available for the second step of the software-reconstruction of the telescope pointing. The attitude reconstruction algorithm is based upon the comparison between star candidates and a dedicated star catalog. The fluxes of stars in catalog are computed from the Hipparcos catalog to simulate the star sensor spectral response.

The reconstruction is achieved using only star sensor data if the gondola spin axis motion is sufficiently slow. That was the case for the Trapani test flight. The precision of the pointing solution is better than 1 arcminute rms for the test flight. For the Kiruna 2001 flight, important speed variation of gondola rotation velocity required to use additional information from gyroscopes and gps to recover a good association between signal and star catalog.

The cryogenics

The focal plane is cooled to 100 mK by means of an open cycle dilution refrigerator. This type of refrigerator has been designed for satellite applications (it will be used on Planck HFI) and Archeops is the first balloon-borne experiment using a dilution refrigerator. The dilution stage is placed in a low temperature box placed on the top of a liquid Helium reservoir at 4.2 K. The top part of this box contains the entrance horns and receives a significant amount of heat power from near infrared radiation (about 500 mW). Exhaust vapours from the helium tank maintain the horns near 7 K. The entrance is protected from radiation by two vapour cooled screens with openings for the input beam. The filters are placed on the horns at 7 K, on the 1.6 K stage (cooled by Joule-Thompson expansion of the dilution mixture) and on the 100 mK stage, just in front of the bolometers. The temperatures of each stage are monitored with thermometers: carbon resistance and NbSi metal insulation transition thermometers.

The bolometers are placed on the 100 mK stage supported by Kevlar cords. The dilution fluids (isotopic pure ^{3}He and ^{4}He) arrive through two small capillary tubes along a heat exchanger with the return mixture. The two capillaries join and the ^{3}He is dissolved into the ^{4}He, cooling down the mixture which is used to cool down the 100 mK plate using a small heat exchanger. We use two extra capillaries of larger diameter (0.5 mm) in order to precool the system by a circulation of ^{4}He gas. These 5 capillary tubes (3 for dilution and 2 for precooling) and the electric wires (9 shielded cables with 12 conductors each) are soldered together, forming the continuous heat exchanger disposed around the 100 mK stage. Input flow is controlled by an electronic flow regulator. The output mixture is pumped with a charcoal pump placed inside the liquid helium (1 liter box filled with charcoal). During pre-launch operations, the output mixture is extracted with an external pump and the dilution stage can stay below 100 mK continuously for months. The hold-time of the cryostat is limited at 48 hours by the liquid helium tank of 20 liters. An electronic regulator is used to maintain constant pressure at one atmosphere in the helium tank.

In order to insure temperature stability of the bolometer a passive filter is used to thermalise the bolometer plate. The open cycle dilution produces large temperature fluctuation ($100\,\mu$K/sqrt(Hz)) which we attenuate with a high specific heat material (HoY). Holmium has a Shottky anomaly around 200 mK which insures the high specific heat. Mixing it with Yttrium helps controlling the conductivity. By cooling down the bolometers through this thermal filter, a stability below a few μK/sqrt(Hz) was obtained during the flight.

The electronics

The bolometers are biased using AC square waves by a capacitive current source. Their output is measured with a differential preamplifier (the first stage uses JFET working at about 120 K) and digitized before demodulation. We use the boxes already designed in preparation for Planck HFI instrument. Each box can manage 6 bolometers and we used 6 of them for a total of 36 channels. All modulations are synchronous and driven by the same clock. This clock is also used for data readout, which is simultaneous for all bolometers and thermometers. Modulation parameters can be controlled by telecommand. Sampling of the raw signal is at 6.51 kHz before demodulation. Demodulation is performed by the EPLD and sampled twice per modulation period. We used a frequency of 76 Hz for modulation and 152 Hz for sampling. The on-board computer uses a transputer T805 and an 4 EPLD Altera 9400 to control all houskeeping measurement, bolometer and star sensor data. The power supply consists of 39 batteries for electronics and satellite telemetry and 36 for the motor. All are 3 V and 36 Ah lithium batteries. With a total power of 150 W, this gives us about 48 h lifetime for the experiment.

The telemetry

After compression, the data are written to a storage module of 2 Gbyte Flash Eprom memory made of 256 circuits of 8 Mbyte each. A dedicated microprocessor is used to write the Eprom. To protect the data in case of bad landing, the data storage module is installed in a sealed box, pressurized at 1 atmosphere. The data are read after retrieval of the gondola. The compressed data are sent via the standard CNES telemetry (400 MHz) at the rate of 108 kbit/sec. This is possible only during the first phase of the flight (about 4 hours) when the balloon is in direct contact with the ground telemetry station. Another telemetry channel using the Inmarsat satellite is used to control the experiment during all the flight. We use the mini-M Inmarsat standard that allows a typical flow rate of 2 kbit/sec. A selected fraction of the data is sent through this channel to control the experiment and commands can be sent to correct all control parameters.

ARCHEOPS FLIGHTS AND FIRST RESULTS

A first flight of the instrument took place in Trapani on July 17th 1999. This test flight used only a few detectors (5) and we got only 4 hours of data during the night. Nevertheless, this flight allowed us to check all the fonctionnalities of the instrument [1].

Flight conditions at Esrange in 2000-2001 Winter

This Winter (December 2000 and January 2001), the polar vortex was not well positioned as the 2 previous winters and we did not get good flight conditions during the campaign. The wind conditions in the stratosphere gave us the possibility of a long flight only on December 1rst, where the Archeops technical flight took place. Later on, we did not get good conditions before January 12th 2001 where we launched Archeops for a 7-8h flight. However, one of the flow-meter controlling the cryostat broke down 1 hour after take off and we had to abort the flight with a landing in Finland and a fast recovery without too much damage on the instrument. The next launch window opened on January 29th for a relatively short duration flight at an altitude of 32 km (too much wind at higher altitude) lower than the nominal 40 km.

The scientific flight

We finally were able to launch Archeops for its first scientific flight on the 29th of January, 2001. Because of the wind conditions we could only use a smaller balloon (150 000 m^3) otherwise the trajectory would have been too much North, endangering the recovery.

We had 7h30 of flight at float, with a temperature of the order of 90 mK on the focal plane almost all the way along. We decided not to fly too close to the Ural mountains: this would have been too risky for the recovery because of strong wind. The experiment was stopped (window closed and motor stopped) about one minute before the separation with the balloon occured thanks to Inmarsat (00h26m30 LT). The gondola reached the ground at a latitude of 62.226 and a longitude of 53.341 (N-E of SyktYvkar).

Preliminary results

The first task after the flight is to reconstruct the pointing using the data from the stellar sensor. The residual error in pointing is given by the distribution of the declination difference between the reconstructed position and the stars. The present precision of the pointing solution for this flight is better than 2 arcminute rms.

To control the accuracy of the reconstructed pointing and extract the in-flight angular resolution of the detectors, their time response, and to calibrate the instrument on point sources, we use the signal of Jupiter measured in the bolometers and the pointing of each detector reconstructed from the Stellar Sensor. Jupiter is a bright source which is seen by one bolometer at a time thanks to Archeops scan strategy, we can therefore easily pinpoint its crossings (we could see it twice during the January 2001 flight).

If the pointing is accurate enough on the corresponding time periods, we can reconstruct the angular resolution as shown on Fig. 2, where the beam of one of the 217 GHz detector is represented as obtained on Jupiter: we found the FWHM of this particular beam to be of the order of 12.5 arcminutes, which includes the bolometer and electronics time response: the deduced optical beam is nominal at 8 arcminutes.

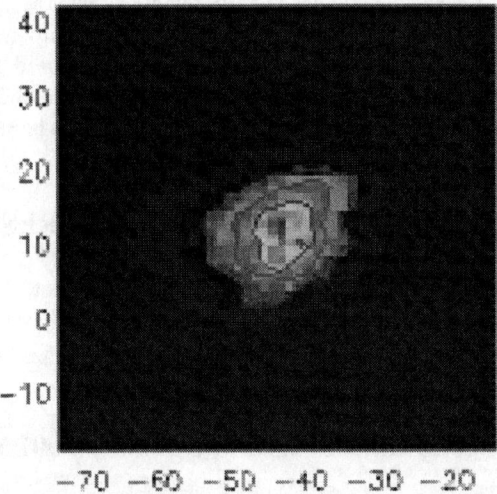

FIGURE 2. Jupiter as observed by one of the bolometers at 217 GHz. Azimuth-Elevation coordinates are in arcminutes. Note that the bolometer time constant spreads the signal a little on the left.

Since the response of the bolometers to Jupiter's signal is a convolution of the intrinsic resolution of the back-to-back horns and the time response of the bolometer, we can also extract a measurement of both parameters: we typically found a time constant of the order of 5 to 15 ms, compatible with measurements on glitches (the intrinsic resolution of the horns being between 5 and 8 arcminutes).

As far as the calibration on point sources is concerned, we make use of the temperature of Jupiter [3] (T = 170 K). Taking into account its solid angle, we have extracted a calibration for each Archeops bolometers.

The 143 and 217 GHz channels are dominated by the cosmic dipole and some extra signal coming from the 10 K back-to-back horn emission (sinusoidal shape). At 545 GHz, the emission from the Galaxy is dominant as well as some atmospheric signal.

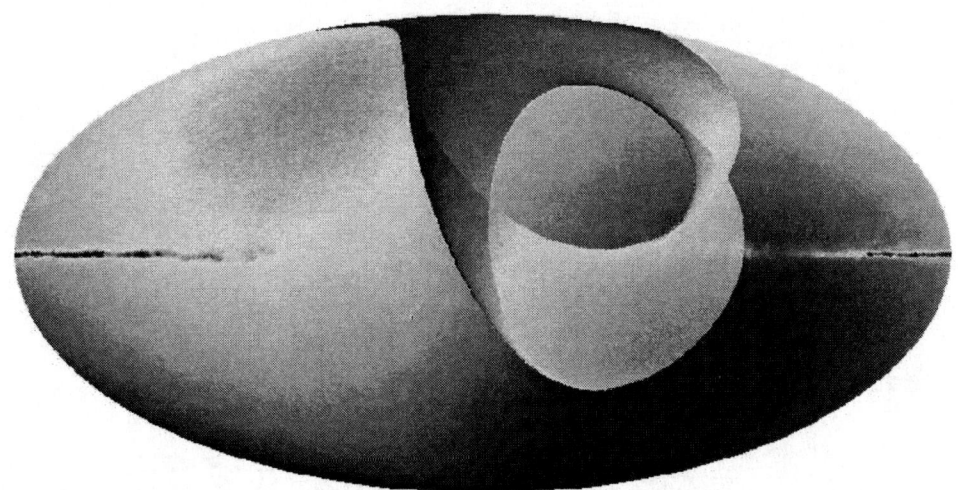

FIGURE 3. Expected 143 GHz map with the CMB dipole, anisotropies and IRAS/DIRBE extrapolation of galactic dust emission ([8]). This is an all-sky Mollweide projection centered on the Galactic anticenter. Overimpression shows the area covered during last Kiruna flight. Revolutions started in the lower right and finished in the upper left. Most of the sky above the galactic plane can be used for CMB anisotropy measurements

Flight at ceiling represents 7.5 hours worth of scientific data taken when the gondola was spinning at 2 rpm. The covered area corresponds to 22 percent of the whole sky. Fig. 3 shows a Mollweide projection in galactic coordinates (centered on the Galactic anticenter) of an extrapolation to the submillimetre of a combined IRAS-DIRBE map [8]. Due to the relative small duration (compared to 24h of one earth rotation), the area where the circles cross each other is very small. As a consequence, we had a relatively poor redundancy during this flight.

The galactic plane is well observed at all frequencies from Perseus to Cygnus regions. Some clouds much below the Galactic plane can easily be identified with their CO and infrared counterparts (Taurus, Pleiades, ...).

With the 353 GHz channels, Archeops will provide the first measurement of galactic polarized emission in this range of frequencies. It is first an important topic in the prospect of foreground removal for Planck-HFI, and is also of great interest to constrain the physics of galactic dust and molecular clouds.

Sensitivities are typically between 50 and $100\mu K_{RJ}$ for one second of integration for one photometric pixel. There are about 8 pixels with a CMB sensitivity between 120 and $200\mu K_{CMB}$ for one second of integration for one photometric pixel. This instrument is therefore very competitive with respect to other designs. Work is in progress to subtract all parasitic signals and extract the CMB fluctuation spectrum. Expected performances for the present flight are shown in Fig. 4. Good detections of the CMB anisotropy spectrum can be expected from low l to beyond the first acoustic peak.

CONCLUSIONS AND FUTURE FLIGHTS

Archeops is a balloon borne experiment dedicated to the measurement of the anisotropies of the Cosmic Microwave Background using the same technology as the Planck satellite will use. We were able to launch the instrument on the 29th of January 2001 at Esrange (Sweden). The flight lasted 7h30 at float, with all detectors working with nominal performances. This is a very important step in validating Planck-HFI concepts (Lamarre et al., this conference). The flight was unusually short due to strong winds in the stratosphere. We are currently analysing the data: reconstructing the pointing using the Stellar Sensor data, calibrating the instrument using point sources (Jupiter) as well as the dipole and the galaxy, and already some results are coming along such as maps of the galactic plane.

Nevertheless, because of the strong winds, the movements of the gondola induced a parasitic signal on bolometers which is different for each frequency. This makes the data analysis more difficult than planned. A flight at higher altitude (with a 400000 or 600000 m3 balloon) should permit to have a more stable experiment, with less atmospheric signal. In addition, to have a better measurement of the Cosmic Microwave Background, a longer flight (a 24 hour flight for instance or two flights with 12 hours each) should provide a larger redundancy on the sky. Expected sensitivity

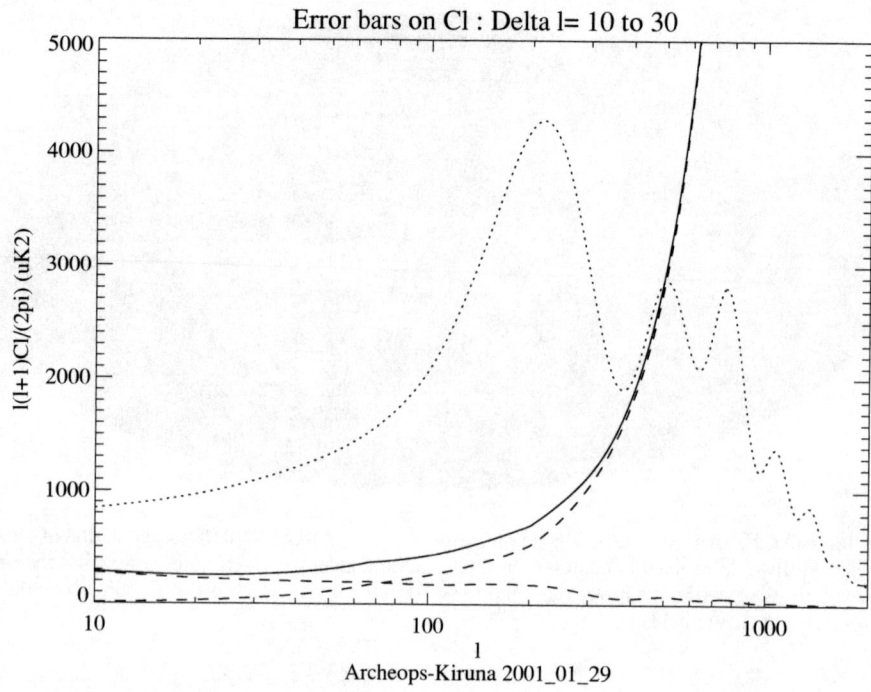

FIGURE 4. Sensitivity expected for the last Kiruna flight. The dotted curve shows a fiducial C_l spectrum for a standard cosmology (in μK^2). The continuous line is the expected 1σ error using the characteristics of the instrument during the flight (flat noise average and responsivity). The first dashed curve (on the left) shows the cosmic variance part of the noise and the second (on the right) shows the effect of detector noise.

shows that significant detections of the C_l spectrum could be achieved between $l = 10$ to 1000. The large area covered by Archeops makes it a unique balloon instrument in lowering the CMB cosmic variance, the most important noise at l below 100 (see Fig. 4). A new campaign is planned for the Winter 2001/2002, with two launch windows: one in December and one in January.

ACKNOWLEDGMENTS

We thank the CNES and Esrange Swedish Facility for their continued support for this project and the flights (technical and scientific) that were realised very smoothly.

REFERENCES

1. Benoît, A., et al. , Astroparticle Physics, in press, astro-ph/0106152, 2001
2. de Bernardis, P., et al. , Nature, 404, 955, 2000
3. Goldin, A. B., et al. , Astrophysical Journal, 488, L61, 1996
4. Hanany, S., et al. , Astrophysical Journal, 343, L3, 2000
5. Hu, W., Sugiyama, N., and Silk, J., Nature, 386, 37, 1997
6. Lee, A. T., et al. , preprint, astro-ph/0104459, 2001
7. Netterfield, C. B., et al. , Astrophysical Journal, submitted, astro-ph/0104460, 2001
8. Schlegel, D. J., Finkbeiner, D. P., Davis, M., Astrophysical Journal, 500, 525, 1998
9. Smoot, G. F., et al. , Astrophysical Journal, 371, L1, 1991

Noise Properties of the BOOMERANG Instrument

F. Piacentini[a], P.A.R. Ade[b], J.J. Bock[c], J.R. Bond[d], J. Borrill[e], A. Boscaleri[f],
K. Coble[g], C.R. Contaldi[d], B.P. Crill[h], P. de Bernardis[a], G. De Troia[a], P. Farese[g],
K. Ganga[i], M. Giacometti[a], E. Hivon[i], V.V. Hristov[h], A. Iacoangeli[a], A.H. Jaffe[l],
W.C. Jones[h], A.E. Lange[h], L. Martinis[m], S. Masi[a], P. Mason[h], P.D. Mauskopf[n],
A. Melchiorri[o], T. Montroy[g], C.B. Netterfield[p], E. Pascale[f], D. Pogosyan[d],
G. Polenta[a], F. Pongetti[q], S. Prunet[d], G. Romeo[q], J.E. Ruhl[g], F. Scaramuzzi[m]

[a] *Dipartimento di Fisica, Università La Sapienza, Roma, P.le A. Moro, 2, 00185, Italy.* [b] *Queen Mary and Westfield College, London, UK.* [c] *Jet Propulsion Laboratory, Pasadena, CA, USA.* [d] *C.I.T.A., University of Toronto, Canada.* [e] *N.E.R.S.C., LBNL, Berkeley, CA, USA.* [f] *IROE-CNR, Firenze, Italy.* [g] *Dept. of Physics, Univ. of California, Santa Barbara, CA, USA.* [h] *California Institute of Technology, Pasadena, CA, USA.* [i] *IPAC, Caltech, Pasadena, CA, USA.* [l] *Department of Astronomy, Space Sciences Lab and Center for Particle Astrophysics, University of CA, Berkeley, CA 94720 USA.* [m] *ENEA, Frascati, Italy.* [n] *Dept. of Physics and Astronomy, Cardiff University, Cardiff CF24 3YB, Wales, UK.* [o] *Nuclear and Astrophysics Laboratory, University of Oxford, Keble Road, Oxford, OX 3RH, UK.* [p] *Depts. of Physics and Astronomy, University of Toronto, Canada.* [q] *Istituto Nazionale di Geofisica, Roma, Italy.*

Abstract. In this paper we report a short description of the BOOMERANG experiment explaining his scientific goal and the technologies implied. We concentrate then on the analysis of the noise properties discussing in particular the scan synchronous noise. Finally we present the calibration technique and the sensitivity of all the channels.

INTRODUCTION

BOOMERANG is a telescope devoted to the detection of structures in the Cosmic Microwave Background (CMB) radiation. The measure of structures in the primordial plasma (anisotropies in the CMB radiation, see [1] for a complete review) was the main goal of the recent observations in cosmology because gives information on the initial conditions that gave rise to the observed Universe. Moreover gives us the capability to tune the cosmological theories measuring the cosmological parameters [2], strongly correlated to the angular power spectrum of the CMB anisotropies (the statistical distribution of the size and the amplitude of the features in the CMB).

The CMB is observed now as Black Body emission at $T = 2.726 \pm 0.005$ [3] which peaks in the microwave frequencies (80 to 250 GHz). The level of the expected fluctuation is of the order of $\Delta T/T < 10^{-5}$, so a good experiment has to be able to detect temperature fluctuations of the order of $30\,\mu K$.

The angular dimensions of the most interesting features goes from a few degrees to a few arcminutes. Fluctuations in this angular range are expected to produce the maximum signal.

THE INSTRUMENT

Given those requirements BOOMERANG and other recent experiments have been developed with great care to make them reliable and to control systematics effects. BOOMERANG [4, 5] succeeded in providing maps of the CMB radiation and the corresponding angular power spectrum with a high signal to noise ratio [6, 7].

The telescope uses 16 bolometric detectors made by a micromesh absorber with an indium bump-bonded NTD

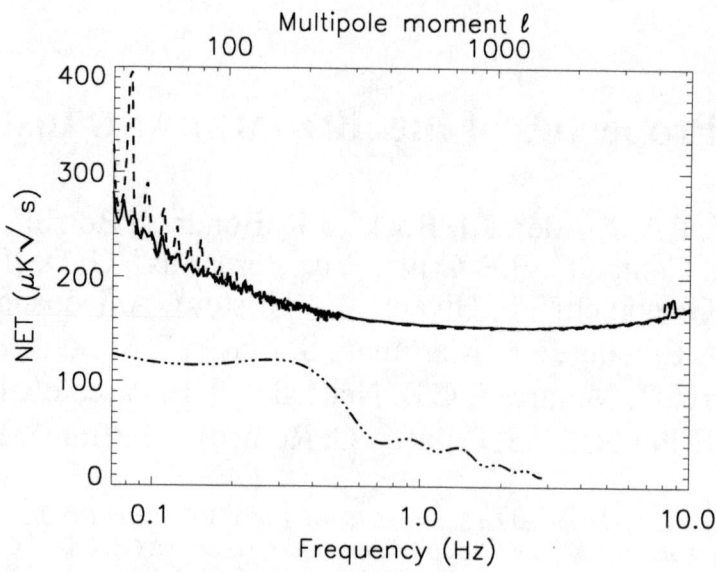

FIGURE 1. Typical power spectrum of the data and of the expected anisotropies for a BOOMERANG channel. The power spectral density of the data is plotted in thermodynamic temperature fluctuations of a 2.73 K blackbody.

Germanium thermistor [8] cooled at 280 μK by a long duration cryogenic system [9, 10]. The microwave frequencies are selected by band pass filters peaked at 90, 150, 240, 410 GHz (with $\sim$ 20% FWHM bandwidth) to map the CMB and to monitor spurious signals from the optics, the residual atmosphere and the galactic foregrounds.

Through the telescope each detector observes the sky in a proper direction. The full payload scans in azimuth with a smoothed triangular waveform (60 degrees peak to peak) so that each detector scans the sky. This movement, combined with the natural sky rotation during the day, determines the sky coverage.

During the scans structures in the sky are converted in temporal frequencies in the signal: anisotropies with a given multipole number are converted in signals with a corresponding electrical frequency in the 1-10 Hz range. The scan speed and the elevation angle, together with the noise and the frequency response of the detectors determine the angular range of sensitivity and the corresponding range in multipole numbers.

The elevation angle ranged during the flight between 40° and 50°. The scan speeds we used ware of 1 and 2°s^{-1}. The choice different scan modes and elevations was one of the several ways to allow control of systematic effects.

NOISE PROPERTIES

In figure 1 we report the power spectrum of the noise of a BOOMERANG channel (150A) for 1 °s^{-1} (continuum line) and 2 °s^{-1} (dashed line). In the same figure the expected angular power spectrum for a typical cosmological model is reported (dot-dashed line). The multipole moment scale on the top corresponds to the frequencies scale on the bottom in the case of 1 °s^{-1} scan speed and 45° of elevation. Comparing the noise power spectrum with the angular power spectrum makes evident how the noise characteristics of the detector are tuned to detect the expected CMB signal. The flat zone of minimum noise is between 0.4 and 3 Hz and the first three peaks in the angular power spectrum is of CMB anisotropies produce signal in the range 0.4-1.5 Hz. The level of this white noise for all the channels is reported in table1 as Noise Equivalent Temperature (NET), together with other performances. The $1/f$ noise knee is at 0.2 Hz, corresponding to an angle in sky of 5 degrees. The rise in the Noise Equivalent Temperature at high frequencies is due to a decrease in sensitivity of the detectors where the transfer function starts to decrease. In fact high frequencies are cut off when the time scale starts to be comparable to the time constant of the detector. This rise in the noise corresponds to angles in the sky of the order of less than 0.1°, smaller than the size of the telescope beam.

The BOOMERANG readout electronics has a major role in the quality of the noise performance achieved. Each bolometer is AC biased with a differential smoothed square wave at $\sim$ 500 Hz. The bias voltage is modulated by the high bolometer impedance to be measured. The signal goes trough a cold J-FETs pair to reduce its impedance and

TABLE 1. Performances of the BOOMERANG channels: time constant, beam size and sensitivity. Only one channel (150A) was used for the first analysis [6] and four channels (150A, 150A1, 150A2, 150B2) were used in the latest analysis [7]. 150B was excluded due to the non stationary noise properties, 150B1 because didn't pass the jack-knife test for internal symmetry. The 90 GHz and the 240 GHz channels will soon be fully analyzed. The 410 GHz channels doesn't contain cosmological information but have been used for dust contamination analysis [11]. The NET is computed at 1 Hz. NET data of the 410 GHz channels are missing because a reliable in-flight calibration for that channels is still in progress.

Channel	Time constant (ms)	Beam size FWHM (arcmin)	NET ($\mu K \sqrt{s}$)
90A	22.5	18 ± 1	145
90B	21.9	18 ± 1	137
150A	10.8	9.2 ± 0.5	130
150B	13.3	9.2 ± 0.5	Variable
150A1	13.3	9.7 ± 0.5	231
150A2	12.0	9.4 ± 0.5	158
150B1	16.3	9.9 ± 0.5	196
150B2	21.2	9.5 ± 0.5	184
240A1	8.9	14 ± 1	221
240A2	7.2	14 ± 1	166
240B1	10.5	14 ± 1	250
240B2	8.8	14 ± 1	792
410A1	4.1	12 ± 1	
410A2	9.9	12 ± 1	
410B1	4.5	12 ± 1	
410B2	4.3	12 ± 1	

its sensitivity to inductive interferences, is amplified by a room temperature low noise differential pre-amplifier and band-pass filtered to remove out-of-band noise. A phase sensitive detector demodulates the signal synchronously with the bias, thus removing any offset and reducing the noise. Subsequent high-pass and low-pass filters get rid of residual $1/f$ and high frequency noise.

The use of AC modulation and demodulation technique reduces $1/f$ noise due to drifts in the bias amplitude, amplifier gain and heath sink temperature. The bias frequency has been selected in order to avoid beating with microphonic frequencies, thus reducing undesired lines in the noise power spectrum.

Low frequency noise and calibration

The noise power spectrum in the $2°s^{-1}$ scan mode has two peaks at the scan frequency and at the first harmonic, evidence for a strong component of Scan-Synchronous Noise (SNN). Those frequencies are well below the frequencies of interest for the sub-degree CMB anisotropy which is the target of this observation. Nevertheless is very important to have a reliable detection of the cosmic dipole signal (well known from COBE/DMR measurements [12]) for calibration. The cosmic dipole signal appears as a large scale linear gradient in the sky region observed and is converted into a signal at exactly the scan frequency by the telescope azimuthal movement.

The effect of the scan synchronous noise is clearly shown and compared to the expected dipole signal in time domain in figure 2. The source of this noise shouldn't be in the sky, because the sky signal doesn't change between fast and slow scan modes. It is probably a systematic effect in the instrument, more effective when the payload moves faster. The gyroscopes don't show significant different pendulations in the two scan modes, excluding the possibility of a signal in the sky excited by larger pendulations. Defining α as the angle in azimuth respect to the center of the scan and binning data from the evaporator thermometer for equal α bins, we detect a tiny coherent evaporator temperature variation in the $2°s^{-1}$ scan mode. The same effect is negligible in the $1°s^{-1}$ scan mode (see figure 3). Thermal effects are thus the first candidates as scan synchronous noise sources.

The calibration of the BOOMERANG telescope [13] is done using the signal from cosmic dipole: low-resolution ($1° \times 1°$) maps are created for the two different scan modes separately; these maps are fit to a dipole model giving the

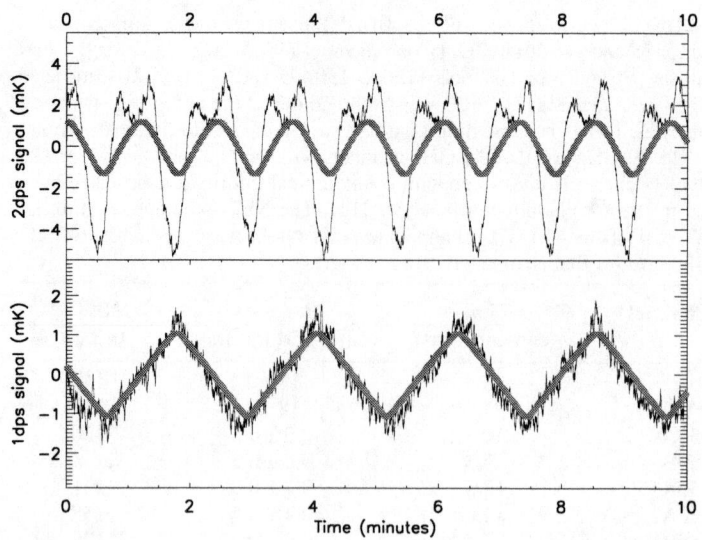

FIGURE 2. Comparison between data (thin line) and dipole simulations (thick light line) for the two different scan modes for the same channel. Data have been low pass filtered in order to display only the low frequency component. The dipole signal expected in each direction of observation has been computed from the COBE-DMR dipole parameters. In the $1°s^{-1}$ data (lower plot) data copy exactly the simulation, showing a very low component of scan synchronous noise. In the $2°s^{-1}$ data, the simulation differs clearly from signal. The discrepancy is due to a strong scan synchronous noise that appears only in the fast scan mode.

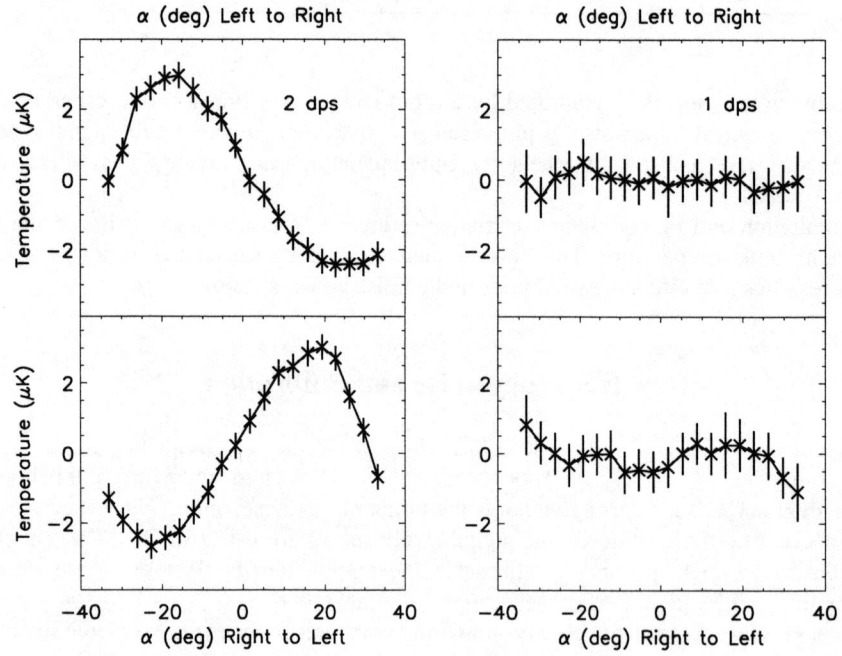

FIGURE 3. Temperature fluctuations of the evaporator binned for equal angles with respect to the center of the scan (α), for different scan modes. The top plots use data in the forward (left to right) part of the scans, the bottom plots in the reverse scans. Left plots are for $2°s^{-1}$ (2dps) and right plots for $1°s^{-1}$ (1dps) scan modes. The fluctuations are scan synchronous in the 2dps scan mode, while no coherence is present in the 1dps mode. Each plot results from integration of 24 hours of data.

calibration constants for the channels sensitive to the CMB. Considering that the dipole signal is scan synchronous, any $1/f$ noise and in particular scan synchronous spurious signal affects the dipole-based calibration. Anyway we know that in the slow scan mode the effect of scan synchronous noise is negligible (see figure 2). Even in the fast scan mode the receivers can be calibrated using the dipole signal, considering that the large scan synchronous noise is substantially reduced in the map-making process, because the scan synchronous signal is not exactly sky synchronous. In other words, making a map most of the scan synchronous noise is averaged out, reducing its effect. Moreover, observing that the scan synchronous noise at low frequencies (90, 150 and 240 GHz) is well correlated to the scan synchronous noise in the 410 GHz channels (where the CMB signal is negligible), a 410 GHz map can be used as a template to remove the residual effect, thus producing a reliable dipole-based calibration for the low frequencies channels in the fast scan mode. The same procedure, applied to the $1°s^{-1}$ scan data, gives a correction of the order of 1%. The two calibration constants, at 1 and $2°s^{-1}$ differ by 10% and this value is chosen as the calibration uncertainty, which is dominated by systematics. Consistency in the two calibrations provides a check for systematic effects, in particular shows that the results don't suffer for incorrect knowledge of the electronics transfer function at low frequency.

An other systematic effect, not directly related to $1/f$ and scan synchronous noise, is the presence in the BOOMERANG maps of faint stripes at nearly constant declination. An equivalent effect is not reproduced in the simulated maps, varies in amplitude and phase between bolometers and is negligible in some of the channels. Several tests have been performed excluding the possibility of a signal in the sky. The effect has been taken into account in the map-making process, filtering it out in Fourier space. The same filter has been applied in any simulated map used in the Monte Carlo Spherical Harmonic Transform technique used to compute the angular power spectrum of anisotropies combining data from different channels [14].

CONCLUSIONS

The noise properties of the BOOMERANG instrument have been studied in detail before science release. Monitoring the full instrument and redundancy in the experimental setup (multi-frequency observations, different scan-speeds, long integration time) allowed control and understanding of spurious effects, thus giving great confidence in the BOOMERANG results [6, 7, 15, 16, 17].

ACKNOWLEDGMENTS

BOOMERANG was supported by PNRA, Universitá di Roma *La Sapienza* and ASI in Italy, by NSF and NASA in USA and by PPARC in the UK. We thanks the 2K1BC conference organization for the very good job they did.

REFERENCES

1. White, M., Scott, D., and Silk, J., *Annu. Rev. Astron. Astrophys.*, **32**, 319–370 (1994).
2. Bond, J. R. *et al.*, *MNRAS*, **291**, L33 (1997).
3. Smoot, G. F. *et al.*, *Ap.J.L.*, **371**, L1 (1991).
4. Piacentini, F. *et al.* (2002), accepted by *Ap.J.Supp.*, astro-ph/0105148.
5. Crill, B. P. *et al.* (2002), in preparation.
6. deBernardis, P. *et al.*, *Nature*, **404**, 955–959 (2000), astro-ph/0004404.
7. Nettefield, C. B. *et al.* (2001), submitted to *Ap.J.L.*, astro-ph/0104460.
8. Mauskopf, P. *et al.*, *Applied Optics*, **36**, 765 (1997).
9. Masi, S. *et al.*, *Cryogenics*, **38**, 319 (1998).
10. Masi, S. *et al.*, *Cryogenics*, **39**, 217–224 (1999).
11. Masi, S. *et al.*, *Ap.J.L.*, **553**, L93–L96 (2001), astro-ph/0101539.
12. Kogut, A. *et al.*, *Ap.J.*, **419**, 1–6 (1993).
13. Crill, B. P., Ph.D. thesis, Caltech, Pasadena, CA (2000).
14. Hivon, E. *et al.* (2001), submitted to *Ap.J.*, astro-ph/0105302.
15. Lange, A. E. *et al.*, *Phys. Rev. D*, **63**, 4 (2001).
16. Jaffe, A. *et al.*, *Phys. Rev. Lett.*, **86**, 3475–3479 (2001), astro-ph/0007333.
17. deBernardis, P. *et al.* (2001), accepted by *Ap.J.*, astro-ph/0105296.

Millimeter and Submillimeter Observations from the Atacama Plateau and High Altitude Balloons

Mark Devlin[1]

University of Pennsylvania
Philadelphia, PA 19104
USA

Abstract. A new generation of ground-based and sub-orbital platforms will be operational in the next few years. These telescopes will operate from high sites in Chile and Antarctica, and airborne platforms where the atmosphere is transparent enough to allow sensitive measurements in the millimeter and submillimeter bands. The telescopes will employ state-of-the-art instrumentation including large format bolometer arrays and spectrometers. I will discuss the results of our observations in the Atacama region of Chile (MAT/TOCO), our future observations on the Balloon-borne Large Aperture Submillimeter Telescope (BLAST) now under construction, and our proposed Atacama Cosmology Telescope (ACT).

THE MAT/TOCO EXPERIMENT IN CHILE

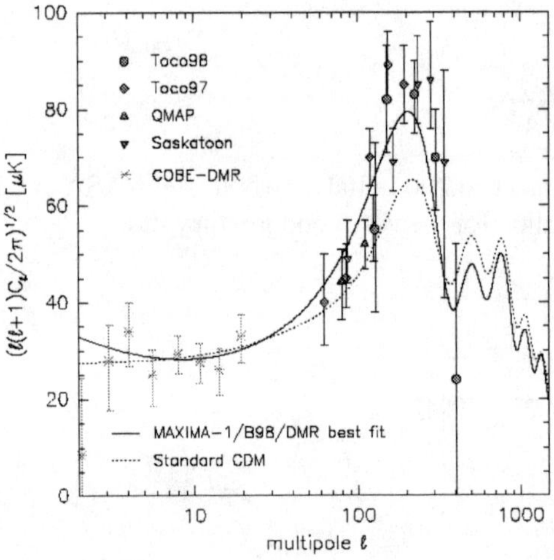

Figure 1 The TOCO power spectrum.

TOCO used the gondola and receiver from QMAP refitted with a mechanical cooler instead of liquid cryogens (Miller, J. Beach et al. 2001) and two SIS-based[2] 144 GHz detector systems to improve the resolution to 0.2°. TOCO employed the Saskatoon-style beam synthesis strategy developed by (Netterfield, Devlin et al. 1997) with eight independent detectors. Instead of observing near the NCP from Saskatoon, CA, we observed near the SCP from the side of Cerro Toco in Northern Chile[3]. At 144 GHz, the atmospheric column density in Saskatoon is too large for anisotropy measurements; a high altitude site such as the Chilean altiplano is required. TOCO operated for two seasons in 1997 and 1998. The primary results and short description of the instrument have been reported (Torbet, Devlin et al. 1998; Miller, Caldwell et al. 1999). The 0.2° resolution allowed us to locate the first peak in the angular spectrum at l~212±14 (Knox and Page 2000). The power spectrum is shown in Figure 1. In the context of the popular adiabatic CDM models, these data were among the first that demonstrated that the universe is geometrically flat.

---

[1] Sloan Fellow

[2] SIS stands for Superconductor-Insulator-Superconductor. The detecting element is a quasi-particle mixer Tucker, R. and M. J. Feldman (1985). "Quantum detection at millimeter wavelengths." Reviews of Modern Physics **57**: 4..

[3] The Cerro Toco site of the Universidad Católica de Chile was made available through the generosity of Prof. Hernan Quintana, Dept. of Astronomy and Astrophysics. It is near the ALMA site.

CP616, *Experimental Cosmology at Millimetre Wavelengths, 2K1BC Workshop,* edited by M. De Petris and M. Gervasi

The TOCO Telescope

The TOCO experiment is shown schematically in Figure 1 [4]. The optics and receiver are contained in the original QMAP balloon gondola. The radar trailer, on which the gondola is mounted, has a separable magnesium base and three legs hold the base off the trailer and stabilize it during observations. The gondola is mounted a 2.5 cm thick flat plate supported by a 1.4 m diameter precision bearing. The plate and gondola are rotated using an on-axis DC motor[5]. During observations of the CMB the telescope is held in a fixed position by a brake that allows the motor to be shut–off preventing it from drawing large currents or oscillating as it seeks the target position in high winds. The motor has a 15 cm diameter hole in the center through which cables and refrigerator hoses pass from the inside of the telescope to the outside. The compressor for the mechanical cryocooler is mounted on the trailer. The azimuth is instrumented with an absolute 17-bit encoder and a 20-bit resolver. The chopping flat sweeps the beam across the sky in a 9° peak-to-peak sinusoidal chop with a frequency of ~ 3 Hz.

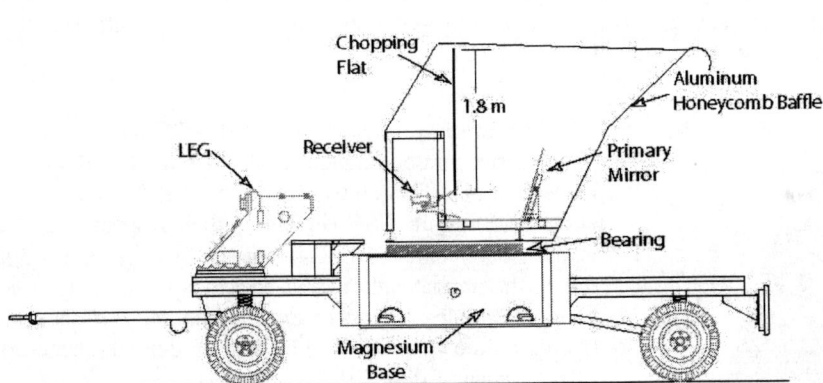

Figure 2

The TOCO Receiver

Radiation from the sky enters the Dewar through a 15.25 cm diameter vacuum window made of 0.56 mm polypropylene and is collected with corrugated feed horns. Three aluminum baffles define the entrance aperture, one attached to the 40 K cold plate, one attached to the Dewar just inside the vacuum window, and one attached to outside of the dewar. Strips of aluminized Mylar connecting the cold feeds to the ambient temperature Dewar wall block RF interference and reduce optical loading on the cold stage. To prevent the formation of frost on the window, warm air is passed through a volume in front of the vacuum window defined by a Saran Wrap-covered aluminum cone attached to the outside of the dewar.

The receiver has 5 feed horns in the focal plane: one at Ka-band (30 GHz), two at Q-band (40 GHz) and two at D-band (144 GHz). Each Ka-band and Q-band horn has two HEMT amplifiers, one at each polarization. The HEMTS and feeds are operated at 40 K. The D horns each have a single SIS amplifier operating in double sideband mode with a total bandpass of 7 GHz. The D horns are polarized and are oriented at 90° with respect to each other. The SIS detectors are operated at 4 K.

THE BALLOON-BORNE LARGE APERTURE SUBMILLIMETER TELESCOPE

BLAST is a "Balloon-borne Large-Aperture Sub-millimeter Telescope" incorporating a 2 m primary mirror and large-format bolometer arrays operating at 250, 350 and 500 μm (Figure 3). Observations will be made from a

---

[4] The Nike Ajax radar trailer was donated to the University of Pennsylvania by Lucent Technologies.

[5] The motor is a Compumotor DR 1100A - 100~Nm torque.

Long Duration Balloon (LDB) platform to be flown from Antarctica. It will provide the first sensitive, large-area (>>10 deg^2), sub-mm survey at these wavelengths.

BLAST will address some of the most important galactic and cosmological questions regarding the formation and evolution of stars, galaxies and clusters. Galactic and extragalactic BLAST surveys will: (i) identify large numbers of high-redshift galaxies; (ii) measure photometric redshifts, rest-frame FIR luminosities and star formation rates thereby constraining the evolutionary history of the galaxies that produce the FIR--sub-mm background; (iii) measure cold pre-stellar sources associated with the earliest stages of star and planet formation; and (iv) make high-resolution maps of diffuse galactic emissions from low to high galactic latitudes. In addition to these goals, the BLAST LDB experiment produce a >>100 deg^2 map of the galactic plane in the sub-mm and a catalogue of 3000—5000 extragalactic sub-mm sources that will serve as a legacy to be followed at other wavelengths and resolutions, including sub-arc imaging with ALMA.

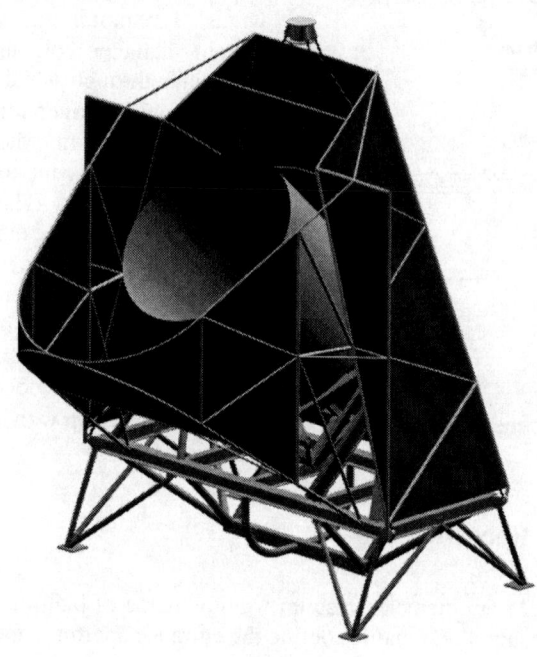

Figure 3 The BLAST Gondola

The primary advantage of BLAST over existing and planned sub-mm bolometer arrays such as SCUBA (Holland, Robson et al. 1999) on the JCMT, SHARC (Hunter, Benford et al. 1996) on the CSO (including their respective upgrades) is its greatly enhanced sensitivity at wavelengths <500 μm due to the dramatically increased atmospheric transmission at balloon altitudes. BLAST complements the Herschel satellite (formerly known as FIRST) by testing identical detectors and filters planned for the SPIRE instrument.

BLAST Science Goals

BLAST will be the first long duration balloon-borne telescope to take advantage of the bolometric focal-plane arrays being developed for *Herschel*. A LDB flight from Antarctica, providing the first surveys at 250, 350 and 500 μm, will significantly extend the wavelength range, sensitivity, and area of existing ground-based extragalactic and galactic surveys. The instrument parameters are given in Table 1. BLAST will conduct unique galactic and extragalactic sub-mm surveys with high spatial resolution and sensitivity. Compared to the pioneering flights of PRONAOS (Lamarre, Giard et al. 1998), BLAST will have an advantage of >100 times the mapping speed. The scientific motivations for BLAST are similar to those of *Herschel* but are achievable within 2 to 4 years with a series of LDB flights.

Using these unique BLAST surveys we expect to achieve the following science goals:

- Conduct a complementary series of wide (shallow) and narrow (confusion-limited) extragalactic 250—500 μm surveys, identifying the galaxy populations responsible for producing the Far-IR and sub-mm backgrounds. BLAST will determine the amplitude of clustering of sub-mm galaxies on scales of 0.1—10 degrees.
- Measure the 250—500 μm spectral energy distributions (SEDs) and colors, from which one can derive rest-frame luminosities and star formation rates (SFRs) for sub-mm selected galaxies.
- Measure the sub-mm source-counts and place the strongest constraints to date on evolutionary models and the global star formation history of starburst galaxies at high-z (see Figure 4)
- Conduct Galactic surveys of the diffuse interstellar emission, molecular clouds, and identify dense, cold pre-stellar (Class 0) cores associated with the earliest stages of star formation.
- Observe solar system objects including the Kuiper-belt objects, planets, and large asteroids.

Extragalactic Surveys at Millimeter and Sub-millimeter Wavelengths

Observations at sub-mm wavelengths of starburst galaxies in the high-z universe have a particular advantage compared to observations in the optical and FIR because a strong negative k-correction enhances the observed sub-mm fluxes. By early 2002, SCUBA and the millimeter camera MAMBO (operating on the IRAM 30 m) will have

Table 1 BLAST Telescope Parameters

Telescope			
Temperature	200 K		
Throughput for each pixel	λ^2		
Bolometers			
Central Wavelengths	250	350	500
Number of Pixels	149	88	43
Beam FWHM (arcseconds)	30	41	59
Field of View for Each Array (arcmin)	6.5×14		
Overall Instrument Transmission	30%		
Filter Widths	~ 30 %		
Observing Efficiency	90%		

completed their first series of extragalactic sub-mm and mm (850 μm — 1.3 mm) surveys. It is important to note that, although covering areas ranging from 0.002 to 0.2 deg^2 (Smail, Ivison et al. 1997; Hughes, Serjeant et al. 1998; Lilly, Eales et al. 1999; Carilli, Owen et al. 2001), these ground-based surveys are hundreds of times smaller than the proposed BLAST surveys (see Table 2).

The following results from the first SCUBA (850 μm) and MAMBO (1.25 mm) surveys have made a significant impact on several cosmological questions, while at the same time demonstrating the necessity for larger area and shorter-wavelength (250—500 μm) sub-mm observations:

Table 2 BLAST sensitivities. Sensitivities for SCUBA and SOFIA are given for comparison.

	250 μm	350 μm	500 μm
Background power [pW]	25.6	18.3	13.5
Background limited NEP [W Hz$^{-0.5}$ x 10^{-17}]	20	14	10
NEFD [mJy √sec]	236	241	239
ΔS (1σ, 1hr) (1 sq. deg.) [mJy]	38	36	36
ΔS (1σ, 6hr) (1 sq. deg.) [mJy]	15.5	14.7	14.6
Comparison with SCUBA (average NEFD) [mJy √sec]	-	1100	1000
Comparison with SOFIA (calculated NEFD) [mJy √sec]	550		

- 30—50% of the sub-mm/FIR background detected by COBE has been resolved into individual sub-mm galaxies. The existing confusion-limited surveys are within a factor of a few in sensitivity of resolving the entire sub-mm/FIR background.

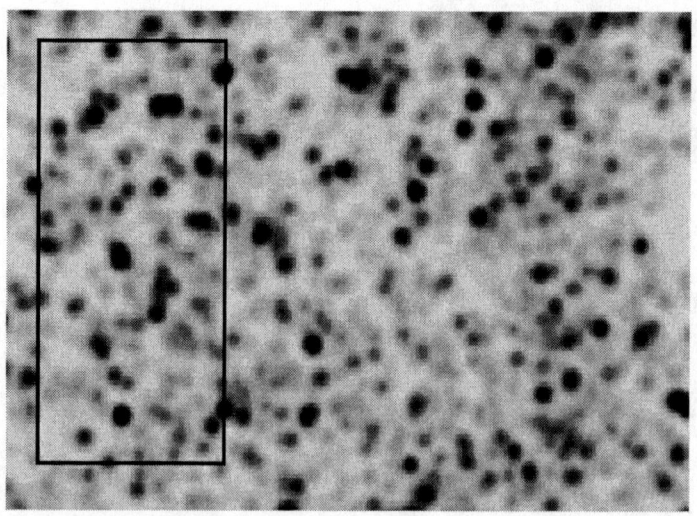

Figure 4 A simulated confusion-limited BLAST 250μm survey with a FWHM resolution of 30' and a 1σ sensitivity of 7 mJy. The assumptions for the evolution and the clustering of submm galaxies are described in Hughes & Gaztanaga 2000. The simulation area covers 0.1 deg^2 and is 1/10th of our smallest proposed survey. The rectangle represents the field of view of the 250 μm array.

- The majority of sub-mm sources appear to be associated with z>>1 Extremely Red Objects (EROs - IR galaxies that are extremely faint or undetected in the optical) and weak radio sources. There is still vigorous debate about the fraction of sources at z≥2.

- The sub-mm source-counts significantly exceed a no-evolution model and require strong evolution out to z~1, but place weak constraints at higher redshifts. The source-counts at bright sub-mm flux densities (N_{850} > 10 mJy) show a large scatter (about a factor of 5) between different surveys which may be the result of clustering on the scale of these surveys.

- The sub-mm surveys appear to find ~5 times the star formation rate that is observed in optical surveys at 2 < z < 4; how much of this discrepancy is due to the effects of dust obscuration and incompleteness in the optical is still being investigated (Adelberger and Steidel 2000).

THE ATACAMA COSMOLOGY TELESCOPE

Exciting recent advances in cosmology have given us a glimpse of new fundamental physics. Observations of galaxy clustering, supernovae, and the cosmic microwave background (CMB), in combination with theoretical advances, have led to a model of a geometrically flat universe of which ~5% is composed of baryons, ~ 30% of cold dark matter, and 65% of a new form of dark energy. Our conception of the contents and fate of the universe is clearer and dramatically different from what it was just five years ago. The scientific goal of ACT (see Figure 5) is to expand our knowledge well beyond the capabilities of the current generation of investigations through a highly interlocking program of innovative observations, sophisticated data analysis, and theory. These efforts will use gravity acting on cosmic length scales to probe fundamental physics.

A revolution in detector technology, namely thousand-element CCD-like bolometric arrays, has opened the door to new discoveries about the physical universe. CMB observations with an instrument based on these detectors can map out the formation of cosmic structure from the high-redshift (z>1000) linear regime through the low-redshift (z<5) non-linear regime where structures form. Optical spectroscopy with a dedicated 10 m class telescope and measurements with modern X-ray satellites enable the determination of the redshifts and masses of hundreds of galaxy clusters, the latest stage of cosmic evolution. Through a combination of these observations: CMB, optical, X-ray, we can probe our picture of the universe to greater depths than with any one set of observations alone and to greater depths than what has been done to date. Our goals are to:

- Map the CMB temperature anisotropy over 100 deg^2 beyond the resolution limits of the *MAP* and *Planck* (launch 2007) satellites. The primary anisotropy constrains the fundamental physics of the infant universe (e.g., inflation) and provides the initial conditions for structure formation; the secondary anisotropies reflect the emergence of structure. Our high resolution maps will be invaluable for fully exploiting the satellite results.
- Find all galaxy clusters with M>4 × 10^{14} M$_{solar}$ in the CMB map region through the Sunyaev-Zel'dovich effect and determine the spectroscopic redshift of over 400 of them. We will also measure the masses of these clusters with X-ray observations and with galaxy velocity dispersions.
- Combine the CMB and cluster measurements to limit or determine the mass of the neutrino to ±0.1 eV. The current cosmological limit on the sum of masses of the light neutrinos is 10 eV.
- Find the equation of state, w=p/ρ, of the "quintessence" or "dark energy" to ±0.1 through the CMB/cluster combination.
- Measure the amplitude of gravitational lensing of the CMB to ~2% to *directly* find the unbiased matter power spectrum on length scales of ~1 Mpc at z~1-2. This technique is independent of and complementary to lower redshift galaxy surveys (which involve bias).
- Detect reionization of the universe from the formation of the first stars at z~10 through the Ostriker-Vishniac effect in the CMB.
- Find all extragalactic mm-wave point sources in 200 deg^2 to a 4σ sensitivity of 1 mJy using our high-resolution, multi-wavelength CMB observations. These data will also be crucial for understanding the foreground contamination of the CMB anisotropy.

The Site

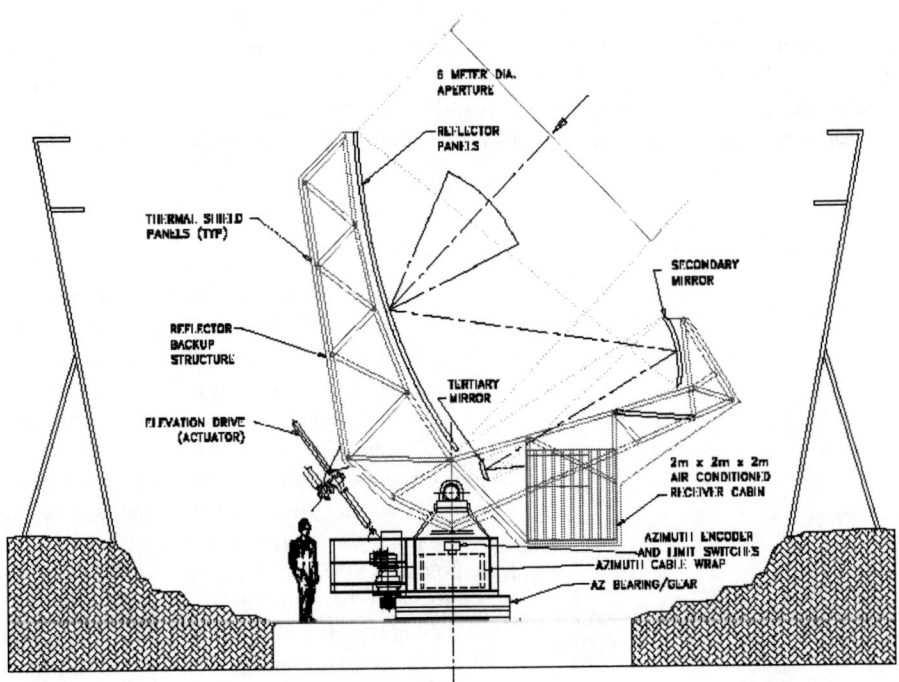

Figure 5 The Atacama Cosmology Telescope

Several groups have expended a great deal of effort evaluating a variety of sites for millimeter and submillimeter observations. The Atacama region of Chile and the South Pole have been identified as the two the premier observing sites in the world. Site selection is a complicated process involving many important parameters. All of these parameters enter into the probability of success for any given experiment. We feel that the Atacama region is the best site for the ACT telescope *based on its particular science goals*.

Atmosphere :

We have evaluated the atmospheric conditions of the Atacama site using experimental and theoretical predictions. NRAO has been monitoring the site since 1995 in preparation for the installation of the ALMA array[6]. The MAT/TOCO experiment in 1997 and 1998 provided a direct measurement of atmospheric quality at the site.

---

[6] See http://www.tuc.nrao.edu/mma/sites/Chajnantor/data.c.html

Theoretical models have assessed the impact of the atmosphere on CMB experiments employing a variety of scan strategies (Lay 2000). A combination of all this information leads to the following conclusions:

- While the altitude of the South Pole site (2850 m) is significantly lower than the Atacama site (5200 m) the overall opacity at the South Pole is less than the Atacama site.

- Well understood seasonal variations at the Atacama site bias the opacity information. When only the six best months at the Atacama site are considered (generally June-December), the opacity at the Atacama site is comparable to the South Pole much of the time. When only the nighttime data is considered it is actually better 25% of the time.

- The power spectrum of the atmospheric noise is a strong function of angular scale. At low l (<500), the atmosphere would dominate over ACT's receiver noise (see Figure 6). However, on smaller angular scales, the atmospheric contribution to the instrument noise is negligible. The MAT/TOCO experiment confirmed these calculations (MAT/TOCO had much higher receiver noise). The conclusion is that if large angular scales are required, it would be worth the effort to go to a site like the South Pole. Otherwise, the Atacama site is more than adequate.

- ACT requires 64 high-quality nights to achieve its minimum science goals. The data and analysis indicate that we should expect to have ~120 nights per 6-month observing season.

Sky Coverage and Scan Strategy:

The Atacama site is at latitude of 23° South. Unlike the South Pole, this site has easy access to over 65% of the sky without compromising the data by scanning near the horizon. This is crucial for required follow-up observations to determine the red-shift of the clusters. It also makes the telescope much more useful as a legacy instrument as it shares sky coverage with many northern and southern hemisphere instruments. In addition, the sky rotation allows the scans to be cross-linked without changing the elevation angle of the telescope. Cross-linking is critical to making large-area maps that are required to achieve our science goals. Similar observations from the South Pole yield scans at constant declination. The elevation of the telescope must be changed to cross-link the scans. This will potentially work for small changes in elevation, but will yield significant atmospheric offsets.

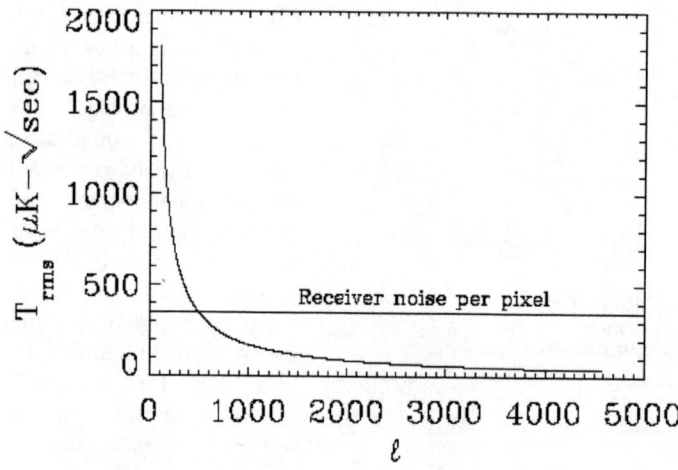

Figure 6 The atmospheric noise as a function of angular scale at the Atacama Site. The noise drops below the receiver noise at angular scales < 2°.

Logistics:

There is little doubt that operations at the South Pole are well-supported by the NSF office of Polar Programs. However, the Atacama site has a clear operational advantage. It is about 1 hour from a town that offers hotels, restaurants and a variety of amenities. It takes only 24 hours to get from a US university to the site. There is no need to have a "winter over" scientist since personnel can be exchanged easily on a regular basis. There is a hospital 100 km from the site so that any problems can be handled without making news headlines across the World (like the recent South Pole evacuations). Finally, it is far less expensive to have experiments installed and maintained because there is significant engineering support available locally.

ACKNOWLEDGMENTS

This work is a summary of the results and research of the many members of the MAT/TOCO, BLAST, and ACT collaborations. In particular the author would like to acknowledge David Hughes, Jeff Klein, and Lyman Page for their contributions directly related to this paper. MAT/TOCO is supported by the NSF (AST97-32960), and the University of Pennsylvania. BLAST is supported by NASA (NRA-99-01-SPA-015).

REFERENCES

Adelberger, K. and C. C. Steidel (2000). "Multiwavelength Observations of Dusty Star Formation at Low and High Redshift." ApJ **544**: 218.

Carilli, C. L., F. Owen, M. Yun, F. Bertoldi, A. Bertarini, K. M. Menten, E. Kreysa and R. Zylka (2001). <u>Deep Millimetre Surveys - Implications for Galaxy Formation and Evolution</u>, World Scientific.

Holland, W. S., E. I. Robson, W. K. Gear, C. R. Cunningham, J. F. Lighfoot, T. Jenness, R. J. Ivison, J. A. Stevens, P. A. R. Ade, M. J. Griffin, W. D. Duncan, J. A. Murphy and D. A. Naylor (1999). "SCUBA: a common-user submillimetre camera operating on the James Clerk Maxwell Telescope'." MNRAS **303**: 659.

Hughes, D., S. Serjeant, J. Dunlop, M. Rowan-Robinson, A. Blain, R. G. Mann, R. Ivison, J. Peacock, A. Efstathiou, W. Gear, S. Oliver, A. Lawrence, M. Longair, P. Goldschmidt and T. Jenness (1998). "High-redshift star formation in the Hubble Deep Field revealed by a submillimetre wavelength survey." Nature **394**: 241.

Hunter, T. R., D. J. Benford and E. Serabyn (1996). "Optical Design of the Submillimeter High Angular Resolution Camera (SHARC)." PASP **108**: 104.

Knox, L. and L. Page (2000). "Characterizing the Peak in the CMB Angular Power Spectrum." **astro-ph/0002162**.

Lamarre, J. M., M. Giard, Pointecouteau, B. E., J.P., , G. Serra, F. Pajot, F. X. Desert, I. Ristorcelli, J. P. Torre, S. Church, N. Coron, J. L. Puget and J. J. Bock (1998). "First Measurement of the Submillimeter Sunyaev-Zeldovich Effect." ApJ **507**: L5.

Lay, O. P. H., N. W. (2000). "The Impact of Atmospheric Fluctuations on Degree-Scale Imaging of the Cosmic Microwave Background." ApJ **543**: 787L.

Lilly, S., S. Eales, W. K. Gear, F. Hammer, O. Le Fevre, D. Crampton, R. Bond and L. Dunne (1999). "The Canada-United Kingdom Deep Submillimeter Survey. II. First Identifications, Redshifts, and Implications for Galaxy Evolution." ApJ **518**: 641.

Miller, A. D., R. Caldwell, M. J. Devlin, W. B. Dorwart, T. Herbig, M. R. Nolta, L. Page, J. Puchalla, E. Torbet and H. T. Tran (1999). "A Measurement of the Angular Power Spectrum of the Cosmic Microwave Background from l=100 to 400." ApJ **524**: L1.

Miller, A. D., J. Beach, S. Bradley, M. J. Devlin, R. Caldwell, H. Chapman, W.B. Dorwart, T. Herbig, D. Jones, G. Monnelly, C. B. Netterfield, M. Nolta, L. A. Page, J. Puchalla, T. Robertson, E. Torbet, H. T. Tran and W. E. Vinje (2001). "The QMAP and MAT/TOCO Experiments for Measuring Anisotropy in the Cosmic Microwave Background." ApJ **Submitted**.

Netterfield, C. B., M. J. Devlin, N. Jarosik, L. Page and E. J. Wollack (1997). "A Measurement of the Angular Power Spectrum of the Anisotropy in the Cosmic Microwave Background." ApJ **474**: L47.

Smail, I., Ivison and A. W. Blain (1997). "A Deep Sub-millimeter Survey of Lensing Clusters: A New Window on Galaxy Formation and Evolution." MNRAS **490**(L5).

Torbet, E. T., M. J. Devlin, W. B. Dorwart, T. Herbig, A. D. Miller, M. R. Nolta, L. Page, J. Puchalla and H. T. Tran (1998). "A Measurement of the Angular Power Spectrum of the Microwave Background Made from the High Chilean Andes." ApJ **521**: L79.

Tucker, R. and M. J. Feldman (1985). "Quantum detection at millimeter wavelengths." Reviews of Modern Physics **57**: 4.

A simple and reliable star sensor for spinning payloads.

F. Nati*, P. de Bernardis*, S. Masi*, D. Yvon† and the Archeops team.

*Dipartimento di Fisica, Università di Roma La Sapienza, Italy
†SPP/DAPNIA/DSM, CEA-Saclay, France

Abstract. We developed a system to reconstruct the attitude of balloon borne spinning experiments, where high accuracy (about 1') and high rotation speed (up to tens of degree per seconds) are required. It is based on a stellar sensor, and gathers togheter hardware simplicity, cheap components, high resolution and sensitivity. It is composed of an optical mirror (diameter 40 cm), an array of 46 fast and sensitive photodiodes, and an ultra-low noise readout electronics. It was designed for the Archeops experiment, a balloon borne millimetric telescope whose goal is to generate high resolution maps of large regions of the sky, to study the temperature anisotropies of the cosmic background radiation.

INTRODUCTION

Sky surveys by means of balloon borne telescopes are actively and effectively pursued in the microwave to infrared and UV to gamma ranges of the EM spectrum ([1], [2], [3], [4], [5]). One of the most important and critical subsystems in these payloads is the attitude control system. This allows to control the pointing of the telescope during the flight, and allows to reconstruct it with sufficient accuracy either in real time or off-line, after the flight.

Among the many possible sensors for attitude control and reconstruction, the star sensors are the most precise, providing resolutions much better than 1 arcmin. The main limits of two-dimensional star sensors are the high data flow and the significant computational power required for extracting the pointing information from the measured image. Moreover, they can work only during the night, or far from Sun direction with appropriate baffles and shields.

We have developed a pointing reconstruction system best suited for fast (tens of degrees/s) spinning payloads, where medium-high accuracy (about 1') is required. Standard systems based on 2-D large format CCD cameras are not optimal in this case, due to the long readout time and to the dilution of star signals over many pixels of the array, which limit the sensitivity and the accuracy of the detection.

Our star sensor is composed of an optical grade parabolic mirror collecting visible flux from the stars and of a linear array of fast and sensitive photodiode detectors, and is complemented by suitable baffles, thermal shields and low noise readout electronics. The array is oriented in the focal plane of the mirror in a direction orthogonal to the scan (the elevation direction in our case). The image of each star sweeps across the photodiode corresponding to its elevation, producing a pulse whose amplitude and timing are accurately measured and recorded. The sequence of detections of all the photodiodes is compared (off-line) to a synthetic one generated by a star catalog, and the pointing parameters are varied to obtain the best fit. Pointing in between detections can be reconstructed by interpolation provided a sufficient star detection rate is achieved.

This system was designed for the Archeops experiment, a balloon borne millimetric telescope whose goal is to build high resolution maps of wide regions of the microwave sky, to measure the temperature anisotropy and polarization of the cosmic microwave background (CMB) radiation ([6]).

The design of the star sensor we have developed gathers together hardware simplicity (both mechanical and electronical) and reliability, low cost and high performance (resolution and sensitivity).

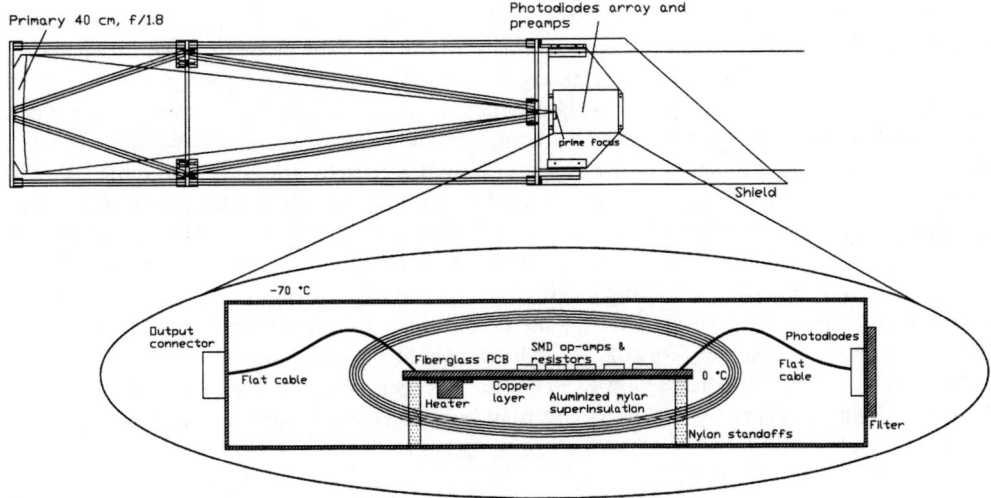

FIGURE 1. Telescope frame for the star sensor and location of the various components and internal details of the electronics box.

DESIGN

The telescope focal plane is composed of a linear array of 46 photodiodes, model Hamamatsu S4111. Each pixel is 4 mm wide (along the azimuth or scan direction) and 1 mm high in the array direction (i.e. along the elevation or cross-scan direction). The angular resolution of the star sensor must be higher than that of the microwave telescope used in the payload; in the case of the last generation of CMB anisotropy experiments the microwave beam is about 5'-10' FWHM. For this reason our target is a 2' resolution of each pixel of the star sensor. To achieve this we choose a mirror with 1.8 m of focal length. If the scan strategy is the one used for Archeops, in one full revolution of the payload the star sensor will produce a map of stars that is a 1.46° thick stripe at about 41° of elevation. Since the spin period of the payload is 2 to 3 rpm, the sampling rate for the signal from the photodiodes will be between 5 ms and 10 ms in order to sample the azimuth with the same resolution as the elevation.

From simulations we see that we can always achieve an accuracy of about 1' in the pointing reconstruction if we have at least a few tens of stars per turn; so we design the telescope in order to detect more than 100 stars in a full revolution. By using the data from the Hipparcos star catalog, we compute the number of stars in the circle sampled in one revolution as a function of the limit magnitude of the instrument. If we can detect magnitude 7 or 8 stars, then we will have more than 100 stars per revolution. The limiting magnitude depends mainly on two parameters: the mirror area and the detector noise. The radiative power of the limit magnitude stars, collected by the mirror surface, must be higher than the NEP of the photodiodes. The photodiodes used in the array S4111 from Hamamatsu have a noise equivalent power of about $2 \cdot 10^{-15} \frac{W}{\sqrt{Hz}}$. With a reasonable total efficiency of 0.5, and in absence of additional noise, we should be able to achieve a magnitude limit of about 8.8, more than enough for our purposes.

IMPLEMENTATION

Optical

The mirror is a glass, optical grade polished, parabolic mirror from Meade, commercially available for optical telescopes. It has a diameter of 400 mm and a focal length of 1830 mm. The off-axis aberrations, dominated by coma, at the focus are smaller than the size of each photodiode pixel. The box containing the photodiodes array and the readout electronics covers only 4% of the mirror area, so the solution with detectors in the prime focus is optimal in terms of simplicity and weight. The mirror and the electronics are mounted on an aluminum frame (Fig 1), wrapped with a black and opaque tissue, with a top shield to prevent stray light from reaching the sensors. The total weight of the system is 38 Kg.

The sensor is a linear array of 46 sensitive photodiodes. The throughput of each diode is $1.5 \cdot 10^{-7} m^2 sr$, so variations in the background will produce a low frequency noise during the rotation af the payload, especially when the Sun rises. This problem has been minimized by carefully reducing the internal reflectivity of the system, and using an optical filter in front of the sensors, so that scattered sunlight is largely reduced. The filter must cut as much as possible scattered sunlight, without loosing too many stars. We computed the percentage of diffused light cut by the filter and the number of stars we loose, as a function of the filter's cutting wavelength. In our simple model, the diffused light comes mainly from Rayleigh diffusion, so it is shifted toward blue frequencies. Instead, we know from stars catalogues that the distribution of temperature of stars has a peak corresponding to the red giants. The two spectra are thus quite different, and a low-pass filter can help to reduce the scattered sunlight while keeping a reasonable stars detection rate. We quantitatively selected the filter by estimating its effect on the Rayleigh diffused light and on the set of all the stars of magnitude less than 7 seen by the telescope during the flight (about 10^4 stars from Hipparcos catalogue). With the filter Schott OG530, featuring a sharp cut-on at $\lambda_{cut} = 530\ nm$, we loose 5% of the stars, while we reduce the diffude sunlight by more than 30%; with the filter Schott OG630, that has its cut-on at 630 nm, we loose about 27% of stars, while the diffused sunlight is reduced by a factor 7: this filter has been used in the latest flights of the sensor.

Electronics

The photodiodes are used in a photovoltaic configuration for minimal noise: any bias on the photodiodes would enhance the dark current, thus reducing the limit magnitude visible by the telescope. The signal from the photodiodes is AC coupled to remove large scale diffuse brightness gradients. The cut-on frequency (1.06 Hz in the first flight) is much lower than the characteristic frequencies of the star signal, but higher than the spin frequency of the payload. In the scan strategy used for the Archeops flights, the payload turns at 2-3 rpm, while a star signal crosses one pixel of 4 mm in a time $T_s = \simeq 30 - 20 ms$. The characteristic frequency of the star signals is of the order of the inverse of this value, about 30-50 Hz, far from both the cut-on frequency and the spin frequency. The filter is a single-pole RC, so the resistance introduces a current noise. In the first flight we used $100 M\Omega$ resistance: this is not the best choice, because the current noise is too high compared to the other contributes from the photodiodes and amplifiers. In the following flights, we increased the value to $1 G\Omega$, therefore the current noise was reduced and the limit magnitude become correspondingly $m_l \simeq 7.3$.

The signal from the photodiodes is amplified by an ultra-low noise operational amplifier (OPA129P) connected as a current to voltage converter. The amplifier features a 1 GΩ feedback resistor with a 0.5 pF capacitor in parallel to get rid of gain peaking. This operational amplifier is selected because of the ultra low bias current ($I_{bias} < 100$ fA); the low noise: 15 nV/$\sqrt{Hz}$ at 10 KHz; the low temperature drift: $< 10\ \mu V/C$ and the extended operating temperature range (from $-40\ °C$ to $+85\ °C$).

Thermal optimization

Stratospheric flights, and in particular the arctic flights in Kiruna, present extreme temperature conditions, between $-60\ °C$ and $-90\ °C$ at float (3 mbar). The mechanics and electronics of the star sensor have been optimized for operating in such conditions. The most critical parts in flight conditions are the operational amplifiers. Their mimimun operating temperature is $-40\ °C$, and they are mounted in the focal plane, so they cannot be shielded with large thermal protections without vignetting significantly the beam. We used 46 amplifiers OPA129 in SMD package to minimize the size of the printed circuit board (PCB). This minimizes radiative emission, and also allows enough space inside the box to wrap the PCB with an aluminized mylar blanket (see Fig. 4B). The PCB is suspended on four teflon standoffs (Fig 1) and small diameter cables are used between the amplifiers PCB and the output connectors, to minimize the thermal conductivity toward the external walls of the box. The lower side of the PCB features a copper ground plane with the double purpose of shielding the amplifiers against parasitic signals and of thermalizing the board. Two heater resistors and one PT100 thermometer are in contact with the copper ground plane to control and monitor the temperature of the boards.

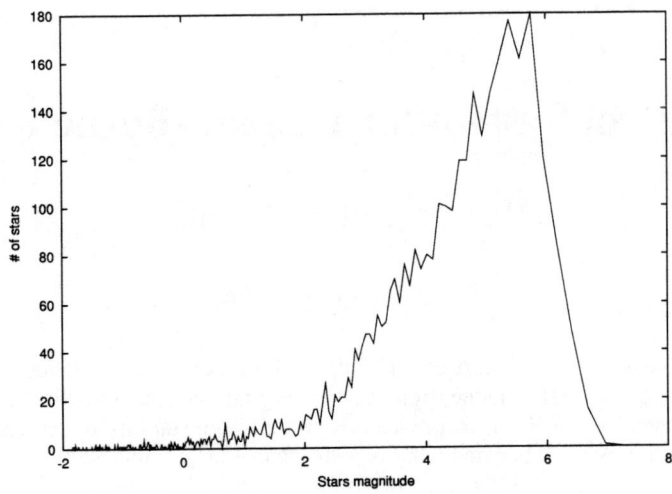

FIGURE 2. Hystogram of amplitudes of star candidates detected in the Kiruna flight (night).

IN-FLIGHT PERFORMANCE

This star sensor has been flown for the first time by the Italian Space Agency from Trapani, for the Archeops test flight in July 1999. Subsequntly it has flown three times by CNES from Kiruna (one flight in April 2000, just for stellar sensor and telemetry testing, and two Archeops flights in January 2001). It worked very well in all these flights, detecting more than hundred star candidates per revolution, allowing the pointing reconstruction with an accuracy of few arcminutes, and with improvements due to small adjustments we made after each flight. The temeperature of the PCB has been mantained around 0 °C during the artic flight. The limit magnitude of the system is ~ 6.5 (Fig. 2)

ACKNOWLEDGMENTS

This program has been supported by the ASI (Italian Space Agency) and the PNRA (Italian National Antarctic Research Program). The Archeops team is defined as in [6].

REFERENCES

1. Lee A. et al. *in 3K Cosmology, AIP 224, astro-ph/9903249*, 1999.
2. Piacentini F. et al. *astro-ph/0105148*, 2001.
3. Devlin M. *astro-ph/0012327*, 2001.
4. Lamarre J. M. et al. *Infrared Phys.*, 277:35, 1994.
5. Grindley J. E. *astro-ph/9712357 and references therein*, 1997.
6. Benoit A. et al. *astro-ph/0106152, Astroparticle Physics, in press*, 2001.

Attitude Control System for Balloon-Borne Experiments

E. Pascale* and A. Boscaleri*

*IROE-CNR, Florence, Italy

Abstract. The Pointing System is a key part of any balloon-borne experiment. It has to achieve the required attitude stabilization of the telescope during the flight, as well as guarantee the post-flight pointing reconstruction. The ACS that has been developed at IROE can reach a high real-time stability in pointing and scanning and it guarantees an attitude solution with an accuracy that is better than 1 arc-minute RMS.

INTRODUCTION

The Attitude Control System (ACS) for stratospheric experiments, designed at IROE [1, 2, 3], was developed for the ARGO [4], TRIP [5], MAX-5 [6], BOOMERANG [7] and SAFIRE [8] experiments. This ACS is able to point the payload in a selected sky direction and to track it or scan over it at a speed of few deg/s in azimuth and with a real-time pointing stability of ~ 1 arc-min rms. Attitude reconstruction in a post-flight processing is calculated from the information provided by several attitude sensors. The accuracy of the solution depends only on the particular choice of the sensor set and can be as good as ~ 0.5 arc-min rms [9]. If necessary, a different choice of the sensor set can dramatically increase this result.

ATTITUDE CONTROL SYSTEM

The Pivot shown in fig. 1 is capable of driving payloads up to 2500 Kg in weight with NASA-NSBF safety factors. It houses two DC torque motors with a torque sensitivity of 1.4 Nm/A and a peak armature current of 13 A, as well as two tachos devoted to control of the angular speed of the flywheels. The mechanical design of the pivot can perform different strategies while pointing sources or scanning large sky regions, according to the interaction of the upper reaction wheel with the flight chain. The hardware that controls the gondola movements is built around a commercial PC 386 CPU board (AMPRO) . Some custom boards interface analog and digital sensors and drive motors using a PWM technique at 28 KHz. The Pointing system is fully digital: it can be controlled by telemetry link and can send data for post processing via different protocols (NRZL or BI-PHASE) to the GSE. An on-board backup system based on a dedicated PC card and custom PCM receiver can store the same down-link matrix. The system is open to accepting any kind of sensor for performing an accurate reconstruction of the telescope's line of sight: magnetometer, sun and star sensors, GPS attitude solution and three axis rate gyroscope. The software takes advantage of a PC architecture and controls the Transfer Function of the motors (PID) for different weigh-inertia moment combinations of the payload. The code also manages any communication link (full duplex) and synchronization with the main scientific equipment.

ATTITUDE SENSORS

Several attitude sensors have been chosen and can be employed in different configuration sets to achieve the desired accuracy in the attitude stabilization and reconstruction.

Magnetometer - It measures azimuth angles from the direction of the local Magnetic Field Vector. Its intrinsic accuracy is ~ 40 arc-sec RMS. It can be used almost everywhere, as long as the payload does not get too close to the Earth's magnetic poles. The main difficulties with this kind of sensor is that its output has to be corrected for the

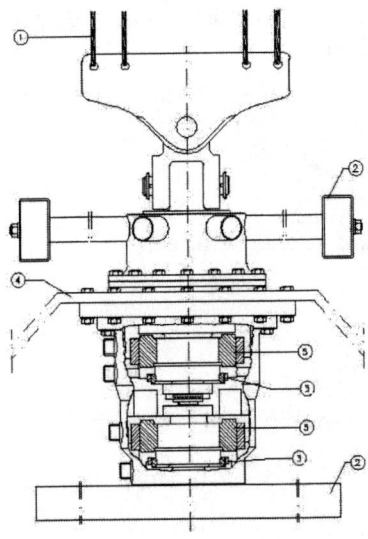

FIGURE 1. The Pivot: 1) Flight Chain, 2) Reaction Wheel, 3) Tachos, 4) Gondola, 5) Motor

local magnetic declination, which is known with a 10% accuracy, and for the pitch and roll of the payload. It is mostly used as a coarse sensor to drive the payload toward positions in which other (finer) sensors can operate.

Precision Fiber Optic Gyros - FOGs are the main sensors for pointing reconstruction. The signal of a 3 axis rate gyro is integrated over one of the available absolute attitude sensors to remove offsets and long-term drifts. We use QVH Industries FOGs, which have a noise of about $5'\sqrt{hour}$.

Star Tracker - The star trackers that we have developed give positional information by measuring the orientation of the payload with respect to the stars. Different classes of star trackers are needed in order to work either during the day or during the night or both, continuously, day and night. At night, a simple bore-sight Cohu 1910 ccd camera with a 50mm f 0.8 lens is able to detect up to mag. 7 stars with an accuracy of $\sim$1 arc-min rms and 33 $msec$ of integration time [9]. The field of view of 6X7 degrees guarantees at least two star detections at once and the full attitude determination.

During the day, the light scattered from the atmosphere at float altitude is less than one hundredth of the light at sea level [10, 11]. This background has to be reduced in order to avoid saturation of the ccd. This is done using lenses with a long enough focal length. Two solutions can be implemented: a servoed or a bore-sight star tracker.

Servoed Star Tracker - A Cohu 1910 camera is mounted on a motorized two axis platform. Each axis is controlled by means of a geared stepper motor and a 16 bit encoder. The camera is used with a 500 mm f 4.6 spotting scope and is high-pass filtered at 715 nm. As can be seen from figure 2b:, the background at float is roughly as bright as a mag. 4 star. The star tracker in this configuration is able to detect stars brighter than the background and to lock on them. As the payload scans and swings, the lock is maintained moving the axes of the motorized platform. The accuracy obtained is given by the encoders (20 arc-sec). Only one star can be locked at one time and no roll information can be obtained. This design has been especially developed for day-time use, but the system can also work at night since the presence of the filter results in a loss of only 0.8 magnitude. The full system can be run up to 33Hz

Bore-sight Star Tracker: The main requirement is to have a large field of view without saturating the ccd with the light coming from the scattered background, so that at least two stars can be detected at one time. Field subtraction is used to extract as many stars as possible from the background: a frame containing only the background signal is subtracted from each frame containing the stars' image on top of the background. Figure 2a shows an example based on the Kodak KAF-3200 CCD sensor (its wider dynamic range allows a significant star's detection over the noise signal). The solid line is the signal from a magnitude 7 star, while the dashed line is the noise after subtraction. With this kind of sensor it is possible to detect stars up to mag. 7 with a 150 mm lens and a field of view of 2-3 degrees, both in azimuth and in elevation. Detection of two stars at one time permits full attitude determination (azimuth, pitch and roll). The field of view can be reduced in order to increase attitude accuracy.

Other kinds of sensors can be used, including attitude GPS (Trimble) and Sun Sensors(ccd or photo-diode based, coarse or fine) [12].

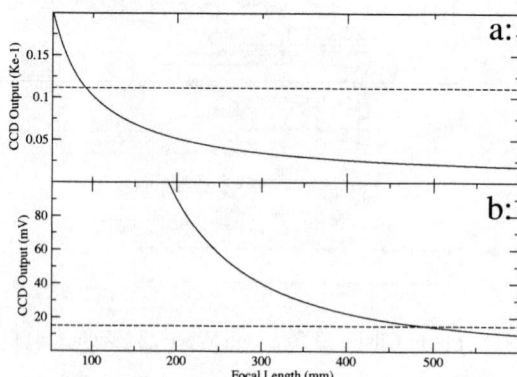

FIGURE 2. a) The signal of the noise on each pixel after frame subtraction (dashed line) is plotted vs. the focal length of the lens. The solid line is the signal of a mag. 7 star. With a 150mm lens it is possible to detect stars up to mag. 7 at float altitude.
b) Signal from the scattered sun light background (dashed line) at float altitude ($\sim$35K) is plotted versus different focal lengths and compared with the signal of mag. 4 star (solid line). All signals are high passed with a 715nm filter.

CONCLUSIONS

The ACS that has been developed is highly configurable. The mechanical parts and the digital control are suitable for driving payloads with different weights and inertia moments and with a real-time stability of $\sim$ 1 arc-min *rms*. The attitude sensors can be configured in different combinations, to match the experiment's requirements. Attitude stabilization and pointing reconstruction are achieved by driving the payload using coarse sensors in positions in which finer sensors (sun sensor, star tracker, etc.) can operate. The choice of a sensor-set also depends on whether the flight is at high or at mid latitudes. In the former case, the flight occurs completely in day-time or during the night, while in the latter case, day-night transitions can occur and special care has to be applied in order to have sensors that are able to work under both conditions. A magnetometer is a good choice for a coarse sensor. Absolute fine sensors have to be used to integrate the gyros. At night, a star-tracker is the easiest solution, while a star-tracker and a sun sensor can both be used during the day. The accuracy of the post-flight attitude solution depends only on the sensors chosen and can be better that 0.5 arc-sec *rms*.

REFERENCES

1. Boscaleri, A., *et al.*,1990, *Spie proceedings*, **1304**, 127.
2. Boscaleri, A., *et al.*, 1990, *Spie proceedings*, **1341**, 58.
3. Boscaleri, A., *et al.*, 1994, *Measurement Science and Technology*, **5**, 190.
4. de Berardis, P., *et al.*, 1993, *New Astron. Astrophys.*, **271**, 683
5. Ventura, G., *et al.*, 1995, **6**, 59
6. Tanaka, S. T., *et al.*, 1996, *Ap. J.*, **568**, L81.
7. de Berardis, P., *et al.*, 2000, *Nature*, **404**, 955
8. Carli, B., *et al.*, 1984, *App. Opt.*, **23**, 2594
9. Piacentini, F., *et al.*, 2001, *Atro-ph/0105148*
10. André, Y., 1990, *Ph. D. Thesys*, Paul Sabatier University, Toulouse, France.
11. Rossi, E., *et al.*, 1989, *IEEE Transactions on Nuclear Science*, **36**, 876.
12. Crill, B. P., *et al.*, 2000, in preparation.

Three Sun Sensors for Stratospheric Balloon Payloads

G. Romeo*, P. de Bernardis†, G. Di Stefano*, S. Masi†, F. Piacentini†, F. Pongetti* and S. Rao*

*Istituto Nazionale di Geofisica - Roma - ITALY
†Università degli Studi di Roma La Sapienza - Roma - ITALY

Abstract. We describe three sun sensors which have been developed for balloon borne experiments. The sensors have different resolutions and sky coverage, and have been developed and used in the BOOMERanG project.

INTRODUCTION.

The measurement of the payload orientation by means of observations of the Sun is an easy way to estimate the attitude of stratospheric balloons payloads flying over the Antarctica during the austral summer. Normally, payload orientation starts with the measurement of the local the magnetic field. In Antarctic flights, even if corrected using the best geomagnetic models, such a measurement cannot be used successfully, because of the strong effects of the solar activity at polar latitudes. Three sun sensor systems have been developed to orientate the BOOMERanG gondola[1],[3],[2]: The coarse system (full sky range) uses an array of cosine detectors. In the fine system (which has no moving parts) a slit projects the sun light on selected pixels of the CCD sensitive surface. The electro-mechanical system uses 4-quadrants photodiodes to point the sun using servomotors, and measures the sun position by means of absolute angular encoders and of the error signals from the photodiodes.

COARSE SYSTEM: A MIXED ANALOG-DIGITAL CIRCUIT RETURNS THE SUN ANGLE.

The coarse system consists of a circular array of cosine detector ('cells' hereafter) and of the related electronics. The principle of operation is explained with the aid of fig.1, where four cells are considered. An arbitrary even number of cells may be used (Boomerang used 6 cells) to improve accuracy. The relative azimuth angle between the sun-light and the payload coincides with the rotation angle of a cell rotating around the vertical axis of the payload, when the cell exhibits the maximum output. The output of the rotating cell is equivalent to the weighted sum of the outputs of the static cells in the circular array, if the weights are periodic functions of time. So our system has no moving parts, but is equivalent to a rotating cell one. In fact, cells are separated by $\pi/2$ angles ($2\pi/n$ in case of n cells) and the signed sum of a couple of opposite cells returns a value proportional to the sine (or cosine) of the angle between sun light and cells axis. An angle generator produces a periodic signal (which represents a virtual angle b rotating from $0°$ to $360°$) and the sine and cosine of this angle multiplies the signal from the cells. The result of the weighted sum may be written as $f = A\cos(a-b)$, where a represents the sun-light angle with respect to the payload. The function f has a maximum when $a = b$. A peak detector connected to the signal f freezes the angle generator output at the right moment, producing a digitized output of a. For improved precision, f was shifted by $90°$, and a 0-crossing detector was used to control the output latch.

The trigonometric functions used to produce signals are stored in non volatile memory and converted to voltage levels using serial DACs. The weighting operation is performed using the linear characteristic illuminance-conductance of CdS photocells. Each cell is biased with a voltage proportional to the weight, and produces a current proportional to the product weight × illuminance. The sum of four current-voltage builds the virtual rotating cell output f. To obtain

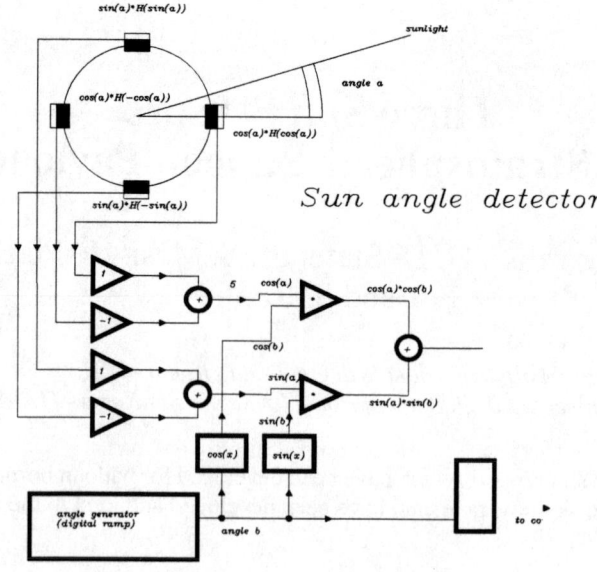

FIGURE 1. Operation of the coarse detector (see text)

a cosine response a diffusion filter has been placed in front of each photocell, and each channel has been electrically trimmed to have the same electrical-optical response. The cells have all been mounted on a shaded common aluminum plate, to avoid errors induced by differential temperature drifts. The lab tests, where it is possible to simulate the sun shining over a completely dark background, exhibited a very good performance (better than 0.5^o) for the 6-cells system. The cosine detectors, however, accept radiation from a solid angle of 2π, which includes bright albedo from the ice during the Antarctic flight. For this reason the in-flight performance was significantly worse than the lab one. The absolute maximum error was anyway contained within 8°, more than enough to enter in the working range of the fine sun sensor.

THE FINE SUN SENSOR

This device uses a 5000 pixel linear CCD array to detect the centroid of the sun image produced by a slit (stenopeic cut) [4], [5], [6]. The slit is orthogonal to the CCD, and is offset by a distance d. Depending on the incidence angle of the sunlight, different regions of the CCD are illuminated. In an ideal detector, a measurement of the location x of the centroid of the illumination pattern with respect to the center of the CCD allows to estimate the incidence angle from the relation $\theta = \arctan x/d$. In the real system effects like internal reflections in the filters and CCD glass and

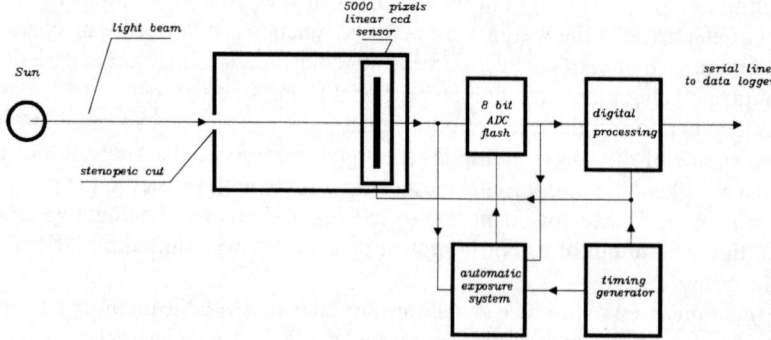

FIGURE 2. Operation of the fine sun sensor (see text)

misalignments of the slit and the CCD modify the previous relationship, and an accurate calibration look-up table is required. An analog automatic exposure system keeps the ADC in range, and a DSP chip performs all the calculations. The system produces 25 measurements per second, and the statistical error for θ is ±0.01° (lab conditions) with 120° of range (azimuth); the statistical error is 0.001° with 45° of range (elevation). The precision of this detector exceeds the precision of 1/5000, related to the number of pixels because the processing uses all the lit pixels and their signal levels to ccompute the centroid. The precision of the system may be improved by placing a diffuser near the CCD. The in-flight noise has been consistent with the lab measurements above, and performance was limited by the precision of the calibration.

THE ELECTRO-MECHANIC SYSTEM

This solution had been originally discarded because it is expensive, and requires moving parts which may be single point failures. It has the advantage, however, of relying on the calibration of the absolute encoders, thus speeding-up significantly the integration with the payload and pre-flight operations. The system sensors are two parallel tubes with a stenopeic hole on the end exposed to the sun, and a 4-quadrants photodiode mounted on the opposite end. The first tube is very short and is used as the coarse sensor to find the sun at system start-up; the second one is longer and is used to lock on the sun and track its position. The tubes are mounted on an alt-azimuth mount on the payload, powered by DC servomotros. The elevation and azimuth angles of the tubes are measured by means of two 16 bit absolute encoders, which offer a resolution of 5/1000 of degree. An even better resolution, and control of perfect locking, is achieved by digitizing the 4-quadrants diode outputs, and by using this information to correct the encoders outputs. Fig 3 shows the block diagram of this system, which will be added to the previous two in the forthcoming flight of BOOMERanG.

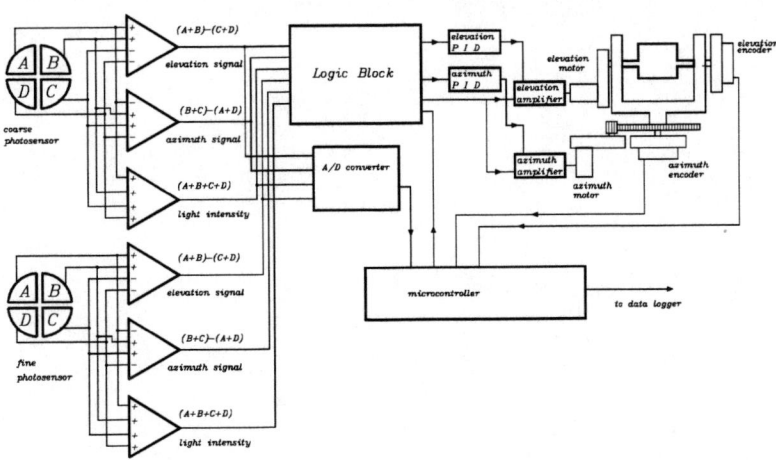

FIGURE 3. Operation of the electro-mechanical system (see text)

ACKNOWLEDGMENTS

This activity is being supported by the Italian Agencies MURST, PNRA and ASI.

REFERENCES

1. de Bernardis, P., et al. 2000, Nature, 404, 955-959.
2. Masi S., et al., 1999, in 3K Cosmology, AIP conf.proc. 476, pg. 237.
3. Piacentini F., et al., these proceedings; see also astro-ph/0105148 (in press on Ap.J.Suppl.)
4. Romeo G., "Peak detector maximizes CCD-sensor range", EDN, August 15, 1996, p.88.
5. Romeo G., Rao S., "Automatic-exposure scheme uses CCD shutter", EDN, July 2 1998, p.86.
6. Romeo G., "Digitization of optical lever instruments", Annali di Geofisica, Vol 43 N. 3, June 2000, pp 545-557

2K1BC Workshop
Experimental Cosmology at millimetre wavelengths

GROUND BASED EXPERIMENTS

First Season Observations with the Degree Angular Scale Interferometer (DASI)

E. M. Leitch[a], J. E. Carlstrom[a], N. W. Halverson[a], J. Kovac[a], C. Pryke[a]
W. L. Holzapfel[b], M. Dragovan[c]

[a] *University of Chicago, 5640 S. Ellis Ave., Chicago IL 60637*
[b] *Jet Propulsion Laboratory, 4800 Oak Grove Drive, Pasadena CA 91109*
[c] *University of California, 426 Le Conte Hall, Berkeley CA 94720*

Abstract. We discuss measurements of anisotropy in the Cosmic Microwave Background (CMB) from the first season of observations with the Degree Angular Scale Interferometer (DASI). The instrument was deployed at the South Pole in the austral summer 1999–2000, and made observations throughout the following austral winter. We have measured the angular power spectrum of the CMB in the range $100 < l < 900$ in nine bands, with fractional uncertainties in the range 10–20%, dominated by sample variance. By comparison of the DASI+COBE-DMR data with a seven-dimensional grid of adiabatic CDM models, we find $\Omega_b h^2 = 0.022^{+0.004}_{-0.003}$, consistent with that derived from measurements of the primordial abundance ratios of the light elements combined with big bang nucleosynthesis theory. Limits on other cosmological parameters are presented.

INTERFEROMETERS AND THE CMB

At its most basic, an interferometer is a correlation device, whose essential function is to multiply signals from pairs of receivers; the time-average of this product for each pair of receivers, known as a *visibility*, is the fundamental data product of an interferometer. Because the time-average of uncorrelated signal products is zero, an interferometer is immune to noise which is uncorrelated between receivers. For any two elements of an interferometer, the multiplication of correlated signals gives rise to a sinusoidal interference pattern on the sky, also called a *fringe pattern*, oriented perpendicular to the baseline vector **b** connecting the antennas (see Fig. 1). This is the effective instrumental response of a single baseline, analogous to the beam of a single-dish telescope. The spacing of the fringe pattern is λ/b, where λ is the observing wavelength and b is the magnitude of the projection of **b** perpendicular to the line of sight.

Each baseline of a multi-element array can therefore be thought of as a matched filter, selecting a specific Fourier component. In the Fourier plane, the spatial frequency **u** selected by a given baseline is $(u, v) = (b_x/\lambda, b_y/\lambda)$, or in terms of multipole moment, $l \simeq 2\pi b/\lambda$ (for $l > 60$, White et al., 1999), shorter baselines measuring larger angular scales, and vice versa. Both real and imaginary components of the Fourier plane can be measured with a complex correlator; the real component is measured by correlating the pair of signals with no relative phase delay phase delay; the imaginary component is measured by correlating the signals with a 90° phase shift introduced into one of the signal paths.

Because they directly sample Fourier components of the sky brightness, interferometers are uniquely suited to measurement of the CMB power spectrum. For observations of the CMB in the Rayleigh-Jeans limit, the visibility in the flat-sky approximation is given by

$$V(\mathbf{u}) = \frac{2k_B T}{\lambda^2} \int_{-\infty}^{\infty} d\mathbf{x}\, A(\mathbf{x}, \lambda) \frac{\Delta T}{T}(\mathbf{x}) e^{-2\pi i \mathbf{u} \cdot \mathbf{x}}, \quad (1)$$

where $A(\mathbf{x}, \lambda)$ is the primary beam. Equivalently, we can write

$$V(\mathbf{u}) = \frac{2k_B T}{\lambda^2} \widetilde{A}(\mathbf{u}, \lambda) * \widetilde{\frac{\Delta T}{T}}(\mathbf{u}), \quad (2)$$

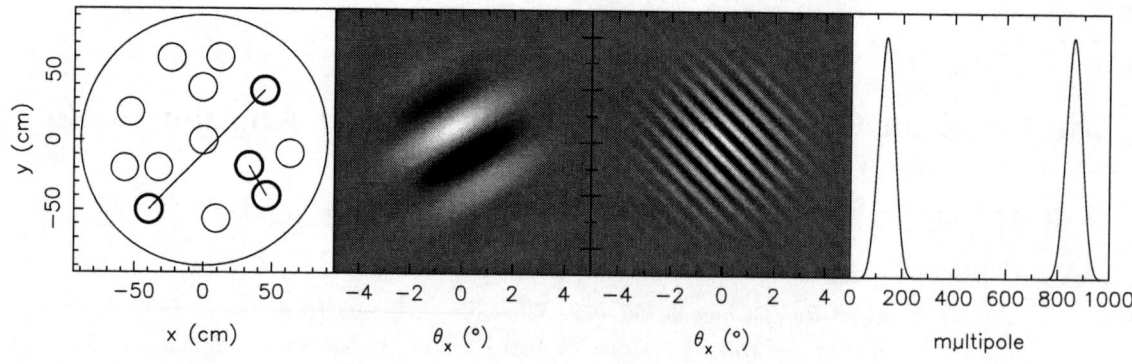

FIGURE 1. (Left panel) The DASI faceplate configuration, with one of the shortest and one of the longest baselines indicated. (Center left) The response on the sky of the the shorter of the two baselines. (Center right) The long baseline response. In each case the fringe pattern is enveloped by the $3°\!\!.4$ FWHM primary beam of the DASI antennas. (Right panel) The l-space window functions for the two baselines at a single frequency.

from which it is clear that the visibility is just the convolution of the aperture autocorrelation function with the transform of the sky brightness $\widetilde{\Delta T}(\mathbf{u})$. Since the CMB power spectrum is directly related to the Fourier transform of the sky brightness, it can be seen that the variance of the visibilities is a direct estimate of the the C_l.

For a pair of infinitesimally small apertures, a given baseline responds to a single l-mode; the effective window function is simply a delta function at $l \simeq 2\pi b/\lambda$. For finite apertures, the l-space response is broadened by the autocorrelation function of the antenna power pattern, which on the sky corresponds to an enveloping of the fringe pattern by the primary beam (as in Eq. 1). The shape of the resulting window function is independent of baseline length (see Fig. 1); for an interferometer, uncertainties in the beam do not lead to resolution-dependent uncertainties in the power spectrum.

THE DASI INSTRUMENT

The Degree Angular Scale Interferometer (DASI) is one of a new generation of ultra-compact microwave interferometers designed to measure anisotropy in the CMB. Other interferometric CMB experiments, including DASI's sister instrument the CBI (Pearson et al., 2000), as well as the VSA (Jones, 1997), are discussed elsewhere in this volume. With 13 elements operating in ten 1-GHz bands from 26–36 GHz, DASI provides dense sampling of the power spectrum from $l \simeq 100$ to $l \simeq 900$, angular scales spanning the first three harmonically related acoustic peaks in $\Omega \sim 1$ cosmologies. A summary of the instrument and observations is given below; for a full description of the experiment, see Leitch et al. (2001).

Telescope Mount

The telescope mount is a standard alt-az design, with antennas affixed to a rigid faceplate which can be separately rotated about its axis. DASI's 13 primary antenna elements are arranged in a three-fold symmetric pattern on the faceplate; the locations of the faceplate slots are optimized to provide nearly uniform sampling over the multipole range probed by DASI. In combination with the three-fold symmetry, faceplate rotation provides a host of consistency checks, permitting discrimination of spurious signals such as cross-talk and contamination from the ground. Because the sky is real, its transform is Hermitian, and each $0°-120°-240°$ triplet of baselines can therefore provide six measurements of the same Fourier component: three from distinct pairs of antennas, and three from their conjugate positions $180°$ away.

With conventional interferometric arrays, in which the array elements are separately mounted, delay lines must be dynamically inserted to compensate for variable geometric phase delays as the antennas track. With a co-planar array such as DASI, the correlator design is greatly simplified, since projected baseline lengths for a co-planar array are independent of the pointing center, and tracking delays are not required.

The 10-GHz correlator, described below, is of an extremely compact design and is mounted to the underside of the faceplate, along with the downconverter and control electronics; because cables carrying

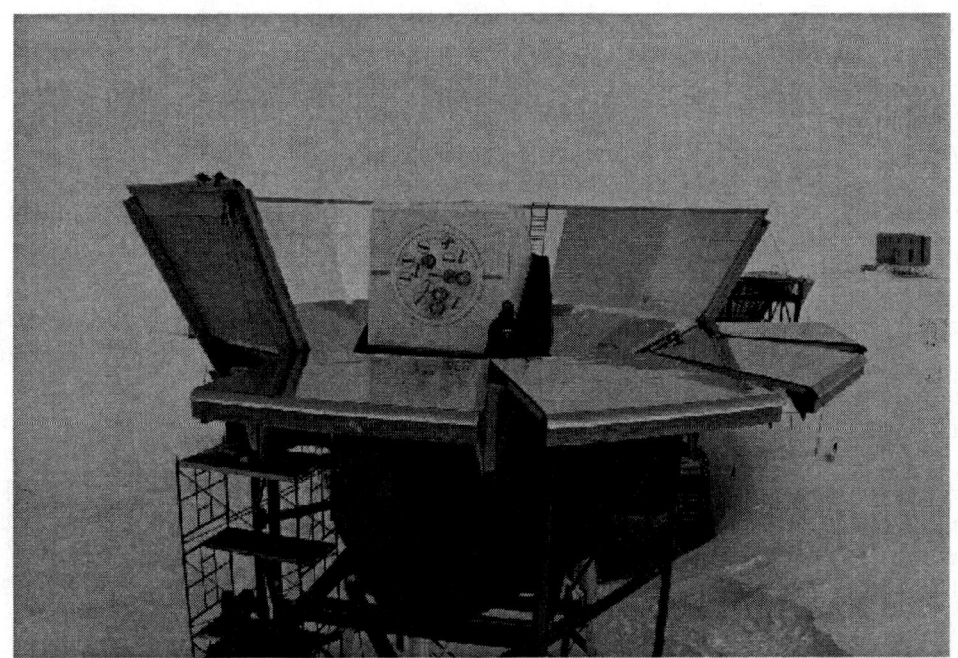

FIGURE 2. The DASI telescope, shown during refitting for polarization observations during the 2000–2001 austral summer. In this photograph, only three of the 13 antennas are mounted. Also visible is the ground shield erected during the 2000–2001 summer season. Four of the panels can be lowered to permit observations of planets and a nearby transmitter.

phase-sensitive signals are fixed to the faceplate, they do not flex with rotation of the instrument, and instrumental phase drifts are less than $10°$ over a period of many weeks.

Receivers

Each antenna consists of a 20-cm aperture, $30°$ semi-flare angle corrugated horn, equipped with a high density polyethylene lens, resulting in a high aperture efficiency, 83.5%, and permitting an extremely compact horn design. At polar temperatures, the lens contributes less than 2.5 K to the system temperature.

By contrast with Cassegrain elements, horns provide unobstructed apertures, with lower sidelobe response and better cross-talk characteristics, important for a compact array in which the elements are nearly touching. Each antenna element is further surrounded by a corrugated shroud, yielding a measured monochromatic crosstalk level of less than -100 dB in laboratory measurements of the horn-lens combination (Halverson and Carlstrom, 2001). To further suppress cross-talk from correlated amplifier noise, the receivers are also equipped with front-end isolators.

The beam pattern, which determines the field of view of the interferometer, has been characterized in range measurements at 26, 30 and 36 GHz and in each case found to agree closely with the predicted pattern, both in the sidelobe response, which is typically -20 dB at the first sidelobe, and in the main beam width, which is $3°.4$ FWHM at 30 GHz (Halverson and Carlstrom, 2001).

Signal Chain

At the heart of each receiver is a cryogenically cooled, 4-stage InP HEMT amplifier operating from 26–36 GHz. These amplifiers were constructed at the University of Chicago, after a design developed at NRAO (Pospiezalski et al., 1995). Receiver temperatures range from 15–26 K at the center of the band, increasing to an average of 30 K at the band edges. These noise temperatures include the HEMTs, isolator and polarizer (all at 10 K), and warm throat, horn and lens. Including CMB and atmosphere, we achieve typical system temperatures of 26 K at band center, for an rms sensitivity per visibility of approximately 60 Jy $s^{1/2}$ in a 1 GHz band.

The 26–36 GHz RF signal from each antenna is mixed down to 2–12 GHz IF band using a local oscillator (LO) tuned to 38 GHz. The IF signal is split into ten 1-GHz wide bands, each of which is further mixed down to 1–2 GHz. The 13 signals at each frequency are fed to one of 10 identical analog correlators (Padin et al., 2000), where the 78 complex multiplications are formed, digitized and integrated for 0.84 s in a digital accumulator. A copy of each input signal is phase-shifted by 90°, and multiplications are performed simultaneously for the real and imaginary Fourier components.

THE SOUTH POLE ENVIRONMENT

The Antarctic plateau is one of the driest deserts on the planet, with annual precipitation averaging less than 8 g/cm^2. The polar atmosphere also has a lower water column density than other sites of the same physical altitude, since cooling flattens the atmosphere above the poles; radiosonde measurements above the Pole indicate that the tropopause occurs between 8 and 9 km, compared to 11–13 km nearer the equator. The precipitable water vapor column above the pole varies between 0.25 and 0.7 mm annually (Chamberlin, 2001).

Opacities were measured with DASI from skydips performed daily during May–November 2000. The mean opacity determined from these data rises from $\tau = 0.012$ to $\tau = 0.023$ over the DASI frequency band (26–36 GHz), with little day-to-day variation; at the lowest frequency, 95% of the measured opacities are < 0.015, while at the highest frequency, 95% are < 0.028. At typical ambient temperatures during the winter ($-60°$ C) these results indicate that over much of the DASI band, the CMB and the atmosphere make roughly equal contributions to our system temperatures.

OBSERVATIONS

During the 2000 season, the ground shields shown in Fig. 2 were not in place, and our observing strategy was driven by the presence of near-field ground contamination. Variation of the ground signal with direction, at a level well above the CMB signal, limits our ability to track single fields over a wide range in azimuth. Repeated tracks over the full azimuth range, however, show little evidence for time variability on periods as long as five days. Observations were therefore divided among 4 constant declination (elevation) rows of 8 fields, on a regular hexagonal grid spaced by 1h in right ascension, and 6° in declination. The grid center was selected to avoid the Galactic plane and to coincide with a global minimum in the IRAS 100 micron map of the southern sky. Each field in a row is observed over the same azimuth range, permitting a constraint on the common ground signal. CMB fields were observed during the period spanning 05 May–07 November 2000, and the data described here comprise 97 days of observation, representing an observing efficiency of better than 85%.

THE POWER SPECTRUM

As discussed in the first section of this contribution, an interferometer makes direct measurements of the Fourier plane, and the angular power spectrum can be extracted from the data without recourse to an image. The calibrated output of the interferometer is the visibility, which is the convolution of the Fourier transform of the sky brightness distribution with the antenna aperture field autocorrelation function (see Eq. 2). A data vector Δ of length $N = 1560 \times 32$ (before data edits) is constructed by combining observations of each visibility for each of the 32 fields. The likelihood function for a set of parameters κ is given by

$$\mathcal{L}_\Delta(\kappa) = \frac{1}{(2\pi)^{N/2} |C(\kappa)|^{1/2}} \exp\left(-\tfrac{1}{2} \Delta^T C(\kappa)^{-1} \Delta\right), \tag{3}$$

where the covariance matrix

$$C(\kappa) = C_T(\kappa) + C_n + \alpha C_C \tag{4}$$

is the sum of the theory, noise, and constraint covariance matrices, described in the next section, and is a function of the parameters κ, in this case the band powers, $l(l+1)C_l/(2\pi)^2$.

We have adopted the iterated quadratic estimator approach of Bond et al. (1998) to find the maximum likelihood values of the angular power spectrum modeled as a piecewise flat (in $l(l+1)C_l$) power spectrum in 9 bands; see Halverson et al. (2001) for a detailed description of this analysis. The quadratic estimator method steps to the peak of the likelihood function by approximating the likelihood near the peak as Gaussian, thereby minimizing the number of computationally intensive matrix inversions required for direct evaluation of the likelihood surface; this method typically converges on the peak in 3–4 iterations.

Maximum likelihood estimates for the nine-band CMB power spectrum are presented in Figure 2. In each of the nine bands, the error is dominated by cosmic (sample) variance. Also shown are band power estimates for an alternate model, in which the bands are shifted with respect to the first model. This is to demonstrate that features in the power spectrum are not artifacts of the data binning, such as anticorrelations between neighboring bands.

Diffuse Foregrounds

Our current understanding of the known Galactic foregrounds implies that the CMB will dominate the anisotropy signal at the frequencies and angular scales to which DASI is sensitive (see, e.g. Tegmark et al., 2000, for a comprehensive review). Nevertheless, we employ an additional check which is independent of the assumptions usually made when extrapolating Galactic foregrounds to microwave frequencies; in the likelihood analysis described above, we use the formalism described in Bond et al. (1998) to marginalize over potentially contaminated modes in the data. The last term in Equation 4, C_C, is a *constraint matrix*, which represents the contribution to the covariance matrix of potential contaminants, for example diffuse foregrounds, or point sources with known positions. A constraint matrix for any foreground can be constructed from a spatial template for that foreground; the matrix is then multiplied by a prefactor α large enough to de-weight the undesired modes without making the covariance matrix singular. Note that we require only the shape of these modes in the datavector to construct the constraint matrix for a foreground, not the absolute scaling; this is equivalent to marginalizing over a mode without knowing its amplitude.

By constructing constraint matrices for synchrotron from low-frequency surveys, for dust from IRAS 100-micron templates and for free-free from H-α images of the DASI fields, we can restrict the contribution of any of these diffuse foregrounds to be less than $\sim 3\%$ of the CMB signal, assuming only that the 26–36 GHz counterparts to these foregrounds have the same spatial distribution as they do at frequencies where the foregrounds have been measured. It is furthermore assumed that each diffuse foreground can be described with a single spectral index constant across a given DASI field. We conclude that dust, free-free and synchrotron emission, as well as emission with *any* spectral index that is spatially correlated with these foregrounds, such as spinning dust grain emission (Draine and Lazarian, 1998), make a negligible contribution to the power spectrum shown in Fig. 3.

Point Sources

As expected for high-frequency, small angular scale experiments, point sources are the dominant astronomical foreground for DASI (see, e.g. Tegmark and Efstathiou, 1996), and the dominant contributor to the measured anisotropy. The method of constraint matrices, described above for projecting out diffuse foregrounds, can be applied to point sources as well, since they produce a unique spatial signature in the visibility data. To null the effect of point sources through this technique, we require only the positions of the sources, *not their fluxes*. Positions for the brightest point sources are determined from the DASI data while positions for point sources below the DASI detection threshold are determined from the PMN catalog (Wright et al., 1994). In all, we constrain every point source from the PMN survey whose 5 GHz flux density, when modulated by the DASI primary beam, exceeds 50 mJy.

For sources below the PMN flux cutoff, we make a statistical correction to the power spectrum. These corrections, given explicitly in Halverson et al. (2001), are determined by Monte Carlo simulation of a point source population with dN/dS from the PMN catalog, and whose spectral index distribution is determined by observation of a complete sample of PMN sources with the OVRO 40-meter telescope at 26–36 GHz (paper in preparation).

CONCLUSIONS

As can be seen in Figure 2, we see strong evidence for both first and second peaks in the angular power spectrum at $l \sim 200$ and $l \sim 550$, respectively, and a rise in power at $l \sim 800$ that is suggestive of a third. The presence of harmonic peaks in the power spectrum is a dramatic confirmation of adiabatic scenarios in

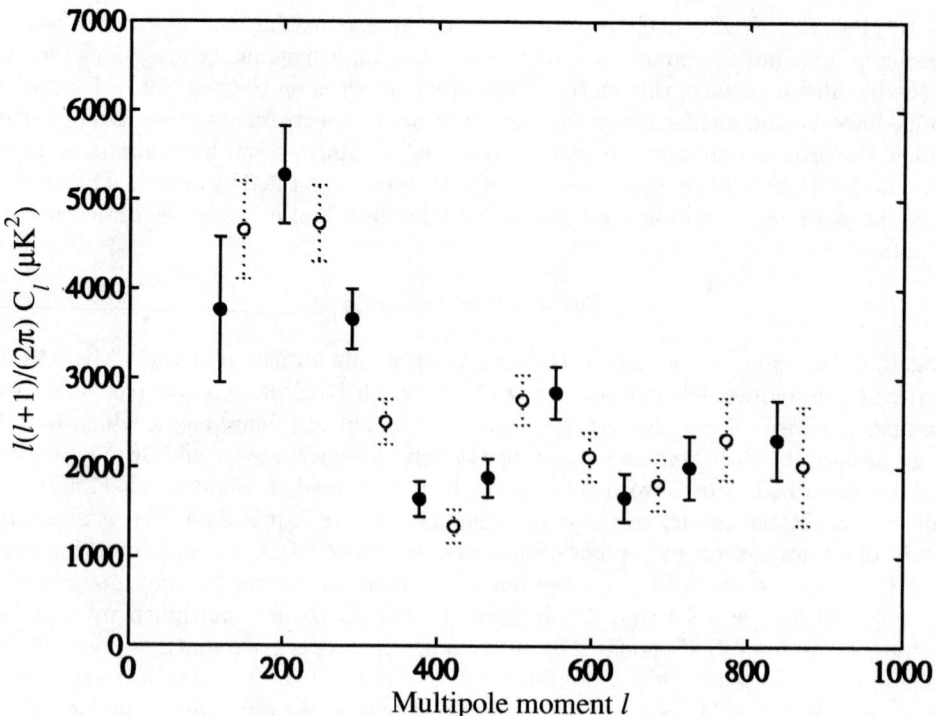

FIGURE 3. The angular power spectrum from the first season of DASI observations, plotted in nine bands (closed circles). We have analyzed the same data in nine bands shifted to the right (open circles). The alternate set of band powers are shown to demonstrate the robustness of the likelihood analysis code. Only the nine bands shown in the primary (closed circle) analysis are used for parameter extraction. Adjacent bands are anticorrelated at the 20% level. Not shown is an additional calibration uncertainty of 8%, expressed as a fractional uncertainty on the C_l band powers, which is completely correlated across all bands due to the combined flux scale and beam uncertainties.

which sub-degree scale anisotropy in the CMB is the result of gravitationally driven acoustic oscillations, seeded by inflation. The rise in power in the region of the predicted third peak moreover strongly supports, from CMB data alone, the presence of dark matter in the Universe.

We have compared the DASI+COBE-DMR data to adiabatic CDM models with initial power law perturbation spectra. Results of this analysis are detailed in Pryke et al. (2001). The best fitting model has an acceptable χ^2 value, indicating that this class of models is a plausible representation of the underlying physics. Adopting conservative priors for the Hubble constant $h > 0.45$ and the optical depth to last scattering, $0.0 \leq \tau_c \leq 0.4$, we find $\Omega_{tot} = 1.04 \pm 0.06$ and $n_s = 1.01^{+0.08}_{-0.06}$, consistent with inflationary predictions. Tighter priors on h and τ_c improve our constraints on Ω_{tot} and n_s respectively, which remain consistent with the theory.

We find that $\Omega_b h^2 = 0.022^{+0.004}_{-0.003}$ and $\Omega_{cdm} h^2 = 0.14 \pm 0.04$, further strengthening the mounting evidence for non-baryonic dark matter. These constraints are only weakly affected by the choice of h and τ_c priors. Setting a strong h prior breaks the $(\Omega_m, \Omega_\Lambda)$ degeneracy such that we constrain $\Omega_m = 0.40 \pm 0.15$ and $\Omega_\Lambda = 0.60 \pm 0.15$— consistent with other recent results.

The current best value for $\Omega_b h^2$ derived from big bang nucleosynthesis (BBN) calculations, in combination with measurements of the primordial abundance ratios of the light elements, is $\Omega_b h^2 = 0.020 \pm 0.002$ (95% confidence, Burles et al. (2001)). The χ^2 of the difference between this and our own value is at the 42% point of the cdf (assuming Gaussian errors on both); the values are hence consistent. Previous CMB analyses have seen little power in the second peak region, and have determined $\Omega_b h^2$ values higher than, and inconsistent with, BBN at the $\sim 3\sigma$ level (Lange et al., 2001; Jaffe et al., 2001).

REFERENCES

1. White, M., Carlstrom, J. E., Dragovan, M. and Holzapfel, W. H., *ApJ*, **514**, 12-24 (1999).
2. Pearson, T. J., Readhead, A. C. S., Padin, S., Cartwright, J. K., Mason, B. S., Myers, S. T., Shepherd, M. C., Sievers, J. L. and Udomprasert, P. S., "The Cosmic Background Imager," in *IAU Symposium 201: New Cosmological Data and the Values of the Fundamental Parameters*, edited by A. Lasenby and A. Wilkinson, ASP, 2000 (astro-ph/0012212).
3. Jones, M. E., "Results from CAT and Prospects for the VSA," in *Microwave Background Anistropies*, proceedings of the XXX1st Recontre de Moriond, 1997, p. 161.
4. Leitch, E. M., Pryke, C., Halverson, N. W., Carlstrom, J. E., Kovac, J., Holzapfel, W. L., Dragovan, M., Cartwright, J. K., Mason, B. M., Padin, S., Pearson, T. J., Readhead, A. C. S. and Shepherd, M. C., *ApJ*, in press (2001).
5. Halverson, N. and Carlstrom, J. E., *IEEE-MTT*, in preparation (2001).
6. Pospiezalski, M. W., Lakatosh, W. J., Nguyen, L. D., Lui, M., Liu, T., Le, M., Thompson, M. A. and Delaney, M. J., *IEEE MTT-S Int. Microwave Symp.*, **1121**, (1995).
7. Padin, S., Cartwright, J. K., Shepherd, M. C., Yamasaki, J. K. and Holzapfel, W. L., *IEEE Trans. Instrum. Meas.*, submitted (2000).
8. Chamberlin, R. A., *Journal of Geophysical Research: Atmospheres*, in press (2001).
9. Bond, J. R., Jaffe, A. H. and Knox, L., *Phys. Rev. D*, **57**, 2117 (1998).
10. Halverson, N. W., Leitch, E. M., Pryke, C., Carlstrom, J. E., Kovac, J., Holzapfel, W. L. and Dragovan, M., *ApJ*, submitted (2001).
11. M. Tegmark, D. J. Eisenstein, W. Hu and A. de Oliveira-Costa, *ApJ*, **530**, 133 (2000).
12. Draine, B. T. and Lazarian, A., *ApJ*, **508**, 157 (1998).
13. Tegmark, M. and Efstathiou, G., *MNRAS*, **281**, 1297 (1996).
14. Wright, A. E., Griffith, M. R., Burke, B. F. and Ekers, R. D., *ApJS*, **91**, 111–308 (1994).
15. Pryke, C., Halverson, N. W., Leitch, E. M., Carlstrom, J. E., Kovac, J., Holzapfel, W. L. and Dragovan, M.", *ApJ*, submitted (2001).
16. Burles, S., Nollett, K. M. and Turner, M. S., *Phys. Rev. D*, **63**, 063512 (2001).
17. Lange, A. E., Ade, P. A. R. Bock, J. J. Bond, J. R. Borrill, J. Boscaleri, A. Coble, K. Crill, B. P. de Bernardis, P. et al., *Phys. Rev. D*, **63**, 042001 (2001).
18. Jaffe, A. H. Ade, P. A. R. Balbi, A. Bock, J. J. Bond, J. R. Borril, J. Boscaleri, A. Coble, K. Crill, B. P. et al., *Phys. Rev. Lett.*, **86**, 3475 (2001).

The Very Small Array

Angela C. Taylor

Astrophysics Group, Cavendish Laboratory, Madingley Road, Cambridge, CB3 0HE, UK

Abstract. The Very Small Array (VSA) is a fourteen-element interferometer designed to study the cosmic microwave background on angular scales of 2.4 to 0.2 degrees (angular multipoles $l = 150$ to 1800). It operates at frequencies between 26 and 36 GHz, with a bandwidth of 1.5 GHz, and is situated at the Teide Observatory, Tenerife. The instrument also incorporates a single-baseline interferometer, with larger collecting area, operating simultaneously with and at the same frequency as the VSA main array. This provides accurate flux measurements of contaminating radio sources in the VSA observations. Since September 2000, the VSA has been making observations of primordial CMB fluctuations. We describe the instrument, observing strategy and current status of the first year of observations.

INTRODUCTION

The Very Small Array (VSA) is a 14-element interferometer designed to make images of the cosmic microwave background (CMB) and to measure its power spectrum over angular scales of 2.4 to 0.2 degrees ($l = 150$ to 1800). The telescope is located at the Teide Observatory, Tenerife at an altitude of 2400 m and, for observations in the region 26–36 GHz, the transparency at the site is approximately 98 percent. There is also negligible correlated emission from the atmosphere.

Interferometers are well-suited to measuring the power spectrum of the CMB since they directly sample the Fourier modes on the sky which can then be converted to a power spectrum. In addition they provide excellent rejection of systematics since only correlated signals are detected, reducing signals such as ground radiation and atmospheric emission. Interferometric systems also offer the opportunity to target a specific range of angular scales on the sky, determined by the spacing of the elements of the array.

In order to achieve constant temperature sensitivity over the full range of angular scales that the VSA is designed to measure, two separate, but scaled, array configurations are used. For measurement of the CMB power spectrum over l-values 150–700, a 'compact array' configuration is used. Each receiver is fitted with a 143 mm diameter antenna and typical baselines in this configuration range from approximately 30 cm to 120 cm. For mapping finer angular scales, the same receivers are re-fitted with larger antennas, 322 mm in diameter. The baselines of this 'extended array' are scaled by the same factor, thus allowing measurements to be made across the whole range of l with constant temperature sensitivity. The specifications of these two arrays are given in Table 1.

The VSA project is a collaboration between the Cavendish Astrophysics Group, Cambridge, Jodrell Bank Obsevatory, Manchester, and the Instituto de Astrofisica de Canarias (IAC), Tenerife. Although the telescope can be operated

TABLE 1. Basic VSA parameters

	Compact Array	Extended Array
Mirror size /mm	143	322
Primary Beam (34 GHz)	4.6°	2.0°
Synthesised Beam (34 GHz)	∼30′	∼11′
l-range	150-700	300-1800
ΔS (28×7 hr) / mJy beam^{-1}	30	6
ΔT (28×7 hr) / μK beam^{-1}	33	33

FIGURE 1. The VSA main array in its enclosure.

remotely, on-site support is provided by the IAC. All three institutions are actively involved in the data analysis.

DESIGN OF THE VSA

The basic design of the VSA is a development of the Cosmic Anisotropy Telescope (CAT) [1], a three-element interferometer which operated at 15 GHz from a sea-level site in Cambridge. The VSA operates over the frequency range 26–36 GHz and has an observing bandwidth of 1.5 GHz. The front-end receivers are cryogenically cooled HEMT amplifiers and provide an overall system temperature of 25–35 K. Each receiver is fitted with a corrugated horn-reflector antenna in which the mirror can be rotated, providing tracking in one dimension. The fourteen VSA receivers are all mounted on a tip-table, providing pointing in a further dimension. Close packing of the array is achieved by mounting each element at 35° to the table. The tracking range of the table in elevation is 0°–70° resulting in a range of accessible declinations of −7° to +63°. To eliminate ground radiation, the complete array is surrounded by a a 3.5 m high octagonal enclosure. Figure 1 shows the VSA main array in its enclosure.

Quasi-independent tracking of the VSA antennas is a key feature that distinguishes the VSA from other interferometric CMB experiments. Since each antenna tracks individually, the astronomical signal path to each antenna varies continuously during an observation. The rate of change of phase (or fringe rate) resulting from this continuous change in path can be calculated for each baseline configuration. The observed complex visibilities are then multiplied by the inverse of this expected fringe rate to give a quasi-constant signal. Since this is equivalent to applying a matched filter, all signals not varying at the expected fringe rate are removed. This rejection of systematics enables us not only to distinguish common systematics such as ground radiation or residual cross-talk between antenna elements but also to filter out the effects of bright sources such as the Sun and the Moon. We have been successful in removing the effects of both the Sun and Moon, even when they are as close as 30° away from the VSA primary beam. The implementation and effectiveness of this filtering technique is discussed further by M. Jones in these proceedings.

A building alongside the main enclosure houses the VSA correlator and control room. The signal from each antenna is down-converted to 8.25–9.75 GHz on the table, with further down-conversion to baseband (0.25–1.75 GHz) before entering the control building. Phase-switching is provided at the first stage of down-conversion. Once inside a screened room, path compensation is provided, in increments of ∼ 7 mm, by a sequence of strip-line elements. Appropriate pairs of signals are then fed by a series of splitters to 182 correlators, providing the real and imaginary components of the

FIGURE 2. One of the 3.7m source-subtraction dishes.

correlated signal from each of the 91 baselines. The outputs of the correlator are sampled every second.

In addition to the VSA main array we operate a separate single-baseline interferometer, with much large collecting area, adjacent to the main enclosure. This 9-metre, north-south baseline operates simultaneously with and at the same frequency as the VSA, and forms part of our source-subtraction strategy described below.

SOURCE SUBTRACTION

Extragalactic radio sources are a major contaminant of CMB observations at centimetre wavelengths. Their population is not well known at frequencies higher than 10 GHz, and many are expected to be variable and/or have rising spectral indices, α (where $S \propto \nu^{-\alpha}$). On the angular scales of interest to primordial CMB work, such sources are also generally unresolved. We deal with this problem by observing all the contaminating radio sources in each VSA field at higher resolution than can be reached using the VSA. To achieve this, we implement a two-stage process.

First, prior to observation with the VSA, we survey all the VSA fields at 15 GHz using the Ryle Telescope (RT) in Cambridge. The RT, which uses five 13 m diameter antennas and gives a resolution of $\sim$30 arcsecs, is used in a raster scanning-mode and reaches an rms noise level of $\sigma = 4$mJy [2]. This allows us to identify all sources above 20 mJy at 15 GHz, and ensures that we find all sources above our source-confusion limit of 80 mJy at 34 GHz, even allowing for a spectral index as steep as -2 between 15 and 34 GHz.

Having identified the contaminating sources in each field, we monitor each source at 34 GHz using a separate single-baseline interferometer working simultaneously with the VSA. This single-baseline interferometer consists of two 3.7m dishes separated by 9 metres on a north-south baseline. Each dish is situated in an enclosure similar to that of the VSA (Fig. 2.) and is fitted with identical horn-reflector feeds and receivers as used on the main array. Every source identified in the 15 GHz survey is monitored daily and its contribution is subsequently subtracted in the uv plane from VSA observations. The monitoring is done simultaneously with and at the same frequency as the VSA observation. This ensures that sources which are variable on time-scales as short as a few days can be subtracted accurately.

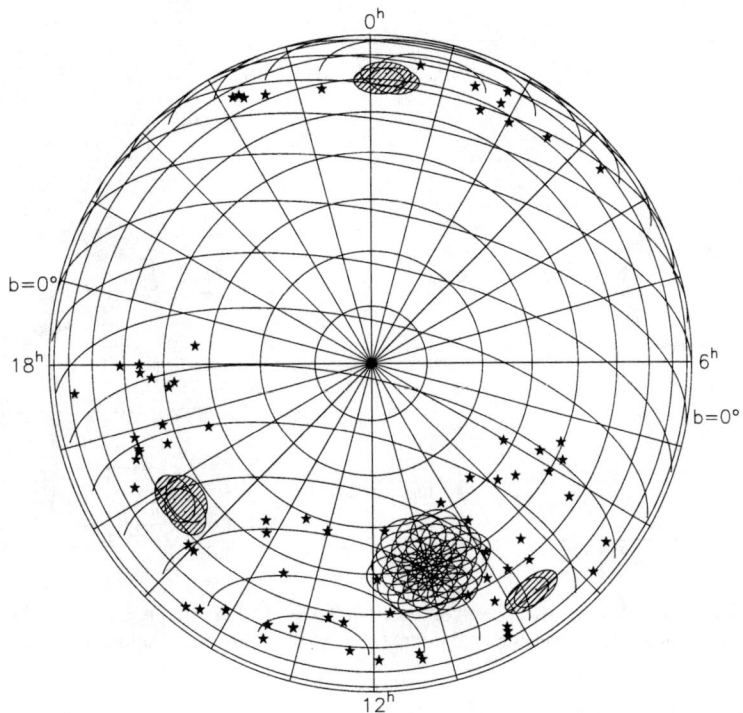

FIGURE 3. Plot of the VSA fields observed during our first year of observations. The hatched regions are fields which were observed using deep mosaicing. The remaining fields were observed as part of a shallow survey. Radio sources predicted to be brighter than 500 mJy are displayed as star symbols. A cut-off at galactic latitude b > 20° has been applied.

OBSERVATIONS

Field Selection

For practical CMB observation, it is important to choose fields which are relatively free from Galactic and extragalactic foregrounds. All the VSA fields are situated at Galactic latitudes greater than 20° and have low Galactic synchrotron and free-free emission, as predicted by the 408-MHz all-sky radio survey of Haslam et. al. (1992)[3]. The dust maps of Finkbeiner et. al. (1994) [4] were used to select fields with relatively low dust contamination. To avoid large-scale structure and clusters we consulted the ROSAT catalogues [5],[6]. More importantly, all fields were chosen to be as free as possible of bright radio sources, since these are the major contaminant of CMB observations at 34 GHz. We used two low-frequency surveys, NVSS [7] at 1.4 GHz and Green Bank [8] at 4.85 GHz to select CMB fields in which there are predicted to be no sources brighter than 500 mJy at 34 GHz. Predictions were made by extrapolating the flux density of every source in the 4.8 GHz catalogue to 34 GHz on the basis of its spectral index between 1.4 and 15 GHz. A further practical consideration which affected the choice of CMB fields was the need to observe all fields for a reasonable length of time, from both Tenerife and Cambridge. This limited the declination range of our fields to +26° – +54°. We also selected fields that are evenly spaced around the sky to enable 24-hour observing. In order to increase the l-resolution of our measurement of the CMB power spectrum, we selected regions of sky where we could mosaic several CMB fields. In each region of sky, mosaiced fields are separated by 2.75°. Our final choice of fields used during the first year of observation is shown in Fig. 3.

Observing strategy

During the first year of VSA observations, we have undertaken two distinct observation programs, each using a compact array. First, we have made deep mosaiced observations of eight fields in three evenly spaced regions of sky (hatched regions in Fig. 3.). Each mosaiced field was observed for $\sim$ 400 hours, reaching a thermal noise of

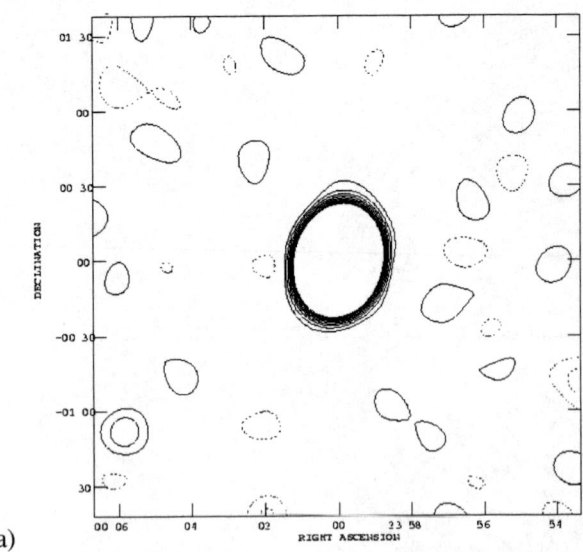 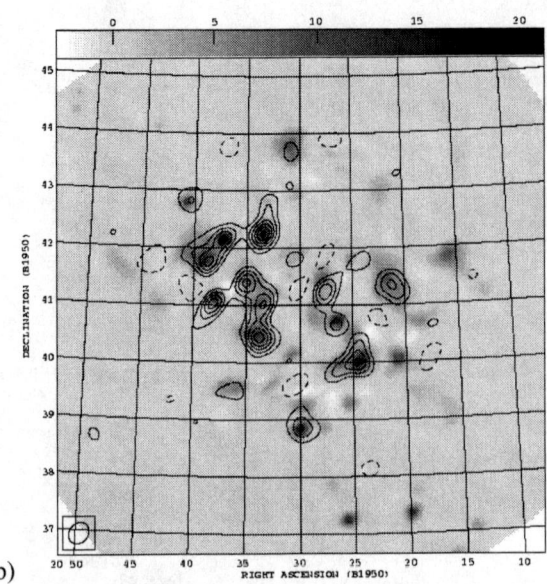

FIGURE 4. (a) An 80-minute observation of Jupiter, contoured at 0.3 Jy. The thermal noise level is 240 Jy (baseline × sec)$^{-1/2}$. (b) Commissioning observation of the Cygnus Loop. Contours (at 3 Jy) are VSA observations at 34 GHz. The grey scale is Green Bank, 4.85 GHz data [9].

approximately 30 mJy. Mosaicing in this way enables us to increase the l-resolution of our measurements whilst also reducing sample variance. However, the time taken for us to survey all fields with the Ryle Telescope prior to any observation with the VSA, has limited the area of sky that we can cover with deep mosaicing in this first year. Consequently, for $l \leq 300$, our measurements of the CMB power spectrum are limited by sample variance. In order to achieve a good estimate of the CMB power spectrum in the region $l \leq 300$, we have now completed the second stage of our observation program; a shallow survey in one area of sky. The shallow survey consists of 2 days observation on each of 30 mosaiced fields (Fig. 3 hexagonal region), and covers an area of approximately 180 sq. degrees. For this shallow survey, our source subtraction strategy no longer limits the area of sky we can observe with the VSA in a given time. Since we are only concerned with low-l observations, where the contribution of point sources to the CMB power spectrum is known to be negligible, the need for prior surveying with the Ryle Telescope is not necessary. Instead, we choose only to monitor sources predicted to be greater than 100 mJy at 34 GHz on the basis of low-frequency survey information.

Calibration

Absolute flux calibration of VSA observations is based on the flux scale of Mason et al. (1999) [10]. Flux calibrations are made each day using one of three primary calibrators – Tau A, Cass A and Jupiter. We also make daily observations of three fainter sources, 3C48, 3C273 and NGC7207 allowing us to check the quality of observations throughout the observing day. The measured flux ratios of these sources to our primary calibrators agree well with those reported by Mason et al., suggesting that the accuracy of our flux calibration is limited by that of Mason et al. to approximately 5 percent.

The overall gain of the telescope is also monitored via a noise injection system. A modulated noise signal is injected into each antenna and is later measured using phase-sensitive detection after the automatic gain control stage of the telescope. The relative contribution of the constant noise source to the total output power from each antenna varies inversely with system temperature, and thus a correction can be made to the overall flux calibration. This system allows us to account for both variations in the gain of the system and for atmospheric attenuation of the astronomical signal. It provides an excellent indication of the weather conditions and is used as a primary indicator for flagging data. For good observing conditions, the gain corrections applied using this system are typically less than a few percent.

Phase calibration of the VSA is also applied on a daily basis using the same three primary calibrators. We find that the VSA is relatively phase stable, with variations of less than 15 degrees per day. A typical calibration

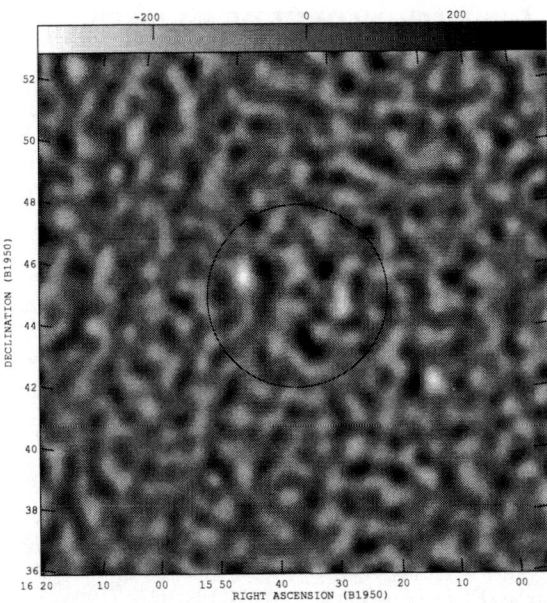

FIGURE 5. An example of a CMB field observed with the VSA. This map is the result of ~ 400 hours of observation. The map has had a gaussian taper applied to enhance the CMB features present within the FWHM (black line) of the VSA primary beam.

observation of Jupiter is shown in Fig. 4(a). This 80-minute observation, made early on in the observing program, also confirms that the telescope is achieving the required sensitivity, with a thermal noise level of approximately 240 Jy/(baseline $\times$ sec)$^{1/2}$.

The pointing accuracy of the VSA is primarily determined by mechanical alignment tolerances, but we frequently make long observations of unresolved calibrators in order to check both the pointing and geometry of the array. Using these observations, and in conjunction with a model of the telescope, we employ a maximum-likelihood technique to simultaneously fit for $\sim$400 parameters. These include the x, y and z co-ordinate of each antenna, correlator gains for each of the 182 correlator channels and the effective observing bandwidth of each baseline. Further observations of offset sources and, for example, the Cygnus Loop, confirm the ability of the VSA to map known structure. Fig 4(b) shows the result of a 90-minute commissioning observation of the Cygnus Loop. Our observed 34 GHz flux contours are overlaid on 15 GHz Green Bank data [9]. As shown in Fig. 4(b), there is good agreement between the structure observed at the two frequencies.

CURRENT STATUS AND CONCLUSIONS

The VSA has been making routine observations of CMB fields in its compact configuration since September 2000. Commissioning and calibration observations confirm that the instrument is working to specification. We have currently completed over 3000 hours of observations, having lost less than 10 percent of our observing time due to bad weather. A typical observation of one of our CMB fields is shown in Fig. 5. This map is the result of ~ 400 hours of observation and, to enhance the CMB features present in the centre, a gaussian taper has been applied (1/e point = 60 λ). The resolution of the map is 31 $\times$ 25 arcmin. The map has not been source-subtracted.

Our observing program using the compact array was completed at the beginning of September 2001, and we are currently re-configuring and upgrading the array ready for observations in its extended configuration. The second season of observations with the new array will begin in October 2001. We will then be able to measure the CMB power spectrum for l-values up to $\sim$ 1800.

ACKNOWLEDGMENTS

The VSA project has involved the work of a large number of people from the three collaborating institutions; the Cavendish Astrophysics Group, Cambridge, Jodrell Bank Obsevatory, Manchester, and the Instituto de Astrofisica de Canarias (IAC), Tenerife.

REFERENCES

1. Robson, M., Yassin, G., Woan, G., Wilson, D. M. A., Scott, P. F., Lasenby, A. N., Kenderdine, S., and Duffett-Smith, P. J., *A&A*, **277**, 314+ (1993).
2. Waldram, E. M., and Pooley, G. G., "Surveying the foreground sources for the VSA with the Ryle telescope at 15 GHz", in *IAU Symposium*, in press, vol. 201.
3. Haslam, C. G. T., Stoffel, H., Salter, C. J., and Wilson, W. E., *A&AS*, **47**, 1+ (1982).
4. Schlegel, D. J., Finkbeiner, D. P., and Davis, M., *ApJ*, **500**, 525+ (1998).
5. Ebeling, H., Edge, A. C., Bohringer, H., Allen, S. W., Crawford, C. S., Fabian, A. C., Voges, W., and Huchra, J. P., *MNRAS*, **301**, 881–914 (1998).
6. Ebeling, H., Voges, W., Bohringer, H., Edge, A. C., Huchra, J. P., and Briel, U. G., *MNRAS*, **281**, 799–829 (1996).
7. Condon, J. J., Cotton, W. D., Greisen, E. W., Yin, Q. F., Perley, R. A., Taylor, G. B., and Broderick, J. J., *AJ*, **115**, 1693–1716 (1998).
8. Gregory, P. C., Scott, W. K., Douglas, K., and Condon, J. J., *ApJS*, **103**, 427+ (1996).
9. Langston, G., Minter, A., D'Addario, L., Eberhardt, K., Koski, K., and Zuber, J., *AJ*, **119**, 2801–2827 (2000).
10. Mason, B. S., Leitch, E. M., Myers, S. T., Cartwright, J. K., and Readhead, A. C. S., *AJ*, **118**, 2908–2918 (1999).

Spurious signals in VSA data

Michael E. Jones

Cavendish Laboratory, Madingley Road, Cambridge, CB3 0HE, United Kingdom

Abstract.
The Very Small Array (VSA) is an interferometer with very small and closely-packed antennas, designed for studies of the cosmic microwave background radiation. During the commissioning of the VSA, an unwanted signal associated with coupling between the closest antennas was discovered. Here we describe the properties of this signal and the measures taken to eliminate it from the data. The same methods can also be used to eliminate the effects of other unwanted signals in the data, such as contamination from the Sun.

INTRODUCTION

The Very Small Array is a 14-element interferometer designed for imaging the cosmic microwave background on degree and sub-degree scales (see Jones & Scott (1998) and Taylor, this volume). Each antenna is a conical horn-reflector in which the reflector can be rotated about the horn axis, providing tracking in one dimension. The antennas are mounted on a table which tips about an east-west axis, providing the other tracking coordinate. In its compact configuration, the reflectors are 143 mm in diameter, and can be placed as close as 190 mm from each other (there is also an extended mode in which 322 mm reflectors are used). This semi-individual tracking results in the path length from the source to the correlator continually changing, so each antenna is provided with a path compensator system, in which different lengths of stripline can be switched into the IF path to compensate for the geometric delay. Since this path is introduced at the IF frequency, rather than the RF frequency, the phase of the signal on each baseline is modulated by $2\pi \times$ path difference $\times (\nu_{RF} - \nu_{LO})$ (the 'fringe rate'). Signals not originating from the pointing direction of the antennas will be modulated at different rates, and can thus be distinguished from astronomical signals.

CHARACTERIZATION OF THE SPURIOUS SIGNAL

During commisioning observations of the VSA both in Cambridge and in Tenerife, a correlated signal was observed in the shortest baseline data (see Figure 1). The properties of this signal can be summarised as follows:

1. The maximum amplitude is about 1 mK.
2. The signal strength drops very rapidly with baseline length.
3. It is strongest on exactly NS or EW baselines, ie when the baseline corresponds with the symmetry axes of the antennas.
4. If the telescope is not tracking, the signal stays constant.
5. The signal is modulated by tracking individual antennas or by changing the relative delays of each antenna, but typically at a much slower rate than the astronomical fringe rate.
6. The signal is *not* modulated by elevation tracking (driving of the table on which all the antennas are fixed).
7. The signal disappears if a large metal sheet is placed between the antennas.
8. Tracking the relative delays produces an amplitude envelope of width roughly equal to the inverse of the IF bandwidth, showing that the signal is broad band.

Several of the points (and especially number 7) indicate that coupling between the antennas must be responsible; the fact that the signal is relatively unchanged when the table is driven suggests that the signal source must be on the table,

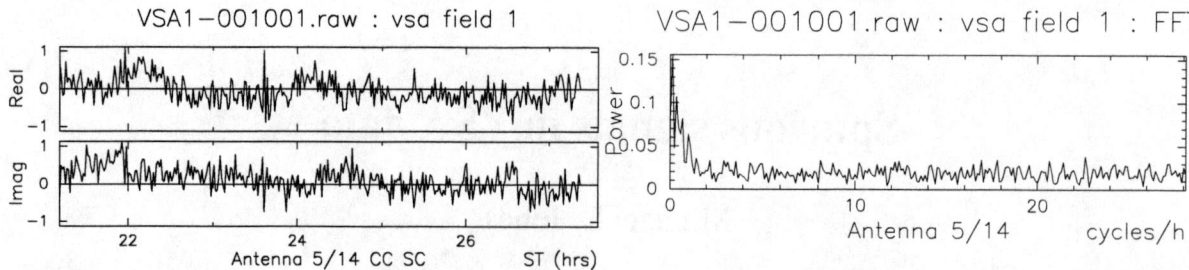

FIGURE 1. Typical spurious signal data on a short VSA baseline, shown both as complex time series data and as a power spectrum.

and is not due, for example, to radiation diffracting over the ground shield. The size of the signal is puzzling, however. Coupling between the antennas was measured in the lab by transmitting a modulated signal from one antenna and detecting it with a phase-sensitive detector connected to an adjacent antenna. Even with the antennas almost touching, a maximum coupling of −100 dB was measured. This would result in any cross-talk signal being attenuated by 50 dB. The measured signal amplitude of ∼ 1 mK therefore implies a correlated signal source of ∼ 100 K, which does not exist anywhere in the telescope system. Experiments in which all room-temperature emissive objects inside the telescope enclosure were removed or shielded failed to have any effect on the measured signal.

REMOVAL OF THE SPURIOUS SIGNAL FROM THE DATA

Despite our lack of success in tracking down the source of the spurious signal, it has been possible to largely remove its effects from the data. This is because the modulation rate of the spurious signal is typically very different from the modulation of the sky signal due to the rotation of the earth. The spurious signal can thus be removed from the time stream for each baseline by a simple fourier filter. Figure 2 shows the power spectrum of the time-ordered data for two baselines in an observation of a bright calibrator. The upper plot shows a baseline for which the fringe rate passes through zero during the observation, the lower plot a baseline where the fringe rate is always high. Both the actual data and the predicted ideal data are shown – the spurious signal can be seen as excess power at low frequency in the lower plot (it is swamped by the calibrator in the upper plot). The shaded regions show the frequencies that are removed by fourier filtering. A Hanning (cosine-edged) filter is used to minimise the introduction of spurious correlations into the CMB data. In the upper plot, data during which the fringe rate of the astronomical signal lies within the range of the filter are subsequently flagged out, as the filtering will affect the astronomical signal as well as the spurious signal. In the lower plot, the two signals are always distinct and no flagging is necessary.

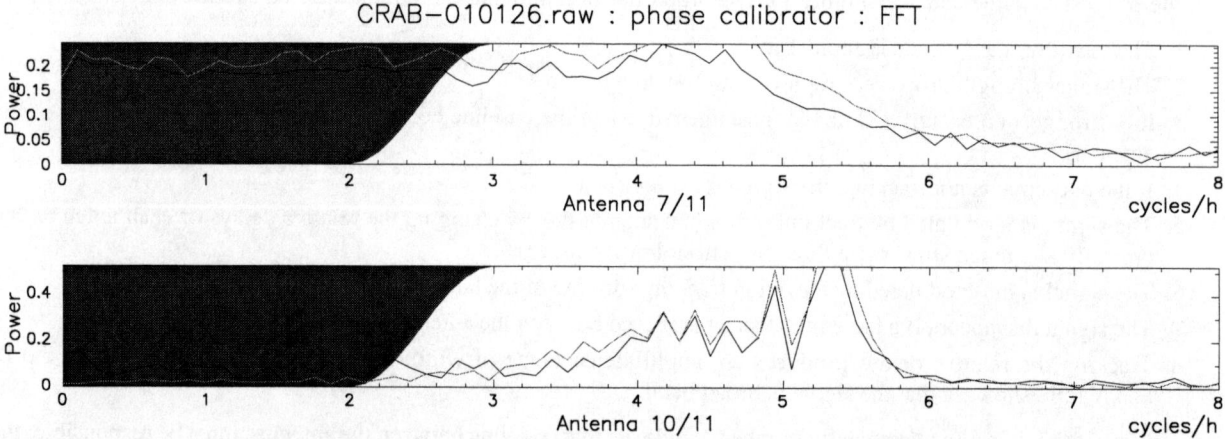

FIGURE 2. Power spectra of data from two VSA baselines for a bright point source. The solid line is the data, the faint line the predicted data for a point source, and the shaded region shows the frequencies removed by the filtering

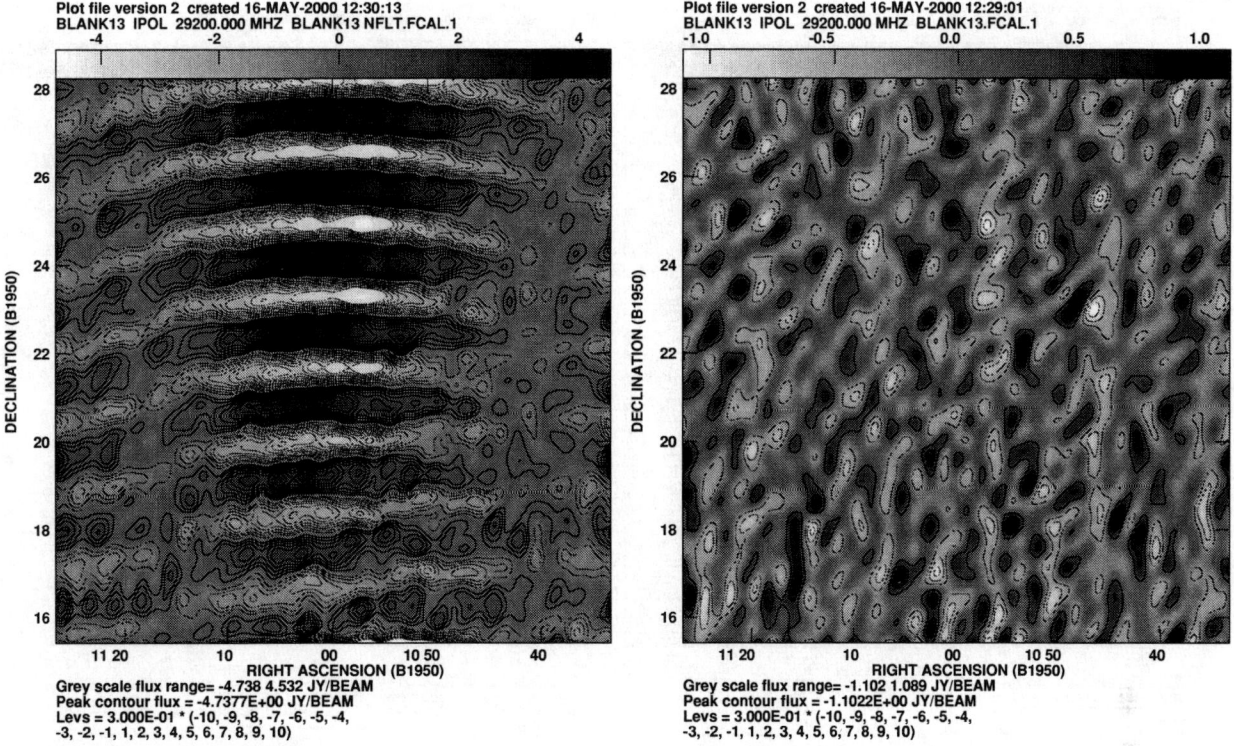

FIGURE 3. Effect of filtering on the map. (Left) Map of a blank area of sky using unfiltered data. (Right) The same map with filtered data. Less than 10 % of the data has been lost in the filtering process, and the noise level is consistent with the expected thermal noise.

The result of this filtering and flagging procedure can be seen in Figure 3, which shows a 6-hour map of a blank area of sky before and after filtering. The noise level on the filtered map is consistent with the expected thermal noise, and contains no spurious correlations. Less than 10 % of the data has been flagged due to overlap in fringe rate with the spurious signal.

FILTERING OF THE SUN AND MOON

The same filtering techniques can be used on any signal whose fringe rate is distinct from the fringe rate of the astronomical signal. For example, the Sun and Moon can produce strong correlated signals even when they are at large angles from the pointing centre; the Sun is about 10^9 times brighter than the smallest CMB signals we are trying to detect, and the maxmimum attenuation of the primary beam is only ~ 60 dB. The signal from the Sun is further attenuated by decoherence due to the large angle from the delay tracking centre, but the residual signal can still be very large. This is a particular problem when the Sun is close to our few bright calibrator sources such as Jupiter or the Crab nebula. However, filtering of the data using a similar technique to that for the spurious signal can reduce the unwanted signal to a low level.

The data are first fringe rotated to the Sun, that is, the phase of the signal from a source at the Sun's position is calculated for every sample and the complex data rotated by the opposite angle. This makes the Sun signal in the data quasi-constant. A high-pass filter is applied, and the data are then fringe rotated back to the field centre. Data when the fringe rates of the Sun and the observed field are within the filter bandwidth of each other are flagged. As an example, Figure 4 shows an observation of Jupiter made when the planet was only 11 degrees away from the Sun, with and without Sun filtering. The filtered data are good enough to act as flux and phase calibrators on most baselines. The same technique can also be used (at rather larger Sun angles) on CMB field data that would otherwise not be usable. This allows the use of daytime data, greatly increasing the observing efficiency of the telescope.

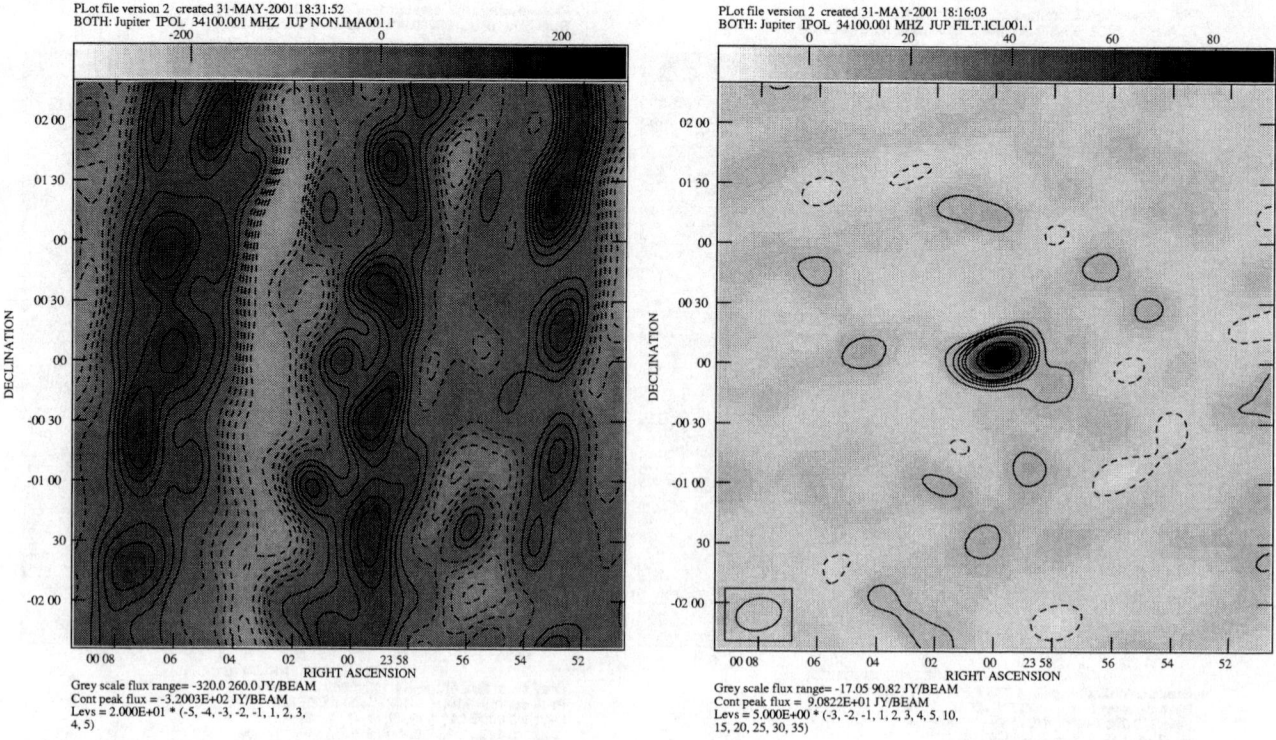

FIGURE 4. Effect of Sun filtering on an observation of Jupiter. (Left) Map using unfiltered data. The Sun is 11° from the map centre; its flux density at 34 GHz is about 10^7 Jy, compared to Jupiter's 100 Jy. (Right) The same map after fringe rotating to the Sun, filtering, and fringe rotating back.

CONCLUSIONS

The shortest baselines of the VSA compact array show a relatively strong spurious correlated signal which is associated with coupling between the antennas, although its origin is unclear. This signal is modulated in a way which is typically quite distinct from the modulation of the astronomical signal due to the rotation of the earth, and can therefore be filtered out with little loss of data. Similar filtering techniques can be used to remove the effect of bright sources such as the Sun, whose fringe rates are also distinct from the fringe rate of the observed field. This allows the observation of calibrator sources even when the Sun is very close, and the observation of CMB fields at sufficiently large angles from the Sun even in daytime.

ACKNOWLEDGMENTS

The VSA is a collaboration between the Cavendish Laboratory, University of Cambridge, Jodrell Bank Observatory, University of Manchester, and the Instituto de Astrofísica de Canarias, Tenerife.

REFERENCES

1. Jones, M. E., Scott, P. F. 1998, in 'Fundamental Parameters in Cosmology', 33rd Rencontres de Moriond, eds Tran Thanh Van, J., Giraud-Herauld, Y., Editions Frontiers

Millimeter and Submillimeter Observations from the South Pole

Antony A. Stark

Smithsonian Astrophysical Observatory, 60 Garden St., Cambridge, MA 02138 USA

Abstract. During the past decade, a year-round observatory has been established at the geographic South Pole by the Center for Astrophysical Research in Antarctica (CARA). CARA has fielded several millimeter- and submillimeter-wave instruments: AST/RO (the Antarctic Submillimeter Telescope and Remote Observatory, a 1.7-m telescope outfitted with a variety of receivers at frequencies from 230 GHz to 810 GHz, including PoleSTAR, a heterodyne spectrometer array), Python (a degree-scale CMB telescope), Viper (a 2-m telescope which has been outfitted with SPARO, a submillimeter-wave bolometric array polarimeter, ACBAR, a multi-channel CMB instrument, and Dos Equis, a HEMT polarimeter), and DASI (the Degree-Angular Scale Interferometer). These instruments have obtained significant results in studies of the interstellar medium and observational cosmology, including detections of the 1° acoustic peak in the CMB and the Sunyaev-Zel'dovich effect. The South Pole environment is unique among observatory sites for unusually low wind speeds, low absolute humidity, and the consistent clarity of the submillimeter sky. The atmosphere is dessicated by cold: at the South Pole's average annual temperature of -49 C, the partial pressure of saturated water vapor is only 1.2% of what it is at 0 C. The low water vapor levels result in exceptionally low values of sky noise. This is crucial for large-scale observations of faint cosmological sources—for such observations the South Pole is unsurpassed.

INTRODUCTION

Development of the geographic South Pole as an astronomical observatory has largely been driven by the desire to measure anisotropy in the Cosmic Microwave Background (CMB) radiation. Observers have actively pursued the measurement of CMB anisotropy because of the theoretical expectation that those measurements would significantly advance our understanding of the Universe [1]. Deep millimeter- and submillimeter-wave observations, of which CMB anisotropy measurements are an extreme example, are often made impossible by opacity and noise due to water vapor in the Earth's atmosphere [2]. For large angular scale measurements, this problem has been successfully finessed by orbital [3] and airborne [4, 5] instruments, but observations of anisotropy on scales less than a few arcminutes require large telescope apertures, which cannot be deployed above the Earth's surface using current technology. This hard truth has led observers to seek new observatory sites having the smallest possible atmospheric water vapor while still being practical locations for the installation and operation of large telescopes. Among the most promising of the new observatories is Amundsen-Scott South Pole Station, the permanent U.S. National Science Foundation base on the Antarctic Plateau.

South Pole Station was originally built for the International Geophysical Year in 1957 and has been in continuous use ever since. Station infrastructure is currently being rebuilt for the second time; the new station is scheduled for completion in 2005. The station supports 50 scientists and staff throughout the Austral winter, with an increase to over 200 during the summer months. From the earliest days, scientific investigations at South Pole Station included observations of weather [6, 7] and atmospheric phenomena such as aurora [8] and ice halos [9]. The South Pole has evolved into an important site for seismology [10], solar astronomy [11, 12], atmospheric studies [13], and cosmic ray physics [14].

Because it is exceptionally cold, the climate at the Pole implies exceptionally dry observing conditions. As air becomes cold, the amount of water vapor it can hold declines dramatically. Air at 0 C can hold 83 times more water vapor than saturated air at the South Pole's average annual temperature of −49 C [15]; together with the relatively high altitude of the Pole (2850 m), this means the water vapor content of the atmosphere above the South Pole is two or three orders of magnitude smaller than it is at most places on the Earth's surface. This has long been known [16], but many years of hard work were needed to realize the potential in the form of new astronomical knowledge.

The EMILIE experiment by the French group Pajot et al. [17] made the first astronomical observations of submillimeter-waves from the South Pole during the Austral summer of 1984-1985. EMILIE was a ground-based single-pixel bolometer dewar operating at $900\mu m$ and fed by a 45 cm off-axis mirror. It had successfully measured the diffuse galactic emission while operating on Mauna Kea in Hawaii in 1982, but the accuracy of the result had been limited by sky noise [18]. Martin A. Pomerantz, a cosmic ray researcher at Bartol Research Institute, and John T. Lynch, the NSF Program Director for Antarctic Aeronomy and Astrophysics [19], encouraged the EMILIE group to relocate their experiment to the South Pole. There they found better observing conditions and were able to make improved measurements of galactic emission [20].

Pomerantz also enabled Mark Dragovan, then a researcher at Bell Laboratories, to attempt CMB anisotropy measurements from the Pole. Dragovan, Stark, and Pernic [21] built a lightweight 1.2 m offset telescope and were able to get it working at the Pole with a single-pixel helium-4 bolometer during several weeks in January 1987. The results were sufficiently encouraging that several CMB groups [22, 23, 24, 25] participated in the "Cucumber" campaign in the Austral summer of 1988-1989, where three Jamesway tents and a generator dedicated to CMB anisotropy measurements were set up at a temporary site 2 km from South Pole Station. These were summer-only campaigns, where instruments were shipped in, assembled, tested, used, disassembled, and shipped out in a single three-month-long summer season. Considerable time and effort were expended in establishing and then demolishing observatory facilities, with little return in observing time. What little observing time was available occurred during the warmest and wettest days of mid-summer.

Permanent, year-round facilities were needed. The Antarctic Submillimeter Telescope and Remote Observatory (AST/RO) [26, 27] is a 1.7 m diameter offset gregorian telescope mounted on a dedicated, permanent observatory building. It was the first radio telescope to operate year-round at South Pole. AST/RO was started in 1989 as an independent project, but in 1991 it became part of a newly-founded National Science Foundation Science and Technology Center, the Center for Astrophysical Research in Antarctica (CARA) [28, see also `http://astro.uchicago.edu/cara`]. CARA has fielded several telescopes in addition to AST/RO: White Dish [29], Python [30, 31, 32, 33, 34] and Viper [35] (Cosmic Microwave Background experiments), DASI [36] (the Degree-Angular Scale Interferometer), and SPIREX [37] (the South Pole Infrared Explorer, a 60-cm telescope, now decommissioned). These facilities are housed in the "Dark Sector", a grouping of buildings which includes the AST/RO building, the Martin A. Pomerantz Observatory building (MAPO) and a new "Dark Sector Laboratory", all located 1 km away from the main base across the aircraft runway in a radio quiet zone.

SITE TESTING

The South Pole is an excellent millimeter- and submillimeter-wave site [39]. It is unique among observatory sites for unusually low wind speeds, the complete absence of rain, and the consistent clarity of the submillimeter sky. Schwerdtfeger [6] has comprehensively reviewed the climate of the Antarctic Plateau and the records of the South Pole meteorology office. Chamberlin [40] has analyzed weather data to determine the precipitable water vapor (PWV) and finds median wintertime PWV values of 0.3 mm over a 37-year period, with little annual variation. *PWV values at South Pole are small, stable, and well-understood.*

Submillimeter-wave atmospheric opacity at South Pole has been measured using skydip techniques. Chamberlin et al. [41] made over 1100 skydip observations at 492 GHz (609 μm) with AST/RO during the 1995 observing season. Even though this frequency is near a strong oxygen line, the opacity was below 0.70 half of the time during the Austral winter and reached values as low as 0.34, better than ever measured at any other ground-based site. From early 1998, the 350 μm band has been continuously monitored at Mauna Kea, Chajnantor, and South Pole by identical tipper instruments developed by S. Radford of NRAO and J. Peterson of Carnegie-Mellon U. and CARA. Results from Mauna Kea and Chajnantor are compared with South Pole in the left panel of Figure 1. *The 350 μm opacity at the South Pole is consistently better than at Mauna Kea or Chajnantor.* The South Pole 25% winter PWV levels have been used to compute values of atmospheric transmittance as a function of wavelength and are plotted in the right panel of Figure 1. For comparison, the transmittance for 25% winter conditions at Chajnantor and Mauna Kea is also shown.

Sky noise is fluctuations in total power or phase of a detector caused by variations in atmospheric emissivity and path length on timescales of order one second. Sky noise causes systematic errors in the measurement of astronomical sources. Lay and Halverson [42] show analytically how sky noise causes observational techniques to fail: fluctuations

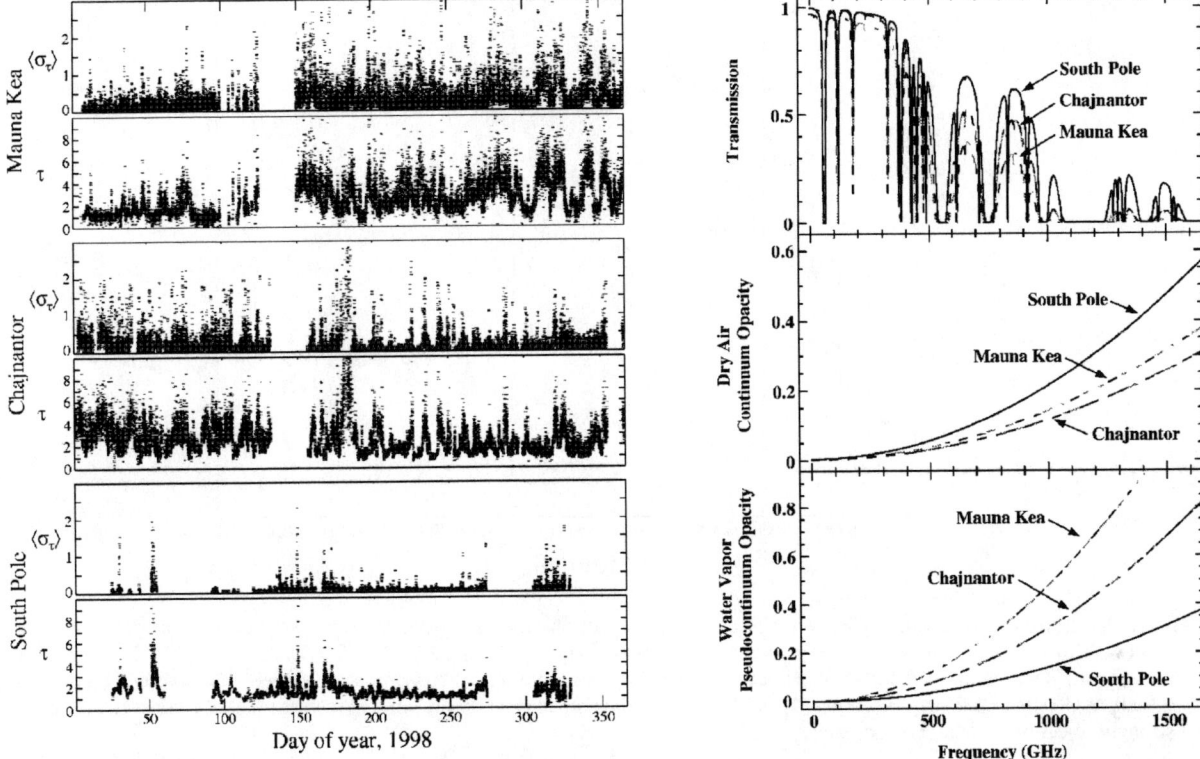

FIGURE 1. Opacity and sky noise at the South Pole. (Left) Measurements at 350μm from three sites. These plots show data from identical NRAO-CMU 350μm broadband tippers located at Mauna Kea, the ALMA site at Chajnantor in Chile, and South Pole during 1998. The upper plot of each pair shows $\langle\sigma_\tau\rangle$, the rms deviation in the opacity τ during a one-hour period—a measure of sky noise on large scales; the lower plot of each pair shows τ, the broadband 350μm opacity. The first 100 days of 1998 on Mauna Kea were exceptionally good for that site, due to a strong El Niño that year. During the best weather at the Pole, $\langle\sigma_\tau\rangle$ was dominated by detector noise rather than sky noise. (Right) Calculated atmospheric transmittance at the three sites. The upper plot is atmospheric transmittance at zenith calculated by J. R. Pardo using the ATM model [38]. The model uses PWV values of 0.2 mm for South Pole, 0.6 mm for Chajnantor and 0.9 mm for Mauna Kea, corresponding to the 25th percentile winter values at each site. Note that at low frequencies, the Chajnantor curve converges with the South Pole curve, an indication that 225 GHz opacity is not a simple predictor of submillimeter wave opacity. The middle and lower plots show calculated values of dry air continuum opacity and water vapor pseudocontinuum opacity for the three sites. Note that unlike the other sites, the opacity at South Pole is dominated by dry air rather than water vapor.

in a component of the data due to sky noise integrates down more slowly than $t^{-1/2}$ and will come to dominate the error during long observations. Sky noise at South Pole is considerably smaller than at other sites, even comparing conditions of the same opacity. The PWV at South Pole is often so low that the opacity is dominated by the *dry air* component [43, 40, cf. Figure 1]; the dry air emissivity and phase error do not vary as strongly or rapidly as the emissivity and phase error due to water vapor. Lay and Halverson [42] have compared the Python experiment at South Pole with the Site Testing Interferometer at Chajnantor [44, 45] and find that the amplitude of the sky noise at South Pole is 10 to 50 times less than that at Chajnantor. *The strength of South Pole as a millimeter and submillimeter site lies in the low sky noise levels routinely obtainable for sources around the South Celestial Pole.*

LOGISTICS

South Pole Station provides logistical support for observatory experiments: room and board for on-site scientific staff, electrical power, network and telephone connections, heavy equipment support, and cargo and personnel transport. The station powerplant provides about 100 kW of power to CARA projects out of a total generating capacity of about 600 kW. Heavy equipment at South Pole Station includes cranes, forklifts, and bulldozers; these can be requisitioned

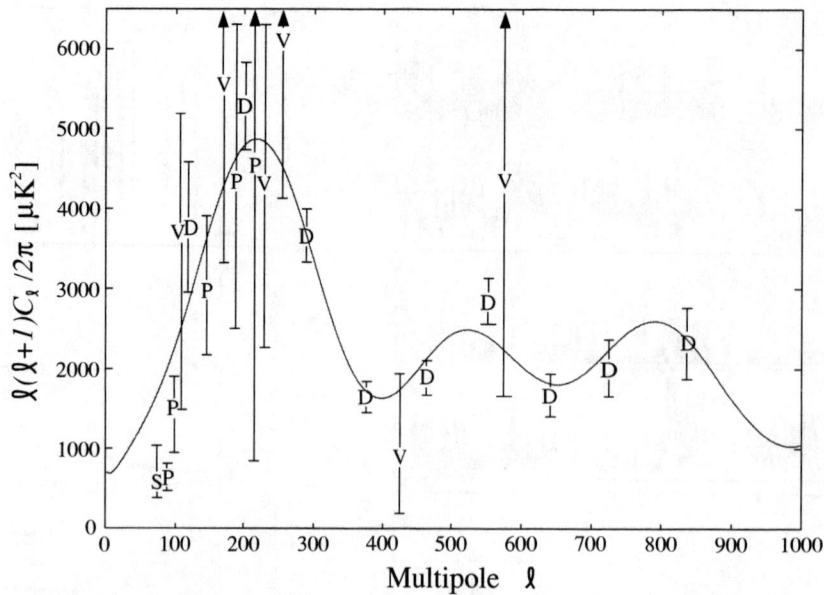

FIGURE 2. Summary of Cosmic Microwave Background Anisotropy measurements from the South Pole. The data shown are: S = U. of California, Santa Barbara [46]; P = Python 1991-1996 [34]; V = Viper 1998 [35]; D = DASI 2000 [47]. The curved line is not a fit to the data, but instead shows the expected values for the cosmological "concordance model": $\Omega_b = 0.05$, $\Omega_{cdm} = 0.35$, $\Omega_\Lambda = 0.60$, $\tau_c = 0$, $n_s = 1.00$, $H_0 = 65\,\mathrm{km\,s^{-1}\,kpc^{-1}}$ [48]. The data are in remarkably good agreement with the concordance model, and strongly reject many other cosmologies.

for scientific use as needed. The station is supplied by over 200 flights each year of LC130 ski-equipped cargo aircraft. Annual cargo capacity is about 3500 tons. Aircraft flights are scheduled only during the period from late October to early February so that the station is inaccessible for as long as nine months of the year. All engineering operations for equipment installation and maintenance are tied to the annual cycle of physical access to the instruments. For quick repairs and upgrades during the Austral summer season, it is possible to send equipment between South Pole and anywhere serviced by commercial express delivery in about five days. During the winter, however, no transport is possible and projects must be designed and managed accordingly.

In summer, there are about 20 CARA people at the Pole at any given time. Each person stays at Pole for a few weeks or months in order to carry out their planned tasks as well as circumstances allow, then they depart to be replaced by another CARA person. Each year there are four or five CARA winter-over scientists, who remain at the observatory for a year.

The receivers used on AST/RO and Viper require about 20 liters of liquid helium per day. Helium also escapes from the station in one or two weather balloons launched each day. The National Science Foundation and its support contractors must supply helium to South Pole, and the most efficient way to transport and supply helium is in liquid form. Before the winter-over period, one or more large (4000 to 12000 liter) storage dewars are brought to South Pole for winter use; some years this supply lasts the entire winter, but in 1996, 2000, and 2001 it did not. The supply of liquid helium has been a chronic problem for South Pole astronomy, but improved facilities in the new station should eliminate single points of failure and provide a more certain supply.

Internet and telephone service to South Pole is provided by a combination of two low-bandwidth satellites, LES-9 and GOES-3, and the high-bandwidth (3 Mbps) NASA Tracking and Data Relay Satellite System TDRS-F1. These satellites are geosynchronous but not geostationary, since their orbits are inclined. Geostationary satellites are always below the horizon and cannot be used. Internet service is intermittent through each 24-hour period because each satellite is visible only during the southernmost part of its orbit; the combination of the three satellites provides an Internet connection for approximately 12 hours within the period 1–16 hr Greenwich LST. The TDRS link helps provide a store-and-forward automatic transfer service for large computer files. The total data communications capability is about 5 Gbytes per day. Additional voice communications are provided by a fourth satellite, ATS-3, and high frequency radio.

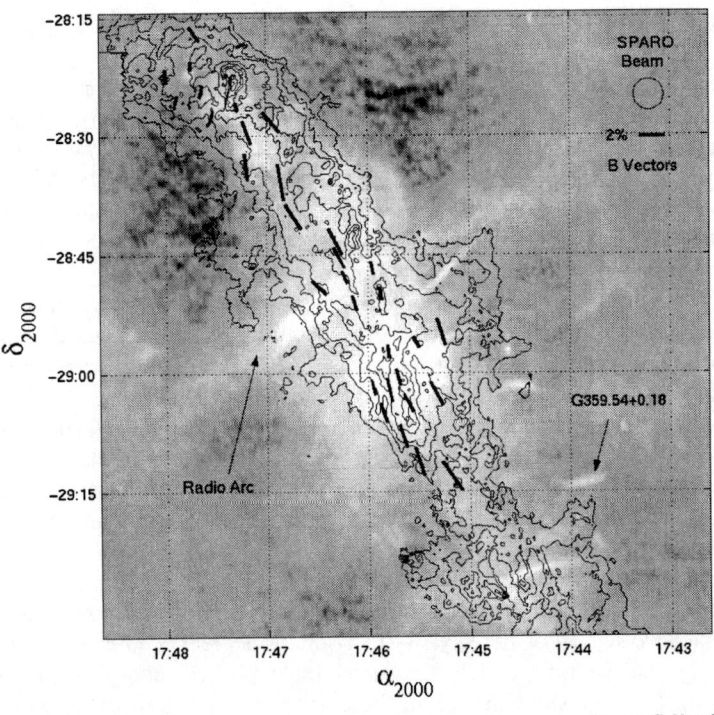

FIGURE 3. SPARO polarimetry results from Novak et al. [49] for the Galactic Center Region (vectors) superposed on radio continuum map by LaRosa et al. [50] (gray scale). The length of each vector gives the measured degree of polarization (see key at upper right), and the orientation of each vector shows the direction of the inferred magnetic field, which is orthogonal to the measured direction of polarization. Black contours show the 800 μm dust emission measured using the JCMT [51]. The radio continuum map shows several of the non-thermal filaments that trace magnetic field lines in hot ionized regions.

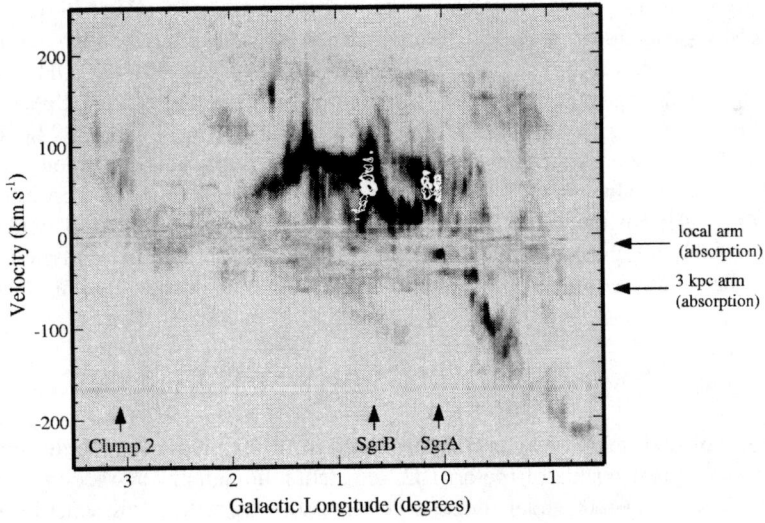

FIGURE 4. AST/RO observations of the Galactic Center Region from Kim et al. [52]. The CO $J = 4 \rightarrow 3$ (greyscale) and $J = 7 \rightarrow 6$ (white contour) lines observed in an $l - v$ strip, sampled every $1'$, at $b = 0°$. These data have been used in conjunction with CO and ^{13}CO $J = 1 \rightarrow 0$ data to determine the the density and temperature in the features shown here [52]. The 300 parsec ring is the rough parallelogram-shape between $l = -1°$ and $l = 1.8°$, and velocities from -200 to $+200 \mathrm{km\,s^{-1}}$. The ring is just below the critical density at which instabilities will cause it to coagulate into a single cloud like Sgr B and Sgr A.

CURRENT INSTRUMENTS AND ILLUSTRATIVE RESULTS

The Degree Angular Scale Interferometer (DASI) [36] is a compact centimeter-wave interferometer designed to image the CMB primary anisotropy and measure its angular power spectrum and polarization at angular scales ranging from two degrees to several arcminutes. As an interferometer, DASI measures CMB power by simultaneous differencing on several scales, measuring the CMB power spectrum directly. DASI was installed on a tower adjacent to the MAPO during the 1999-2000 Austral summer and has had two successful winter seasons so far. The first season results are a stunning confirmation of the "concordance" cosmological model, which has a flat geometry and significant contributions to the total stress-energy from dark matter and dark energy [47, 48].

Viper is a 2.1 meter off-axis telescope designed to allow measurements of low contrast astronomical sources. It is mounted on another tower at the opposite end of the MAPO from DASI. Viper is used with a variety of instruments: Dos Equis, a CMB polarization receiver operating at 7 mm, SPARO, a bolometric array polarimeter operating at 450 μm, and ACBAR, a multi-wavelength bolometer array used to map the CMB.

ACBAR is a 16-element bolometer array operating at 300 mK. It was designed specifically for observations of CMB anisotropy and the Sunyaev-Zel'dovich effect (SZE). It was installed on the Viper telescope during the Austral summer of 2000-2001 and was successfully operated throughout most of the Austral winter of 2001. ACBAR has made high-quality maps of SZE in several nearby clusters of galaxies and has made significant measurements of anisotropy on the scale of degrees to arcminutes.

The Submillimeter Polarimeter for Antarctic Remote Observing (SPARO) is a 9-pixel polarimetric imager operating at 450 μm. It was operational on the Viper telescope during the early Austral winter of 2000. Novak et al. [49] mapped the polarization of an extended region of the sky ($\sim$ 0.25 square degrees) centered approximately on the Galactic Center. Their results imply that, within the Galactic Center molecular gas complex, the toroidal component of the magnetic field is dominant. The data show that all of the existing observations of large-scale magnetic fields in the Galactic Center are basically consistent with the "magnetic outflow" model of Uchida et al. [53]. This magnetodynamic model was developed in order to explain the Galactic Center Radio Lobe, a limb-brightened radio structure that extends up to one degree above the plane and may represent a gas outflow from the Galactic Center.

AST/RO is general-purpose 1.7 m diameter telescope [26, 27] for astronomy and aeronomy studies at wavelengths between 200 and 2000 μm. It is used primarily for spectroscopic studies of neutral atomic carbon and carbon monoxide in the interstellar medium of the Milky Way and the Magellanic Clouds (see reference list at http://cfa-www.harvard.edu/~adair/AST_RO). Five heterodyne receivers and three acousto-optical spectrometers are currently installed: (1) a 230 GHz SIS receiver [54], (2) a 450–495 GHz SIS quasi-optical receiver [55, 56], (3) a 450–495 GHz SIS waveguide receiver [57, 58], which can be used simultaneously with (4) a 800-820 GHz fixed-tuned SIS waveguide mixer receiver [59], and (5) the PoleSTAR array—four 800-820 GHz fixed-tuned SIS waveguide mixer receivers [60, 61, see http://soral.as.arizona.edu/pole-star]. Spectral lines observed with AST/RO include: CO $J = 7 \rightarrow 6$, CO $J = 4 \rightarrow 3$, CO $J = 2 \rightarrow 1$, HDO $J = 1_{0,1} \rightarrow 0_{0,0}$, [C I] $^3P_1 \rightarrow ^3P_0$, [C I] $^3P_2 \rightarrow ^3P_1$, and [^{13}C I] $^3P_2 \rightarrow ^3P_1$. There are four currently available acousto-optical spectrometers (AOS) [62]: two low resolution spectrometers with a bandwidth of 1 GHz, an array AOS having four low resolution spectrometer channels with a bandwidth of 1 GHz for the PoleSTAR array, and one high-resolution AOS with 60 MHz bandwidth. Two new instruments for AST/RO are under development: TREND, a 1.5 THz heterodyne receiver [63, 64] and SPIFI, an imaging Fabry-Perot interferometer [65]. AST/RO observing time is open to all astronomers on a proposal basis.

THE FUTURE 8-M TELESCOPE

An 8 m diameter off-axis telescope has been proposed for the South Pole [66]. It will be equipped with a large field of view [67] that will feed a state-of-the-art 1027-element bolometer array receiver. The initial science goal will be a large SZE survey covering 4000 square degrees at $1.3'$ resolution with 10μK sensitivity at a wavelength of 2 mm. The SZE survey will find all galaxy clusters above a mass limit of 3.5×10^{14} M$_\odot$ regardless of redshift; should clusters exist at redshifts much higher than currently predicted, they will be found by the SZE survey, but missed in even the deepest X-ray observations planned. It is expected that an unbiased sample of approximately 20,000 clusters will be found, with over 1,000 at redshifts greater than one. The sample will provide sufficient statistics to use the density of clusters to determine the equation of state of the dark energy component of the Universe.

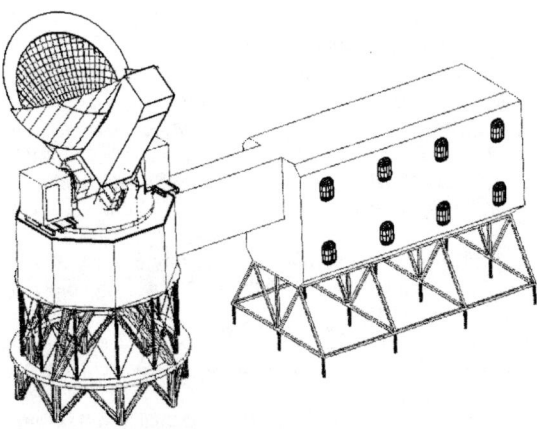

FIGURE 5. A sketch of the proposed 8-m South Pole Telescope and tower mounted to the side of the new Dark Sector Laboratory. A guard ring surrounding the primary and shielding around the lower part of the secondary arm support region minimize contamination from scattered radiation. The guard ring allows the full aperture to be illuminated without excessive noise from spillover. Like the DASI and AST/RO telescopes, the environment of all critical and moving components will be kept warm, and therefore access will not require the winterover staff to work in the extreme cold conditions.

ACKNOWLEDGMENTS

We thank Jeff Peterson of CMU and Simon Radford of NRAO for the data shown in the left panel of Figure 1. We thank Juan R. Pardo of Caltech for discussions on atmospheric modeling and for carrying out the calculations shown in the right panel of Figure 1. This work was supported in part by the Center for Astrophysical Research in Antarctica and the NSF under Cooperative Agreement OPP89-20223.

REFERENCES

1. Peebles, P. J. E., *Physical Cosmology*, Princeton University Press, 1971.
2. Radford, S. J. E., Boynton, P. E., Ulich, B. L., Partridge, R. B., Schommer, R. A., Stark, A. A., Wilson, R. W., and Murray, S. S., *Astrophys. J.*, **300**, 159 (1986).
3. Smoot, G. F., et al., *Astrophys. J.*, **396**, L1 (1992).
4. de Bernardis, P., Ade, P. A. R., Bock, J. J., Bond, J. R., Borrill, J., Boscaleri, A., Coble, K., Crill, B. P., De Gasperis, G., Farese, P. C., Ferreira, P. G., Ganga, K., Giacometti, M., Hivon, E., Hristov, V. V., Iacoangeli, A., Jaffe, A. H., Lange, A. E., Martinis, L., Masi, S., Mason, P. V., Mauskopf, P. D., Melchiorri, A., Miglio, L., Montroy, T., Netterfield, C. B., Pascale, E., Piacentini, F., Pogosyan, D., Prunet, S., Rao, S., Romeo, G., Ruhl, J. E., Scaramuzzi, F., Sforna, D., and Vittorio, N., *Nature*, **404**, 955 (2000).
5. Hanany, S., Ade, P., Balbi, A., Bock, J., Borrill, J., Boscaleri, A., de Bernardis, P., Ferreira, P. G., Hristov, V. V., Jaffe, A. H., Lange, A. E., Lee, A. T., Mauskopf, P. D., Netterfield, C. B., Oh, S., Pascale, E., Rabii, B., Richards, P. L., Smoot, G. F., Stompor, R., Winant, C. D., and Wu, J. H. P., *Astrophys. J. Lett.*, **545**, L5 (2000).
6. Schwerdtfeger, W., *Weather and Climate of the Antarctic*, Elsevier, Amsterdam, 1984.
7. Warren, S. G., "Antarctica", in *Encyclopedia of Weather and Climate*, edited by S. H. Schneider, Oxford Univ. Press., New York, 1996, vol. 1, p. 32.
8. Landolt, A. U., *Proc. Ast. Soc. Pac.*, **70**, 581 (1958).
9. Ohtake, T., *Antarctic. J. US*, **13**, 174 (1978).
10. Roult, G., and Rouland, D., *Physics of the Earth and Planetary Interiors*, **84**, 15 (1994).
11. Harvey, J., "Solar Observing Conditions at the South Pole", in *Astrophysics in Antarctica*, edited by D. J. Mullan, M. A. Pomerantz, and T. Stanev, AIP Press, New York, 1989, p. 227.
12. Libbrecht, K. G., and Woodard, M. F., *Science*, **5016**, 152 (1991).
13. Fan, S. M., Blaine, T. L., and Sarmiento, J. L., *Tellus Series B-Chemical and Physical Meteorology*, **51**, 863 (1999).
14. Smith, N. J. T., et al., *Nucl. Inst. & Meth. in Physics Res.*, **276**, 622–7 (1989).
15. Goff, J. A., and Gratch, S., *Trans. Amer. Soc. Heat. and Vent. Eng.*, **52**, 95 (1946).
16. Smythe, W. D., and Jackson, B. V., *Applied Optics*, **16**, 2041 (1977).
17. Pajot, F., Gispert, R., Lamarre, J. M., Peyturaux, R., Pomerantz, M. A., Puget, J.-L., Serra, G., Maurel, C., Pfeiffer, R., and Renault, J. C., *Astron. Astroph.*, **223**, 107 (1989).

18. Pajot, F., Gispert, R., Lamarre, J. M., Peyturaux, R., Puget, J.-L., Serra, G., Coron, N., Dambier, G., Leblanc, J., Moalic, J. P., Renault, J. C., and Vitry, R., *Astron. Astroph.*, **154**, 55 (1986).
19. Lynch, J. T., "Astronomy & Astrophysics in the U. S. Antarctic Program", in *Astrophysics from Antarctica*, edited by G. Novak and R. H. Landsberg, ASP, San Francisco, 1998, vol. 141 of *ASP Conf. Ser.*, p. 54.
20. Pajot, F., Gispert, R., Lamarre, J. M., Peyturaux, R., and Pomerantz, M. A., *Astron. Astroph.*, **223**, 107–111 (1989).
21. Dragovan, M., Stark, A. A., Pernic, R., and Pomerantz, M. A., *Appl. Opt.*, **29**, 463 (1990).
22. Dragovan, M., Platt, S. R., Pernic, R. J., and Stark, A. A., "South Pole Submillimeter Isotropy Measurements of the Cosmic Microwave Background", in *Astrophysics in Antarctica*, edited by D. J. Mullan, M. A. Pomerantz, and T. Stanev, AIP Press, New York, 1990, p. 97.
23. Gaier, T., Schuster, J., and Lubin, P., "Cosmic Background Anisotropy Studies at 10° Angular Scales with a HEMT Radiometer", in *Astrophysics in Antarctica*, edited by D. J. Mullan, M. A. Pomerantz, and T. Stanev, AIP Press, New York, 1989, p. 84.
24. Meinhold, P. R., Lubin, P. M., Chingcuanco, A. O., Schuster, J. A., and Seiffert, M., "South Pole Studies of the Anisotropy of the Cosmic Microwave Background at One Degree", in *Astrophysics in Antarctica*, edited by D. J. Mullan, M. A. Pomerantz, and T. Stanev, AIP Press, New York, 1989, p. 88.
25. Peterson, J., "Millimeter and Sub-millimeter Photometry from Antarctica", in *Astrophysics in Antarctica*, edited by D. J. Mullan, M. A. Pomerantz, and T. Stanev, AIP Press, New York, 1989, p. 116.
26. Stark, A. A., Chamberlin, R. A., Cheng, J., Ingalls, J., and Wright, G., *Rev. Sci. Instr.*, **68**, 2200 (1997).
27. Stark, A. A., Bally, J., Balm, S. P., Bania, T. M., Bolatto, A. D., Chamberlin, R. A., Engargiola, G., Huang, M., Ingalls, J. G., Jacobs, K., Jackson, J. M., Kooi, J. W., Lane, A. P., Lo, K.-Y., Marks, R. D., Martin, C. L., Mumma, D., Ojha, R., Schieder, R., Staguhn, J., Stutzki, J., Walker, C. K., Wilson, R. W., Wright, G. A., Zhang, X., Zimmermann, P., and Zimmermann, R., *Proc. Ast. Soc. Pac.*, **113**, 567 (2001).
28. Novak, G., and Landsberg, R. H., editors, *Astrophysics from Antarctica*, ASP Conf. Ser. 141, San Francisco: ASP, 1998.
29. Tucker, G. S., Griffin, G. S., Nguyen, H. T., and Peterson, J. S., *Astrophys. J. Lett.*, **419**, L45 (1993).
30. Dragovan, M., Ruhl, J., Novak, G., Platt, S. R., Crone, B., Pernic, R., and Peterson, J., *Astrophys. J. Lett.*, **427**, L67 (1994).
31. Alvarez, D. L., *Measurements of the Anisotropy in the Microwave Background on Multiple Angular Scales with the Python Telescope*, Ph.D. thesis, Princeton University (1995).
32. Ruhl, J. E., Dragovan, M., Platt, S. R., Kovac, J., and Novak, G., *Astrophys. J.*, **453**, L1 (1995).
33. Platt, S. R., Kovac, J., Dragovan, M., Peterson, J. B., and Ruhl, J. E., *Astrophys. J. Lett.*, **475**, L1 (1997).
34. Coble, K., Dragovan, M., Kovac, J., Halverson, N. W., Holzapfel, W. L., Knox, L., Dodelson, S., Ganga, K., Alvarez, D., Peterson, J. B., Griffin, G., Newcomb, M., Miller, K., Platt, S. R., and Novak, G., *Astrophys. J. Lett.*, **519**, L5 (1999).
35. Peterson, J. B., Griffin, G. S., Newcomb, M. G., Alvarez, D. L., Cantalupo, C. M., Morgan, D., Miller, K. W., Ganga, K., Pernic, D., and Thoma, M., *Astrophys. J. Lett.*, **532** (2000).
36. Leitch, E. M., Pryke, C., Halverson, N. W., Kovac, J., Davidson, G., LaRoque, S., Schartman, E., Yamasaki, J., Carlstrom, J. E., Holzapfel, W. L., Dragovan, M., Cartwright, J. K., Mason, B. S., Padin, S., Pearson, T. J., Shepherd, M. C., and Readhead, A. C. S., *Astrophys. J.* (2001), in press, `astro-ph/0104488`.
37. Nguyen, H. T., Rauscher, B. J., Severson, S. A., Hereld, M., Harper, D. A., Lowenstein, R. F., Morozek, F., and Pernic, R. J., *Proc. Ast. Soc. Pac.*, **108**, 718 (1996).
38. Pardo, J. R., Cernicharo, J., and Serabyn, E., *IEEE Trans. Antennas and Propagation* (2001), in press.
39. Lane, A. P., "Submillimeter Transmission at South Pole", in *Astrophysics from Antarctica*, edited by G. Novak and R. H. Landsberg, ASP, San Francisco, 1998, vol. 141 of *ASP Conf. Ser.*, p. 289.
40. Chamberlin, R. A., "Comparisons of saturated water vapor column from radiosonde, and mm and submm radiometric opacities at the South Pole", in *Characterization and Atmospheric Transparency in the mm/submm Range*, edited by Vernin, Munoz-Tunon, and Benkhaldoun, Astr. Soc. of the Pacific, 2001, vol. IAU SITE2000 of *ASP Conference Series*.
41. Chamberlin, R. A., Lane, A. P., and Stark, A. A., *Astrophys. J.*, **476**, 428 (1997).
42. Lay, O. P., and Halverson, N. W., *Astrophys. J.*, **543**, 787 (2000).
43. Chamberlin, R. A., and Bally, J., *Int. J. Infrared and Millimeter Waves*, **16**, 907 (1995).
44. Radford, S. J. E., Reiland, G., and Shillue, B., *PASP*, **108**, 441 (1996).
45. Holdaway, M. A., Radford, S. J. E., Owen, F. N., and Foster, S. M., Fast switching phase calibration: Effectiveness at Mauna Kea and Chajnantor, Millimeter Array Technical Memo 139, NRAO (1995).
46. Ganga, K., Ratra, B., Gundersen, J. O., and Sugiyama, N., *Astrophys. J.*, **484**, 7 (1997).
47. Halverson, N. W., Leitch, E. M., Pryke, C., Kovac, J., Carlstrom, J. E., Holzapfel, W. L., Dragovan, M., Cartwright, J. K., Mason, B. S., Padin, S., Pearson, T. J., Shepherd, M. C., and Readhead, A. C. S., *Astrophys. J.* (2001), in press, `astro-ph/0104489`.
48. Pryke, C., Halverson, N. W., Leitch, E. M., Kovac, J., Carlstrom, J. E., Holzapfel, W. L., and Dragovan, M., *Astrophys. J.* (2001), in press, `astro-ph/0104490`.
49. Novak, G., Chuss, D. T., Renbarger, T., Peterson, J. B., Griffin, G. S., Newcomb, M. G., Loewenstein, R. F., Pernic, D., and Dotson, J. L., *Astrophys. J. Lett.* (2001), submitted, `astro-ph/0109074`.
50. LaRosa, T. N., Kassim, N. E., Lazio, T. J. W., and Hyman, S. D., *Astron. J.*, **119** (2000).
51. Pierce-Price, D., Richer, J. S., Greaves, J. S., Holland, W. S., Jenness, T., Lasenby, A. N., White, G. J., Matthews, H. E., Ward-Thompson, D., Dent, W. R. F., Zylka, R., Mezger, P., Hasegawa, T., Oka, T., Omont, A., and Gilmore, G., *Astrophys. J. Lett.*, **545** (2000).
52. Kim, S., Martin, C. L., Stark, A. A., and Lane, A. P., *American Astronomical Society Meeting 197, BAAS*, **32**, 4.04 (2000).

53. Uchida, Y., Shibata, K., and Sofue, Y., *Nature*, **317**, 699 (1985).
54. Kooi, J. W., Man, C., Phillips, T. G., Bumble, B., and LeDuc, H. G., *IEEE trans. Microwaves Theory and Techniques*, **40**, 812 (1992).
55. Engargiola, G., Zmuidzinas, J., and Lo, K.-Y., *Rev. Sci. Instr.*, **65**, 1833 (1994).
56. Zmuidzinas, J., and LeDuc, H. G., *IEEE Trans. Microwave Theory Tech.*, **40**, 1797 (1992).
57. Walker, C. K., Kooi, J. W., Chan, W., LeDuc, H. G., Schaffer, P. L., Carlstrom, J. E., and Phillips, T. G., *Int. J. Infrared and Millimeter Waves*, **13**, 785 (1992).
58. Kooi, J. W., Chan, M. S., Bumble, B., LeDuc, H. G., Schaffer, P. L., and Phillips, T. G., *Int. J. IR and MM Waves*, **16** (1995).
59. Honingh, C. E., Hass, S., Hottgenroth, K., Jacobs, J., and Stutzki, J., *IEEE Trans. Appl. Superconductivity*, **7**, 2582 (1997).
60. Walker, C., Groppi, C., Hungerford, A., Kulesa, K., C. an d Jacobs, Graf, U., Schieder, R., Martin, C., and Kooi, J., "PoleSTAR: A 4-channel 810 GHz Array Receiver for AST/RO", in *Twelfth Intern. Symp. Space THz Technology*, Twelfth Intern. Symp. Space THz Technology, NASA JPL, 2001, in press.
61. Groppi, C., Walker, C., Hungerford, A., Kulesa, C., Jacobs, K., and Kooi, J., "PoleSTAR: An 810 GHz Array Receiver for AST/RO", in *Imaging at Radio Through Submillimeter Wavelengths*, edited by J. G. Mangum and S. J. E. Radford, ASP Conference Series, San Francisco, 2000, vol. 217, p. 48.
62. Schieder, R., Tolls, V., and Winnewisser, G., *Exp. Astron.*, **1**, 101 (1989).
63. Gerecht, E., Musante, C. F., Zhuang, Y., Yngvesson, K. S., Goyette, T., Dickinson, J., Waldman, J., Yagoubov, P. A., Gol'tsman, G. N., Voronov, B. M., and Gershenzon, E. M., *IEEE Trans.*, **MTT-47**, 2519 (1999).
64. Yngvesson, K. S., Musante, C. F., Ji, M., Rodriguez, F., Zhuang, Y., Gerecht, E., Coulombe, M., Dickinson, J., Goyette, T., Waldman, J., Walker, C. K., Stark, A. A., and Lane, A. P., "Terahertz Receiver with NbN HEB Device (TREND) - A Low-Noise Receiver User Instrument for AST/RO at the South Pole", Twelfth Intern. Symp. Space THz Technology, 2001, in press.
65. Swain, M. R., Bradford, C. M., Stacey, G. J., Bolatto, A. D., Jackson, J. M., Savage, M., and Davidson, J. A., *SPIE*, **3354**, 480 (1998).
66. NRC, *Astronomy and Astrophysics in the New Millennium*, National Research Council, National Academy Press, 2001.
67. Stark, A. A., "Design Considerations for Large Detector Arrays on Submillimeter-wave Telescopes", in *Radio Telescopes*, edited by H. R. Butcher, 2000, vol. 4015 of *Proceedings of SPIE*, p. 434.

MAD-4-MITO, a Multi Array of Detectors for ground-based mm/submm SZ observations

L. Lamagna*, M. De Petris*, F. Melchiorri*, E. Battistelli*, M. De Grazia*, G. Luzzi*, A. Orlando* and G. Savini*

*Experimental Cosmology Group, Dipartimento di Fisica, Università di Roma "La Sapienza"
P.le A. Moro, 2 - 00185 ROMA (Italy)

Abstract. The last few years have seen a large development of mm technology and ultra-sensitive detectors devoted to microwave astronomy and astrophysics. The possibility to deal with large numbers of these detectors assembled into multi–pixel imaging systems has greatly improved the performance of microwave observations, even from ground–based stations, especially combining the power of multi–band detectors with their new imaging capabilities. Hereafter, we will present the development of a multi–pixel solution devoted to Sunyaev–Zel'dovich observations from ground–based telescopes, that is going to be operated from the Millimetre and Infrared Testagrigia Observatory.

INTRODUCTION

The Experimental Cosmology Group at the University of Rome "La Sapienza" has long been active in the development of instrumentation devoted to microwave astronomy from ground–based stations and balloon–borne platforms. In particular, observational campaigns have been performed from the Testa Grigia Observatory (3500m a.s.l.) searching for Sunyaev Zel'dovich effect on nearby clusters of galaxies, with special regard for the Coma cluster (A1656). Thanks to the experience accumulated in dealing with the site characteristics and the general issues of such measurements, and due to the importance of the SZ effect as a cosmological probe and a unique source of information on cluster physics, the instrument is going to be upgraded from a single–pixel, 4–band photometer into a multi–band bolometer array. It is planned to become operational by the end of 2002, when it will be employed to perform a systematic arcminute–scale search for the effect over a sample of nearby clusters.

THE SUNYAEV ZEL'DOVICH EFFECT AS A PURE ESTIMATOR ON THE COSMOLOGICAL DISTANCE LADDER

The full power of SZ measurements in providing systematic–free and redshift independent estimations of cosmological parameters has been widely explored in recent papers [1], as well as demonstrated from a series of experimental results from interferometric arrays and single dish observations in the radio region [2].

SZ effect [3] arises from inverse Compton scattering of CMB photons over a population of hot electrons present in galaxy clusters atmospheres, and is separated into a *thermal* component, due to the effect of velocity distribution in the hot gas, and a kinematic component, arising from the bulk motion of the hot gas along the line of sight. While the latter has not yet been object of a systematic search (apart form few ground–based observational campaigns from the SUZIE group), the thermal SZ effect has already been observed from a wide variety of instruments, mainly operating in the radio region with single dish or interferometric techniques. Combining SZ measurements over a cluster of galaxies with accurate X–ray surface brightness information provided from satellite platforms yields an estimate of the cluster angular diameter distance, so that a sampling of the redshift–distance Hubble diagram is possible. This can bring out an estimate of the main cosmological parameters (H_0 and Ω_m) along with information on the total mass of the cluster. In particular, the possibility to extract the Hubble constant H_0 from a combination of SZ and X–ray data makes this

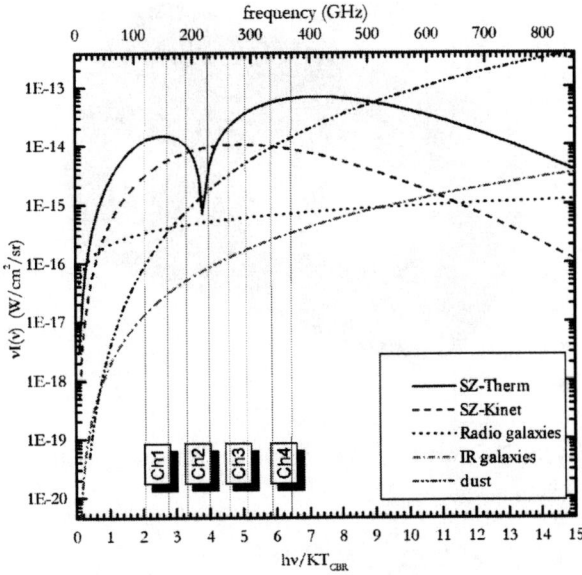

FIGURE 1. Main astrophysical contributions to signal in the 4 bands selected for the MAD experiment.

kind of measurements the natural complement to the mainly Ω-sensitive CMB power spectrum information.

SZ observation program from MITO

The Millimetre and Infrared Testagrigia Observatory has been devoted to SZ search for the last 3 years, during which the main issues of SZ detection (at instrumental as well as at data reduction level) have been studied and cleared with the aid of a wide field ($\sim 17'$) single pixel photometer operating in 4 bands, chosen to match the main features of thermal SZ spectrum with the highest transparency windows of the atmospheric emission spectrum. The Coma cluster ($A1656$) has been used as a benchmark for testing instrument and observation strategies (drift–scans with azimuthal 3–field modulation have been performed recently), along with software simulation capabilities and data reduction techniques. The latter, in particular, are based on the combination of spectral and spatial cross–correlation of sky signals (which include both CMB data and atmospheric noise) in the spectral bands [4, 5].

THE NEW MAD-4-MITO

MAD (Multi Array of Detectors) is an experiment designed to operate in the 4 MITO bands ($140, 220, 270$ and $350 GHz$). These have been chosen to match the best atmospheric transparency spectral windows and, at the same time, to exploit the full potential of thermal SZ effect spectral signature (see fig. 1). The highest frequency channel has been implemented to monitor the foreground contribution of galactic dust emission, which becomes more significant with increasing frequencies and decreasing galactic latitudes. Each channel is designed to receive radiation from the focal plane after free propagation in the optics system, and then it pixelizes the focal plane image into nine regions of $\sim 4'$ dimensions, arranged in a 9×9 bolometer array (see below). Thus, pixelization is performed only after splitting the focal plane into 4 identical spectrally defined beams, instead of employing optically insulated beam guides directly coupled to a single focal plane image. This ensures uniform band selection on each pixel, and should provide good efficiency to the whole system due to the few optical elements deployed between the telescope focal plane and each detector array.

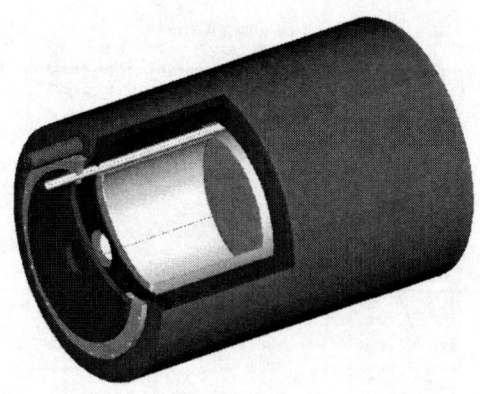

FIGURE 2. MAD cryostat layout. Short optical paths constrain the detector work area in the upper part of the dewar.

Telescope

MAD is designed to operate at the 2.6m telescope of the Testa Grigia site, located at 3500m a.s.l. on the Alps, just along the Italian/Swiss border line. The station, famous for the good atmospheric conditions and low water vapour content (less than 1mm during Winter months, with minimum peaks in December/January), ensures ideal observing conditions for microwave experiments, and has been used by the Rome group along the years as operating base for its main single-pixel SZ-devoted experiment [6], and as a benchmark site for the Diabolo cryostat [7]. The telescope has an aberration-free field of view of 20', which will be fully pixelized from the MAD optic system. Recent improvement in the alignment of the different optical components (especially the wobbling subreflector) and baffling at the focal plane level, ensure good control of modulated spurious signals, which may propagate through the demodulation system as slowly variable offsets, with time frequencies similar to those of the astrophysical signal coming from drift–scanned SZ sources, and thus potentially hard to discriminate on the basis of pure Fourier analysis.

Cryostat

As any other component of the experiment, MAD's cryostat has been designed to ensure long operation times with minimal servicing, enabling us to concentrate on observations and treatment of newly acquired data from calibration and SZ sources while still performing observations at the telescope.

The cryostat (fig. 2) is designed to have an upper radiation window, with the work area limited to a few tens of cm immediately below. This is necessary to assemble the minimum number of optical elements in front of the detectors and still have the possibility to operate in 4 different bands: the system is characterized from extremely small optical paths from the telescope focal plane to the four detector arrays (fig. 3) and thus needs to be concentrated in the upper part of the cryostat. This brought to the need of a good thermalization system, with the helium tank directly below the optics box, and the full height nitrogen tank providing $77K$ radiation shielding by itself. The goal performance of the cryostat, whose assembling is still on its way, is to have a fully thermalized system at $4K$ for $7-8$ days of continuous operation, without any need for cryogen refilling. With the heat inputs from radiation window, internal tank supports, and wiring for 40 detectors, it should need about $30l$ for each of the liquid cryogens to reach this performance. Due to internal layout and tank shaping, the cryostat still remains very compact, with an external height of less than 70cm.

Two-stage $He^3 - He^4$ refrigerator

Detector cool–down to the operating temperature of 300mK will be made possible by means of a two stage $He^3 - He^4$ adsorption refrigerator. Each of the two parts of the system is made of a self–contained condensation

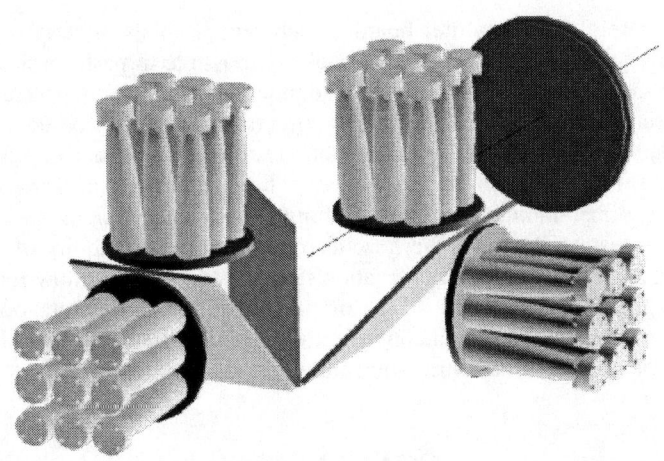

FIGURE 3. Sketch of detector layout in the MAD optics box. The circular shape indicates the position of cryostat radiation window

chamber, directly connected to the cold stage, a heat exchanger, thermally connected with the main He^4 bath, and a cryopumping chamber filled with the proper amount of active charcoal, which may be thermally controlled using an externally powered heater and an electro–mechanical heat switch. The system uses cryopumping on He^4 chamber to bring the cold stage below He^3 condensation temperature, then the second cryopump lowers the liquid He^3 (and thus the cold stage) temperature to the operating level of ~ 300mK. The same refrigerator has been successfully used with the present MITO cryostat, reaching 290mK limit temperature and more than 80 hours of continuous operation with a total heat input of $65\mu W$ on the cold stage [8]. One of the main advantages of this design is the total self-consistency of the fridge operation, since it doesn't need pumping on the main He^4 bath with external devices.

Detectors and Optics

MAD's detectors, built at Haller-Beeman, Inc., are standard composite bolometers made of a sapphire radiation absorber and an NTD Ge thermistor. They are suspended within a heat trap, optimized for the 4 operating wavelengths, by means of $30\mu m$ nylon fibers that keep the absorber firm at half-wavelength height and keep the whole system response fast (expected time constant, still to measure at this time, is $\tau \simeq 8 \div 12ms$). Electrical responsivity is about $1.5MV/W$.

Each of the 4 arrays is arranged in a 3×3 layout (fig. 3), with a tenth bolometer used as a blind monitor of system noise and channel cross-talk. Since the whole MITO telescope corrected focal plane area undergoes such pixelization, single pixel angular resolution is $\sim 4.5'$. The spectral selection is performed over the unpixelized image, directly below the focal plane, and the beam is splitted into four images by means of low pass mesh filters operating at 45 incidence. Single bands are then selected from band-pass filters directly above the multi–mode Winston concentrators that couple detectors with the spectrally selected image. We think that this solution will ensure good optical efficiency to the system, especially if compared to closed light–pipe solutions, but we are aware of the problems that can arise from the need for large filters with uniform performance in order to have pixel–independent spectral selection. The focal plane layout, provided that the sky image is kept fixed on each array through the sidereal motion, allows for continuous detection over the desired direction and the nearest ones, and, above all, allows for simultaneous observation of the pixelized sky region in the different bands: this will ensure low systematics arising from time–delayed signal detection over the pixels and the different spectral regions.

Readout electronics and data reduction

Each detector is DC coupled with a bias circuit and a differential readout circuit, designed to keep low levels of microphonics and e.m. spurious signals. Low output impedance is achieved by buffering the two outputs of each detector through a dual JFET amplifier, set to a common drain (*i.e.* unity gain) configuration. Finally, the signal is amplified from a warm differential preamplifier board, which sends signals directly to data acquisition hardware. Proper filtering at lowest frequencies and anti–aliasing makes it possible to perform data acquisition at few tens of samples/second with 16 bit resolution. We will employ a commercial software controlled ADC board, for which fast monitoring software has been developed. Since the default observation strategy is based on signal modulation through wobbling of the telescope subreflector, on-the-fly demodulation has to be performed in order to extract the information content from each signal. This feature has been included in the acquisition software, together with specific self-calibraton functions, such as offset–monitoring, phase shifting of the different signals with respect to the modulation reference, and others still to test on the field. We have also included the possibility of offline access to modulated data, since we plan to try many other different observation strategies that would allow for multi–mode demodulation over the same modulated signal, such as triangle wave azimuthal scanning or even total power unmodulated scanning, depending on the final detector performance and contingent weather conditions that could make it possible to remove the bulk of atmospheric noise by pixel cross-correlation alone.

CONCLUSIONS

We have shown the main characteristics of the MAD project, a multi–pixel 4–band detector designed for SZ search from ground–based stations, that will become operational from the Testa Grigia station within next Winter. The project headlines have been designed on the basis of the experience accumulated in the last 3 years with the single-pixel instrument that is still operational at the MITO telescope, and has already allowed for SZ detection over the Coma cluster. With the new instrument, we plan to perform an extensive observation program to measure the SZ effect over few tens of clusters and thus map the Hubble diagram from SZ detection over arcminute scales. This would also bring out many candidate SZ sources to include in the upcoming OLIMPO balloon–borne experiment observation program that is undergoing planning and design in our group.

ACKNOWLEDGMENTS

This project is being funded from MURST and University of Rome "La Sapienza". We also wish to thank P. de Bernardis and S. Masi for continuous support during our work.

REFERENCES

1. Rephaeli, Y., "The SZ effect: current status and future prospects", in *2K1BC Workshop - Experimental Cosmology at Millimetre Waves*, edited by M. D. Petris and M. Gervasi, AIP Conference Proceedings, American Institute of Physics, New York, 2001.
2. J.E. Carlstrom, *et al.*, "The Sunyaev–Zel'dovich Effect: Results and Future Prospects", in *Constructing the Universe with Clusters of Galaxies*, edited by F. Durret and G. Gerbal, IAP conference, 2001.
3. Sunyaev, R., and Zel'dovich, Y., *Comm. Astrophys. Space Phys.*, **4**, 173 (1972).
4. M. De Petris, *et al.*, *ApJ Letters* (2001, in press).
5. L. Lamagna, *et al.*, "SZ Effect Detection towards the Coma Cluster with the MITO experiment", in *Proceedings of the Ninth Marcel Grossman Meeting*, World Scientific, 2001.
6. M. De Petris, *et al.*, *New Astronomy*, **1**, 121 (1996).
7. A. Benoit, *et al.*, *Astron. Astrophys. Suppl. Ser.*, **141**, 523–532 (2000).
8. T. Maiani, *et al.*, *Cryogenics*, **39**, 459 (1999).

SZ surveys with the Arcminute MicroKelvin Imager

Christian Holler

Cavendish Laboratory, Madingley Road, Cambridge, CB3 0HE, United Kingdom

Abstract. The Arcminute MicroKelvin Imager (AMI) is an instrument currently under construction in Cambridge designed to produce a survey of galaxy clusters via the Sunyaev-Zel'dovich effect. It consists of two interferometric arrays, both operating at 12–18 GHz; one of ten 3.7-m antennas to provide good temperature sensitivity to arcminute-scale structures, and one of eight 13-m antennas (the present Ryle Telescope) to provide flux sensitivity for removing contaminating radio sources. The telescope is due to be observing by 2003, and will produce a public survey of galaxy clusters, as well as being available to guest observers.

INTRODUCTION

It is now widely recognised that an important next step in cosmology is to conduct a survey of galaxy clusters via the SZ effect. Galaxy clusters are the largest gravitationally bound objects in the universe, and by pointing out the peaks of the initial density field, they provide a sensitive indicator of the growth of structure and the parameters that control it. They are also large enough to provide a fair sample of the material consitituents of the universe. The promise of SZ surveys is that they can in principle select clusters over a very large redshift range with well-understood and physically meaningful selection criteria [1].

Although the first detections of the SZ effect were made using single-dish radiometers, the majority of the detections made so far have been using interferometers. Interferometers offer several significant advantages over total power measurements: they are not susceptible to scan-synchronous systematics; they can use the rotation of the earth to modulate the sky signal in a way that is distinguishable from other signals such as crosstalk or groundspill; they are insensitive to most atmospheric emission; and they can simultaneously measure, and separate, the SZ effect and point source contamination.

To observe the SZ effect efficiently, however, the interferometer must have baselines which are well matched to the angular scale of the cluster. Most clusters at moderate to high redshift have angular sizes of a few arcminutes, and hence most of their power in the fourier plane is on scales of a few hundred to a thousand wavelengths (fig. 1). The interferometers that have provided most of the SZ detections to date, the Ryle Telescope in Cambridge and the OVRO and BIMA arrays in California, have minimum baselines of at least 500 wavelengths (limited by the size of the antennas), and thus resolve out most of the flux density of the cluster. For efficient SZ surveys, telescopes with smaller antennas and hence shorter minimum baselines are required. This also has the beneficial effect of increasing the instantaneous field of view of the telescope, further enhancing the survey speed.

AMI DESIGN

The design requirements for an SZ survey instrument are: maximum sensitivity, hence low system temperature and high bandwidth; shortest baselines of $\sim 200\lambda$ (confusion from primary CMB anisotropies becomes a problem at shorter baselines); and ability to cope with foregrounds. The first question is the choice of observing frequency—this controls all subsequent aspects of the design.

Contributions to the system temperature of a telescope due to the atmosphere and from typical HEMT recievers give a clear advantage of operating at low frequency, even from a relatively poor site (fig.2). However, this is offset by the temperature contribution of a source of given flux density, which falls as approximately ν^{-2}. Since extragalactic point sources are the main contaminant for SZ observations, this is an important consideration. Several strategies are

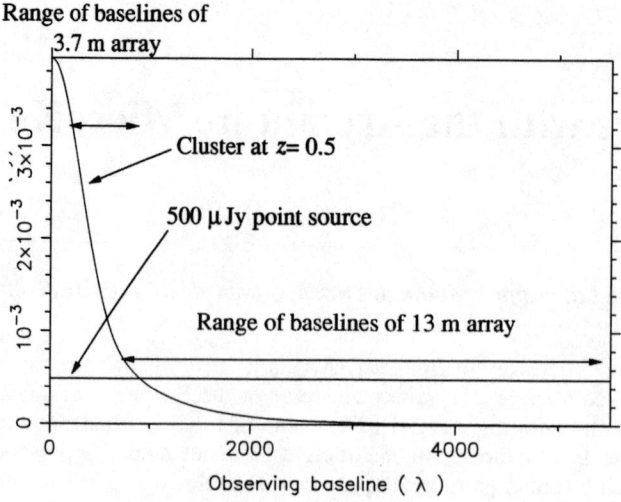

FIGURE 1. The 15 GHz flux density due to a massive cluster at $z \sim 0.5$ as a function of interferometer baseline. Also shown are the ranges of baselines of the two arrays of AMI, and the flux density of a typical confusing radio source.

possible. One method is to operate at high frequency, $\nu \sim 100$ GHz, where source contamination is minimised, and to offset the increased system temperature by going to a high, dry site and maximising the number of antennas and the bandwidth. An alternative is to exploit the low system temperature available at low frequency, but to provide sufficient flux sensitivity at higher resolution to detect and then remove the point sources. This is the approach we adopt with AMI, since we have the large antennas of the Ryle Telescope available to do source subtraction. Operating at 15 GHz from Cambridge it is possible to achieve total system temperatures below 25 K, with the cost and logistical advantages of working only a few km from our home institution.

With the wavelength fixed at $\lambda \simeq 2$ cm, the remaining design parameters are fairly constrained. The available bandwith is the K_a waveguide band of 12–18 GHz; the minimum required baseline of $\sim 200\lambda$ fixes the antenna size at ~ 4 m. We use 3.7-m diameter antennas, for which single-piece spun reflectors are available at low cost. The number of

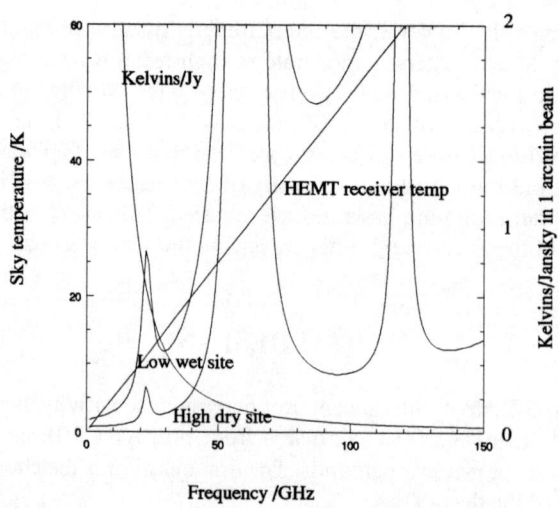

FIGURE 2. Contributions to system temperature as a function of frequency; the atmospheric emission temperature from a sea level site such as Cambridge and a high dry site such as Mauna Kea are shown, along with typical noise temperature of a HEMT amplifier; the total system noise is the sum of the amplifier and sky noise (plus other contributions from losses, spillover, the CMB etc.). The other line (and scale on right) shows the flux/temperature conversion for an unresolved source in a 1 arcmin beam.

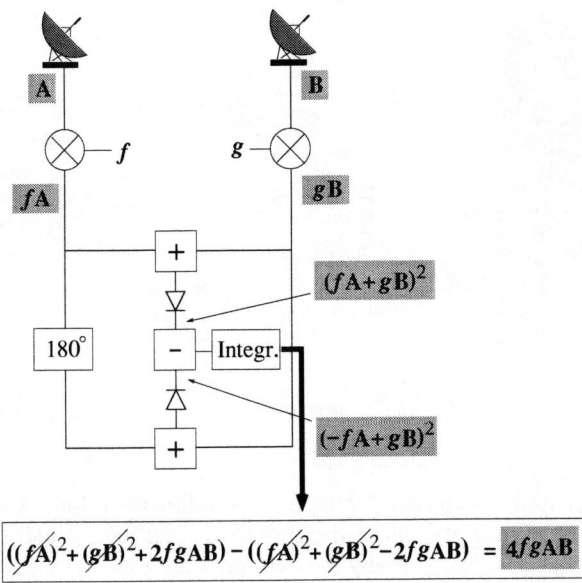

FIGURE 3. Phase-switched add-and-square type correlation with 180° invertion. The signal is synchronously detected with the product switching function fg.

these antennas is fixed by the flux sensitivity available for source subtraction from the Ryle Telescope and the 15 GHz source counts to about ten, ie a collecting area equal to about one RT antenna.

THE CORRELATOR

To maximize sensitivity it is important to use as large a bandwidth as possible. In order to maximize the AMI bandwidth while minimizing the cost of both the correlator and the IF system, we have been developing a correlator to cover the entire available 6 GHz bandwidth in a single IF channel. This design combines the recent cheap availability of the required active components (amplifiers, switches, detectors, voltage-variable attenuators) at frequencies above 10 GHz with the relative ease of design of passive components (splitters, phase-shift networks) at low *fractional* bandwidth. The correlator thus uses an IF band of 6–12 GHz, with frequency resolution provided by an 8-channel multiple-lag design. In order to measure amplitude and phase of the signal a so called "complex correlator" is used. A broadband 90° phase shift in the 'sin' arm produces the imaginary terms. An entire multi-channel correlator for one baseline is integrated on to a single substrate. The correlation scheme used is the phase-switched add-and-square type introduced by Ryle [2], also used in both CAT and the VSA, which uses single detector diodes as the multiplying elements (fig.3). Phase switching the signals with two orthogonal functions f and g and a second correlation with a broadband 180° phaseshift in one arm is used to suppress any noise and spurious signals picked up between the antennas and the correlator.

CURRENT STATE OF DEVELOPMENT

For cost reasons it was decided to carry out the design completely in printed circuit techniques and not to buy additional components such as broadband splitters and hybrids. Therefore these had to be developed seperately. Broadband in-phase splitting is obtained through a component called the "Wilkinson divider", which is well treated in the literature [3]. The broadband split with 90° phase difference between the outputs is realised through a Wilkinson divider and two tight coupled lines with different length and coupling (fig.4) [4]. The results show the performance as simulated (grey curve) and measured (black curve). The transmission over the range 5–13 GHz is very flat and the phase error in the same range is less than 5°. Fig.5 shows the concept of producing a 180° phase difference between

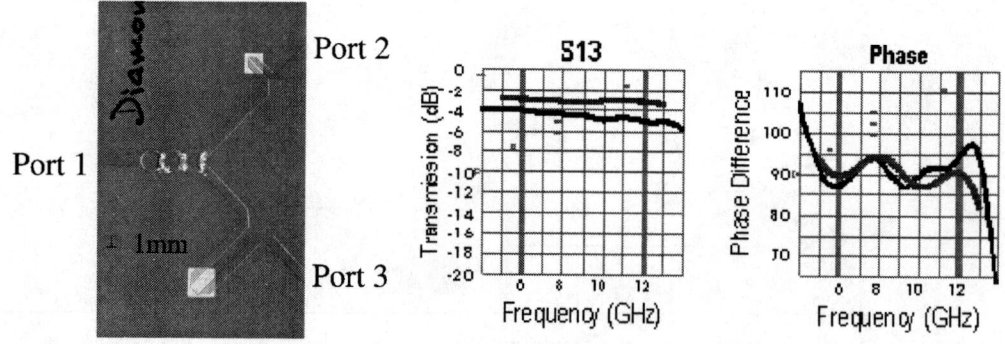

FIGURE 4. 90° phase shifter and its performance (grey: simulation, black: measurement).

two in-phase signals. This is realised through combining microstripline with slotline techniques [5]. The electric field lines of an incoming signal couple from the microstrip line to the slot line, which is etched into the ground plane on the other side of the dielectric substrate. Thereby the field lines are rotated by 90° in space. When coupling back from the slotline to the microstripline, the field lines can either be rotated back by −90° (left picture) or forward by an additional 90° (right picture). Together these two procedures can create a phase difference of 180°, which is independent of frequency in the first order. Results again show that the performance over the range 5–13GHz is very satisfactory (fig.6).

This frequency range also creates difficulties in the design of the diode detector circuit. The need for very low junction capacitance diodes results in its very small size (300μm) and matching the diode to the 50Ω circuit without losing too much sensitivity results in very narrow microstriplines. The results in fig.7 show the reflection coefficient of this detector circuit. It is very acceptable for most of the range 5–13GHz.

CONCLUSION

AMI thus consists of two arrays, one of eight 13-m antennas and one of ten 3.7-m antennas, using the same 12–18 GHz receivers and 6–12 GHz correlator. The small array provides most of the sensitivity to the SZ effect and the large array most of the sensitivity to point sources, although there is some overlap. The flux sensitivities are 2 mJy s$^{-1/2}$ and 20 mJy s$^{-1/2}$ for the large and small arrays, in 6 and 20 arcmin fields of view respectively. The temperature sensitivity depends on the exact array configuration used; a representative number would be about 15 μK in a week, in a 1.5 arcminute beam and 20 arcmin field of view.

AMI itself is now largely funded and under construction; first results are expected by mid-2003. In recognition of the large amount of follow-up work in other wavebands that will result from the AMI surveys, we are proposing to

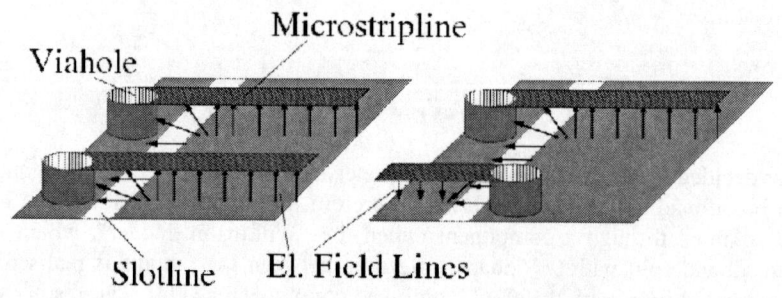

FIGURE 5. Concept of a 180° phase shifter.

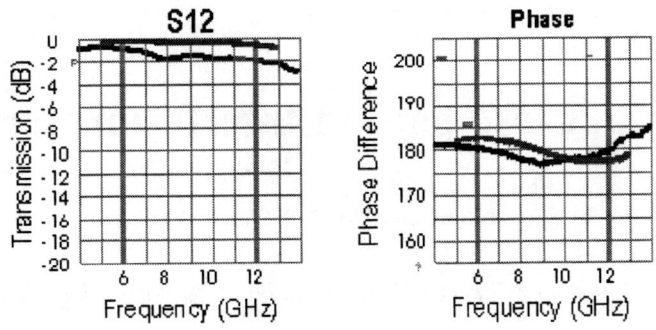

FIGURE 6. Performance of the 180° phase shifter; (grey: simulation, black: measurement).

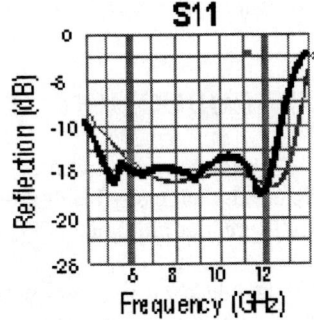

FIGURE 7. Performance of the diode matching circuit; (grey: simulation, black: measurement).

make these surveys public as soon as possible after the data are taken. Also, a significant fraction of observing time on AMI will be open to guest observers – details will be published when the telescope is nearer completion.

ACKNOWLEDGMENTS

AMI is supported by the Particle Physics and Astronomy Research Council and the Department of Physics, University of Cambridge

REFERENCES

1. Kneissl, R., Jones, M. E., Saunders, R., Eke, V. R., Lasenby, A. N., Grainge, K., and Cotter, G., *MNRAS*, **in press** (2001).
2. Ryle, M., *Proc. Roy. Soc. A.*, **211**, 351 (1952).
3. Li, C. Q., Li, S. H., and Bosisio, R. G., *Microwave Journal*, **Nov**, 125–135 (1984).
4. Schiffmann, B. M., *IRE Trans. on Microwave Theory and Techniques*, **April**, 232–237 (1958).
5. Gupta, K. C., Garg, R., and Bahl, I. J., *Microstrip Lines and Slotlines*, Arctech House Inc., 1996, p. 324.

Cross-talk in Close-packed Interferometer Arrays

Ravi Subrahmanyan

Australia Telescope National Facility (CSIRO), Locked bag 194, Narrabri, NSW 2390, Australia

Abstract. Close-packed antenna configurations are important for high-surface-brightness observations of the CMB. The visibilities measured in baselines formed between close antenna elements — in particular, between shadowed elements — of interferometers are often observed to be corrupted. I show data characterizing the spurious cross-talk between antennas of the Australia Telescope Compact Array (ATCA).

INTRODUCTION

Many of the large and medium angular scale CMB anisotropy measurements to date have been made with total power radiometers using differencing techniques to cancel the large sky brightness which exceeds the anisotopy amplitude by a factor about 10^5. For example, the COBE differential radiometers [1] and the Boomerang bolometers [2] were both successful instruments of this class.

Interferometric telescopes, like the CAT [3] and DASI [4], have successfully reported measurements of the anisotropy on degree scales. On small arcmin angular scales, interferometric arrays have placed useful limits [5] and even reported detections [6]. Interferometric methods have definite advantages for measurements of the extremely small CMB temperature anisotropies: they do not respond to the mean sky background and atmospheric emission. Therefore, they do not require precision differencing techniques and consequently the receivers forming the arms of the interferometers do not require to be built with the same stability.

In practice, interferometers do not always give a zero response in the absence of sky anisotropy. This is particularly the case when the interferometer elements are close packed and, unfortunately, high sensitivity to surface brightness fluctuations in the sky requires small baselines. This spurious response is present in the CBI [7], DASI [4] and VSA [8]; their cause is not well understood and their removal has necessitated the adoption of differential measurement techniques in these interferometric observations as well. I herein refer to this spurious unwanted component, which is present in the interferometer response and is not owing to sky anisotropy, as 'cross talk'.

The Australia Telescope Compact Array (ATCA; The Australia Telescope [9]) has been used in an ultra-compact configuration for searches for arcmin scale anisotropy [5]. Five antennas of the array which are reconfigurable on a 3 km East-West railtrack were used for this experiment. The antennas are 22 m in diameter and have Cassegrain type on-axis optics. The reflectors are shaped and the panels forming the main reflector are solid. During early observations with the array it was noticed that the response in baselines between close antennas was corrupted when one of the antenna apertures was in the geometric shadow of an adjacent antenna. The closest allowed baseline in the ATCA is 30.6 m and observations of fields in the southern sky which are at declinations south of about $-50°$ will never involve shadowed baselines; therefore, deep observations for CMB anisotropies with the ATCA were made in fields at these declinations and cross talk was avoided. Nevertheless, with a view to understanding the origin of the cross talk at the ATCA, an effort has been made to characterize the cross talk and these data are being presented here.

CROSS-TALK BETWEEN CLOSE ATCA ANTENNAS

When observing an unresolved calibrator source with the interferometers formed between the antennas of the ATCA, the calibrated visibility amplitudes are expected to be identical on all baselines. However, if the aperture of one of the antennas is shadowed, it is expected that the correlation coefficient would drop on all baselines with the shadowed antenna. In the geometric optics approximation, the amplitude drop would increase with the extent of geometric

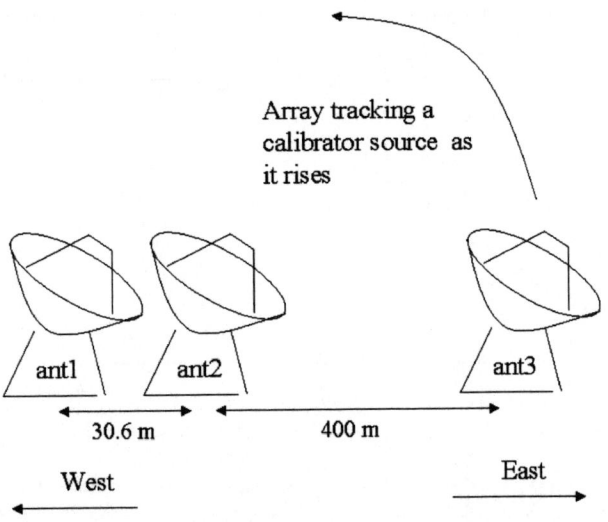

FIGURE 1. Array configuration for the cross talk measurements

shadowing; however, we would expect the drop to be identical on all baselines with the shadowed antenna.

Three ATCA antennas, referred to as ant1, ant2 and ant3, were located on the East-West railtrack so that the ant1-ant2 spacing was 30.6 m and ant3 was located several hundred metres to the east. The configuration is depicted in Fig. 1. An unresolved continuum source at a declination −4° was observed down the track toward East as it rose in the sky. Ant1 was shadowed by ant2 initially while the projected baseline ant1-ant2 was less than the antenna diameter of 22 m. As the source elevation increased, the geometric shadowing progressively reduced and at higher elevations, when the projected baseline exceeded 22 m, there was no geometric shadowing.

The amplitudes on the ant1-ant2 and ant1-ant3 baselines are shown in panel (a) of Fig. 2 versus the projected length of the ant1-ant2 baseline. These data were obtained at 4800 MHz and in a continuum band 128-MHz wide. Contrary to expectations, the amplitude seen in the short ant1-ant2 baseline — between the shadowed and shadowing antennas — deviates from the longer baseline visibility amplitude in the regime where there is geometric shadowing: the ant1-ant2 baseline is corrupted.

The same array configuration was used and a 'blank' sky field at the same declination was tracked over a similar hour angle range. An 'unexpected' and 'unwanted' response is seen on the ant1-ant2 baseline alone in the regime where geometric shadowing occurs. This response is shown in panel (b) of Fig. 2. There is clearly an additive cross

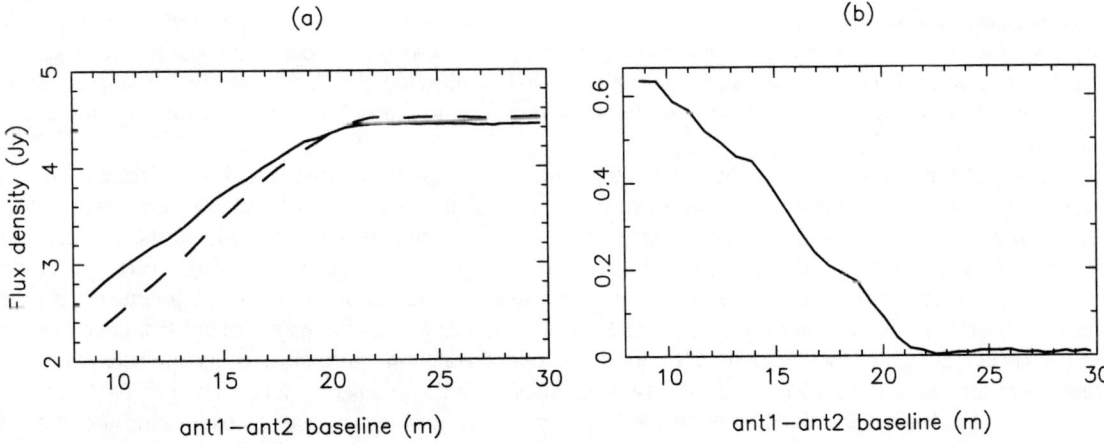

FIGURE 2. Visibility amplitude on the ant1-ant2 and ant1-ant3 baselines versus the projected length of the ant1-ant2 baseline. Panel (a) shows amplitudes while tracking an unresolved source, ant1-ant2 is shown as a continuous line and ant1-ant3 is the dashed line. Panel (b) shows the ant1-ant2 amplitude while tracking a blank field which has the same declination.

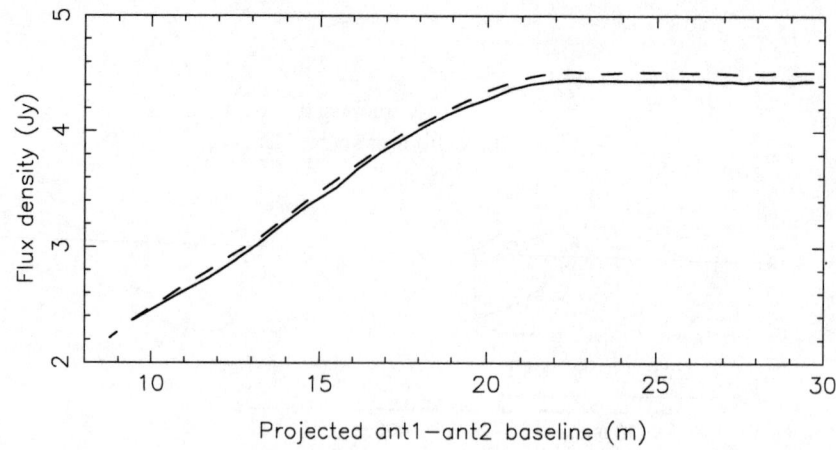

FIGURE 3. Visibility amplitudes on the ant1-ant2 and ant1-ant3 baselines while tracking the unresolved source. Here the ant1-ant2 visibilites have been corrected for cross talk by subtracting the visibilities measured on the blank field.

talk between the shadowing and shadowed antennas.

When the cross talk visibilities that are measured on the blank field are vectorially subtracted from the ant1-ant2 visibilities on the calibrator source, the amplitude and phase on the ant1-ant2 baseline matches, within the errors, the complex visibility on the ant1-ant3 baseline (see Fig. 3). It has been observed that the cross talk component has a phase which varies with frequency; however, a complex subtraction of the cross talk component makes the ant1-ant2 visibilities the same as that on ant1-ant3 and other long baselines. The implication is that the dominant cross talk between ATCA antennas in shadowed regimes is an additive component.

CHARACTERISTICS OF THE ADDITIVE CROSS TALK

The additive cross talk component, in units of flux density as well as in units of antenna temperature, is seen to increase with wavelength. Roughly, the flux density and temperature scale as the wavelength at cm wavelengths (1-10 GHz).

Continuous wave (CW) signals are usually present in the antenna electronics as part of the local oscillator modules in the conversion chains. Cross talk could originate in a leakage of these tone-like signals between close antennas. An examination of the cross talk between ATCA antennas with the correlator in a spectral-line mode has shown that the cross talk here is a continuum signal and is not tone-like.

The signals in the ATCA antennas are effectively sampled at Nyquist rate (twice the bandwidth) and the ATCA correlator measures the correlation coefficient over a range of positive and negative lags/delays. Using 128 MHz bandwidth signals sampled at 3.9 ns intervals, the lag spectrum of the cross talk component, measured over ± 32 lag steps, is shown in panel (a) of Fig. 4. Using a narrower 4 MHz bandwidth, the lag spectrum measured over a larger range of delays is shown in panel (b) of the same figure. The δ-function form lag spectrum implies that the cross talk is a continuum type of signal.

If the source of the cross talk is in the near field, it may be possible for this signal to propagate into the two feed horns in the two nearby antennas via multiple propagation paths as a result of multiple scatterings off antenna structural elements like the subreflector support structures. However, there is only one spike in the lag spectrum over the range $\pm 128\ \mu s$ (panel b of Fig. 4): this implies that the cross talk component has a single propagation delay in its path to the two receiver feed horns. While observing toward a sky direction with any antenna pair, the geometric free-space propagation and electronics instrumental delays for signals arriving at the correlator from that sky direction are compensated. Therefore, the appearance of the cross-talk δ-function spike at zero lag indicates that the cross talk component arrives at the feed horns of the two antennas with the geometric delay. The precise location of the δ-function spike has been determined to be within ± 10 cm of the geometric delay by examining the phase versus frequency behaviour of the Fourier transform of the lag spectrum shown in panel (a) of Fig. 4.

The amplitude of the cross talk increases with the extent of geometric shadowing. The cross talk flux density was measured with the 30.6 m ant1-ant2 baseline observing toward a common sky direction: the elevation was increased keeping the antenna azimuth fixed due East, and the variation in the cross talk flux density versus the projected

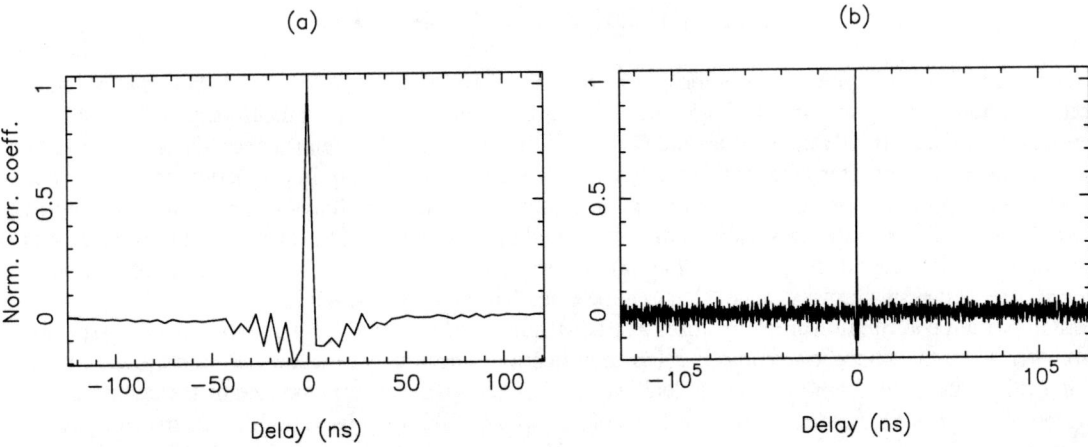

FIGURE 4. Normalized correlation coefficient of the cross talk component versus delay. Panel (a) is with 3.9 ns resolution and over a small delay range, panel (b) is with 125 ns resolution and over a large delay range.

baseline is shown in Fig. 5. Measurements of the cross talk amplitude in different azimuth directions has shown that this amplitude depends only on the projected radial separation between the antenna centers, *i.e.*, the cross talk amplitude depends only on the extent of geometric shadowing and not on the orientation of the baseline with respect to the ground-sky horizon.

Cross talk is often traced to the leakage of front-end receiver noise from one antenna into a close adjacent antenna. This is more likely if there is no isolator in the signal path between the feed horn and the first low noise amplifier. In the case of the cross talk seen at the ATCA in shadowed configurations, the location of the cross talk related δ-function close to the zero in the lag spectrum rules out receiver noise as the source of the cross talk. There are calibration noise diodes which inject noise into the horns: the cross talk was unchanged when they were switched off ruling out the possibility that these noise sources cause the cross talk.

Close antennas may respond to the ground radiation in their near fields. However, if the source of the cross talk in the ATCA is external to the antenna structures and their electronics, the closeness of the cross talk delay to the geometric path delay indicates that the source of the cross talk is in the sky direction toward which the antennas point.

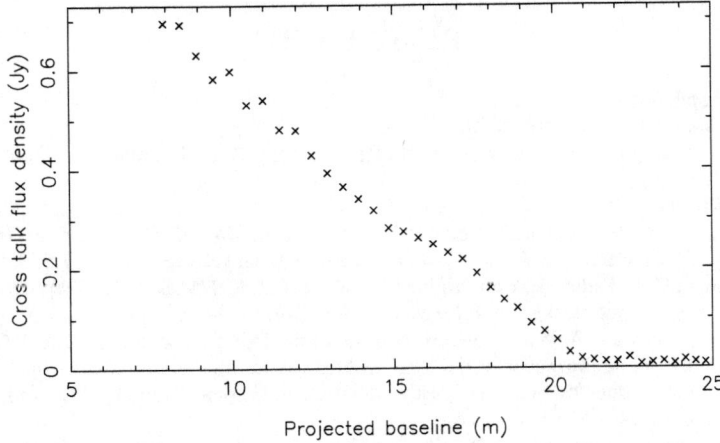

FIGURE 5. Variation of cross talk flux density on the ant1-ant2 baseline as a function of the projected baseline length.

SOURCE OF THE CROSS TALK

The evidence suggests a source which is located in the sky direction toward which the antennas point. From this direction, the antennas receive radiation from the atmosphere as well as from the uniform sky background.

At cm wavelengths (1-10 GHz), D^2/λ for the $D = 22$ m diameter ATCA antennas exceeds about 1.6 km and most of the atmospheric is in the near field. When the antennas are close and overlap in projection, there may be common atmospheric emission in the beams of the two antennas and this may be, perhaps in part, contributing to the cross talk. Correlated variations in the cross talk amplitude picked up in orthogonal linear polarizations have been noticed at low elevations and at higher frequencies (9 GHz), and these have not been accompanied by corresponding phase fluctuations. This suggests that 'sky noise' [10] may be contributing to the cross talk.

A more important contribution probably arises as a consequence of the effective generation of 'zero-spacings' in the interferometer when the adjacent antennas overlap in projection. These may be generated either owing to diffraction at the rim of the antenna in front or owing to leakage through the main reflector surface of the antenna in front. The coupling between the fields incident on the apertures of the two antennas may be viewed as arising from a scattering of the field incident on the aperture of the antenna in front. When zero-spacings are generated, the interferometer would respond to the uniform background sky and atmosphere: both these would appear as a stable continuum cross talk and with a delay appropriate for a source at the pointing center.

WORK IN PROGRESS

Arvind Deshpande (Raman Research Institute, India) and I are currently working on examining the spatial distribution and coherence properties of the cross talk component across the apertures of the shadowed antennas. Experimentally, holography-type measurements [11, 12] are being made with one of ant1 or ant2 stationary and the other scanning in azimuth and elevation about the nominal pointing direction. The data is being modelled using Geometric optics and Physical optics propagation of emission from plausible source models.

ACKNOWLEDGMENTS

The Australia Telescope is funded by the Commonwealth of Australia for operation as a National Facility managed by CSIRO.

REFERENCES

1. Smoot, G.F., et al., *ApJ* **360**, 685-695 (1990).
2. de Bernardis, P., et al., *Nature* **404**, 955-959 (2000).
3. Robson, M., Yassin, G., Woan, G., Wilson, D.M.A., Scott, P.F., Lasenby, A.N., Kenderdine, S., and Duffett-Smith, P.J., *A&A* **277**, 314-320 (1993).
4. Padin, S., et al., *ApJ* **549**, L1-L5 (2001).
5. Subrahmanyan, R., Kesteven, M.J., Ekers, R.D., Sinclair, M., and Silk, J., *MNRAS* **315**, 808-822 (2000).
6. Dawson, K.S., Holzapfel, W.L., Carlstrom, J.E., Joy, M., LaRoque, S.J., and Reese, E.D., *ApJ* **553**, L1-L4 (2001).
7. Pearson, T.J., Readhead, A.C.S., Padin, S., Cartwright, J.K., Mason, B.S., Myers, S.T., Shepherd, M.C., Sievers, J.L., and Udomprasert, P.S., "The Cosmic Background Imager," in *New Cosmological Data and the Values of the Fundamental Parameters*, edited by A. Lasenby and A. Wilkinson, proceedings of the IAU Symposium 201, ASP Conf. Ser. , 2000, pp.2.
8. Jones, M.E., "Results from Cat and Prospects for the Vsa," in *Microwave Background Anisotropies*, proceedings of the XVIth Moriond Astrophysics Meeting, edited by Francois R. Bouchet, Richard Gispert, Bruno Guilderdoni, and Jean Tran Thanh Van, Gif-sur-Yvette: Editions Frontieres, 1997, 161.
9. *The Australia Telescope, special issue of J. Electr. Electron. Eng. Aust.*, The Institution of Engineers, Australia and The Institution of Radio and Electronics Engineers, Australia, **12**, June 1992.
10. Holdaway, M.A., Owen, F.N., and Emerson, D.T., *NRAO Millimeter Array Memo* 137 (1995).
11. Bennett, J.C., Anderson, A.P., McInnes, P.A., and Whitaker, A.J.T, *IEEE Trans. Ant. Prop.* **24**, 295-303 (1976).
12. Scott, P.F., and Ryle, M., *MNRAS* **178**, 539-545 (1977).

Results from the First Engineering Run of BOLOCAM and Plans for the Future

P. Mauskopf*, P. Ade*, J. Bock[†], S. Edgington**, S. Golwala**, A. Goldin[‡], J. Glenn[§], D. Haig*, V. Hristov**, B. Knowles[§], A. Lange**, H. Nguyen[‡] and B. Rownd[§]

*Cardiff University - Cardiff - Wales - UK
[†]JPL and Caltech - Pasadena - CA
**Caltech - Pasadena - CA
[‡]JPL - Pasadena - CA
[§]University of Colorado - Boulder - CO

Abstract. We present results from the engineering run of the BOLOCAM receiver at the Caltech Submillimeter Observatory in May, 2000 and simluations of the sensitivity of the completed BOLOCAM receiver to astrophysical sources. BOLOCAM is a 144 element array of bolometers designed to operate in the millimeter wave atmospheric windows at 150, 220, and 280 GHz. These bands are well suited to studies of the Sunyaev-Zel'dovich effect in distant clusters of galaxies as well as measurements of the emission of interstellar dust.

INTRODUCTION.

BOLOCAM is one of a new generation of cameras designed for imaging and surveys at millimeter and sub-millimeter wavelengths. Submillimeter and millimeter wave cameras mounted on the JCMT [1] and on the IRAM 30 meter telescope [2] have begun to perform large area surveys containing sources corresponding to dust emission from distant galaxies (e.g. [3],[4]). At lower frequencies, interferometric arrays have begun to perform surveys of the Sunyaev-Zel'dovich effect in distant clusters [5],[6]. These measurements probe the history of galaxy and cluster formation since the intensity of signals from both dusty galaxies and the SZ effect diminish less rapidly with redshift than optical or X-ray signals. We plan to continue this effort with BOLOCAM observations of large areas of blank sky regions selected to have complementary data at many other wavelengths.

INSTRUMENT

The combination of several new technologies in BOLOCAM allows new methods of observation that have the potential for low systematic errors and offer sensitivity to signals at a wide variety of angular scales. In BOLOCAM, the detectors and electronics are designed to have no intrinsic excess noise in integrations lasting up to one minute in time. While integrating, the dominant source of noise is from changes in atmospheric emission which must be removed by subtracting the average signal of all of the detectors in the array. This limits BOLOCAM's sensitivity to flux from sources larger than the size of the array on the sky. Because the optics only have to be designed to give diffraction limited performance at a single position of the secondary mirror, the available field of view is maximized and can be significantly larger than the typical chop throw of a secondary mirrow. This also allows the maximum possible number of detectors in the focal plane.

The design of BOLOCAM is described in detail elsewhere [7]. Here, we present a summary of the performance of the system during an engineering run at the Caltech Submillimeter Observatory (CSO) in May, 2000.

Detectors

The detectors in BOLOCAM are 'spider-web' type bolometers fabricated as a monolithic array on a three inch silicon wafer. Different configurations of these devices have been designed for use in a variety of experiments including SuZIE [9], BOOMERANG [10], and HERSCHEL (SPIRE) [11]. For the BOLOCAM engineering run, we used a prototype of the 144 element bolometer array that contained a total of 62 working bolometers.

Cryogenics

For ground-based operation, a base temperature of 300 mK for the bolometers is sufficient to approach photon-limited noise performance. The BOLOCAM array is cooled to a base temperature of 256 mK using a closed cycled adsorption He^3/He^4 refrigerator. The fridge contains 3 independent volumes, each consisting of a space for collecting condensed liquid and a cryopump, one charged with He^4 and the other two with He^3 gas. The He4 refrigerator is used to condense the gas in the two He^3 refrigerators during the cool-down cycle. One of the He^3 fridges, the InterCooler (IC) is designed to have a maximal hold time as a function of thermal input by allowing the temperature of the He3 liquid to be almost unconstrained. This fridge incorporates a long pump tube and heat exchanger that efficiently extracts the heat of enthalpy from the evaporating He^3 gas and serves as athermal buffer to the other fridge. In operation it reaches a base temperature of about 400-500 mK. The second He^3 fridge, the UltraCooler (UC) has a shorter pump tube that is thermally shorted in the middle to the IC cold head to minimize parasitic heat loads and obtain a minimum base temperature.

One advantage of the BOLOCAM fridge is the ability to cycle it without pumping on the main He^4 bath. The total fridge cycle takes 1.5 hours and is fully electrical and automated at the telescope. The hold time of the fridge is limited by the hold time of the IC buffer fridge and was measured to be over 24 hours when BOLOCAM was mounted on the CSO (the fridge was cycled once per day and was always still cold at the beginning of the cycle). Details of the design and performance of the fridge are given in [8].

The cryostat is a standard LHe/LN_2 cryostat from Precision Cryogenics. The volume of the LHe and LN_2 tanks are 15 litres each. The consumption of LHe at the CSO during the engineering run was 10 litres per day. We expect to be able to reduce the consumption by at least a factor of 2 in the future.

Electronics

The readout electronics are designed to give detector noise limited signals in a frequency band from 0.015-20 Hz. The readout circuit is based on the circuit used in the BOOMERANG instrument. This readout scheme has been field tested by BOOMERANG and found to have good performance over the entire frequency band. In adapting the BOOMERANG design for the larger number of bolometers in BOLOCAM, we used surface mount components to minimize the space needed for the electronics. We also changed the first warm preamplifier from an AD624 to INA103 which has a lower white noise specification ($1\,nV/\sqrt{Hz}$ vs. $4\,nV/\sqrt{Hz}$). Laboratory tests of the BOLOCAM electronics after the engineering run showed that the surface mount components used in the BOLOCAM preamplifier boards as well as the INA103 amplifiers have a higher temperature coefficient than the components used in the BOOMERANG electronics. This leads to a large 1/f noise that is not completely correlated from one amplifier channel to another and therefore cannot be removed. This was observed during the engineering run as an excess noise below 1 Hz over the detector noise of between a factor of 2-4. Selecting surface mount components with lower temperature coefficient and using the AD624 chips gives back the expected performance.

Optics

The optical design for BOLOCAM provides an unobstructed 9 arminute field of view diffraction limited at 1 mm wavelength, the maximum size allowed by apertures in the back structure of the CSO. The reimaging optics and receiver are mounted together in a box attached to the back structure of the telescope near the Cassegrain focus. Two flat mirrors fold and direct the incoming radiation from the telescope onto a 36×36 cm ellipsoidal mirror that reimages the Cassegrain focus to a point inside the BOLOCAM cryostat and reimages the primary mirror onto a 4 K aperture directly behind the entrance window of the cryostat. This 3.5 cm diameter aperture is undersized in order to define the illumination pattern of the bolometers on the primary mirror and control spillover. Behind this 4 K Lyot stop, there is a single lens, 10 cm in diameter, of high density polyethylene, with a focal length equal to its distance from the Lyot stop. This lens acts to flatten the overall field curvature of the reimaging optics while ensuring that the detector illumination overlaps as much as possible on the primary mirror. A monolithic array of 151 hexagonally packed conical horns with entrance aperture of 5 mm each is placed at the image plane.

The illumination pattern of the conical horns is larger than the size of the Lyot stop. Therefore, a combination of the horn geometry and the 4 K cold aperture contribute to the definition of the illumination pattern of the detectors on the primary mirror and on the sky.

The filter bands are determined by a combination of the high pass wave guide cutoff from the section of circular waveguide at the exit aperture of the conical feed horns and an 80 mm diameter low pass filter placed in front of the horn array. Additional blocking filters are placed at the 4 K aperture and on the 77 K shield. An initial test of the filters and horns with five bolometers gave an average optical efficiency for the system of 19%. However, the prototype array used in the engineering run described in this paper had an average optical efficiency of <7% mounted in the same configuration using the same horns and filters. The reason for the poor efficiency of the prototype array is most likely due to poor metallization of the absorber on the bolometers.

Scan Strategy

During the engineering run, all observations were made using drift scans, with the telescope fixed in azimuth and elevation while the earth's rotation caused the source to move across the field of view. This type of scan minimizes sources of environmental signal modulations from moving optical components or sidelobe pickup from the ground. However, it also causes the astrophysical signals to change very slowly and therefore relies on the stability of the detectors and electronics. For a point source observation with a Gaussian beam with FWHM = 40", the 3dB point of the signal bandwidth is at a frequency of 80 mHz.

OBSERVATIONS

BOLOCAM was first mounted on the CSO for five nights from May 18 - 22, 2000. The first two nights were devoted to setup and commissioning tests while the last three nights were devoted to test observations. Data was obtained in observations of Uranus, 3C273, W58, G34.3, DR21, and Abell 1835, all in drift scan mode. Here we present preliminary maps of Uranus, DR21, and Abell 1835.

DATA ANALYSIS

Merging

The signals from the BOLOCAM bolometers are sampled at 50 Hz by a PC-based data acquisition system. At the same time, the pointing data from the telescope computer are written to a different file at a rate of 100 Hz. In both data streams there are common TTL signals used for communication between systems on BOLOCAM and the telescope computer and for synchronization of the data streams. These TTL signals include signals from the telescope that switch from high to low when the telescope changes from tracking mode to drifting mode. These signals are the primary ones used to synchronize the data streams and are also used to mask out data from the cleaning and map making processes when the telescope is slewing.

Both the raw bolometer data files and the telescope data files are written to disk once per minute. A merging routine combines these files once every several minutes into a single time stream that contains all of the receiver and telescope data.

Cleaning

After the data has been combined into a single file, we perform a deconvolution of the electronics transfer functions and remove common mode noise in the detectors which is dominated by common-mode sky temperature fluctuations. To do this, we first separate the data in the time stream into individual drift scans using the TTL signals from the telescope. Then, we remove spikes in the data and deconvolve scan by scan. The bolometer time constants are comparable to the 20 Hz bandwidth of the readout electronics, so cosmic ray events only last a few samples and are easily removed. We remove an average of 2% of the data in a given bolometer.

After despiking, we remove the common mode signals from across the array. The output voltage from each detector can be described by:

$$V_i(t) = \frac{dR_i}{dP} * (F(\theta_i(t)) + A(t)) + \delta I_{bias}(t) R_0 + \frac{\delta G_i(t)}{G} I_{bias} R_0 + e_i(t) \qquad (1)$$

where $F(\theta)$ is the flux on the sky as a function of angle, $A(t)$ is the common mode emission of the atmosphere + telescope as a function of time, e_i is the detector noise, R_i is the bolometer resistance, G is the electronics gain, and I_{bias} is the common bias current. The five terms in equation xx are: i) sky signal $\propto F(\theta_i(t))$, ii) sky noise $\propto A(t)$, iii) bias noise $\propto \delta I_{bias}(t)$, iv) electronics gain noise $\propto \frac{\delta G_i(t)}{G}$ and, v) detector noise. Terms (ii) and (iii) are common mode and can be removed by cleaning.

In addition, to these components, we find a component of the signal that is correlated within each hextant or group of 24 channels all going through the same preamplifier boards, cryogenics cables, JFET boards, etc., but not correlated from one hextant to another. This component of noise is most likely from gain variations in the preamplifier boards due to temperature variations. To remove it, we make an average of the signals within each hextant and remove that from the individual bolometers within the hextant instead of removing the average over the entire array. The resulting cleaned data has about a factor of 1.5 less noise at low frequencies. Finally, we also either remove a 2nd order polynomial from each scan or reapply a high pass filter to the time ordered data in order to reduce the effect of remaining 1/f noise in the time stream.

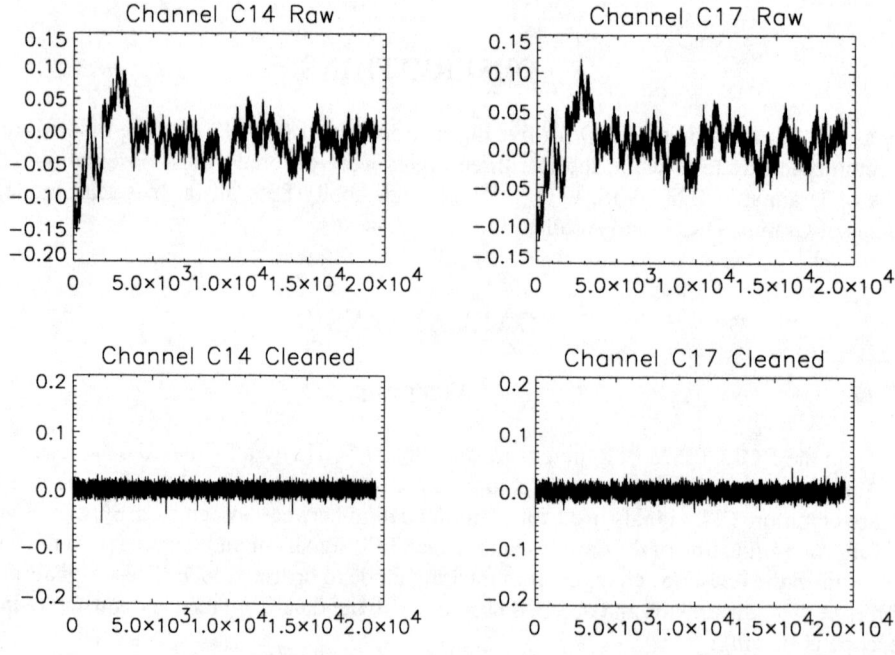

FIGURE 1. BOLOCAM data before and after cleaning

In observations where there is a bright source present, the removal of a polynomial or the application of a high pass filter creates shadows, or regions of "negative flux" around the source. In order to avoid this, we allow the option of running a source removal module that masks out all data within 2 beam FWHM around specified sources in the subsequent cleaning and low frequency noise removal modules. This eliminates the shoadowing effect.

Mapping

After the data has been cleaned, we make binned maps on the sky from the time stream, using the pointing data from the telescope and the data from the rotator. We first calibrate the raw time stream of each detector using the individual detector responsivities measured in observations of Uranus. We then average data from all of the detectors in each bin, weighted by the noise integrated in a band from 100-200 mHz. We compute a weighted mean and variance from the data in each bin.

RESULTS

Figure 2 shows a map of one observation of the planet Uranus. The map is made from a total of 114000 samples from each of 62 bolometers, corresponding to a total observation time of 2280 seconds. Of this total observing time, a total of 66373 samples from each bolometer are actually used in the map, an observing efficiency of 58%. The map is pixellized in 5" × 5" pixels and covers a total area of 200 sq. arcmins. The map was made in a raster scan mode, the telescope stepping in declination between each drift scan with a step size of 5". The rotator was running during the observation to maintain the same array rotation with respect to parallactic angle at all times.

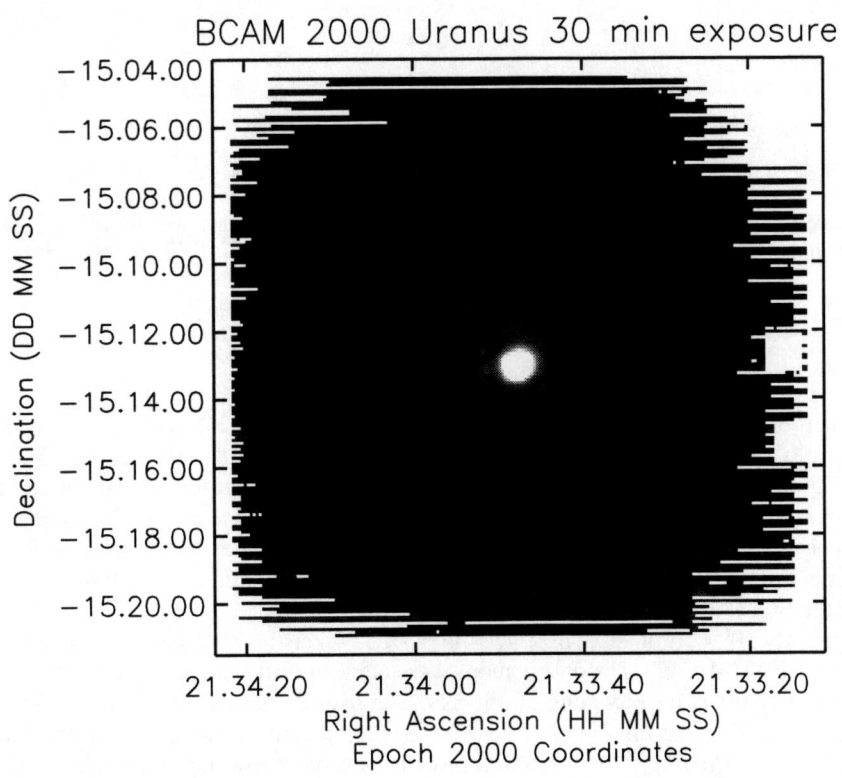

FIGURE 2. Map from one 30 minute observation of Uranus

From the map we derive the following average beam parameters: solid angle= 0.43 sq. arcmins, beam FWHM= 38" in agreement with the expected values from the optical design. The flux from Uranus in the BOLOCAM 1.4 mm band is approximately 30 Jy [12]. There is an Airy ring clearly visible in the map and in the radial profile plot (Figure 3) with an amplitude equal to 1% of the peak amplitude. In addition, there is a negative going feature at a radius of about 5σ from the beam center with an amplitude of $< 0.3\%$ of the peak. The error per 5"x5" pixel in the center of the map is approximately 80 mJy, corresponding to an error of 12 mJy/beam, and a signal to noise ratio in the central 5" × 5" pixel of almost 500.

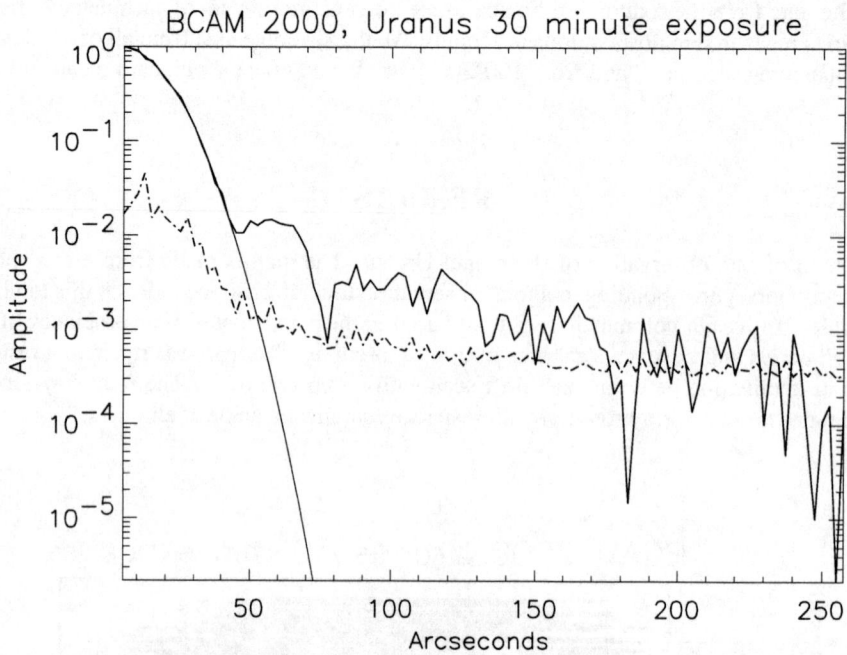

FIGURE 3. Plot of radially averaged flux from Uranus map. The solid line is the raw data overlaid with a Gaussian profile with FWHM=38". The dashed line is the error measured from the variance of the pixels in the radially averaged bin.

A1835

Figure 4 shows a map of the region around the cluster Abell 1835 (z=0.2523). This cluster is one of the most luminous X-ray emitting clusters in the sky with a total X-ray luminosity of 3.8×10^{45} erg/s [15], a core radius of $\theta_{core} = 0.22'$ for a King model with $\beta = 0.66$. It is one of the most studied clusters of galaxies in the wavelength region from the radio through the submillimeter. It has a strong SZ thermal signal with a measured peak y-parameter of 4.6×10^{-4} [14] and also has a strong central radio source with a flux of 3 mJy at 30 GHz [16] and a central sub-mm dust source with a central flux of 20 mJy at 450 um [13]. We integrated on this region for a total of 3 hours during the last night (May 22) of the BOLOCAM engineering run.

The map is approximately $10' \times 16'$ in size, pixellized in 5" × 5" pixels and has been smoothed with a Gaussian smoothing function with a FWHM= 40". The observations were made in drift scan mode with three declination steps of 5" each in order to evenly fill in the sampling an the sky. The map is made from a total of 564000 samples of which 339685 are actually used in the final map or 60% observing efficiency. The error per pixel in the smoothed map is 5 mJy= $100\mu K_{R-J} = 300\mu K_{CMB}$. We perform a fit of the unsmoothed map to the cluster profile smoothed to the BOLOCAM beam size and find a most likely value of the peak amplitude of the SZ signal at 1.4 mm of $-170 \pm 340 \mu K_{CMB}$, which corresponds to a flux in a BOLOCAM beam centered on the cluster of -1.5 ± 3 mJy.

SENSITIVITY

We calculate our sensitivity in several different ways. We calculate the sensitivity of each detector to sky signal as a function of frequency.

$$S(\text{mJy}/\sqrt{\text{Hz}}) = \text{PSD}(V/\sqrt{\text{Hz}}) \text{Jy}_{\text{Uranus}}/V_{\text{Uranus}} \qquad (2)$$

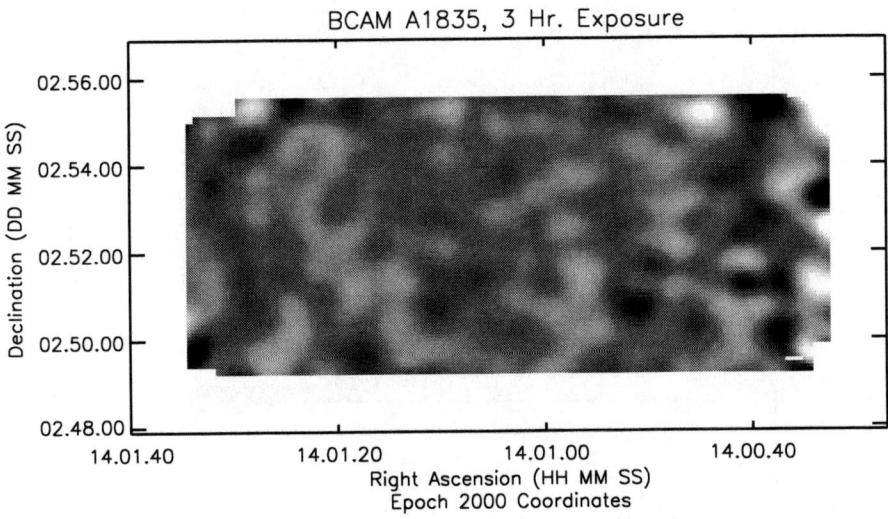

FIGURE 4. Map of the region of sky near the cluster Abell 1835

Where the PSDs are calculated from the cleaned data and calibrated by the peak signal from each bolometer on observations of Uranus. Figure 5 shows a plot of the sensitivity of a good bolometer vs. frequency. As described earlier, there is excess noise from the prototype electronics at low frequencies that remains after cleaning and therefore is not correlated from bolometer to bolometer. The white noise of the detector is about 50 mJy/$\sqrt{Hz}$ while the noise at the effective frequency of the drift scan observations is 100-200 mJy/$\sqrt{Hz}$. This is a factor of 4-8 times worse than the optimum sensitivity for a single pixel. The factors contributing to this loss in sensitivity are:

1. Optical Efficiency of prototype wafer: 2-3
2. Excess 1/f noise in electronics: 2-4

We also calculate the overall sensitivity of the array from the noise in the maps in terms of a mapping speed. Mapping speed is defined as the solid angle that it is possible to map down to a given noise level in a given amount of time (ref. Glenn). This is given by:

$$M(\text{arcmin}^2/\text{mJy}^2/\text{Hr}) = 3600(\text{sec}/\text{Hr})N_{det}\varepsilon_{int}\Omega_{beam}(\text{arcmin}^2)/S_{det}^2(\text{mJy},\sqrt{\text{sec}}) \quad (3)$$

where ε_{int} is the efficiency of integration time. For the Abell 1835 observation, we find an effective mapping speed of the array of:

$$M = 0.7 \text{arcmin}^2/\text{mJy}^2/\text{Hr} \quad (4)$$

CHANGES

Since the May, 2000 observing run, we have made several changes to the instrument. First, we have fabricated a new wafer which has over 100 good bolometers for which have measured optical efficiencies about 2.5 times higher than the detectors in the prototype array. Second, we have modified the readout electronics and eliminated the excess 1/f noise so that the readouts are now consistent with white noise down to frequencies < 30 mHz. Finally, we have changed the filters and horn array to observe at a wavelength of 2.1 mm rather than 1.4 mm. Currently we are in the final stages of preparation for a second engineering run at the CSO in December, 2001. Taking these improvements into account, we expect to achieve a mapping speed of:

$$M \simeq 100 \text{arcmin}^2/\text{mJy}^2/\text{Hr} \quad (5)$$

or

$$M \simeq 100 \text{arcmin}^2/(35\mu K)^2/\text{Hr} \quad (6)$$

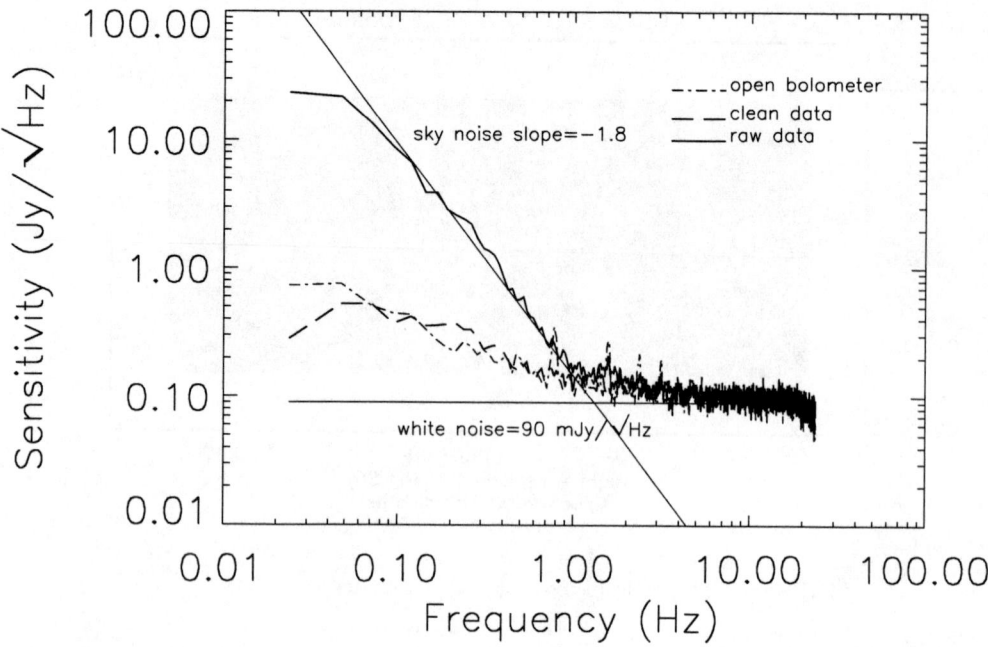

FIGURE 5. PSDs of BOLOCAM channel before and after cleaning, with a PSD of an open channel showing only electronics noise

including observing overheads during the upcoming run. With this mapping speed, we can map over one square degree to an error of less than 20 μK per beam in 50 hours of observation and to 12μK per beam in 100 hours.

SIMULATIONS

We have used the estimated mapping speeds above to perform simulations of observations with BOLOCAM during the observing run this December. Figure 6 shows maps of a 1° × 1° degree region of sky with and without detector noise added assuming 100 hours of integration time. The sky signal is a combination of primary and secondary CMB anisotropies. The primary anisotropy signal is generated from a Λ-CDM model that fits the current CMB power spectrum data at lower multipoles. The secondary SZ anisotropies are taken from hydrodynamical simulations for flat-Λ cosmology. The overall rms from the sky is 20μK of which the SZ signal contributes an rms of 12.5μK and the primary CMB anisotropy on angular scales from the BOLOCAM beam size of 1' to the 8' size of the array contributes an rms of 15.5μK. The noise rms in 1' pixels is 12μK. There are 8 clusters in the map with a signal to noise ratio greater than 4.

CONCLUSIONS

We have presented some preliminary results from the first run of the BOLOCAM instrument at the CSO in May, 2000. During 5 nights of observation, the instrument was successfully integrated with the telescope and observations were made of various galactic and extragalactic sources. Since this engineering run, a new wafer of bolometers for the focal plane has been produced with over 100 working detectors and the nominal optical coupling to incoming radiation, which is a factor of 2-3 improvement in coupling over the engineering wafer. In addition, the readout electronics have been upgraded to eliminate excess 1/f noise present during the engineering run. Finally, a full software pipeline has been developed and used to analyze the data taken from the engineering run. These improvements will be tested in detail during the next engineering run of BOLOCAM on the CSO this December.

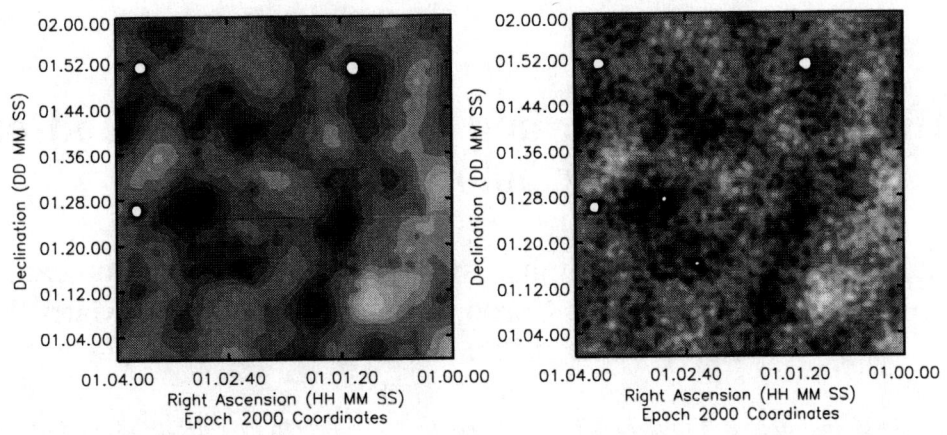

FIGURE 6. Simulation of 1° × 1° field at 150 GHz with and without simulated instrument noise

ACKNOWLEDGMENTS

This activity is being supported by the US Agencies NASA and NSF and by PPARC, and by the Leverhulme Trust in the UK.

REFERENCES

1. Holland, W., et al., 1999, MNRAS, 303, 659
2. Kreysa, E., et al., 1998, Proc. SPIE Vol. 3357, p. 319-325, Advanced Technology MMW, Radio, and Terahertz Telescopes, Thomas G. Phillips; Ed.
3. Eales, S., et al., 2000, ApJ, 120, 2244-2268
4. Bertoldi, F., 2000, A&A, 360, 92-98
5. Joy, Marshall, et al., 2001, ApJ, 551, L1-L4
6. Jones, M., et al., 2001, MNRAS (submitted), astro-ph/0103046
7. Glenn, J., et al., 1998, Proc. SPIE Vol. 3357, p. 326-334,
8. Bhatia, R., et al., 2000, Cryogenics, 40, 685-691
9. Mauskopf, P. D., et al., 1997, Appl. Opt., 36, p. 765
10. , Bock, J., et al., "Infrared Bolometers With Silicon Nitride Micromesh Absorbers", Proceedings of "Submillimetre and Far-Infrared Space Instrumentation", 30th ESLAB Symposium 24-26 September 1996, ESTEC, Noordwijk, Netherlands.
11. Bock, J., et al., 1998, Proc. SPIE Vol. 3357, p. 297-304, Advanced Technology MMW, Radio, and Terahertz Telescopes, Thomas G. Phillips; Ed.
12. Griffin, M. J. and Orton, G. S., 1993, Icarus, 105, 537
13. Edge A., et al., 1999,MNRAS, 306, 599-606.
14. Mauskopf P., et al., 2000, Ap.J., 538, 506-516.
15. Allen, S., et al. 1996, MNRAS, 283, 263-281.
16. Cooray A., 1999, New A., 4, 377-388. Advanced Technology MMW, Radio, and Terahertz Telescopes, Thomas G. Phillips; Ed.

The Diabolo photometer and the future of ground-based millimetric bolometer devices

Désert, F.-X. *, Benoît, A.[†], Camus, Ph.[†], Giard, M.[**], Pointecouteau, E.[**], Aghanim, N.[‡], Bernard, J.-P.[‡], Coron, N.[‡], Lamarre, J.-M.[‡], Marty, Ph.[‡], Delabrouille, J.[§] and Soglasnova, V.[¶]

Laboratoire d'Astrophysique de l'Observatoire de Grenoble, 414 rue de la Piscine, BP53, F–38041 Grenoble Cedex 9, France
[†]*Centre de Recherche sur les Très Basses Températures, 25 Avenue des Martyrs BP166, F–38042 Grenoble Cedex 9, France*
[**]*Centre d'Étude Spatiale des Rayonnements, 9 avenue du Colonel Roche, BP 4346, F–31029 Toulouse Cedex France*
[‡]*Institut d'Astrophysique Spatiale, Bât. 121, Université Paris XI, F–91405 Orsay Cedex, France*
[§]*Physique Corpusculaire et Cosmologie, College de France, 11 pl. Marcelin Berthelot, F-75231 Paris Cedex 5*
[¶]*Space Research Institute, Astrospace Center, Academy of Science of Russia, Profsoyuznaja St. 84/32 117810 Moscow, Russia*

Abstract. The millimetric atmospheric windows at 1 and 2 mm are interesting targets for cosmological studies. Two broad areas appear leading this field: 1) the search for high redshift star-forming galaxies and 2) the measurement of Sunyaev–Zel'dovich (SZ) effect in clusters of galaxies at all redshifts. The Diabolo photometer is a dual-channel photometer working at 1.2 and 2.1 mm and dedicated to high angular resolution measurements of the Sunyaev–Zel'dovich effect towards distant clusters. It uses 2 by 3 bolometers cooled down to 0.1 K with a compact open dilution cryostat. The high resolution is provided by the IRAM 30m telescope. The result of several Winter campaigns are reported here, including the first millimetric map of the SZ effect that was obtained by Pointecouteau et al. (2001) [13] on RXJ1347-1145, the non-detection of a millimetric counterpart to the radio decrement towards PC1643+4631 and 2 mm number count upper limits. We discuss limitations in ground-based single-dish millimetre observations, namely sky noise and the number of detectors. We advocate the use of fully sampled arrays of (100 to 1000) bolometers as a big step forward in the millimetre continuum science. Efforts in France are briefly mentionned.

INTRODUCTION

The atmospheric windows at 1 and 2 mm wavelengths constitute a large opening for ground–based cosmological studies. Continuum observations on large single–dish telescopes (IRAM 30 m, JCMT, CSO, SEST, ...) have already provided outstanding results in that respect. Whereas the search for high redshift galaxies has proved very successful in the near past mostly at 1.2 and 0.8 mm (this conference), we would like here to also emphasize the usufulness of millimetre SZ measurements in the 2.1 mm window by showing the results that have been achieved with the Diabolo instrument[1]. The IRAM 30 m millimeter telescope at Pico Veleta (Spain) provides the highest angular resolution on SZ effect with the combination of the size of the telescope and the operating wavelength, namely about 20 arcsecond at 2 mm. This can be very important for the study of high redshift clusters of galaxies which may not be fully virialized.

One can note that both (sub)millimetric flux of galaxies and the SZ effect brightness (although not for the same reason) share the property of being rather insensitive to their redshift. Hence, number counts can be much more

---

[1] More details on the experiment can be found at http://www-laog.obs.ujf-grenoble.fr/desert/diabolo/diabolo.html

sensitive to the luminosity function than to distance effect. The high redshift population of objects can stick out more easily than in other wavelength domains. In particular, when confusion is close, this can be a very important positive leverage to extract the early population from the low redshift crowd.

THE DIABOLO INSTRUMENT

FIGURE 1. One of the 2 arrays of 3 bolometers used in Diabolo. The Winston cones are arranged in a close-packed triangular configuration.

FIGURE 2. Diabolo cryostat in the Nasmisth cabin of the 30m. One can see shock absorbers (black springs) around the cryostat to damp vibrations coming from cryocoolers in the same cabin. The electronics box is on the upper right.

Diabolo is a dual-channel photometer with 0.1 K bolometers cooled by a space-compatible dilution fridge of the same type as what will be flown on Planck-HFI (Lamarre et al., this conference). It is described in length by Benoît et al. (2001) [2]. The AC square bias electronics to read the bolometers is described by Gaertner et al. (1997) [6].

It now contains two small arrays of 3 bolometers each of which has a Winston cone at its entrance aperture (Fig. 1). With a beam splitter, both arrays simultaneously measure the sky brightness resp. at 1.2 mm and at 2.1 mm. This is essential to spectrally separate sky noise from the SZ effect (see below). The FWHM of the beam is 22 arcseconds when the photometer is installed on the IRAM 30m telescope at Pico Veleta (Spain). Fig. 2 shows the cryostat at the Nasmith focus.

In 1995 and 1996, we performed ON-OFF (target) along with ON-OFF (blank-sky) on selected clusters of galaxy (Désert et al. 1998 [3]) to make first detections and to check for systematics. Since then, we have done small raster maps where the telescope is held fixed in local coordinates and the Earth rotation makes a drift subscan at constant declination. A map is made by repeating those subscans at different declination.

The sensitivity is below 1×10^{-4} (about 1 mJy/beam) for the comptonisation parameter y at 1σ in one hour of integration and after sky noise is subtracted (see below).

SCIENTIFIC RESULTS

The SZ effect

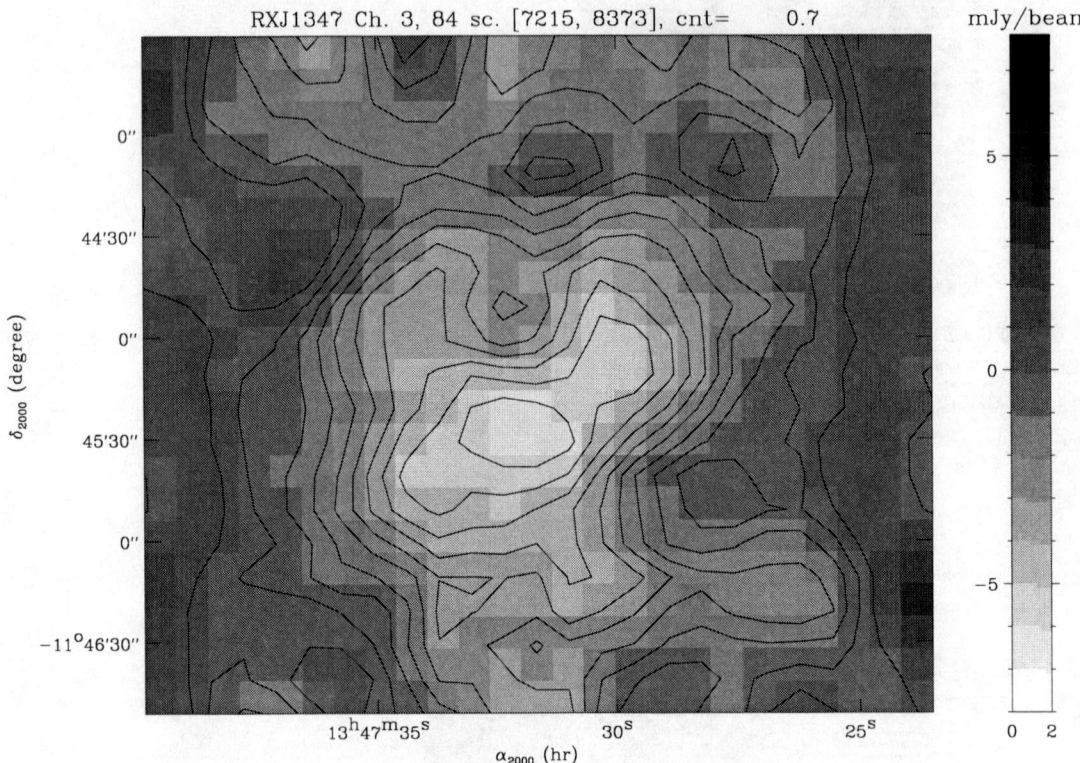

FIGURE 3. 2.1 mm map obtained with the Diabolo instrument in 1999 of the cluster of galaxies RXJ1347-1145. It corresponds to the coaddition of 84 independant rasters. The grey scale is from white (negative brightness) to black (positive brightness). Contours are in units of 0.7 mJy per beam (1σ level) from -9 to 0. Pixel size is 10 arcseconds. A smoothing by 3 pixels was applied.

The SZ effect is clearly detected in one of the most X-ray luminous clusters, RXJ1347-1145, at a redshift of 0.45. The map shown in Fig. 3 is obtained after coadding 16 hours of rasters taken in January 1999. First results were described by Pointecouteau et al. (1999) [12] and these observations are analysed at length by Pointecouteau et al. (2001) [13]. A new mapping algorithm is used here in order to deal with the effect of wobbling (Marty et al. 2001) [11]. A projected gas mass can be almost directly deduced from these observations $1.1 \pm 0.1 \times 10^{14} M_\odot$ within an angular radius of $\theta = 74''$ in agreement with X-ray expected gas mass. The SZ effect is the strongest ever detected ($y = 7 \times 10^{-4}$). This is accomplished with a high signal to noise (about 20).

Other clusters have been mapped with the same experiment and will be reported by Marty et al. (2001) [11].

Dark clusters

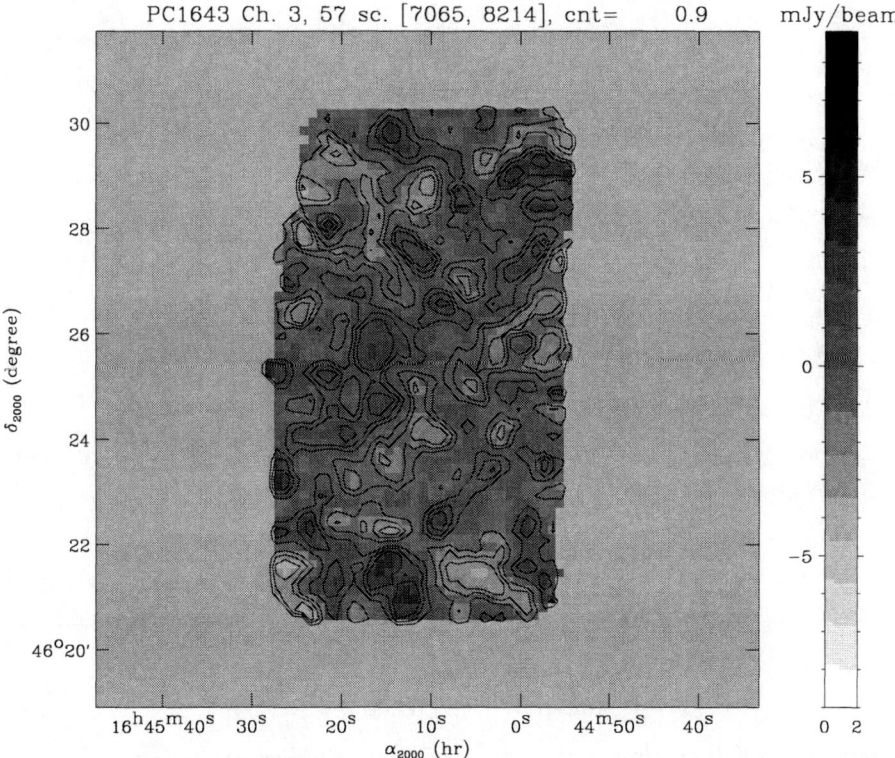

FIGURE 4. 2.1 mm Diabolo map in the field of PC1643+4631. The grey scale is from white (negative brightness) to black (positive brightness). Contours are in units of 0.9 mJy per beam (1σ level for the applied 30 arcsecond smoothing). The center of the Ryle decrement is at 16h45m11.2+46d24'56"

A strong SZ effect has been detected with the radio Ryle interferometer at a position near the pair of quasars PC1643+4631 by Jones et al. (1997) [10]. This brightness decrement observed at 15 GHz could not be confirmed by other experiments like BIMA at 28.5 GHz [9] or SuZie at 2 mm [7]. Here we wanted to map a sufficiently large map so as not to miss any decrement that could have been mispositioned, especially in declination, and have high resolution as well so as not to miss any relatively compact source. About 18 hours were spent in January 1999 providing our deepest field ever observed at 2.1 mm. 57 maps of 1200 seconds each can be coadded in order to have a typical sensitivity of 1.5 mJy (1σ) for each of the 20 arcsecond pixel making up the final 4 by 9 arcmin map. Fig. 4 shows this final map. No strong SZ effect is detected in this map. To set a preliminary upper limit, we have computed the integrated flux inside varying radii for a given central position. Fig. 5 shows that the absolute flux is never larger than 15 mJy in the 1 to 2 arcminute radius range (the optimum range for our 2.5 arcmin wobbling amplitude), whatever the center declination is chosen. We can safely exclude sources with an absolute flux larger than 20 mJy. The SZ effect expected with the minimum parameters advocated by Jones et al. [10] (core radius of 1 arcmin, $\Delta T/T = 2 \times 10^{-4}$) is -35 mJy at 2.1 mm, which is not observed. If the Ryle decrement were due to a kinetic SZ effect, as would arise, for example, from a bubble of matter ionized by early quasars (Aghanim et al. [1]), then our spectral leverage implies a flux twice larger (-70 mJy) which is clearly not observed. Analysis of the PC1643 field in terms of CMB anisotropies should also be reported soon.

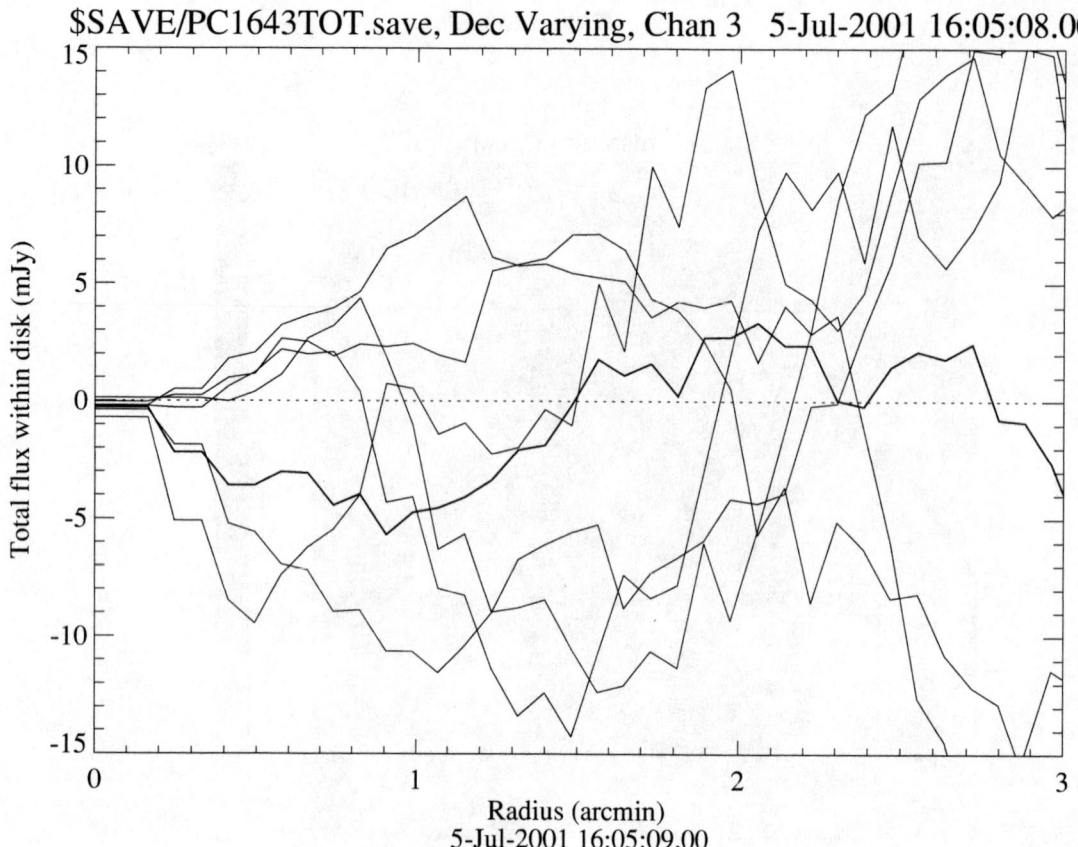

FIGURE 5. Integrated 2.1 mm flux radial profile obtained from the previous map. Each line corresponds to a central position with a constant right ascension but a declination varying by steps of 1 arcminute across the map.

2 mm source counts

From the previous deep survey, we also analysed the map for the presence of point sources. No sources could be detected at the level of 6 mJy in an area of 34 arcmin^2. A 2σ upper limit on integral number counts is thus 400 deg^{-2} at the flux limit of 6 mJy. This limit, which is above 850 μm and 1.2 mm counts, could be improved with a larger observing time and/or next generation of bolometer instrument (see below). These constraints may prove very useful for the knowledge of high redshift galaxies.

Sky noise

One of the main limitations in ground-based (sub)millimeter observations is sky noise. This is due to inhomogeneous water vapour layers travelling above the telescope. This fluctuating emission produdes a spatially and temporally variable noise degrading the performance of millimetre continuum measurements. Two methods are used to counteract this noise. The spatial method is used when one wants to observe point sources with an array of bolometers, thereby one subtracts the average signal from neighbouring pixels (Kreysa, this conference). For extended sources this is insufficient. With a dual channel instrument like Diabolo we were able to perform the spectral method, whereby the SZ signal having a spectrum very different from the water vapour emission, a simultaneous measurement at 2 wavelengths (namely 1.2 and 2.1 mm) allows one to subtract the sky noise induced map at all spatial scales. Fig 6 illustrates this method. In this case, the gain in signal to noise has proven not to be dramatic (30 to 50%). However the statistics of the signal is much improved in that the remaining (hopefully) detector noise is closer to Gaussian, hence improving the quality of SZ detections (*e.g.* [3]).

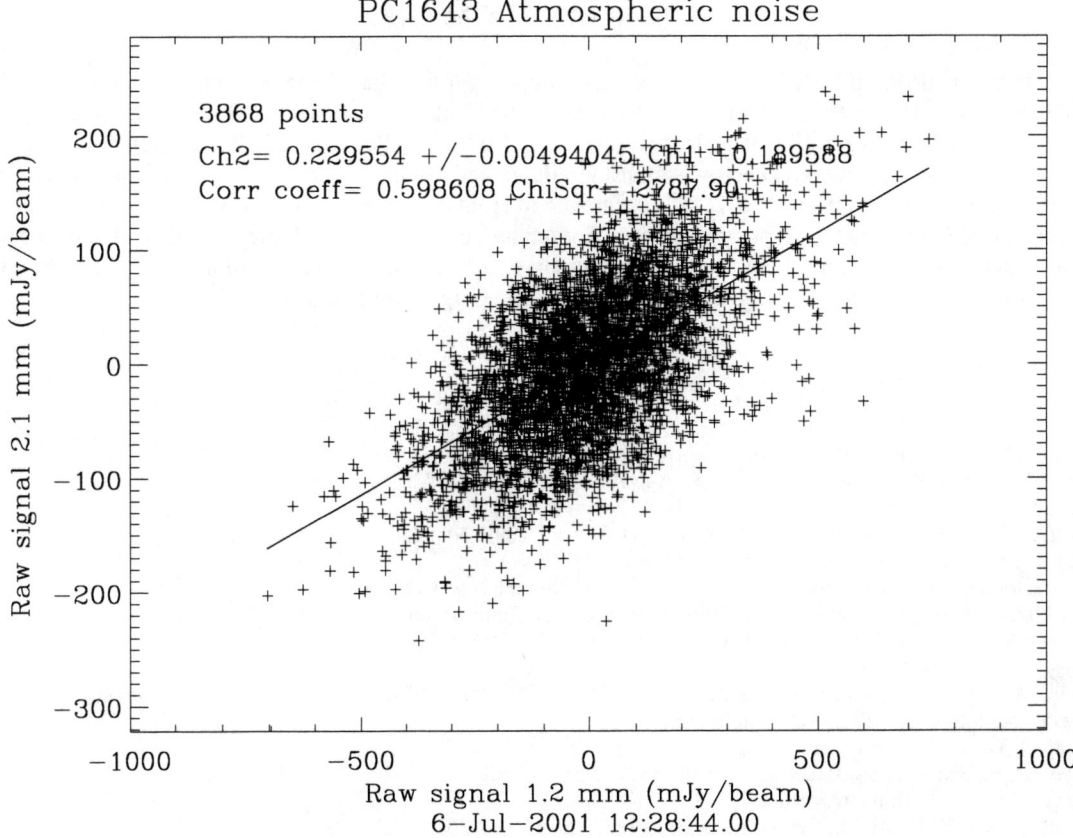

FIGURE 6. Illustration of millimetre sky noise. Raw signal correlation between two bolometers during one scan made on the "empty" field PC1643. The signal at 1.2 mm (containing almost no SZ effect) provides a template on which the 2.1 mm (SZ) signal can be decorrelated. The slope is compatible with the spectrum expected from water vapour emission.

THE STEP FORWARD IN MILLIMETER CONTINUUM ASTROPHYSICS

The previous results have shown examples of the importance of the large surveys at 1.2 but also 2.1 mm wavelengths with high resolution. The 4 most important (extragalactic) reasons at 2.1 mm are the study of the SZ effect in conjontion with X-ray observatory data, the study of secondary anisotropies (the CMB is flat on these small angular scales), the detection of primordial galaxies, and the mapping of external galaxies. To achieve that goal, this conference has seen many projects and realisations of cameras with many bolometers packed together with individual horns. On the other hand, we wish to advocate the use of filled arrays of bolometers fully sampling the available focal plane of large millimetre dishes (see details in [4]). Indeed, there are three reasons for this new design to be competitive: the sky noise may be better handled (no instantenous holes in the observed field of view), the confusion noise, nearly reached even with large telescopes, can be better tackled and the efficiency of photon gathering is optimised. The challenge is clearly the optical behaviour of such new cameras (the background is a million time larger than the objects to be detected), adapting a multiwavelength operation (should we keep dichroic or use wavelength sensitive piled up arrays?), multiplexing arrays of several thousand pixels. Such cameras can be the workhorse for the ground-based follow-up of large surveys made by the next generation space instruments like SIRTF, Herschel and Planck. Developments in France follow from the CEA/Leti design of a submillimeter camera for Herschel (initially for SPIRE [8] and now for PACS) and other advance in NbSi technology [5]. Prototypes are currently being built and tested to qualify these new designs.

ACKNOWLEDGMENTS

We thank INSU, MESR, and PNC for their continued support for the Diabolo experiment as well as IRAM for the observing logistics. The instrument could not have been made operational at the 30 m IRAM telescope without extensive testing on smaller millimeter telescopes, namely MITO (de Petris et al., this conference) and POM2, a 2.5 m dish on Plateau de Bure operated by Bernard Fouilleux and Gilles Duvert (LAOG). We thank Marco de Petris and co-organisers for this very lively 2K1BC conference in front of the wonderful Cervinio. Finally, we pay tribute to the memory of Guy Serra who was a pionneer in French and European submillimetre study of the diffuse galactic emission. Although he was very much involved in balloon experiments, he managed to help us in the alignment (by -15 deg.C) of the Diabolo instrument in the early days at Testa Grigia observatory (1994).

REFERENCES

1. Aghanim, N., Désert, F. X., Puget, J.-L., Gispert, R., 1996, A & A, 311, 1
2. Benoît, A., Zagury, F., Coron, N., et al., 2000, A & A Suppl. Ser., 141, 523
3. Désert, F.-X., Benoît, A., Gaertner, S., et al., 1998, NewA 3, 655
4. Désert, F.-X., & Benoît, A., 1999, IRAM Newsletter, January 1999, Nr. 38, http://iram.fr/ARN/jan99/jan99.html, "The case for a bolometric millimetre camera at the IRAM 30m telescope", astro-ph/9901414
5. Dumoulin, L., Berge, L., Lesueur, et al., 1993, Journal of Low Temperature Physics, 93, 301
6. Gaertner, S., Benoît, A., Lamarre, J.-M., et al., 1997, A&A Suppl. Ser. 126, 151
7. Ganga, K., et al., 2000, Astro. Lett. & Comm., 37, 303
8. Griffin, M. J., Vigroux, L. G., Swinyard, B. M., 1998, Proc. SPIE Vol. 3357, p. 404-413, Advanced Technology MMW, Radio, and Terahertz Telescopes, Thomas G. Phillips (Ed.)
9. Holzapfel, W.L., Carlstrom, J.E., Grego, L., et al., 2000, ApJ, 539, 67
10. Jones, M. E., Saunders, R., Baker, J. C., et al., 1997, ApJ, 479, L1
11. Marty, P., et al., 2001, in preparation
12. Pointecouteau, E., Giard, M., Benoît, A., et al., 1999, ApJ, 519, L115
13. Pointecouteau, E., Giard, M., Benoît, A., et al., 2001, ApJ, 552, 42
14. Schindler, S., Hattori, M., Neumann, D. M., & Boehringer, H., 1997, A&A, 317, 646

MASTER: Millimetre And Sub-millimetre Triple hEterodyne Receiver

E. S. Battistelli*, G. Boella*, F. Cavaliere*, M. Gervasi*, A. Passerini*, G. Sironi*, M. Zannoni*, D. Andreone[†], L. Brunetti[†], V. Lacquaniti[†], S. Maggi[†], R. Steni[†], E. Natale**, G. Tofani**, E. Bava[‡], U. Pisani[§], J. R. Thorpe[¶] and M. De Petris[||]

*Università degli Studi di Milano Bicocca - Milan - ITALY
[†]Istituto Elettrotecnico Nazionale Galileo Ferraris - Turin - ITALY
**Osservatorio Astrofisico di Arcetri - Florence - ITALY
[‡]Department of Electronic - Politecnico - Milan - ITALY
[§]Department of Electronic - Politecnico - Turin - ITALY
[¶]Department of Electronic and Electrical Engineering - University of Leeds - Leeds - UK
[||]Department of Physics - University La Sapienza - Rome - ITALY

Abstract. We will describe a system of three heterodyne receivers. Its mixers are based on SIS tunnel junctions which allow a very sensible downconversion of the detectable signal from 94 GHz, 225 GHz and 345 GHz to 1.5 GHz. This will allow the detection of molecular rotational transition lines from diffuse molecular clouds. Current status of a 94 GHz receiver, prototype of MASTER, will also be described. Technical design and cryogenic problems solution will be shown and we will focus our attention on the optical coupling technique based on gaussian beam analysis.

THE RECEIVER

Master is a system of three heterodyne receivers working at 94GHz, 225GHz and 345GHz. Cryogenic mixers are realized using both waveguide and quasi-optical technology and are based on Nb/AlOx/Nb SIS (Superconductor/Insulator/Superconductor) tunnel junction. The high non linearity of their characteristic (I vs V) (see Figure 1) is used in order to efficiently downconvert the frequency of the signal we want to detect, into an intermediate frequency (IF) of 1.5GHz [4].

We are already assembling a 94GHz waveguide radiometer, prototype of Master and the main characteristics of the

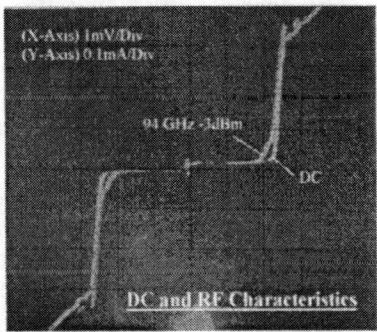

FIGURE 1. I vs V characteristics

triple receiver are being defined.

Two Gunn Oscillators and two multipliers are used to produce the three Local Oscillator signals and tuning is achieved with aid of mechanical backshort. A first amplification stadium is achieved at 4K using HEMT amplifiers.

FIGURE 2. 94GHz waveguide receiver

TABLE 1. MASTER's characteristics

	94GHz	225GHz	345GHz
IF	1.5GHz	1.5GHz	1.5GHz
Inst. bandwidth	600MHz	600MHz	600MHz
DSB Noise Temp.	100/130K	120/150K	140/170K
Total Bandwidth	10GHz	20GHz	35GHz
V_{bias}	2.1/2.3mV	2.1/2.3mV	2.1/2.3mV

In the waveguide configuration, radiation from Local Oscillators and from the sky is collected by 7° corrugated feed horns geometrically scaled with respect to the wavelengths while in the open structure configuration, a hyperhemispherical quartz lens is used to couple the local-oscillator signal and radio-frequency: lens synthesis a dielectric half-space realizing broadband 70-400 GHz Open-Structure Mixers (useful for high frequency detections) [5].

Cryogenic technic are of course needed for noise considerations but also in order to reach temperatures below Niobium critical temperature. The 94GHz radiometer is housed in a standard Nitrogen/Helium dewar while a hybrid cryocooler/liquid Helium dewar will be used for Master. Calibrations are made using Eccosorb AN-72 at 77K and at room temperature. The problem of eliminating the contribution of the Cooper pairs has been solved by applying a magnetic field around the Mixers.

Every single receiver of Master will be able to take measurements within a total bandwidth of approximately 10% of the central frequency and, through changing the LO frequency, it will be able to give always the same IF. A bias is needed in order to make the junctions work where the non-linearity is stronger (on the knee of the characteristic). A bandpass filter determines the instantaneous bandwidth.

The estimated MASTER's characteristics are reported in table 1 [1].

ASTOPHYSICAL OBSERVATIONS

The main scientific goal of MASTER is the observation of lines associated to molecular rotational transitions and atomic hyperfine transitions. This allow the investigation of star formation processes, for instance through the detection of emission lines of the CO molecule, used as a track of H_2. Another interesting topic would be the search of high redshift lines whose intensity is determined by the CBR, in order to look for a link between T_{CBR} and redshift z:

$$T_{CBR}(z) = T_{CBR}(0)(1+z) \tag{1}$$

Even if the instantaneous bandwidth is narrow we could think to attempt continuous measurements. For instance distortions of the CMBR spectrum due to inverse Compton scattering when the CBR is scattered by a hot electron gas in a cluster of galaxy could be studied (S-Z effect) [6].

OBSERVATIONAL SITES AND OPTICAL COUPLING

Astrophysical observations at millimetre and sub-millimetre wavelengths need dry high altitude sites because water vapour is the dominant source of opacity. Master's frequencies have been chosen considering the atmospheric transmission: the higher and the lower ones take into account the atmospheric windows at these frequencies while the middle one will be used in order to study atmospheric contributions. Observations with Master will be taken from MITO telescope at Testagrigia observatory on Italian Alps (3480m a.s.l.)[3]. Hopefully observations will follow from Dome C site on Antarctic Plateau depending on the reflectors availability.

Optical coupling between receivers and telescopes is one of the key point for sensitive detection and clear understanding of astronomical signals. This problem has been studied through Gaussian Beam analysis which gives wavelength-depending corrections to geometrical optics and introduces diffractional effects. Using this technique has been possible to find a frequency independent configuration for a perfect coupling between the 94GHz radiometer and MITO telescope [2]. This can be done by refocusing on the receiver the radiation from the telescope by using two ellipsoidal mirrors (preferred comparing to lenses because are less absorbing). Analogue configurations are possible in coupling Master with any other telescopes.

ACKNOWLEDGMENTS

This activity is being supported by the Italian Agencies MURST, CNR, PNRA and CSNA.

REFERENCES

1. E. S. Battistelli et al., MASTER: Three heterodyne receiver for millimeter and sub-millimeter wave astronomy, Proc. Of the IX Convegno GIFCO, Lecce May 2000, vol.68, S. Aiello and A. Blanco Eds., p. 245 (2000)
2. Ta-Shing Chu, An imaging beam wave guide feed, IEEE transition and propagation, Vol. AP-31 n4 p. 614-619,(1983)
3. M. De Petris et al., MITO: the 2.6-m millimeter telescope at Testagrigia; New Astronomy, Vol.1, p. 121-132 (1996)
4. V. Lacquaniti, G. Sironi et al., Development of SIS Junction for Astrophysical observation, Proc. 30th ESLAB Symp., Submillimetre and Far-Infrared Space Instrument, 24-26 September 1996, ESTEC, Noordwijk, The Netherlands, ESA SP-338 (1996)
5. J.R. Thorpe et al., Highly sensitive heterodyne receivers for submillimeter-wave astronomy, Proc. of Work-shop on Astronomy and Astrophysics at submillimeter wavelenght, Rome Dec. 1999, vol.66, M. Candidi et al. Editors, p.113 (1999)
6. M. Zannoni et al., MASTER: a triple heterodyne receiver for astronomy in the millimeter and sub-millimeter domain, Proc. of the Third International Workshop on Astrophysics at Dome C 28 - 29 June 2001 Hobart, Tasmania (Australia)

MBI: Millimetre-Wave Bolometric Interferometer

S. Ali [1], P. Rossinot [1], L. Piccirillo [1], W. K. Gear [1], P. Mauskopf [1], P. Ade [1], V. Haynes [1], P. Timbie [2]

[1] *Dept of Physics & Astronomy, Cardiff University, Cardiff,CF243YB,UK*
[2] *Dept of Physics, University of Wisconsin, Madison, WI, USA*

Abstract. We present the design of the prototype of a millimetre-wave bolometeric interferometer (MBI). This interferometer uses two arrays bolometers as detectors. The combination of high sensitivity bolometers and interferometric imaging appears to be well suited for precision measurements in observational cosmology.

INTRODUCTION

We describe a new instrument for observations of faint astrophysical sources at millimeter and sub-millimeter wavelengths: an interferometer, which uses sensitive bolometers as detectors. An interferometer has multiple advantages over a single telescope system, such as:

- Direct measurement of the Fourier transform (visibility) of the sky brightness in 2-D
- Reduced effects of atmospheric fluctuations in ground-based observations
- High angular resolution without large, expensive single dish
- Rapid mechanical chopping not required
- Bolometers time constants can be long and
- Reduced side-lobe pickup.

We are building an instrument to exploit these advantages of interferometry to image the CMB polarization [4], image the S-Z effects in clusters of galaxies [1,3] and also to look for primordial galaxies [2] at sub-mm wavelengths.

INSTRUMENT

Two Element Interferometer

In a simple 2-element radio interferometer, signals from two telescopes aimed at the same point in the sky are multiplied (correlated) so that the sky temperature is sampled with an interference pattern with a single spatial frequency. The output of the multiplying interferometer is the visibility. To recover the full phase information, complex correlators are used to measure simultaneously both the in-phase and quadrature phase components of the visibility. In an interferometer that uses incoherent detectors, such as an optical interferometer, the electric field wavefronts from two telescopes are added and then squared in a detector (an "adding" interferometer as opposed to a "multiplying" interferometer [5]). The result is a constant term proportional to the intensity plus an interference term. The constant term is an offset that we propose to remove by modulating the length of the baseline D by a few wavelengths at a frequency of ~1 Hz. Phase-sensitive detection at this modulation frequency recovers both the in-phase and quadrature phase interference terms and reduces susceptibility to low-frequency drifts (1/f noise) in the bolometer and readout electronics. We recover the same visibility as for the multiplying interferometer.

2-Element Prototype Interferometer

The baseline of the prototype MBI is formed by two identical flat mirrors, which direct their beams to two Cassegrain telescopes. The telescopes couple the beams from the flat mirrors to a cryogenic beam combiner and detector system housed inside a 4K cryostat. The detector system includes two separate 9 pixel bolometric arrays. The detectors are spider-web bolometers cooled to ~ 280 mK by ^{3}He refrigerator.

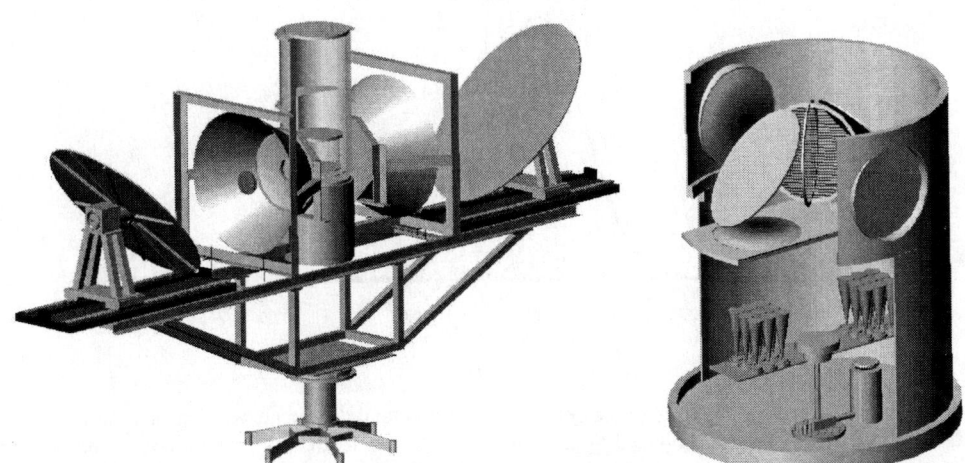

Figure 1. Prototype 2-element interferometer. Two flat mirrors form the baseline, which can be varied from 1 to 4 meters. Wave fronts reflected from these two mirrors are directed to small (0.5 m diameter) telescopes attached to a cryostat, which houses cold optics and the detectors. The figure on the right shows the details of the quasi-optical beam-combiner and the bolometric detector arrays. Two crossed polarizers, constructed from a lithograph of 0.2 micron thick copper wires on a poly-propylene substrate, form the beam-splitter.

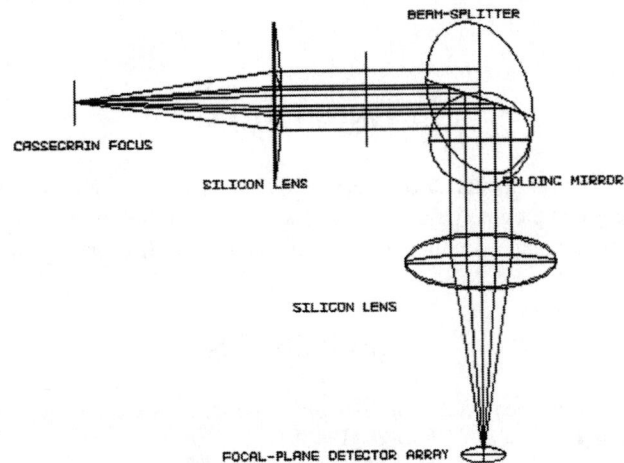

FIGURE 2. Schematic of one arm of the interferometer: the Cassegrain telescope and the beam-combiner. The incoming beam is collimated by the first silicon lens on to the orthogonal metal grid beam splitters. The vertical and horizontal beam splitters reflect and transmit the horizontally and vertically polarized component of the beam respectively, which comes from one arm of the interferometer, and mix with the corresponding component of the incoming beam from the other arm. After interference the resulting beam is directed towards the detector array by the folding mirror and another silicon lens. For clarity only the central pixel is shown.

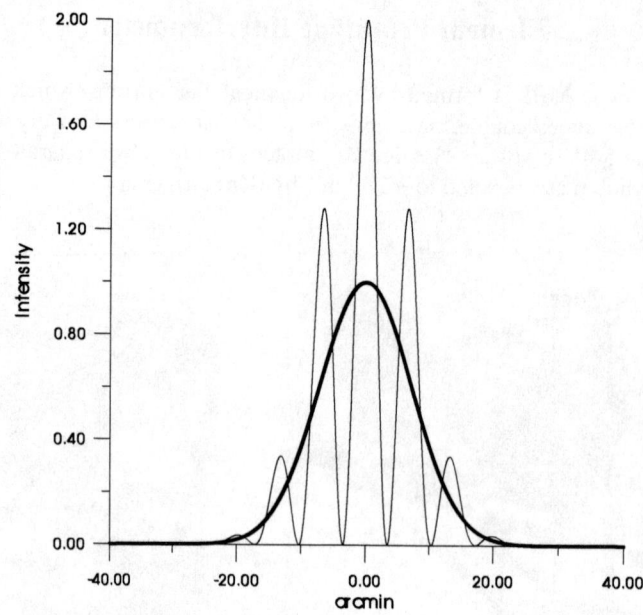

FIGURE 3. Simulated Gaussian beam response of a single 50 cm diameter antenna element working at 2 mm wavelength. Superimposed is the interferometer fringe pattern produced by a single baseline of 1 metre. By moving one of the flat mirrors back and forth +/- 3 λ (corresponding to +/- 6mm) – the whole complex visibility function can then be measured. The Fourier Transform of the complex visibility function gives a map of the sky brightness.

If the bolometers detect a single at 2 mm wavelength with a bandwidth of $\Delta\nu = 20\%$ and optical efficiency 0.3, we estimate the sensitivity of each bolometer to be NET $\approx$ 370 μK/$\sqrt{Hz}$. For MBI with two 9-bolometer arrays we expect 5 mK·$\sqrt{s}$. noise per pixel in the synthesized image. For a good site we estimate that approximately 37 days are needed to integrate an image of 324 pixels down to 3 μK noise per pixel. This performance is comparable to a filled aperture 6 m telescope.

CONCLUSION

We plan to test the prototype in Tenerife by April 2002 and later take the interferometer to better observing sites, e.g. South Pole for future campaigns. Later we propose to build a 3-element longer baseline interferometer so we can probe more points in the uv plane simultaneously and can achieve higher angular resolution.

REFERENCES

1. Birkinshaw, M. "The Sunyaev-Zel'dovich Effect," in *AIP Conf. Proc.* 476: 3K cosmology, 1999, pp 298.
2. Bond, J. R., Carr, B. J., and Hogan, C. J., *ApJ*, **306**, 428 (1986).
3. Carlstrom, J. E., Joy, M., and Grego, L., *Ap.J*, **456**, L75, (1996).
4. Hu, W. and White, M., *New Astronomy*, **2**, 323, (1997).
5. Rohlfs, K. and Wilson, T. L., **Tools of Radio Astronomy**, 2d ed., Springer, New York, 1996.

Fastscanning: a new observing technique for bolometer arrays on ground based telescopes

L.A. Reichertz, B. Weferling, W. Esch, and E. Kreysa

Max-Planck-Institut für Radioastronomie, Auf dem Hügel 69, 53121 Bonn, Germany

Abstract. The use of a wobbling secondary mirror to suppress atmospheric noise restricts the scan velocity in mapping modes and has other disadvantages. The fastscanning observing technique allows observations with bolometer arrays from ground based telescopes without the need of a wobbling secondary mirror. Therefore mapping of large sky areas can be done in a shorter time than usual which is especially useful in surveys to search for new sources. We present here the basic principle of this method.

INTRODUCTION

Noise from fluctuations of the atmospheric emission in astronomical observations from ground based mm/submm telescopes is normally removed by dual beam techniques: A common method is to use a wobbling secondary mirror which alternatively points the beam to two adjacent positions on the sky. After phase sensitive detection, spatial and temporal emission fluctuations should cancel, at least to first order. Although this method is widely used on many telescopes, there are some disadvantages: The required precisely moving mirror mechanism allows only low modulation frequencies and this limits the scan velocity in mapping modes. Furthermore, the usually fixed wobbling direction restricts the coordinate system of observing modes and any asymmetries in the mirror movements can lead to large offsets, due to different optical paths for each beam. Last but not least, a wobbling secondary mirror is not available on every telescope where continuum observations are of interest. Here we describe a new observing technique that we have tested with the MPIfR bolometer array MAMBO [1] where a wobbling secondary mirror is no longer necessary.

THE FASTSCANNING PRINCIPLE

In the fastscanning mode the secondary mirror stays fixed and the atmospheric signal is removed by taking advantage of the fact that in detector arrays all beams are receiving atmospheric emission simultaneously. The atmospheric contribution is then removed during the data reduction by correlation analysis of all the detector pixels. The scan velocity has to be relatively fast in this method for the following reason. Thermal drifts and electronic 1/f noise make AC coupled amplifiers for the bolometer readout necessary. In the absence of the wobbling mirror, the signal is not modulated. Therefore, a scanning telescope is needed to convert the spatial frequencies of the sky into the frequency band, which is defined by the combination of the AC coupling and the bolometer response function. We call this frequency band the bolometer filter function. It can be obtained from the equation of the high pass filter of the amplifier

$$A(\nu) = \frac{1}{1 + 1/(i2\pi\nu\tau_{amp})} \quad (1)$$

and the equation of the frequency dependence of the bolometer resonse $S(\nu)$. Using a simplified bolometer model (e.g. Richards [2]): with an effective thermal time constant τ_{bol} the bolometer response $S(\nu)$ follows the equation of a low pass filter

$$S(\nu) = S(0) \frac{1}{1 + i2\pi\nu\tau_{bol}}. \quad (2)$$

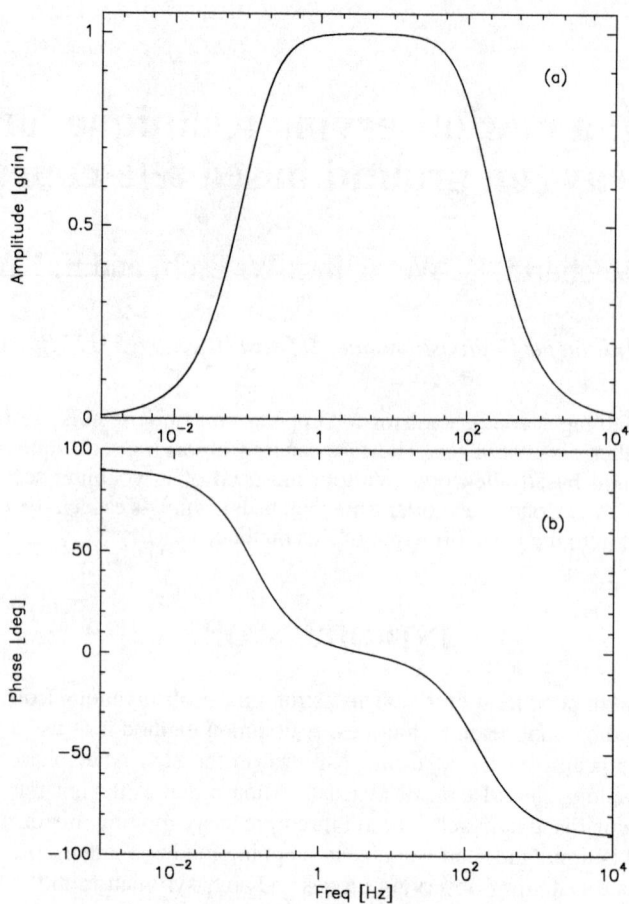

FIGURE 1. Amplitude and phase of the bolometer filter function [3].

Hence, the total bolometer filter function is given by the expression

$$G(\nu) = \frac{1}{(1 + 1/(i2\pi\nu\tau_{\text{amp}}))(1 + i2\pi\nu\tau_{\text{bol}})}. \tag{3}$$

For MAMBO the time constant of the AC coupling is $\tau_{\text{amp}} = 1.32$ sec, corresponding to a 3 dB roll-off at 0.12 Hz, and the bolometer time constant is $\tau_{\text{bol}} = 1.1$ msec. Figure 1 shows this complex function as a plot of amplitude and phase versus frequency. Any signal that is applied to the detector system is convolved with this filter function. Deconvolution of the acquired signals using this function allows the reconstruction of the original photometric signals.

FIRST TEST RESULTS

We investigated the fastscanning technique at the IRAM 30 m telescope on Pico Veleta in Spain. The bolometer signal frequencies generated by scanning a celestial point are relatively low (e.g. 99% is below 5 Hz). Signal losses due to the attenuation of the lowest frequencies, can be calculated as a function of scanning velocity, beamwidth and source extension [3]. A compromise between integration time per scan and signal loss has to be made. In our first test with the 30 m telescope at 250 GHz (beamwidth 10.5 arcsec FWHM), we chose a scanning velocity of 40 arcsec/sec, corresponding to a signal loss of about 7 %. Figure 1 shows the result of a scan on Saturn. Saturn had a 250 GHz flux of 635 Jy at that time. Such a strong source is useful to investigate the effect of the bolometer filter function and the deconvolution. In this figure the signal of the central array channel is shown. The diagram on the left shows the raw

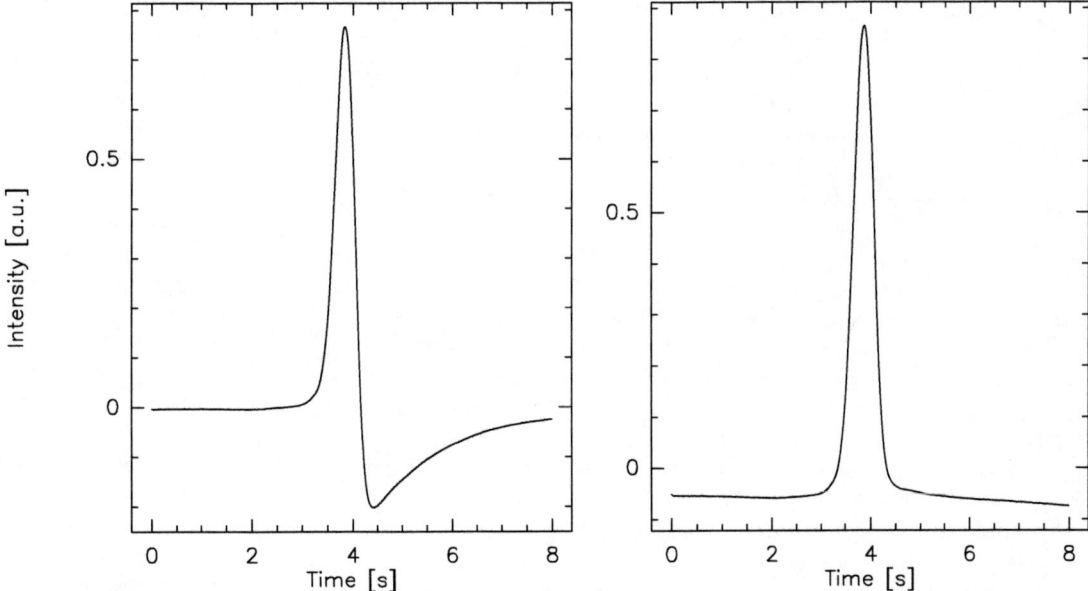

FIGURE 2. Experimental result of a scan on Saturn, scan velocity 40 arcsec/sec, left: raw signal of the center channel, right: after the deconvolution. Saturn had a 250 GHz flux of 635 Jy at that time [3].

signal without deconvolving the data by the bolometer filter function and the diagram on the right shows the result after the raw data were deconvolved. One can see that the distortion of the signal due to the filter function is removed, the original signal from the telescope beam on a point-like source is obtained as expected. In the fastscanning technique, the scanning velocity can be 10 times faster than in the wobbling mode. Maps of large sky areas can be done in a much shorter time now. We found a very high correlation of the atmospheric noise in all channels, and so this noise could be removed efficiently. More details about fastscanning can be found in [3] and [4].

CONCLUSIONS

- With arrays of bolometers it is possible to perform millimeter observations without the need of a wobbling secondary mirror, and therefore maps of large regions of the sky can be done in much shorter time than before.
- A fastscanning telescope is necessary to generate AC signals with a sufficient frequency bandwidth.
- Determining the bolometer filter function and deconvolving the data with this function allows the reconstruction of the original photometric signal.

REFERENCES

1. Kreysa, E., Gemünd, H.-P., Gromke, J., Haslam, C.G.T, Reichertz, L.A., Haller, E.E., Beeman, J.W., Hansen, V., Sievers, A., Zylka, R., *Infrared Phys. Techn.* **40**, 191 (1999).
2. Richards, P.L., *J. Appl. Phys.* **76 (1)**, 1 (1994).
3. Reichertz, L.A., Weferling, B., Esch, W., Kreysa, E., *submitted to A&A* (2001).
4. Weferling, B., Reichertz, L.A., Schmid-Burgk, J., Kreysa, E., *submitted to A&A* (2001).

2K1BC Workshop
Experimental Cosmology at millimetre wavelengths

POLARIZATION EXPERIMENTS

Polarization Observations with the Cosmic Background Imager

John K. Cartwright[1], Anthony C. S. Readhead[1], Martin C. Shepherd[1], Steve Padin[1], Timothy J. Pearson[1], Greg B. Taylor[2]

[1]*Department of Astronomy, California Institute of Technology, Pasadena, CA 91125*
[2]*National Radio Astronomy Observatory, Socorro, NM 87801-0387*

Abstract. We describe polarization observations of the CMBR with the Cosmic Background Imager, a 13 element interferometer which operates in the 26-36 GHz band from Llano de Chajnantour in northern Chile. The array consists of 90-cm Cassegrain antennas mounted on a steerable platform which can be rotated about the optical axis to facilitate polarization observations. The CBI employs single mode circularly polarized receivers which sample multipoles from $\ell\sim400$ to $\ell\sim4250$. The instrumental polarization of the CBI was calibrated with 3C279, a bright polarized point source which was monitored with the VLA.

INTRODUCTION

The Cosmic Microwave Background Radiation provides a unique means of testing many aspects of the Standard Model of the early universe. All variations agree that the CMBR is the redshifted radiation from the initial plasma, and that as such, it contains clues about the fundamental characteristics of the universe: Ω_0, Ω_b, Ω_Λ, h_0, n, τ, & T/S [1]. This information resides in the spatial fluctuations of the intensity and polarization of the CMBR. The past decade has seen the emergence of low noise detector technologies which are propelling us into an new era of precision measurements of the characteristics of the CMBR. Recent measurements of the intensity fluctuations on degree scales have provided direct evidence for a flat universe [2,3,4]. In contrast to the intensity fluctuations, polarization anisotropies are sufficiently small to have eluded detection thus far [5].

THE COSMIC BACKGROUND IMAGER

The CBI is a 13 element interferometer which operates in the 26-36 GHz band. The array consists of 90-cm Cassegrain antennas mounted on a single, fully steerable platform (fig. 1). In addition to employing standard alt-az axes, the antenna platform can rotate about the telescope boresight, and as we will see, this feature facilitates polarization observations. The platform allows a range of configurations for the telescopes, permitting observations of anisotropies on $400<\ell<4250$ scales. A downconverter splits the 26-36 GHz band into 10 channels, and the spectral information provided by these channels aids foreground rejection. An important feature of the CBI is its sensitivity; the low noise HEMT amplifiers in the receiver frontends typically have $T_n\sim25$ K, which enables detections of $\delta T\sim20\mu K$ intensity fluctuations in a single night. This sensitivity puts a detection of the polarization predicted by standard models of $\delta p\sim2\mu K$ within reach. We deployed the CBI at the Chajnantor site in northern Chile in fall of 1999, and routine observations have been underway since January 2000. The initial CBI observations demonstrated a significant decrease in C_ℓ between bins centered on $\ell\sim600$ and $\ell\sim1200$ [6]; we are augmenting this result by pushing this range to higher ℓ and improving the resolution in ℓ. The CBI web site (http://www.astro.caltech.edu/~tjp/CBI) provides details about our scientific program and the instrument.

Polarization Observations

The CBI employs single mode circularly polarized receivers. To implement a polarization detection effort in parallel with the intensity observations which constitute the CBI's primary mission, we configured 12 receivers for LCP and one receiver for RCP; the resulting array consisted of 66 intensity (LL) and 12 cross polarized (RL) baselines, all spanning $400<\ell<4250$. A single interferometer baseline measures a *visibility*, which is the Fourier transform of the intensity distribution on the sky. The cross polarized visibility RL, for example, at a point in the aperture plane (u,v) is given by,

$$RL(u,v) = \int\int A(x,y)P(x,y)e^{-2\pi i(ux+vy)}dxdy$$

Figure 1. The Cosmic Background Imager at an altitude of 5080m on the Chajnantor site in northern Chile. The 13 Cassegrain antennas have cylindrical shields which reject crosstalk between the antennas. In this picture, the array is in the initial sparse configuration; in April 2000, the antennas were reconfigured to emphasize shorter baselines. The clamshell dome, seen open at the base of the telescope, is surrounded by shipping containers which contain living spaces, a control room, and laboratory and machine shop facilities. Twin diesel generators provides power for the facility, and a cell phone with an amplified link ties the site to the project base camp in the nearby town of San Pedro de Atacama.

where $A(\mathbf{x})$ is the beam pattern on the sky and $P(\mathbf{x}) = Q(\mathbf{x}) + iU(\mathbf{x})$. Generally, we may write,

$$\begin{pmatrix} LL(\mathbf{u}) & RL(\mathbf{u}) \\ LR(\mathbf{u}) & RR(\mathbf{u}) \end{pmatrix} = \begin{pmatrix} \tilde{I}(\mathbf{u}) - \tilde{V}(\mathbf{u}) & \tilde{Q}(\mathbf{u}) + i\tilde{U}(\mathbf{u}) \\ \tilde{Q}(\mathbf{u}) - i\tilde{U}(\mathbf{u}) & \tilde{I}(\mathbf{u}) + \tilde{V}(\mathbf{u}) \end{pmatrix}$$

The CBI directly measures $LL(\mathbf{u})$ and $RL(\mathbf{u})$. For extended sources, $\tilde{Q}(\mathbf{u})$ and $\tilde{U}(\mathbf{u})$ are complex, and they cannot be obtained without both $LR(\mathbf{u})$ and $RL(\mathbf{u})$. We know, however, that $LR(\mathbf{u}) = RL^*(-\mathbf{u})$ [7]; rotation of the antenna platform can be used to switch from $(\mathbf{u})$ to $(-\mathbf{u})$. Thus, with the aid of rotation, we can obtain both $\tilde{Q}(\mathbf{u})$ and $\tilde{U}(\mathbf{u})$ from a system which employs only single mode receivers. $\tilde{Q}(\mathbf{u})$ and $\tilde{U}(\mathbf{u})$ are Fourier transforms of the spatial fluctuations $Q(\mathbf{x})$ and $U(\mathbf{x})$ on the sky, and thus provide direct measurements of the polarized power spectrum.

Polarization Calibration

We committed ~15% of each observing session to a variety of calibration observations. We selected 3C279, a bright extragalactic radio source, to serve as our primary polarization calibrator. With $I_\nu \sim 25$ Jy and $p_\nu \sim 10\% I_\nu$ at 31 GHz, and no discernible extended emission, 3C279 permits quick, simple calibrations. 3C279 is variable, however, so it was monitored throughout the polarization campaign with the VLA at 22.46 GHz and 43.34 GHz.

The plane wave incident on the interferometer can be expressed in terms of RCP and LCP components:

$$E(\mathbf{x}, \nu; t) = E_R(\mathbf{x}, \nu; t)e^{i\phi} + E_L(\mathbf{x}, \nu; t)e^{-i\phi}$$

The factor of $e^{\pm i\phi}$ reflects the fact that the baseline orientation advances or retards the phase of the circularly polarized components of the wavefront, depending on the mode. The position of the baseline in the aperture plane (u, v) determines ϕ: $\phi = \tan^{-1}[v/u]$. Because the baselines are fixed to the deck, we will regard the baseline orientation and deck position as interchangable to within an offset determined by the array geometry.

An ideal circularly polarized receiver responds to only a single mode of circular polarization. In practice, however, effects such as bandpass errors and optical irregularities contaminate a pure CP visibility of one mode with the orthogonal mode; this error is characterized by the leakage term ϵ. Consider two imperfect receivers (j, k) which combine to form a cross polarized baseline. The signals at the receiver outputs are simply voltages:

$$V_R(\mathbf{u}, \nu; t) = g_j\left[\tilde{E}_R(\mathbf{u}, \nu; t)e^{i\phi} + \epsilon_j \tilde{E}_L(\mathbf{u}, \nu; t)e^{-i\phi}\right] \qquad (j \Rightarrow RCP)$$

$$V_L(\mathbf{u}, \nu; t) = g_k\left[\tilde{E}_L(\mathbf{u}, \nu; t)e^{-i\phi} + \epsilon_k \tilde{E}_R(\mathbf{u}, \nu; t)e^{i\phi}\right] \qquad (k \Rightarrow LCP)$$

The correlator computes the visibility v_{RL}, which is the time-averaged complex product of V_R and V_L:

$$v_{RL} = g_j g_k^* \left[\langle \tilde{E}_R \tilde{E}_L^* \rangle e^{2i\phi} + \epsilon_k^* \langle \tilde{E}_R \tilde{E}_R^* \rangle + \epsilon_j \langle \tilde{E}_L \tilde{E}_L^* \rangle + \epsilon_j \epsilon_k^* \langle \tilde{E}_L \tilde{E}_R^* \rangle e^{-2i\phi}\right]$$

On letting Stokes V=0, we find:

$$v_{RL}(\mathbf{u}, \nu) = g_j g_k^* \left[\tilde{p}(\mathbf{u}, \nu)e^{2i\phi} + \tilde{I}(\mathbf{u}, \nu)(\epsilon_j + \epsilon_k^*) + \epsilon_j \epsilon_k^* \tilde{p}^*(\mathbf{u}, \nu)e^{-2i\phi}\right]$$

We can make some assumptions to simplify this expression. For typical sources, $p \sim 0.1 I$, and for the CBI, $\epsilon \sim 10\%$, so that $p:\epsilon I:\epsilon^2 p^*$ scale as $0.1:0.1:10^{-3}$. We therefore ignore second order terms in ϵ. In addition, we have a leakage term ϵ_j for each of the 13 antennas, but we have only 12 cross polarized baselines; we need only solve for the sum of the two terms associated with each baseline. We therefore regard the leakages (ϵ_j, ϵ_k) associated with a pair of antennas as a baseline-based parameter ϵ_{jk}. We will do the same with the gain, letting $G_{jk} = g_j g_k^*$. Then,

$$v_{RL}(\mathbf{u}, \nu) = G_{jk;\nu}\left[\tilde{p}(\mathbf{u}, \nu)e^{2i\phi} + \epsilon_{jk;\nu}\tilde{I}(\mathbf{u}, \nu)\right]$$

The goal of polarization calibration is to determine the gains G_{jk} and the leakages ϵ_{jk} for each of the ten CBI bands ν.

Our polarization calibration procedure capitalizes on the fact that deck rotation modulates the source polarization term $\tilde{p}(\mathbf{u}, \nu)e^{2i\phi}$ relative to the instrumental polarization $\epsilon \tilde{I}(\mathbf{u}, \nu)$. The calibration routine combines multi-deck angle observations of polarization calibrators with values for $\tilde{I}(\mathbf{u}, \nu)$ and $\tilde{p}(\mathbf{u}, \nu)$ supplied by external measurements to solve

Figure 2. CBI instrumental polarization comparison, RX8–RX12

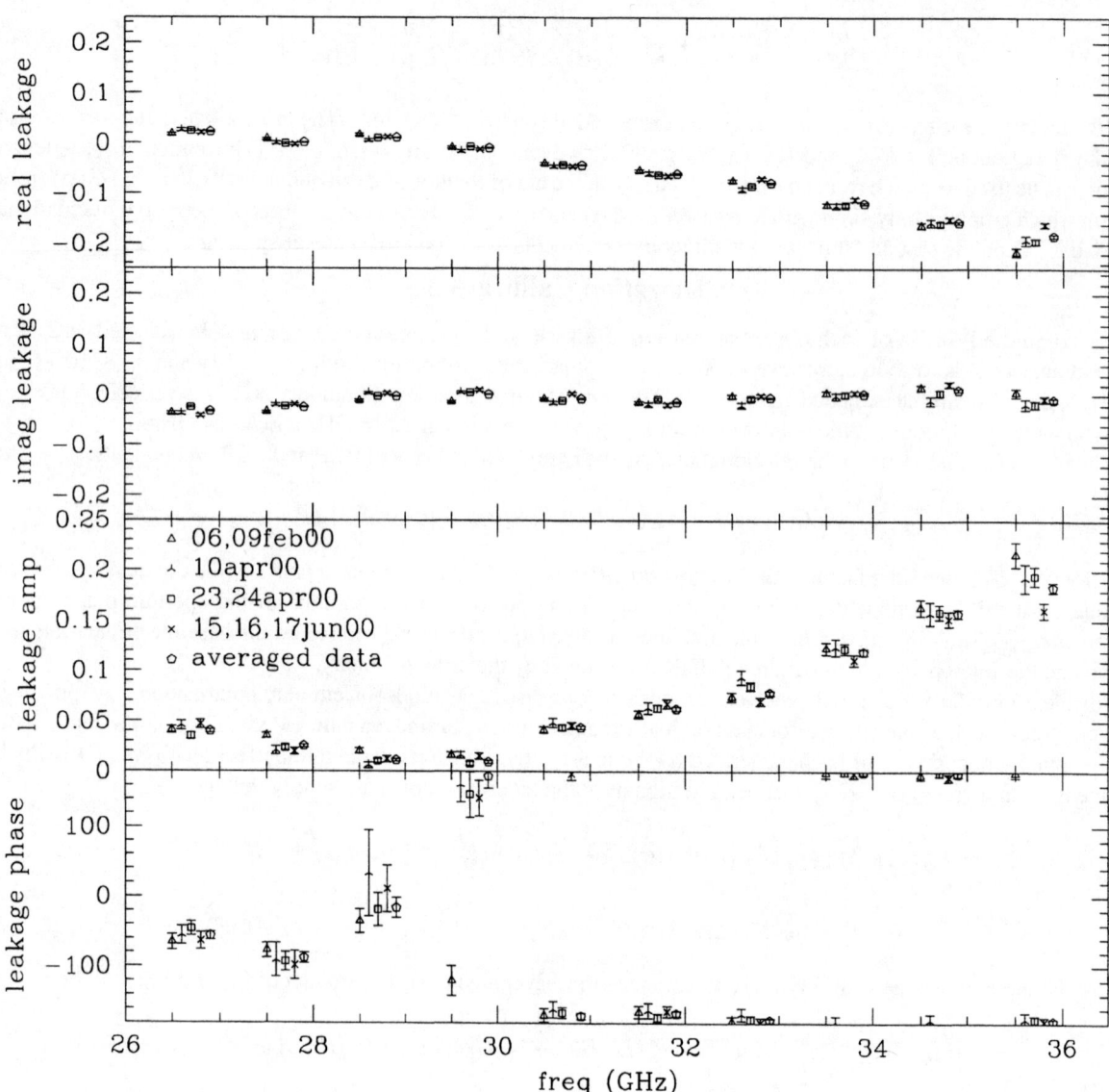

Figure 2. Instrumental polarization for baseline RX8-RX12. The top pair of figures provides a comparison of the real and imaginary components of the leakage terms for dates spanning Feb-Jun 2000. The data are collected into 10 bands as a function of frequency, which is shown on the x-axis. The points within each group are observations on different dates, where the dates are averaged as shown in the third frame. The comparison between the 10apr00 and 23,24apr00 data, for example, is notable because these points span a reconfiguration of the array; the instrumental polarization remained constant throughout the reconfiguration. The bottom pair of frames shows the corresponding amplitude and phase for the instrumental polarization. The leakage terms for other baselines have similar shapes; the large instrumental polarization at the band edges arises from simple bandpass errors in the quarter wave plates which define the mode of circular polarization for the receivers.

for $G_{jk;\nu}$ and $\epsilon_{jk;\nu}$. We merge values for I_ν from the CBI with those for $m=p/I$ and position angle χ from the VLA, and supply these as inputs for the polarization calibration. The uncertainty on the VLA data is $\sim$6% for both m and χ for both bands, and the absolute uncertainty for I_ν from the CBI is 5%.

We performed several measurements of the instrumental polarization over the Feb-Jun 2000 period to establish the stability of the instrumental polarization. Figure 2 compares eight deep measurements of the instrumental polarization for one baseline for this period; for clarity, the data are grouped into four sets of dates. The excellent agreement between 10apr00 and 23/24apr00 is notable because these observations bracket a period during which the CBI array was reconfigured to emphasize shorter baselines. The rise in instrumental polarization at high frequencies stems from simple bandpass errors in the quarter wave plates which define the receiver polarizations. These figures indicate the typical reproducability of the instrumental polarization. A χ^2 analysis of the individual dates relative to the sample mean demonstrated that the leakage terms are in good agreement. While the instrumental polarization can be large, this comparison shows that it remains constant over long timescales.

Efforts to observe extended sources such as the CMBR require an understanding of the polarization characteristics across the entire primary beam. The off-axis characteristics of the beams were determined from measurements of the instrumental polarization at the beam half-power points in the four cardinal directions. A χ^2 analysis of the four sets of leakage terms measured at the half-power points relative to that at the beam center demonstrated the uniformity of the polarization characteristics across the primary beams of all the receivers.

The polarization observations included sources of known polarization to confirm that the system was working as desired. We observed 3C273 throughout the polarization campaign with both the CBI and the VLA; typically the CBI recovered the VLA values for the fractional polarization and polarization angle of 3C273 to within 10%. To assess the CBI's mapping capabilities, we observed several extended sources, such as supernova remnants; in all cases these observations agreed qualitatively with maps in the literature. The mapping observations, coupled with the 3C273 monitoring observations, provided great confidence in the polarization capabilities of the instrument.

CMBR Observations

The CBI polarization observations spanned two periods: Jan-May 2000 and Aug-Oct 2000. The short baselines suffered from ground spillover, so to reject this contamination, we observed fields in lead/trail pairs separated by 8^m in right ascension and differenced the pairs of fields on a point-by-point basis in the aperture domain. We collected $\sim$75 baseline hours of data on a lead/trail pair centered at $\alpha=08^h48^m$, $\delta=-3°10'$, and $\sim$250 baseline-hours on a pair centered at $\alpha=20^h52^m$, $\delta=-3°30'$. The analyses of these data are in progress.

CONCLUSION

The Cosmic Background Imager is well-suited to polarization observations of the CMBR. The high sensitivity of the instrument puts a polarization detection effort within reach, and the stability of the instrument enables the long integrations required for deep polarization observations without numerous time-consuming instrumental polarization calibrations. The VLA monitoring campaign facilitates a robust calibration program, and the CBI's imaging capabilities permit tests of the system with extended sources of known polarization; all of the checks contained therein provide great confidence in the polarization capabilities of the CBI.

ACKNOWLEDGEMENTS

This work was made possible by NSF grant AST-9802989, and the generous support from the California Institute of Technology, Ronald and Maxine Linde, and Cecil and Sally Drinkward.

REFERENCES

1. Kamionkowski, M., & Kosowski, A., Ann. Rev. Nucl. Part. Sci. 49 (1999) 77
2. De Bernardis, P., et al., Nature 404 (2000) 955
3. Hanany, S., et al., ApJ 545 (2000) L5
4. Halverson N. W., et al., astro-ph/0104489 (2001)
5. Staggs, S. T., Gunderson, J. O., & Church, S. E., astro-ph/9904062 (1999)
6. Padin, S., et al., ApJ 549 (2001) L1
7. Conway, R. G., and Kronberg, P. P., M.N.R.A.S. 142 (1969) 11

The BaR-SPOrt Experiment: the Science

E. Carretti*, G. Bernardi*, S. Cecchini*, S. Cortiglioni*, C. Macculi*, E. Morelli*, C. Sbarra*, G. Ventura*, J. Monari†, S. Poppi†, G. Boella**, S. Bonometto**, M. Gervasi**, G. Sironi**, M. Tucci**, M. Zannoni**, M. Baralis‡, O. Peverini‡, R. Tascone‡, R. Fabbri§, V. Natale¶, M. Bruscoli∥, A. Boscaleri††, E. Pascale†† and L. Nicastro‡‡

*Istituto Te.S.R.E./C.N.R., via P. Gobetti 101, I–40129 Bologna
†I.R.A./C.N.R., via P. Gobetti 101, I–40129 Bologna
**Dip. di Fisica, Univ. di Milano - Bicocca, P.za della Scienza 3, I–20126 Milano
‡I.R.I.T.I./C.N.R., c.so Duca degli Abruzzi 24, I–10129 Torino
§Dip. di Fisica, Univ. di Firenze, Via Sansone 1, I–50019 Sesto Fiorentino, Firenze
¶C.A.I.S.M.I./C.N.R., Largo E. Fermi 5, I–50125 Firenze
∥Dip. di Astronomia, Univ. di Firenze, Largo E. Fermi 5, I–50125 Firenze
††I.R.O.E./C.N.R., Via Panciatichi 64, I–50127 Firenze
‡‡I.F.C.A.I./C.N.R., via U. La Malfa 153, I–90146 Palermo

Abstract. BaR-SPOrt is a balloon–borne experiment in the microwave range (32-90 GHz) aimed at studying the polarization of the CMB and of the diffused Galactic Background. Here we present the main scientific goals as well as the observing strategies of the project.

INTRODUCTION

The Cosmic Microwave Background (CMB) is a powerful tool to understand origin and evolution of the Universe. The CMB looks like a Black Body at 2.725 K [1] almost isotropic and unpolarized: any detection of deviations from its ideal behaviour allows the estimate of cosmological parameters [2, 3, 4]. Very small temperature anisotropies have been detected at both large [5, 6] and small [7, 8, 9] angular scales, but only upper limits on the CMB polarization (CMBP) have been set up to now. In particular, the information contained in the CMB Polarization can solve the

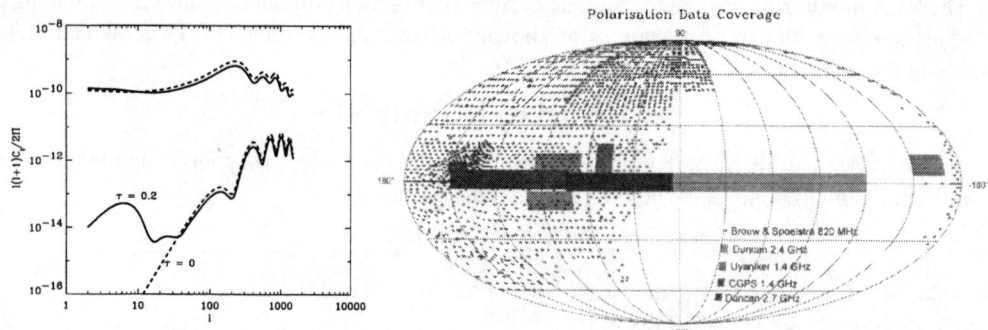

FIGURE 1. **Left:** Anisotropy and E-mode power spectra. Two $\Omega_\Lambda = 0.7$ cosmological models which differ only for the reionization optical depth τ are shown. **Right:** Sky coverage of the existing continuous polarization surveys in Galactic coordinates (The Galactic centre is in the middle).

degeneracies among cosmological parameters that CMB anisotropy alone is not able to remove [3]. Figure 1 presents the comparison between temperature and polarization (E-mode) power spectra for two cosmological models which differ only for the optical depth τ of the re-ionized medium in the dark ages. It is clear that the E-mode spectrum is much more sensitive to τ than the temperature one and that this new information is found at large angular scales ($l < 20$, i.e. $\theta > 10°$). The CMBP brings important information also at subdegree angular scales. For instance, the *coherent* primordial fluctuations foreseen by the inflation produce a well defined Doppler peak pattern: the peaks in the T and E power spectra are alternate. Thus, the detection of the CMBP at subdegree scales allows a test of the inflationary model [10]. Finally, the CMBP allows the separation between the scalar and the tensorial components of the primordial fluctuations providing a way to disentangle among different inflationary models [11]. Figure 1 shows also that the polarized emission peaks at the sub-degree scales, suggesting where it would be easier to make a first detection.

In spite of its importance, the CMBP predicted level is very low (few μK at sub-degree scales and less than a 1 μK at large ones). Table 1 shows the present upper limits: it is clear that the first detection is at least one order of magnitude far away.

TABLE 1. Present upper limits of the CMBP

Frequency (GHz)	Beam	Sky Coverage	Upper Limit	Reference
4.0	15°	scattered	300 mK	[19]
100-600	1°.5 - 40°	GC	3-0.3 mK	[20]
9.3	15°	$\delta = +40°$	1.8 mK	[21]
33	15°	$\delta \in (-37°, +63°)$	180 μK	[22]
5.0	18" - 160"	$\delta = +80°$	4.2 mK - 120 μK	[23]
26-36	1°.2	NCP	30 μK	[24]
26-36	1°.4	NCP	18 μK	[25]
33	7°	SCP	267 μK	[26]
8.7	6'	$\delta = -50°$	16 μK	[27]
90	0°.24	NCP	13 μK	[28]
26-36	7°	$\delta = +43°$	10 μK	[29]

Surveys of the polarized diffuse emission in the microwave range are crucial also for the study of the Galactic contribution. In fact, the Galactic background, beside its intrinsic interest, acts as a foreground for CMB experiments and only its accurate knowledge will allow clear measurements of CMB features (see [12] and [13] and references therein). Our Galaxy is featured by a smooth linearly polarized background emission, carrying information on the Galactic structure. Up to now, observations were carried out only at frequencies up to 2.7 GHz [14, 15, 16, 17, 18], where the Galactic emission results to be dominated by synchrotron. Such observations either are widely undersampled [14] or cover narrow stripes around the Galactic Plane [15, 16, 17, 18] (see Figure 1) calling for further observations.

SCIENTIFIC GOALS

The BaR-SPOrt (Balloon-borne Radiometers for Sky Polarization Observations) project, together with other CMBP experiments (for a review see [30], the conclusion remarks in this volume and several contributions in this proceedings), is aimed at filling the gap in polarization observations at microwave frequencies (32 and 90 GHz) by observing sky patches of about $20° \times 20°$. The main efforts will be concentrated in measuring the CMBP, though the Galactic synchrotron emission, especially at 30 GHz, will be studied as well.

Polarization at large scales can be detected only by all-sky surveys which, due to the low emission level of the signal, require a very stable environment to allow all-sky scans. Such investigations are left to space missions, like MAP, PLANCK and SPOrt. Ground-based and ballon-borne experiments, limited to small sky patches, can concentrate their efforts to the sub-degree scales attempting a first CMBP detection. Furthermore, Carretti et al. [31] have shown that ground-based instruments can be limited by the atmospheric emission. The instrumental polarization can correlate the unpolarized atmospheric signal, whose fluctuations can then degrade the expected sensitivity. Ballon-borne experiments do not suffer from this problem thanks to the low level of residual atmosphere at their flight altitude.

As a result, BaR-SPOrt has the following main features:

- a balloon experiment to minimize the atmospheric effects;
- a small sky patch ($20° \times 20°$) as observing target to minimize the instability induced by environmental fluctuations over a single scan (like thermal fluctuations and spillover changes);

- two frequency channels at 32 and 90 GHz to match the best band for CMBP observation (90 GHz) and to check the Galactic synchrotron contribution.
- sub-degree angular resolution to attempt the first CMBP detection where the polarization signal peaks.

Due to the low expected signal, much care has to be spent to optimize the instrument design with respect to the systematics generation, observing time efficiency and long term stability. Sharing the know-how of the SPOrt project, the following choices were operated for BaR-SPOrt (for details see also the companion paper [32] in this volume):

- correlation polarimeters to improve the stability;
- correlation of the two circularly polarized component E_L and E_R that provides directly and simultaneously the detection of both Q and U (100% observing time efficiency).
- on axis optics in order to minimize the spurious polarization induced by the f pattern (a combination of co-polar and cross-polar pattern, see [31] for details) and the CMB anisotropy at the beam scales: with such a configuration, 40 dB of cross-polarization allows a contamination $< 0.3\ \mu K$ (10 times smaller than the expected CMBP signal);
- high OMT insulation (~ 60 dB) [33] which is the major responsible for Q and U offset generation in a correlation polarimeter [31]: such a OMT insulation allows offset values as low as 50 mK, making the radiometer very stable (knee frequency $f_{\text{knee}} \sim 10^{-4}$, about 10^6 times lower than that of the RF amplifiers).

As a result the spurious signal generated by the CMB anisotropy is negligibile and the $1/f$ part of the noise is low enough to be recovered by destriping techniques. The scanning strategy of BaR-SPOrt is based on 1 min horizontal scans of the patch to be observed, with small changes (1/3 of the beam) of the vertical pointing every several scans. The crossing between the scans, as required by destriping algorithms, is ensured by the tilting of the patch during its daily motion on the sky. The destriping technique consists of an iterative procedure studied for the SPOrt experiment. Figure 2 presents the correlation function of Q in the worst case of $f_{\text{knee}} = 10^{-2}$ Hz and shows how the destriping is able to remove the $1/f$ noise contribution.

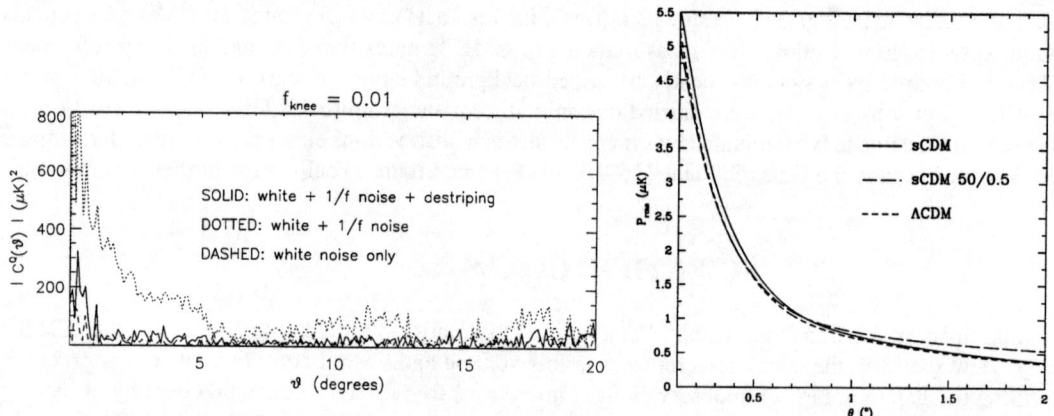

FIGURE 2. **Left:** Correlation function of Q for a simulated BaR-SPOrt experiment. The effect of the $1/f$ noise before and after the destriping is shown ($f_{\text{knee}} = 10^{-2}$ Hz). For comparison, the white noise case is also shown. **Right:** expected P_{rms} for different cosmological models (sCDM, ΛCDM and a sCDM with re-ionization at $z_{ri}=50$ and a re-ionization fraction of 0.5) versus the angular resolution θ.

Table 2 reports the sensitivity foreseen for 14 days of observing time. The pixel sensitivity does not allow the building of CMBP maps, but the full-patch sensitivity should be enough to detect the mean polarized signal $P_{rms} = \sqrt{\langle Q^2 + U^2 \rangle}$. In fact, Figure 2 shows the expected CMBP P_{rms} with respect to the angular resolution (HPBW) of the instrument for different cosmological models: at sub-degree angular scales P_{rms} is rather independent of the cosmological model and comparing it with the full-patch sensitivity (which is roughly the sensitivity for a flat spectrum analysis) it appares that BaR-SPOrt should be able to provide a detection at both 32 and 90 GHz.

TABLE 2. BaR-SPOrt main characteristics: σ_{1s} is the sensitivity in 1 second, σ_{PX} and σ_{FP} are the final sensitivity per pixel and the full patch rms sensitivity, respectively, considering a two weeks flight duration.

Frequencies (GHz)	Bandwidth	HPBW	$\sigma_{1s}[\text{mKs}^{1/2}]$	$\sigma_{PX}[\mu K]$	$\sigma_{FP}[\mu K]$
32	10%	0.5°	0.5	18	0.4
90	10%	0.2°	0.7	64	0.6

OBSERVING STRATEGIES

Both Diffuse Galactic synchrotron and dust emission can pollute the CMBP signal at low and high frequency, respectively. Therefore, an important role is played by the selection of the target patch to be observed. Ideally, we would look for an area where these two contaminants are negligible at both 32 and 90 GHz. At present the most probable launch site is Antarctica, but also an Artic flight is being considered, due to the availability of launch facilities for example in Norway and Sweden: thus, target patches in both the northern and southern sky are considered. Due to the lack of polarized data, especially at high Galactic latitude, we use unpolarized survey at low ($\ll$ 32 GHz) and high ($\gg$ 90 GHz) frequencies to estimate the foreground contributions.

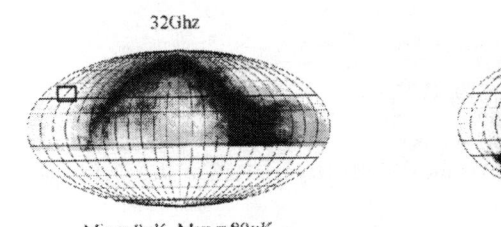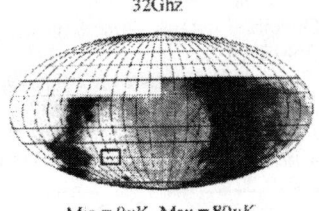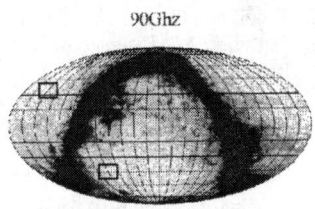

FIGURE 3. Reich & Reich [34] (left) and Rhodes/HartRAO [35] (middle) maps scaled up to 32 GHz together with DIRBE [37] map (right) scaled down to 90 GHz. The two possible target patches for the BaR-SPOrt experiment are superimposed to the maps. The maps are in Celestial coordinates.

The synchrotron emission level has been evaluated starting from the 1.4 GHz Reich & Reich map [34] for the northern sky and the 2.3 GHz Rhodes/HartRAO survey [35] for the southern emisphere. The total intensity was scaled up to 32 GHz by the power law $T_S \propto \nu^{-3.0}$ [36] and the polarization percentage was assumed to be 10%. Figure 3 (left and middle) shows the result for the channel where the synchrotron emission should be more important. There are two possibile target patches (for the northern and southern sky, respectively) where the synchrotron emission is very low: the rms fluctuation of the polarized signal ΔT_S (32 GHz) should allow us to search for CMBP even at 32 GHz. In particular, location and estimate of the polarized emission level are the following:

- RA = 10 h; DEC = +35° $\Rightarrow$ ΔT_S (32 GHz) < 0.3 μK
- RA = 5 h; DEC = -45° $\Rightarrow$ ΔT_S (32 GHz) < 1 μK

One should note that these levels are only upper limits and the real situation might be even better. As a matter of fact, the same results can be obtained simply by scaling the noise of the original map, suggesting the estimated rms fluctuations in the two patches should be noise rather than Galactic signal. At 90 GHz the situation is even better so that we can expect that these two patches are free from the Galactic synchrotron emission.

A similar study has been done also for the dust emission T_D. In this case we started from the DIRBE map at 1250 GHz [37] and scaled it down to 90 GHz. The frequency behaviour was assumed to be $T_D \propto \nu^{-2.7}/(\exp^{h\nu/kT} - 1)$ [13] with a polarization level 5%. The result is shown in Figure 3 (right): in the two regions selected with the synchrotron analysis the fluctuation of the dust polarized emission is

$$\Delta T_D(90\,\text{GHz}) < 0.15 \mu K. \tag{1}$$

Therefore, the two identified regions are very good targets to search for the CMBP signal.

ACKNOWLEDGMENTS

Authors wish to thank J. L. Jonas for providing us the data of the Rhodes/HartRAO survey. We wish also thank the organizers of the 2K1BC workshop for such a succesful meeting. We acknowledge use of CMBFAST and HEALPix packages for performing our analysis. BaR-SPOrt is an ASI funded project.

REFERENCES

1. Mather, J. C., Fixsen, D. J., Shafer, R. A., Mosier, C., Wilkinson, D. T., *ApJ*, 512, 511-520, 1999.
2. Jungman, G., Kamionkowski, M., Kosowsky, A., Spergel, D. N., *PRD*, 54, 1332-1344, 1996.
3. Zaldarriaga, M., Spergel, D. N., Seljak, U., *ApJ*, 488, 1, 1997
4. Efstathiou, G., Bond, J. R., *MNRAS*, 304, 75, 1999.
5. Smoot, G.F., et al., *ApJ*, 396, L1, 1991.
6. Bennet, C.L., et al., *ApJ*, 464, L1, 1996.
7. De Bernardis, P., et al., *Nature*, 404, 955, 2000.
8. Hanany, S., et al., *ApJ*, 545, L5, 2000.
9. Miller, A. D., et al., *ApJ*, 524, L1, 1999.
10. Kosowsky, A., *NewAR*, 43, 157-168, 1999.
11. Kamionkowski, M., Kosowsky, A., *PRD*, 57, 685, 1998.
12. Tucci, M., Carretti, E., Cecchini, S., Fabbri, R., Orsini, M., Pierpaoli, E., *NewA*, 5, 181, 2000.
13. Tegmark, M., Eisenstein, D. J., Hu, W., de Oliveira–Costa, A., *ApJ*, 530, 133, 2000.
14. Brouw, W. N., Spoelstra, T. A. Th., *A&AS*, 26, 129, 1976.
15. Duncan, A. R., Haynes, R. F., Jones, K. L., Stewart, R. T., *MNRAS*, 291, 279, 1997.
16. Duncan, A. R., Reich, P., Reich, W., Fürst, E., 1999, *A&A*, 350, 447, 1999.
17. Uyaniker, B., Fürst, E., Reich, W., Reich, P., Wielebinski, R., *A&AS*, 138, 31, 1999.
18. Gaensler, B. M., Dickey, J. M., McClure–Griffiths, N. M., Green, A. J., Wieringa, M. H., Haynes, R. F., *ApJ*, 549, 959, 2001.
19. Penzias, A. A., and Wilson, R. W., *ApJ*, 142, 419-221, 1965.
20. Caderni, N., et al., *Phys. Rev. D*, 17, 1901-1907, 1978.
21. Nanos, G. N., *ApJ*, 232, 341-347, 1979.
22. Lubin, P. M., and Smoot, G. F., *ApJ*, 245, 1-17, 1981.
23. Partridge, R. B., et al., *Nature*, 331, 146-147, 1988.
24. Wollack, E. J., et al., *ApJ*, 419, L49-L52, 1993.
25. Netterfield, C. B., et al., *ApJ*, 445, L69-L72, 1995.
26. Sironi, G., Boella, G., Bonelli, G., Brunetti, L., Cavaliere, F., Gervasi, M., Giardino, G., Passerini, A., *NewA*, 3, 1-13, 1998.
27. Subrahmanyan R. et al., *MNRAS*, 315, 808-822, 2000.
28. Hedman M. M. et al., *ApJ*, 548, L111-L114, 2001.
29. Keating, B. G., et al., submitted to *ApJL*, 2001 (*astro-ph/0107013*).
30. Staggs, S.T., Gundersen, J. O., Church, S. E., in Microwave Foregrounds, ed. A. de Oliveira-Costa & M. Tegmark, ASP Conference Series, 181, 299, 1999
31. Carretti, E., Tascone, R., Cortiglioni, S., Monari, J., Orsini, M., *NewA*, 6, 173-187, 2001 (*astro-ph/0103318*).
32. Macculi, C., et al., *this volume*, 2001.
33. Tascone, R., et al., *this volume*, 2001.
34. Reich, P., Reich, W., *A&AS*, 63, 205-288, 1986.
35. Jonas, J. L., Baart, E. E., Nicolson, G. D., *MNRAS*, 297, 977-989, 1998.
36. Platania, P., Bensadoun, M., Bersanelli, M., de Amici, G., Kogut, A., Levin, S., Maino, D., Smoot, G. F., *ApJ*, 505, 473, 1998.
37. Hauser, M. G., et al., *ApJ*, 508, 25-43, 1998.

BaR-SPOrt: a technical overview

Macculi C.*, Bernardi G.*, Carretti E.*, Cecchini S.*, Cortiglioni S.*, Morelli E.*, Sbarra C.*, Ventura G.*, Monari J.[†], Poppi S.[†], Boella G.**, Bonometto S.**, Gervasi M.**, Sironi G.**, Tucci M.**, Zannoni M.**, Baralis M.[‡], Peverini O.[‡], Tascone R.[‡], Fabbri R.[§], Natale V.[¶], Bruscoli M.[‖], Boscaleri A.[††], Pascale E.[††] and Nicastro L.[‡‡]

I.Te.S.R.E./C.N.R., via P. Gobetti 101, I–40129 Bologna
[†]*I.R.A./C.N.R., via P. Gobetti 101, I–40129 Bologna*
**Dip. di Fisica, Univ. di Milano - Bicocca, P.za della Scienza 3, I–20126 Milano*
[‡]*I.R.I.T.I./C.N.R., c.so Duca degli Abruzzi 24, I–10129 Torino*
[§]*Dip. di Fisica, Univ. di Firenze, Via Sansone 1, I–50019 Sesto Fiorentino, Firenze*
[¶]*C.A.I.S.M.I./C.N.R., Largo E. Fermi 5, I–50125 Firenze*
[‖]*Dip. di Astronomia, Univ. di Firenze, Largo E. Fermi 5, I–50125 Firenze*
[††]*I.R.O.E./C.N.R., Via Panciatichi 64, I–50127 Firenze*
[‡‡]*I.F.C.A.I./C.N.R., via U. La Malfa 153, I–90146 Palermo*

Abstract. BaR-SPOrt is a project to measure the linearly polarised emission of 20°x20° sky patches from a stratospheric balloon, at 32 GHz and 90 GHz. It consists of correlation polarimeters for direct measurements of the Q and U Stokes parameters, coupled to an optics providing a beam of 0°.5 (32 GHz) and 0°.2 (90 GHz). The instrument design is described. Particular emphasis is put on the hardware solutions adopted to reduce the systematic effects in high sensitivity polarisation measurements.

INTRODUCTION

BaR-SPOrt (Balloon-borne Radiometer for Sky Polarisation Observations) is a balloon experiment housing correlation microwave polarimeters (32 & 90 GHz) for the direct measurement of the Q and U Stokes parameters with HPBW=0°.5 & 0°.2. It shares most of the SPOrt know-how [1, 2], aimed at studying the polarisation of the Cosmic Microwave Background (CMB) as well as the diffused Galactic Background. The polarised component of the CMB can be related to cosmological parameters, providing information about the nature of the primordial fluctuations and the re-ionization era, thereby allowing to discriminate among different cosmological models. The study of the linearly polarised emission is fundamental to understand the physical processes taking place in our Galaxy [3]. The Galactic emission represents also a foreground signal to be carefully removed from the measurements of CMB experiments. BaR-SPOrt is an ASI (Italian Space Agency) funded experiment which should be ready to fly by the end of 2003. The first launch, which might be from Antarctica, will carry only the 32 GHz channel. The 90 GHz will be implemented as a second step. BaR-SPOrt scientific goals are the building of linear polarisation maps of sky patches and the tentative detection of the linearly polarised component of the CMB radiation, which is expected to be at μK level. The main features of the BaR-SPOrt experiment are summarized in Table 1.

TABLE 1. BaR-SPOrt main characteristics: σ_{1s} is the instantaneous sensitivity, σ_{PX} and σ_{FP} are the final per pixel and the full patch rms sensitivities for a flight of two weeks.

Frequencies (GHz)	Bandwidth	Beam	σ_{1s}[mKs$^{1/2}$]	$\sigma_{PX}[\mu K]$	$\sigma_{FP}[\mu K]$
32	10%	0.5°	0.5	18	0.4
90	10%	0.2°	0.7	64	0.6

BAR-SPORT INSTRUMENT DESIGN

A brief overview

The detection of a signal as low as that expected from CMB polarisation ($\leq 1\ \mu K$) asks for extremely sensitive and stable radiometers. The BaR-SPOrt instrument design has been developed to minimize instrumental effects and to reduce 1/f noise, thereby increasing the long term instrument stability. Particular care has been taken in the realisation of the antenna system to control the spurious polarisation [4]. Correlation techniques are widely adopted in high sensitivity measurements because of their capability to reduce the effects of gain fluctuations. Residual instabilities are usually recovered by using destriping tecniques [5, 6, 7, 8], which require the radiometer to be stable over a single scan period ($30 \div 60$ s for BaR-SPOrt). Starting from the correlation radiometer sensitivity equation:

$$\Delta T_{rms} = \sqrt{\frac{k^2 T_{sys}^2}{B\tau} + T_{offset}^2 \left(\frac{\Delta G}{G}\right)^2 + \Delta T_{offset}^2} \qquad (1)$$

care has been taken to reduce the system noise temperature (T_{sys}), the gain fluctuations ($\Delta G/G$), the instrumental offsets (T_{offset}) and the offset fluctuations (ΔT_{offset}), by means of correlation and modulation techniques.
The main instrumental characteristics are (see also Table 1):

- direct amplification architecture: no down conversion to avoid possible phase error;
- low cross-polarisation optics providing HPBW of $0°.5$ ($0°.2$) at 32 GHz (90 GHz);
- correlation unit based on a custom design waveguide Hybrid Phase Discriminator (HPD), with unpolarised component rejection equal to 30 dB [9];
- custom design Orthomode Transducer (OMT) with high isolation between channels (> 60 dB) to limit contaminations from the unpolarised component [4, 9];
- phase modulation (lock-in system) and correlation providing > 70 dB of total rejection to the unpolarised component;
- a cryostat (see Figure 1) to cool ($T < 80.0 \pm 0.1$ K) LNAs, circulators, the polariser and the OMT by a closed loop cryocooler. The horn, at present designed to be at 300 K, might be cooled as well. A thermal shield, stabilised at temperature $T \cong 300.0 \pm 0.1$ K, is foreseen to increase the thermal stability;
- custom design internal calibrator to inject reference polarised signals.

Considering the parameters of the antenna system shown in Table 2, the physical temperatures of its devices and the noise temperature of the HEMT (30 K), the expected system temperature for the 32 GHz channel is: $T_{sys} \simeq 40K$.

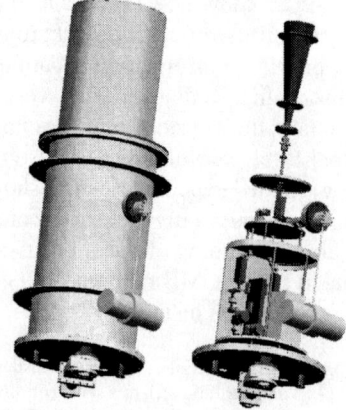

FIGURE 1. The outer and inner parts of the cryostat.

TABLE 2. Goal attenuation parameters of the 32 GHz antenna system.

Feed horn [dB]	Polariser [dB]	OMT [dB]
0.05	0.1	0.1

The polarimeter

The most peculiar features of the BaR-SPOrt polarimeter are: the direct and simultaneous measurement of Q and U, by correlating the two circularly polarised components of the radiation; the adoption of cryogenic Low Noise Amplifiers (LNA); the implementation of a *new concept* waveguide device (HPD) for on board analog correlation. The electric scheme of the BaR-SPOrt radiometer is shown in Figure 2. The choice of correlating the circular components is due to the expression of the Stokes parameters with respect to the basis on which the electric field is decomposed [3]:

$$\begin{aligned}
I &\propto \langle |\vec{E}_x|^2\rangle + \langle |\vec{E}_y|^2\rangle & I &\propto \langle |\vec{E}_L|^2\rangle + \langle |\vec{E}_R|^2\rangle \\
Q &\propto \langle |\vec{E}_x|^2\rangle - \langle |\vec{E}_y|^2\rangle & Q &\propto \langle |\vec{E}_L||\vec{E}_R|\cos(\delta_c)\rangle \\
U &\propto \langle |\vec{E}_x||\vec{E}_y|\cos(\delta)\rangle & U &\propto \langle |\vec{E}_L||\vec{E}_R|\sin(\delta_c)\rangle \\
V &\propto \langle |\vec{E}_x||\vec{E}_y|\sin(\delta)\rangle & V &\propto \langle |\vec{E}_L|^2\rangle - \langle |\vec{E}_R|^2\rangle
\end{aligned} \quad (2)$$

where I describes the total power, Q and U the linear polarisation, V the circular polarisation, $\vec{E}_x, \vec{E}_y$ and $\vec{E}_L, \vec{E}_R$ are the linearly and circularly polarised components of the electric field, respectively, and δ and δ_c are the phase differences of the two linear and circular components, respectively. Thus, simultaneous Q and U outputs cannot be obtained by correlating the linear components. Figure 2 shows how a phase delay of 90° between the linear components, inserted by the polariser, and a rotation of 45° between the reference frame of the OMT and the polariser can transform the linear components into circular ones, then fed to the HPD.

The use of cryogenic LNAs is necessary to reduce the system noise temperature. The BaR-SPOrt radiometers have two LNA amplification stages: the first one is cryogenic ($\sim$ 80K) and the second one is warm ($\sim$ 300K). A mechanical

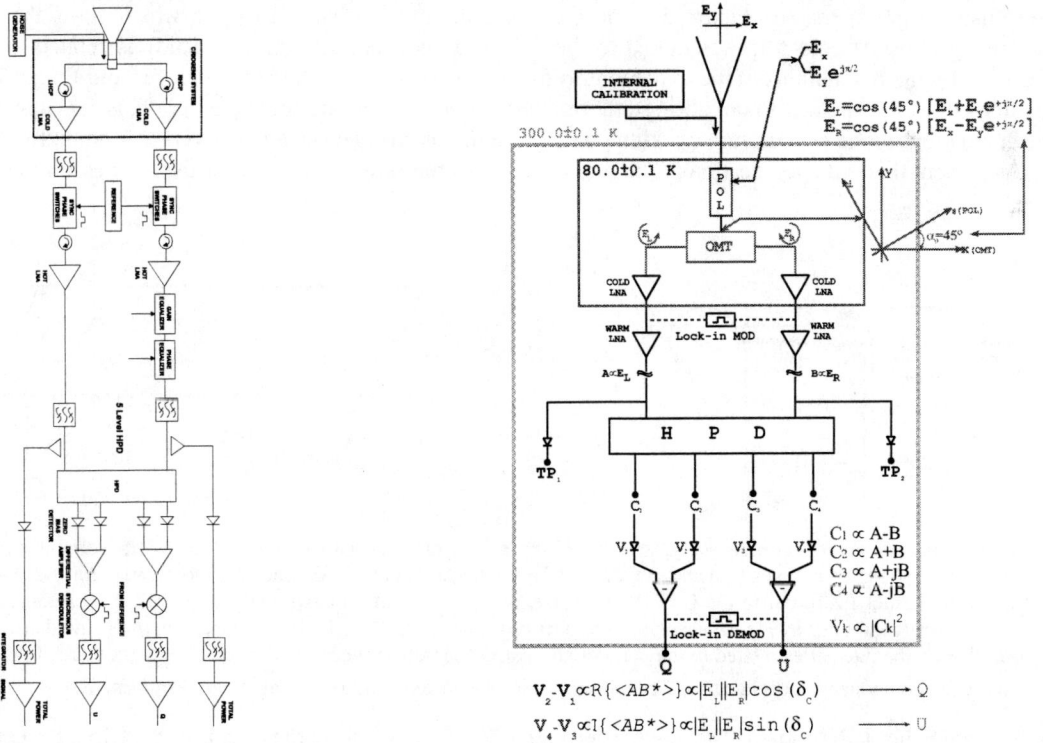

FIGURE 2. Electric scheme (left) and operational scheme (right) of BaR-SPOrt polarimeters.

Stirling cryocooler with closed loop control will provide the cooling down to 80 K with stability of 0.1 K. Since the BaR-SPOrt performances strongly depend on the temperature stability, particular care has been put in the thermal design. As already mentioned, an active temperature control is planned also for the warm parts.

The correlation unit is based on the HPD [1], a passive microwave circuit that processes the signal in order to have four outputs proportional to:

$$\vec{E}_R - \vec{E}_L \quad \vec{E}_R + \vec{E}_L \quad \vec{E}_R + j\vec{E}_L \quad \vec{E}_R - j\vec{E}_L \tag{3}$$

After square law detection the four HPD outputs become (Figure 2):

$$\begin{aligned} V_1 &\propto [|\vec{E}_L|^2 + |\vec{E}_R|^2 - 2\Re(\vec{E}_L \cdot \vec{E}_R^*)] \\ V_2 &\propto [|\vec{E}_L|^2 + |\vec{E}_R|^2 + 2\Re(\vec{E}_L \cdot \vec{E}_R^*)] \\ V_3 &\propto [|\vec{E}_L|^2 + |\vec{E}_R|^2 + 2\Im(\vec{E}_L \cdot \vec{E}_R^*)] \\ V_4 &\propto [|\vec{E}_L|^2 + |\vec{E}_R|^2 - 2\Im(\vec{E}_L \cdot \vec{E}_R^*)] \end{aligned} \tag{4}$$

which are properly differentiated to get as final outputs the two quantities:

$$\begin{aligned} V_2 - V_1 &\propto |\vec{E}_L||\vec{E}_R|\cos(\delta_c) \longrightarrow Q \\ V_4 - V_3 &\propto |\vec{E}_L||\vec{E}_R|\sin(\delta_c) \longrightarrow U \end{aligned} \tag{5}$$

After integration, these provide time averaged values proportional to the Q and U Stokes parameters. One of the most important HPD specifications is the rejection of the unpolarised component of the signal, mainly due to the CMB and to the system noise temperature.

Thermal analysis and vacuum window

The request for high precision and stability instruments, necessary to measure faint polarisations, translates also in an *ad hoc* thermal design. Models are currently being developed to understand the effects of the thermal instability caused by the closed loop cryocooler (CLC) as well as the spurious polarisation induced by the dielectric vacuum window (usually polymers, see Figure 3). The CLC is a thermo-mechanical device which cools the components to low temperature. Its cold finger is linked to the cold box by a copper rod. The study is related to the thermal noise caused by the fluctuations of the cryocooler refrigeration power, which in turn induce Cold Finger Temperature Fluctuations (CFTF). The maximum allowed thermal fluctuations, over one scanning period, is 100 mK. Since CFTF induce fluctuations to the temperature working point, the study of this thermal noise is very important for the definition of the instrument thermal specifications. The model provides the power spectrum of the temperature fluctuations of

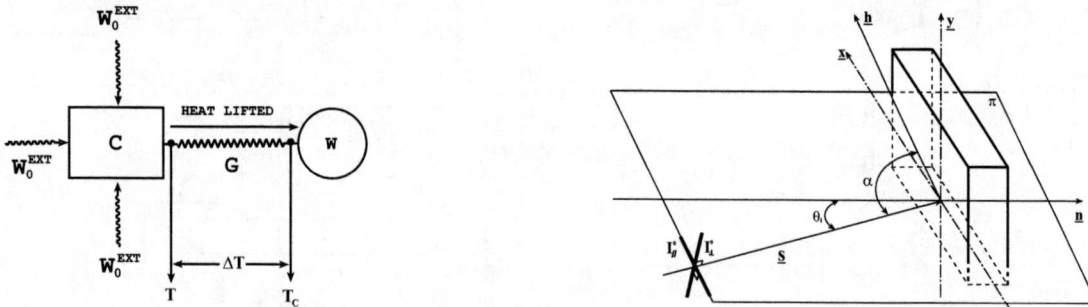

FIGURE 3. The cryocooler thermal analysis (left): C represents the thermal capacity of the warm component (WC), G the conductance between the cryocooler (W) and WC, T and T_C the temperature of WC and W, respectively, ΔT the temperature gap, W_0^{EXT} the thermal input EXTernal to the C-G-W sub-systems. The vacuum window analysis (right): $\underline{h}$ represents the alignment direction of the structural units inside the sheet, $\underline{n}$ is the normal to the sheet, $\underline{S}$ the incidence direction, π the incidence plane defined by $\underline{n}$ and $\underline{h}$, $I_\parallel^0$ e $I_\perp^0$ the intensities related to the parallel and normal vibration modes with respect to the π plane, θ_i is the incidence angle, $\underline{x}$ and $\underline{y}$ the components individuating the laboratory reference frame and α the angle between $\underline{h}$ and $\underline{S}$.

the device which has to be cooled. The power spectrum $P_T^{WC}(f)$ depends on the details of the CLC, the conductance G

between the cryocooler and the Warm Component (WC), and the thermal capacity C of the WC (see 6 and [10, 11]):

$$P_T^{WC}(f) = (1+\frac{\alpha}{G})^2 \frac{P_T^{CF}(f)}{1+(\frac{f}{f_C})^2} \tag{6}$$

where $P_T^{CF}(f)$ is the power spectrum of the cold finger temperature, $f_C = \frac{G}{2\pi C}$ is the cut–off frequency and α is the slope of the cryocooler calibration curve $W = \alpha T_C + \beta$ (in the temperature working range, the trend of the refrigeration power W is linear in T_C). The consistency of this model will be tested in the laboratory.

The manufacturing of the dielectric material making the vacuum window creates, inside it, chains of polymers aligned along some directions (see Figure 3). Those chains induce birefringence ($\Re\{n_\| - n_\perp\} \neq 0$ where $\|$ and $\perp$ individuate the principal directions with respect to the π plane) and dichroism ($\Im\{n_\| - n_\perp\} \neq 0$), as in uniaxial crystals (see [12, 13, 14, 15, 16, 17, 18, 19]), though their origin is quite different, being linked to the manufacturing and not intrinsic. The optical anisotropy then induces spurious polarisation and signal depolarisation, superimposed to the usual trend of isotropic dielectrics (analysis performed by Fresnel's coefficients). In particular, also in the best case of isotropic diffuse radiation, a residual (spurious polarisation) is produced when integrating the signal over the antenna pattern, because of the breaking of the azimuthal symmetry originating from the polymer chain alignment. On the contrary, the same integration should provide a null spurious signal, in case of isotropic dielectrics which have azimuthal symmetry. The model and the measurements should verify the different trends between isotropic and anisotropic dielectric.

In conclusion, the BaR-SPOrt instrument represents an opportunity to test *state of the art* technological solutions that are the base also for more ambitious CMBP space–projects [1, 2, 4]. Moreover, by improving technologies for high sensitivity polarisation measurements at microwaves and millimetric wavelengths, it will play an important role to promote instrumental developments useful for ground applications (radioastronomy).

ACKNOWLEDGMENTS

Authors wish to thank the organizers of the 2K1BC Workshop for providing a great opportunity to show the BaR-SPOrt experiment.

REFERENCES

1. Cortiglioni S. et al., The SPOrt Project: an Experimental Overview, in proceedings of the *International Conference on 3K Cosmology EC-TMR, Roma 1998*, edited by L. Maiani, F. Melchiorri and N. Vittorio, AIP Conference Proceedings 476, 1999, pp.186-193.
2. Fabbri R. et al.,The SPOrt Project: Cosmological and Astrophysical Goals, in proceedings of the *International Conference on 3K Cosmology EC-TMR, Roma 1998*, edited by L. Maiani, F. Melchiorri and N. Vittorio, AIP Conference Proceedings 476, 1999, pp.194-203.
3. Kraus D., *Radio Astronomy*, 2nd edition, Cygnus-Quasar Books.
4. Carretti E., Tascone R., Cortiglioni S., Monari J., Orsini M., *NewA*, 6, 173–187, 2001 (*astro-ph*/0103318).
5. Delabrouille J., *A&ASS*, 127, 555–567, 1998.
6. Wright E. L., *astro-ph*/9612006.
7. Carretti E., et al., 2001 in preparation.
8. Revenu B. et al., *A&AS*, 142, 499, 2000.
9. Tascone R. et al. in the same 2k1BC Conference Proceeding, Cervinia(Ao), 2001.
10. P. de Bernardis, Dispense del corso di Laboratorio di Fisica 2 indirizzo Astrofisica, Universita' di Roma La Sapienza, A.A. 1992 - 1993.
11. Macculi C., Thermal fluctuations induced by cryocooler temperature fluctuations, *BaR-SPOrt Int. Tech. memo*, 01–01, 2001.
12. J. R. White, Origin and Measurement of Internal Stress in Plastics, Polymer Testing, 4, 165–191, 1984.
13. B. E. Read et al., Birefringence Techniques for the Assessment of Orientation, Polymer Testing, 4, 143–164, 1984.
14. Yu N. Gnedin and N. A. Silant'ev, Basic Mechanisms of Light Polarization in Cosmic Media, *Astrophysics and Space Physics Reviews*, edited by R. A. Sunyaev, Vol. 10, 1997.
15. E. D. Palik, Handbook of optical constants of solid II, Academic Press, 1991.
16. Max Born and Emil Wolf, *Principles of Optics*, Pergamon Press, 1970.
17. M. Nigro and C. Voci, *Problemi di Fisica Generale: Elettromagnetismo-Ottica*, Edizioni Libreria Cortina, Padova 1993.
18. J. D. Jackson, *Classical Electrodynamics*, John Wiley & Sons, Inc., 1975.
19. Macculi C., Spurious polarisation from dielectrics, *BaR-SPOrt Int. Tech. memo*, 2000.

Millimeter wave passive devices for measurements of the polarized sky emission

R. Tascone*, D. Trinchero*, M. Baralis*, O. A. Peverini*, A. Olivieri*, E. Carretti[†] and S. Cortiglioni[†]

*IRITI-CNR, Politecnico di Torino, Torino, Italy
[†]ITESRE-CNR, Bologna, Italy

Abstract. The polarized sky emission can be detected by measuring its Stokes parameters through a correlation unit, usually called Hybrid Phase Discriminator (HPD). In the millimeter wave range and for rather large bandwidths, heterodyne receivers are not applicable, and the correlation units have to work in the frequency band of the radiometer. The SPOrt (Sky Polarization Observatory) project will investigate the polarized radiation by means four radiometers at the frequencies 22, 32, 60 and 90 GHz within a 10% bandwidth. This contribution deals with the study of a new configuration of waveguide HPD's that will be mounted on these radiometers. The configuration presents a high degree of sensitivity for the detection of linearly polarized radiation and is suitable for those technologies that offer a high degree of accuracy. Moreover, the set up is particularly robust with respect to the mechanical uncertainties, which are more significant in the higher frequency bands.

INTRODUCTION

The Astrophysical project SPOrt (Sky Polarisation Observatory) [1] was selected by ESA and financially supported by ASI. The aims of this project are the construction of multifrequency polarization maps of the galactic emission and the Cosmic Microwave Background investigation. The instrumentation will consist of four radiometers, which cover the frequency range 20-90 GHz and will be mounted on the International Space Station (ISS). The linearly polarized emission will be detected by measuring the Q and U Stokes parameters which correspond to

$$\begin{aligned} Q &= |E_x|^2 - |E_y|^2 \\ U &= 2\Re\{E_x E_y^\star\} \end{aligned} \qquad (1)$$

These parameters will be detected by means of correlation units (HPD) whose input signals come from an antenna with double circular polarization. The antenna consists of a corrugated horn, a polarizer which converts the circular polarizations into the linear ones and an ortho-mode transducer (OMT) [2]. In the case of a linearly polarized emission the two outputs of the OMT, under ideal conditions, are:

$$\begin{aligned} A(\omega) &= \tfrac{1}{\sqrt{2}} |\underline{E}| \exp(+j\theta) \\ B(\omega) &= \tfrac{1}{\sqrt{2}} |\underline{E}| \exp(-j\theta) \end{aligned} \qquad (2)$$

where $\underline{E}$ is the spectral distribution of the electric field of the emission and θ is the angle with a principal direction of the polarizer. Through the correlation product

$$AB^\star = \frac{1}{2} |\underline{E}|^2 \exp(2j\theta) \qquad (3)$$

it is possible to evaluate the Q and U Stokes parameters of the polarized emission. Of course, the process must present a high level of rejection for the non polarized components. In the past, the correlation was performed after

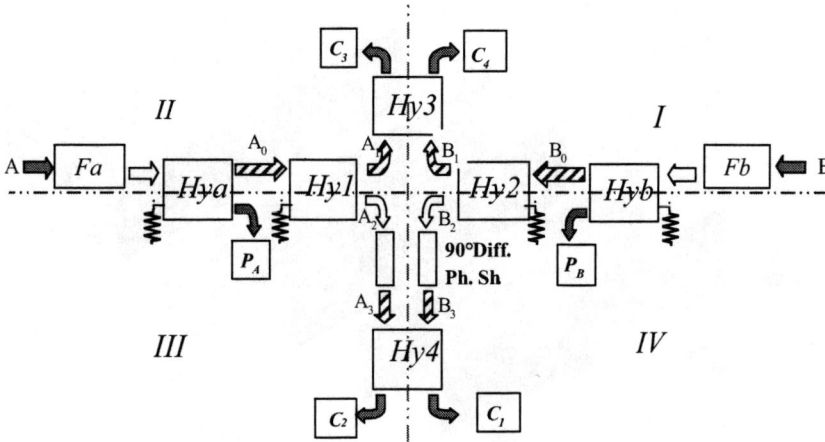

FIGURE 1. Scheme of the HPD where the two input filters and the two directional couplers for monitoring the total power are integrated.

a down conversion by using a local oscillator [3]. Apart from stability problems, this is possible only for narrow band investigations. In the SPOrt project the bandwidths are of 10% centered at 22, 32, 60 and 90 GHz. Obviously, the absolute bandwidths are so large that it is not possible to process the signals at lower frequencies, hence the correlation must be performed at the antenna frequency.

This paper deals with a new configuration of waveguide Hybrid Phase Discriminator which yields at its four output ports the sum and the difference in phase and quadrature of the two input signals. The analog operations are performed at the antenna frequency by means of 3 dB directional couplers and phase shifters. Moreover, two input waveguide filters for the definition of the operative band, and two additional directional couplers for monitoring the total power of the emission are integrated in the HPD. The configuration of this device extends along five levels using E-plane and H-plane discontinuities to form the various components. In this way, a very compact configuration has been obtained which was called FL - HPD (Five Level Hybrid Phase Discriminators, patent pending)

HYBRID PHASE DISCRIMINATORS

The scheme of the correlation units (HPD) used in the radiometers is shown in Fig. 1. Besides the correlator itself, it contains two waveguide filters, with a high rejection in the stop band with the purpose to define the measuring band, and two 3 dB directional couplers to monitor the signal level in the two channels. With reference to Fig 1, the two input signals A and B are filtered by the two 13 cavity waveguide filters (Fa and Fb), subsequently a fraction of the two signals is drawn off by two 3 dB directional couplers (Hya and Hyb) and detected by two diodes with quadratic characteristic. The remaining fraction (A_0 and B_0) are divided into identical parts along two branches. For this operation, two 3 dB directional couplers ($Hy1$ and $Hy2$) are used instead of two power splitters because a high level of decoupling between the two branches is required. In the upper branch the signals A_1 and B_1 are combined by the 3 dB coupler $Hy3$ to produce the output signals C_3 and C_4, proportional to $A + jB$ and $A - jB$, respectively. In the lower branch, before the combination performed by the 3 dB coupler $Hy4$, the signal B_2 undergoes a 90° phase shift with respect to the signal $A2$. In this way, the output signals C_1 and C_2 are proportional to $A - B$ and $A + B$, respectively. The detection of the signals C_k, with $k = 1, \cdots, 4$, by quadratic characteristic diodes and the subsequent amplification by means of two differential amplifiers yields the real and imaginary parts of the average of the correlation product $<AB^\star>$. If the signals A and B are the outputs of an antenna with double circular polarization, the real and imaginary parts of $<AB^\star>$ correspond to the Q and U Stokes parameters of a linearly polarized radiation, respectively.

$$Q = <|A+B|^2 - |A-B|^2>$$
$$U = <|A+jB|^2 - |A-jB|^2>$$
(4)

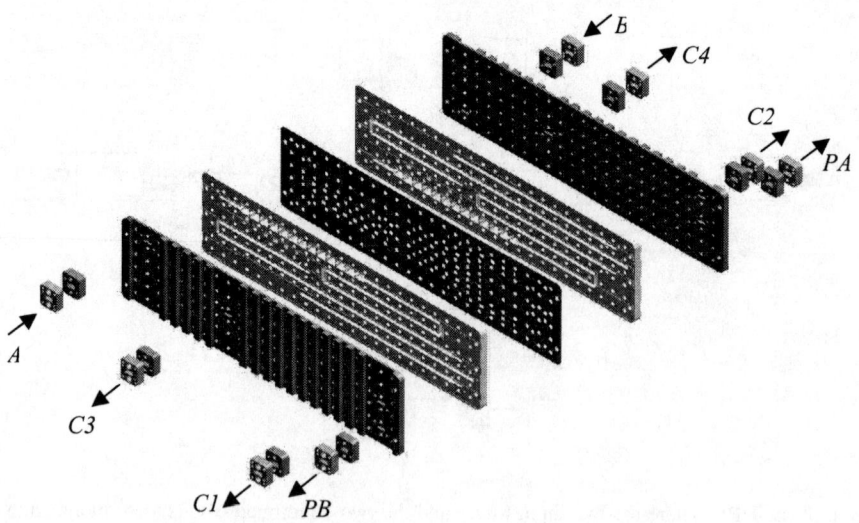

FIGURE 2. Cutaway perspective of the Hybrid Phase Discriminator FL-HPD. The two covers, the two plates containing the waveguides and the central coupling layer are visible.

In (4) the quantities $|A|^2$ and $|B|^2$ are eliminated by cancellation. Moreover, the level of these two quantities is practically defined by the non polarized component, which can be 40-60 dB higher than the polarized one. Hence, the device must present a very high rejection for the auto-correlation terms. This can be obtained by imposing very severe specification to the various components, even if the admitted error for the detection of the Stokes parameters can be within 10%. The problem was attacked from two sides: First, the development of specialized synthesis and analysis techniques able to guarantee a high level of accuracy in such a way that experimental tuning is not necessary. Second, the study of a compact configuration particularly robust with respect to the mechanical uncertainties.

CONFIGURATION OF THE FL-HPD

The device was designed in rectangular waveguide. The dimensions of the internal waveguides are chosen in order to minimize the dispersion effects of the directional couplers within the corresponding operative bands. By means of appropriative waveguide transitions, integrated in the device, the external connections use standard rectangular waveguides: WR42 for K band (22GHz), WR28 for Ka band (32 GHz), WR15 for Q band (60 GHz) and WR10 for W band (90 GHz).

The direct implementation of the circuit scheme of Fig. 1 involves a cross-like geometrical configuration, which requires volume for the K band realization, and with accuracy problems especially for the higher bands. It is important to define a geometrical configuration whose manufacturing process can be done in a unique phase, by avoiding movements of the sample during the manufacturing. In this way, positioning errors are eliminated. Moreover, within the limits of the mechanical uncertainties, it is important to guarantee a high level of symmetry in order to obtain a high rejection for the non polarized radiation. The selected configuration, called FL-HPD (Five Level Hybrid Phase Discriminator) is shown in Fig. 2. It is a *sandwich* structure, closed by two covers where the input and output standard flanges are placed. The waveguides are obtained within two plates of constant thickness separated by a 0.1 mm thick layer placed in the middle of the device (the thickness of this layer is the same for all the bands). The directional couplers are obtained by coupling two parallel waveguide lengths by means of H-plane rectangular apertures realized on the central layer. The phase shifter consists of a cascade of H-plane stubs obtained by rectangular holes on the adjacent waveguide plate. In order to maintain a high level of integration, the two 13 cavity waveguide filters are obtained by a cascade of E-plane discontinuities.All components form a snake-like geometry which expands on five levels in the E-plane. All the mechanical parts present only through holes, hence they can be manufactured by wire electric discharge machines. Moreover, apart from a 180° rotation, the two waveguide plates are identical and can be manufactured at the same time by overlapping the two pieces. The matched loads of the directional couplers are inside the FL-HPD and are made of ECOSORB MF-190 material. Two kinds of matched transitions are present in the

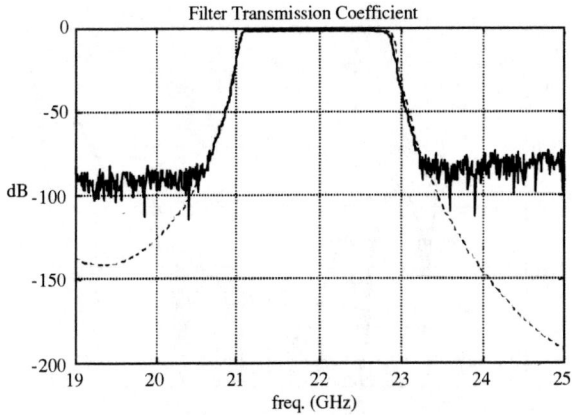
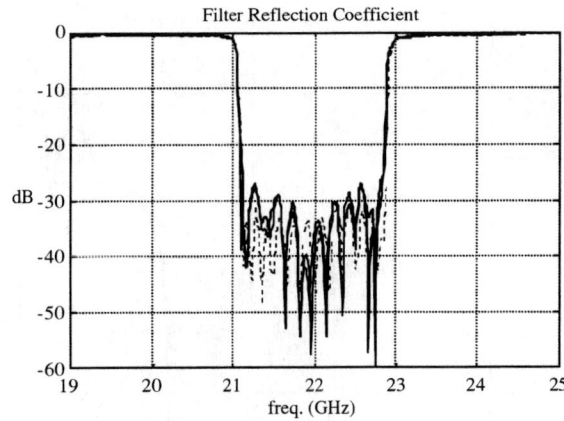

FIGURE 3. Frequency response of the E-plane filter with two L-shaped junctions used in the K band FL-HPD. Measured data, continuous lines; predicted data, dashed lines.

device: L-shaped junctions, to obtain the eight lateral input/output ports in standard waveguide; C-shaped junctions to connect different levels in the waveguide plates. If the scheme of Fig. 1 is divided into four quadrants as indicated in the same figure, the parts corresponding to the *I* and *III* quadrants are realized on the left waveguide plate of Fig. 2, whereas those corresponding to the *II* and *IV* quadrants are realized on the right one. With reference to Fig. 1, the shaded arrows correspond to the L-shaped junctions, whereas the dashed arrows are related to the C-shaped junctions. The white ones represent direct connections.

The filters were synthesized by using the method presented in [4]. A similar technique was used for the design of the various transitions, since they can be seen as filters where the geometry of the input and output ports are assigned. For the other components the technique was modified. In particular, in the case of the directional couplers, the synthesis was carried out in the sum and difference mode basis by controlling the phase difference between the transmission coefficients of these two modes and the reflection coefficient of the sum mode. Similar technique was adopted for the design of the phase shifter.

The analysis was carried out by applying the method of moments with weighted Gegenbauer polynomials as basis functions to represent the aperture field distribution with the right edge conditions. For each kind of discontinuity the projection of the basis functions onto the modal sets were evaluated in the spectral domain by exploiting the analytical knowledge of the Fourier transform of the basis functions. The material losses were taken into account by introducing the relevant impedance boundary conditions in the moment method application. In this way, the loss phenomena were accurately modelled, including also the losses produced by reactive fields excited by the discontinuities.

RESULTS

Before proceeding to the realization of the devices, each component was manufactured to experimentally verify the accuracy of the analysis tools developed for this activity. As an example, measured and predicted data for the K band filter with two L-shaped junctions are compared in Fig. 3. In the same way, Fig. 4 shows the measured and predicted scattering parameters of the 3 dB directional couplers used in the K-band HPD. In the device four L-shaped junctions are present to allow the connection with the standard WR42 waveguide. The reflection coefficient at the all four ports is lower than -35 dB in the band of interest. The transmission coefficients at the uncoupled ports is about -40 dB. The Transfer function ratio between the bar and the cross transmissions for the four input ports is almost flat at level of about 0 dB in the operative band. Subsequently, two prototypes of the complete HPD in K and Ka bands were designed and manufactured. The scattering parameters of the 8 port-devices were measured. In Fig. 5 the reflection and transmission scattering coefficients are reported. In order to evaluate the performance of the FL-HPD the measurements of the scattering parameters were elaborated to obtain the transfer functions which yield the Stokes parameters. If an ideal behavior of the diodes and of the differential amplifiers is assumed, it is possible to define a spectral distribution of the Stokes parameters whose integration yields the relevant data. To this end consider the

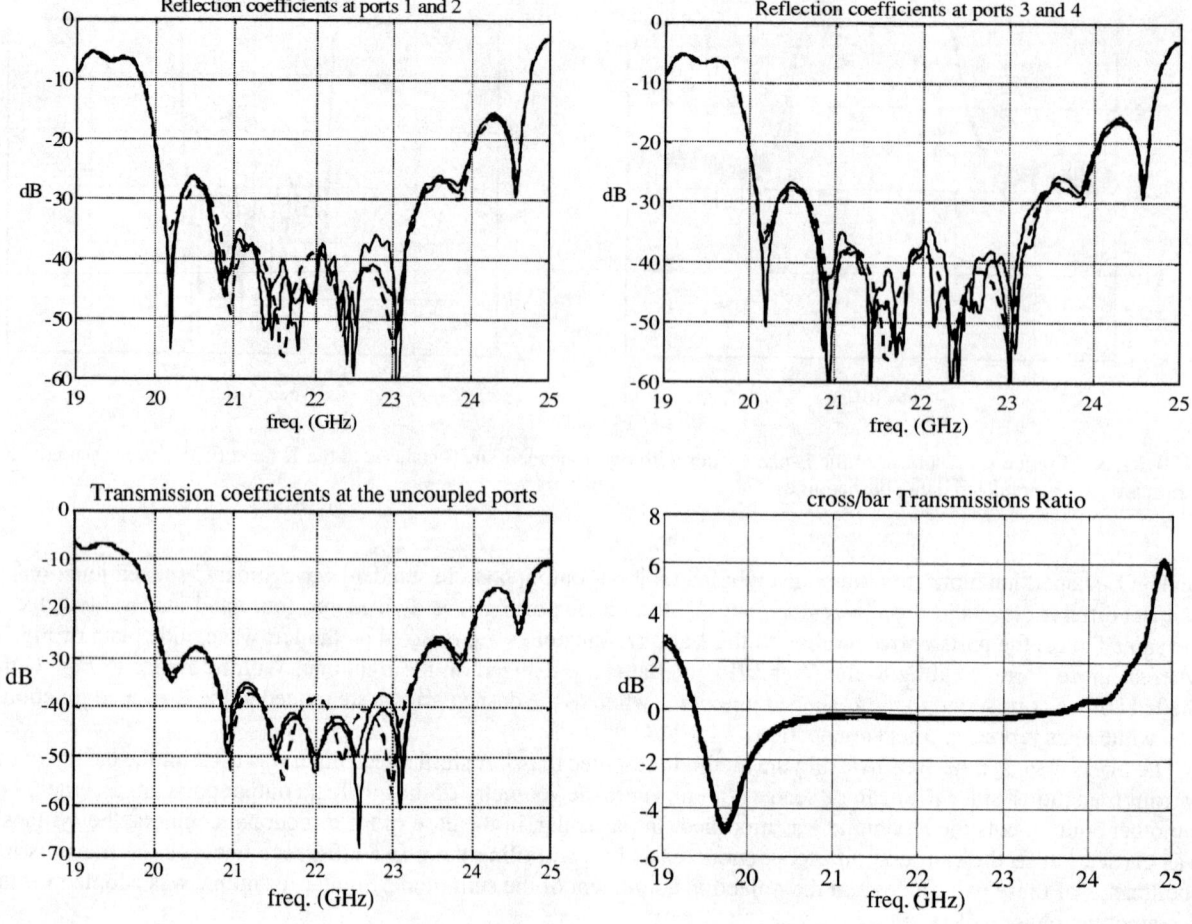

FIGURE 4. Scattering parameters of the 3 dB directional coupler with four L-shaped junctions used in the K band FL-HPD. Measured data, continuous lines; predicted data, dashed lines.

quantities:

$$C_k = |S_{ka}A + S_{kb}B|^2 \text{ with } k = 1,2,3,4 \quad (5)$$

where, with reference to Fig. 1, S_{ka} and S_{kb} are the scattering parameters of the FL-HPD. By subtracting C_1 from C_2 and C_4 from C_3, one obtains:

$$\begin{aligned} Q_m &= |C_2|^2 - |C_1|^2 = \\ &= H_{qq}\Re\{AB^\star\} + H_{qu}\Im\{AB^\star\} + \\ &+ H_{qa}|A|^2 + H_{qb}|B|^2 \end{aligned} \quad (6)$$

$$\begin{aligned} U_m &= |C_3|^2 - |C_4|^2 = \\ &= H_{uq}\Re\{AB^\star\} + H_{uu}\Im\{AB^\star\} + \\ &+ H_{ua}|A|^2 + H_{ub}|B|^2 \end{aligned} \quad (7)$$

where

$$H_{qq} = 2\Re\{S_{2a}S_{2b}^\star - S_{1a}S_{1b}^\star\} \quad (8)$$

$$H_{qu} = -2\Im\{S_{2a}S_{2b}^\star - S_{1a}S_{1b}^\star\} \quad (9)$$

$$H_{uq} = 2\Re\{S_{3a}S_{3b}^\star - S_{4a}S_{4b}^\star\} \quad (10)$$

$$H_{uu} = -2\Im\{S_{3a}S_{3b}^\star - S_{4a}S_{4b}^\star\} \quad (11)$$

$$H_{qa} = |S_{2a}|^2 - |S_{1a}|^2 \quad (12)$$

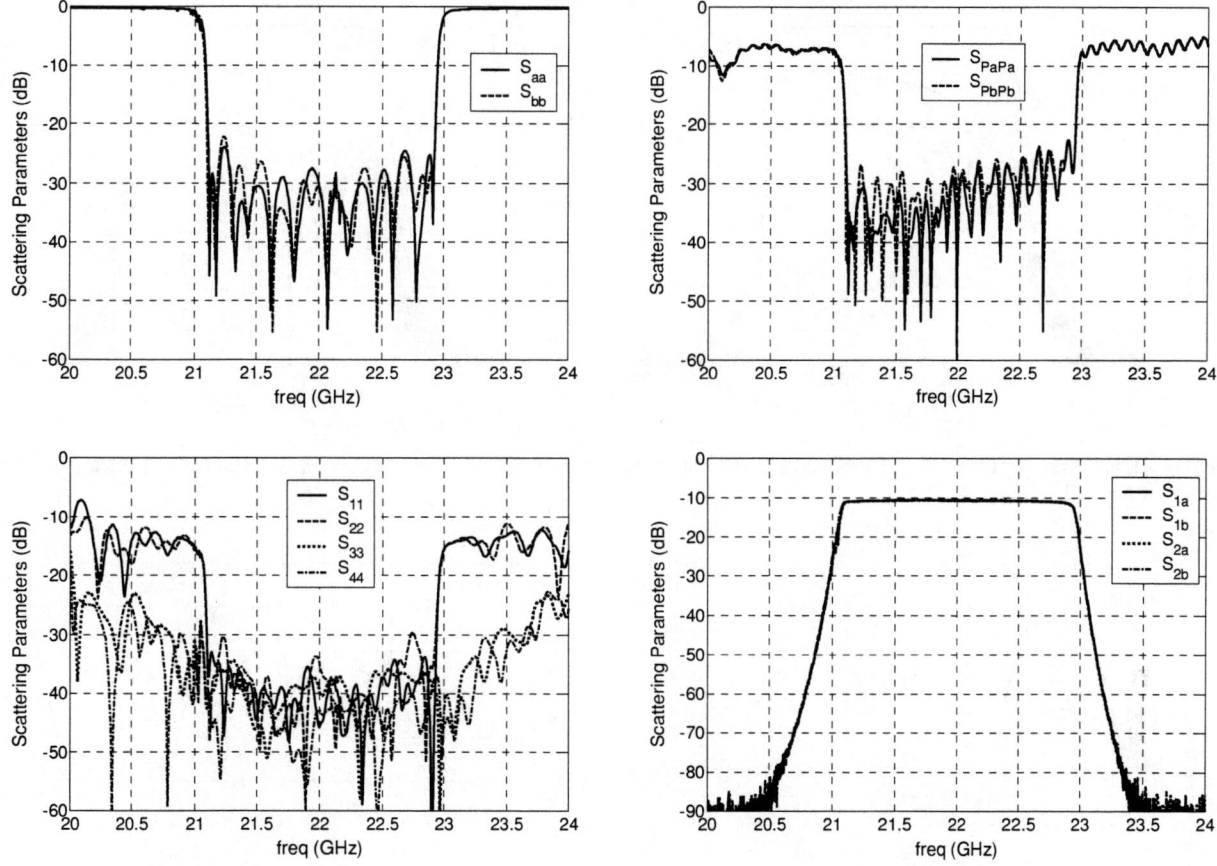

FIGURE 5. Measured reflection and transmission scattering coefficients of the K band FL-HPD.

$$H_{qb} = |S_{2b}|^2 - |S_{1b}|^2 \qquad (13)$$

$$H_{ua} = |S_{3a}|^2 - |S_{4a}|^2 \qquad (14)$$

$$H_{ub} = |S_{3b}|^2 - |S_{4b}|^2 \qquad (15)$$

The eight transfer functions defined by the previous equations are shown in Fig. 6. These plots refer to the Ka band FL-HPD. One can observe that the rejection for the auto correlation terms is about 30 dB. Moreover, by their integration the detection error of the linearly polarized radiation is better than 0.28 dB for the amplitude and 1.83 deg, with an offset of 0.54 deg, for the direction. Fig. 7 shows the amplitude and phase error as a function of the polarization angle.

ACKNOWLEDGMENTS

The authors would like to thank M. Franciotti for the prototype manufacturing.

REFERENCES

1. S. Cortiglioni, S. Cecchini, M. Orsini, G. Boella, M. Gervasi, G. Sironi, R. Fabbri, J. Monari, A. Orfei, Ng. Kin-Wang, L. Nicastro, U. Pisani, R. Tascone, L. Popa, I. A. Strukov, " Sky Polarization Observatory (SPOrt): a Project for the International Space Station", *Proc. of ESA Workshop on Space Exploration and Resources Exploitation* , Cagliari, 20-22 Oct 1998
2. E.Carretti, R. Tascone, S. Cortiglioni, J. Monari, M.Orsini, "Limits Due to Instrumental Polarisation in CMB Experiments at Microwave Wavelengths", *New Astronomy, Elsevier Science B.V.*, Vol. 6/3, pp. 173-188, May 2001.

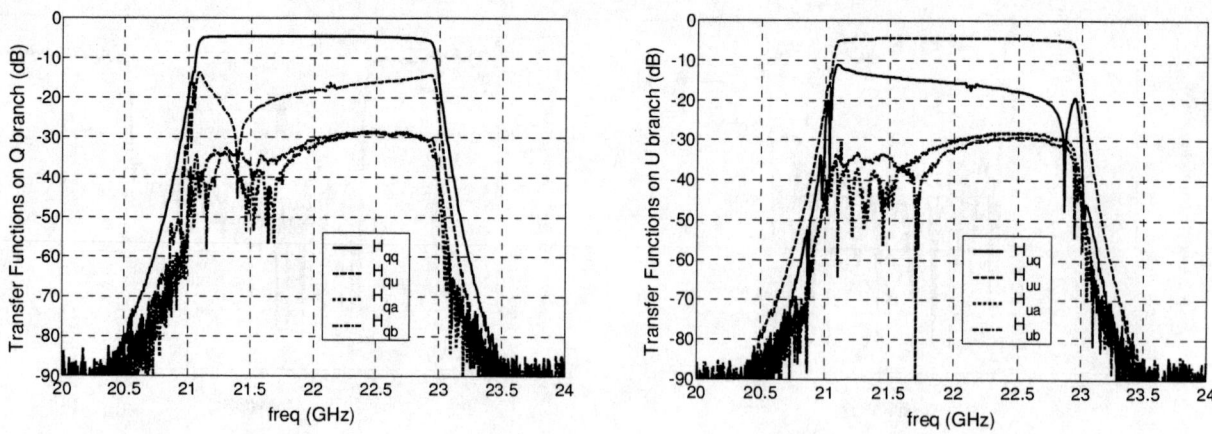

FIGURE 6. Transfer functions corresponding to the evaluation of the Q and U Stokes parameters obtained from the measurements of the K band FL-HPD prototype.

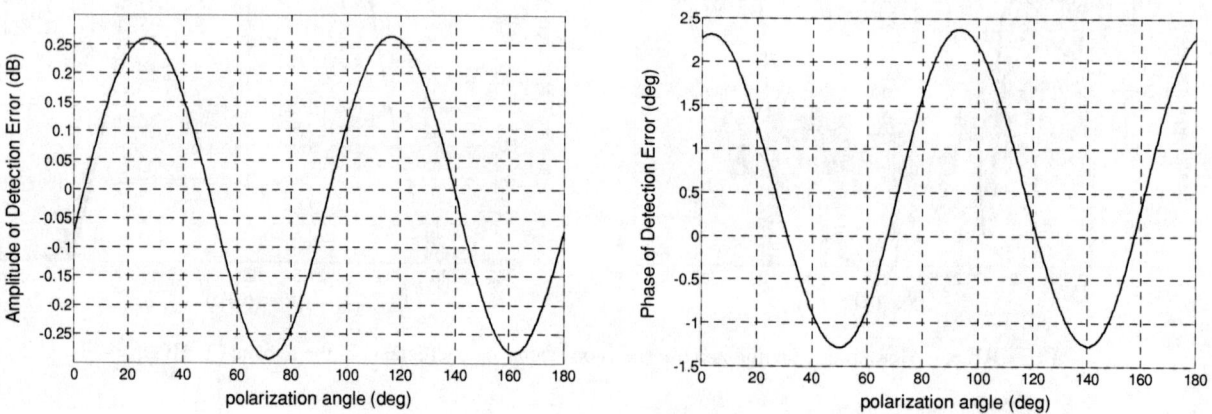

FIGURE 7. Detection error as a function of the polarization angle for a linearly polarized radiation. The angle is detected with an error less the 1.83 degrees and an offset of 0.54 degrees. As for the amplitude, the maximum error is 0.28 dB.

3. G. Sironi, G. Boella, G. Monelli, L. Brunetti, F. Cavaliere, M. Gervasi, G. Giardino, A. Passerini, "A 33Ghz polarimeter for observations of the Cosmic Microwave Background", *New Astronomy, Elsevier Science B.V.*, Vol. 3, pp.1-13, 1998.
4. R.Tascone P.Savi, D. Trinchero, R. Orta, "Scattering Matrix Approach for the Design of Microwave Filters", *IEEE Trans. Microwave Theory Tech.*, Vol.48, No. 3,pp 423-430, March 2000.

MITO-Pol, a polarimeter for the Testa Grigia Observatory

R. Maoli*, G. Savini* and F. Melchiorri*

*Experimental Cosmology Group, Dipartimento di Fisica, Università di Roma "La Sapienza"
P.le A. Moro, 2 - 00185 ROMA (Italy)*

Abstract. After a short review of the existing upper limits on CMB polarization and planned experiments, we present the solution adopted to resolve polarized signals with MITO-Pol, a polarimeter that will operate at the focal plane of MITO telescope. The first MITO-Pol campaign is planned for Winter 2002–2003. The main goals of this instrument are search for calibration sources and investigation of the polarization field around the CMB anisotropy peaks.

INTRODUCTION

The interest in CMB polarization originated in the seventies due to the possible detection of a huge quadrupole anisotropy (at the level of a few mK) claimed by Princeton Group (Yuma experiment). Rees noted [1] that an anisotropic universe could have a quadrupole anisotropy even larger than the dipole and would be characterized by a significant polarization of CMB. As a consequence Florence group attempted to detect the polarized signals through a balloon borne experiment (the data have been published and discussed several years later [2]).

This was the first far infrared experiment devoted to the study of the degree of polarization of the infrared sky. It posed an upper limit of $3 \cdot 10^{-4}$ to any polarized component of CMB at angular scales ranging from 0.5 degrees up to 90 degrees, thereby excluding the possibility that the anisotropy of the universe was at the level claimed before. Also, the experiment posed the first upper limit to the degree of polarization of the galactic emission in the $400 - 2000\,\mu m$ range. At the angular scale of 0.5 degrees the Galactic Center appeared not polarized with an upper limit of 10^{-4} in the shorter wavelength range. It is curious to note that this is still one of the most stringent upper limits to galactic contamination.

The Florence experiment (see fig.1) employed a rotating analyser in front of a Ge:Ga composite bolometer operating at $1.2\,K$, at the focal plane of a $30\,cm$ Cassegrain metallic telescope. The experiment clearly indicated the basic limitations arising from such a kind of modulation system. The residual atmospheric emission, polarized by the rotating analyser, produced a large offset due to the residual spurious polarization of the optics: the signal was subtracted by tilting the entire instrument at several elevations and fitting the signals with a secant law.

The results of the experiment were re-analysed in 1982 [3] and for the first time it was suggested that an universal magnetic field could be detectable through the Faraday rotation of the polarization plane at different wavelengths. Also it was pointed out that a primordial re-ionization could produce a huge increase in the degree of polarization.

PAST EXPERIMENTS IN POLARIZATION

Few upper limits are available at about the same large angular scale from other experiments performed at the same period. Nanos [4] obtained an upper limit of $1.5\,mK$ at $15°$ with a $3.2\,cm$ Faraday-switched polarimeter. Berkeley group [5][6][7] used a dual-mode corrugated horn with a Faraday rotation switch at $\lambda = 9\,mm$. Its best upper limit for the linear polarization of the CMB was $0.2\,mK$ at a $7°$ scale.

Other upper limits are available at small angular scales while little or nothing exists at $0.1 - 1$ degree, where we expect the largest signal. This situation is the consequence of the secondary role played by the observation of CMB polarization until a couple of years ago. This measurement has been rarely the main goal of an experiment, therefore it was obtained essentially as a by-product in CMB anisotropy observations.

FIGURE 1. The Florence experiment.

At small angular scales interferometers give access to all the four Stokes parameters. Using the VLA, Partridge and collaborators [8] [9][10] improved upper limits for the CMB polarization down to $30\mu K$ at an angular scale of 1 arcmin.

At intermediate scales, telescopes have to be used, implying an increasing difficulty to control the spurious polarization of the instrument. Before the HEMT development, telescopes were usually coupled with bolometers. These devices are intrinsically insensitive to the polarization. It is necessary to employ an optical element (i.e. wire-grid, half-wave plate etc.) that halves the sensitivity of the experiment to the CMB anisotropies. In this sense the choice of observing CMB polarization is alternative (or at least penalizing) regarding the CMB anisotropy observation.

The development of HEMT amplifiers and their availability to higher frequencies has improved the situation. The only upper limits to the CMB polarization at intermediate scales comes from the Saskatoon experiment [11] where an off–axis parabola is coupled to HEMT amplifiers at $30 GHz$. Even if polarization was not the main goal of this experiment, an upper limit of $25 mK$ at $1.44°$ was obtained.

PRESENT AND FUTURE OF POLARIZATION

After BOOMERANG [12] and MAXIMA [13] results, the situation is completely changed. CMB polarization became one of the main goals of observational cosmology and many polarization-dedicated experiment started to be planned.

In table 1 we reported the main characteristics of planned CMB polarization experiments. At small angular scales interferometric techniques are employed in the CBI and ATCA experiments while for POLATRON a $5.5 m$ Cassegrain telescope is coupled with spider web bolometers. In the future ALMA interferometer [14] will also play an important role.

The two "large scale" experiments, POLAR and MILANO, have also an intermediate scale version that coupled the feed horn with a $2.6 m$ Cassegrain telescope. MITO-Pol and PIQUE are two other ground-based experiment aiming to the CMB polarization detection at angular scales of 10–30 arcmin.

Always at intermediate scales three balloon borne experiment are planned: the polarization-dedicated version of BOOMERANG and MAXIMA and the Bar–SPOrt Experiment.

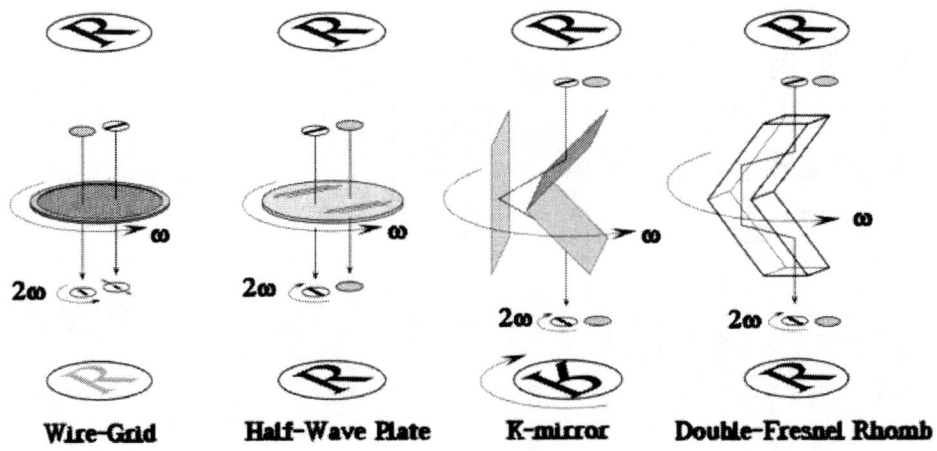

FIGURE 2. Different modulating techniques.

Three different space–based experiments will observe the CMB polarization in the next few years: MAP, Planck and SPOrt.

The polarized signal to detect is usually of a magnitude of a few percent of any relevant signal from an astrophysical or cosmological source, so it is fundamental to separate the incoming polarized radiation from the non-polarized foregrounds.

The separation must take place without producing any further polarization of radiation on the incoming signal. In the millimeter band, if bolometric receivers are used, this is obtained with a modulating technique applied on the polarized signal either by using a rotating wire-grid [17], a half-wave plate, a K-mirror or (in our case) a double-Fresnel rhomb (see fig.2). The signal is then extracted by de-modulating with a lock-in amplifier.

If we consider a polarized signal I_p and a non polarized signal $I_0 >> I_p$ we see that due to its nature the output of the signal after the modulation depends on which of the previous methods has been adopted. In the case of a rotating polarizer, the output signal will be totally polarized and only the polarized component will have a frequency dependence (see fig.2). Yet, even the slightest mis-alignment of the axis rotation of the polarizer with the optical axis would produce a residual polarized signal with the same rotating frequency dependence. Intensity in this case will be distributed (where distribution is referred to the intensity of radiation collected in two orthogonal modes of polarization -see fig.4-), in the way that previously non-polarized radiation can be efficiently removed by computing the difference between the two signals from the detectors that collect radiation split by a 45° wire-grid used as a beam-splitter for the two orthogonal modes of polarization. For this reason, care must be taken in the choice of radiation collectors to avoid differences in spurious polarization.

Other methods are preferred for they leave the non-polarized signal untouched (except for an attenuation constant in the case of the double-Fresnel rhomb or half-wave plate). In the case of the K-mirror [16], this modulation technique has two major disadvantages: the first being that it implies the rotation of the image as well as of the polarized component, and if adopting an array of detectors this excludes its employment; the second is in the mechanical constraint, any slight mis-alignment can result in a spurious polarized signal due to spurious polarization of radiation that is reflected by the plane mirrors with an angle that differs from 0° or 90° (this would negate the effort to detect very faint signals, like the polarized cosmological sources).

MITO-POL – A POLARIZATION DEDICATED INSTRUMENT

The experimental setup that we intend to employ at the MITO telescope[15] has been chosen between the different solutions described before.

We prefer the use of a half-wave plate or a double-Fresnel rhomb (see fig.3), for they achieve the rotation of only the polarized radiation, leaving the rest untouched. The rhomb can be employed if the right material is adopted. Most

TABLE 1. Ground and balloon-based Polarization experiments...

Name outlet	Frequency/ bandwidth (GHz)	Angular scale	Sensitivity	Sky coverage	Detectors	Polarimeter	Telescope	Upper limit
CBI *	26-36	3′ – 30′	20 (μK/night)	-	HEMT	Interferometer	13 × 90cm	-
ATCA †	8.7	2′	-	6 pix	HEMT	Interferometer	5 × 22m	$\leq 11 \mu K$
POLATRON **	90/20	2.5′	0.7 ($mK/\sqrt{Hz}$)	850 pix/6 months	Bolom.	OMT	5.5m OVRO	-
PolKa ‡	350	10″ – 1′	-	-	Bolom.	rotating analyzer	IRAM	$\leq 300 \mu K$
MITO-Pol §	150-350	5′	1 ($mK \cdot \sqrt{s}$)	10° × 10°	HEMT	Fresnel-rhomb+MartinPupplet	(see MITO)	-
PIQUE ¶	40,90/16	14′	2 ($mK \cdot \sqrt{s}$)	~25 pix	HEMT	OMT	1.4m	$\leq 10 \mu K$
MilanoPol ‖	33 / 1.5	7° – 14°	1 ($mK/\sqrt{Hz}$)	-	HEMT	correlator	-	-
MilanoPol2 ††	33 / 1.5	15′	1 ($mK/\sqrt{Hz}$)	-	HEMT	correlator	MITO	-
POLAR ‡‡	26-36,90-100	7°	1 ($mK/\sqrt{Hz}$)	1844°	HEMT	OMT	-	$\leq 10 \mu K$
COMPASS §§	26-36,90-100	10′-20′	1 ($mK \cdot \sqrt{s}$)	-	HEMT	OMT	(see MITO)	-
MAXIPOL ¶¶	150,240,410	10′	.041 ($mK \cdot \sqrt{s}$)	10^6 pix	Bolom.	grid.pol.+HWP	1.3m	-
Boomerang2K ***	90,150,240,410	10′	200 ($\mu K/\sqrt{Hz}$)	10^6 pix	Bolom.	polarization abs.	1.3m	-
BAR-Sport †††	32,90	30′,12′	.5 , .7 ($mK \cdot \sqrt{s}$)	20° × 20°	HEMT	OMT	-	-
MAP ‡‡‡	22,30,40,60,90	13′ – 56′	35 ($\mu K/pix$)	full sky	HEMT	OMT	1.4m × 1.6m	-
Planck §§§	30-100,143,217,545	5′ – 33′	6 ($\mu K/pix$)	full sky	HEMT + Bolom.	-	1.5m	-
SPOrt Project ¶¶¶	22,32,60,90/10%	7°	1 ($mK \cdot \sqrt{s}$)	82% of the sky	HEMT	OMT	-	-

* see J.K. Cartwright et al. contribution to these Proceedings
† Subrahmanyan R., et al., Mon. Not. R. Soc., **315**, 808 (2000)
** Philhour B.J., et al., astro-ph 0106543
‡ see G. Siringo et al. contribution
§
¶ Hedman M.M., et al., Ap. J. Letters, **548**, L111 (2001)
‖ see M. Zannoni et al. contribution
†† see M. Zannoni et al. contribution
‡‡ see B.G. Keating et al. contribution
§§ see L. Piccirillo et al. contribution
¶¶ see P.L. Richards contribution
*** see S. Masi et al. contribution
††† see E. Carretti et al. and C. Macculi et al. contributions
‡‡‡ MAP home page: http://map.gsfc.nasa.gov
§§§ see N. Mandolesi et al. and J.M. Lamarre et al. contributions
¶¶¶ SPOrt home page: http://sport.tesre.bo.cnr.it/

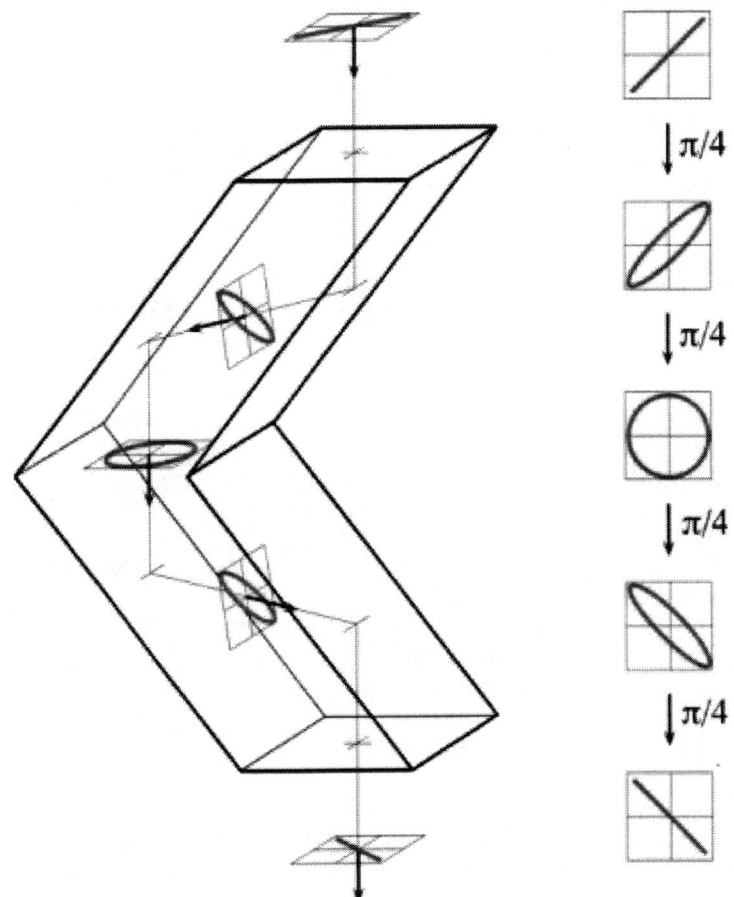

FIGURE 3. Double-Fresnel rhomb.

important is the constant refraction index in the spectral range that we intend to adopt, for slight variations of the latter determine a rotation (like in the half-wave plate) of a fraction of the linearly polarized radiation, decreasing the signal. We have chosen HDPE (High Density Polyethylene) that fulfils this requirement and moreover, is easily machinable. Unfortunately the optical path enclosed in the double-rhomb implies an absorbing coefficient due to non-unity of the rhomb transmittance coefficient. On the other hand the employment of a half-wave plate, preferable for transmittance reasons, implies the choice of a narrower bandwidth.

To achieve spectral information while efficiently rejecting residual unpolarized radiation of our instrument, we have interposed a Martin-Puplett [18] interferometer in the stage between the modulator and the final polarizing beam-splitter (see fig. 4). In this way, we can obtain the interference pattern due to coupling between the phase-shift of the modulated-polarized signal and the presence of the subsequent fixed polarizer [19].

The interferogram is related by the Fourier transform to the spectral distribution of the radiation. Fourier-transform spectroscopy allows us to have a "frequency multiplexing" due to the collection of all the spectral signal during the integration time of the detectors, and not only that of a small frequency interval [21]. In this way the effective integration time coincides with the entire interferogram procedure interval, therefore maximizing the signal-to-noise ratio.

If we consider a system that has its focal plane at the exit of this setup (as to make use of an array of pixels), its geometry would be strongly constrained by dimensions and disposition of optics. In our case, maximization of signal has been pursued, with a single pixel configuration, through ray-tracing of the setup [20]. Presently, the optics are such that the image of the primary mirror is focused on the entrance of the Winston cones.

Employment of such a setup in other experiments with arrays sets limits on its efficiency with the f-number. The choice lies between transforming the beam in a larger f-number one (thus losing efficiency in the process), or

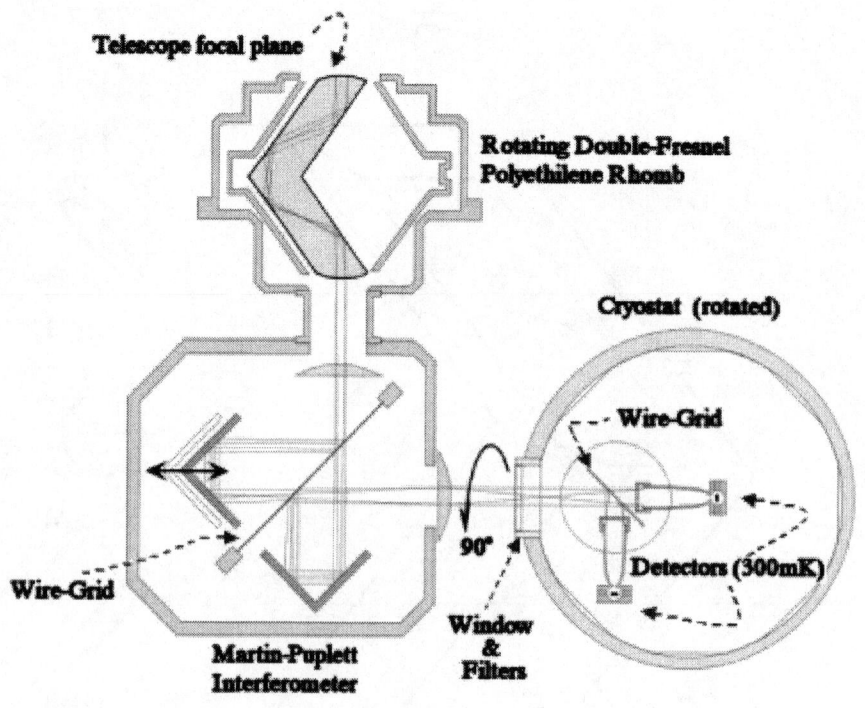

FIGURE 4. MITO-Pol setup.

constructing a setup with the focal plane halfway through the optical path. Re-imaging in any case is most problematic if we consider that at first order, there is a different optical path for the two split beams.

CONCLUSIONS

Experiments aiming at the measurement of the CMB polarization (and not only at the establishment of upper limits) have to face the problem of a poor knowledge of the polarization properties of the foregrounds. While these properties could be disregarded when instrument sensitivities were at a level of $50-100\mu K$, thanks to a careful choice of the sky region and of the observational frequencies, this is not feasible when the expected signal is $\sim 1-5\mu K$. Therefore other experiments whose goals are the measurement the foreground polarization are needed.

Another primary necessity for a successful CMB polarization experiment is the availability of calibration sources. If CMB dipole is a good calibrator for anisotropy experiments, this is not the case for the polarization ones; other calibration sources are scarce and not well known. The search for these sources can also be a first target for polarization experiments.

Finally, anisotropy experiments started to produce maps of the sky. Therefore it is possible to study the structure of the polarization field around the peaks of the CMB anisotropies. Theoretical models foresee that polarization can be as high as 30–45 % [22] and the expected signal at a level of $\sim 100\mu K$.

MITO-Pol will be put at the focal plane of the $2.6m$ MITO telescope in Winter 2002–2003. For its angular resolution ($\sim 5'$) and sensitivity, MITO-Pol is very well suited for observation of these three intermediate targets, of primary importance for the study of the CMB polarization.

ACKNOWLEDGMENTS

The realization of MITO polarimeter has been part of the thesis of G. Siringo, L. Terenzi, L. Moretti and of the PhD thesis of G. Pisano.

REFERENCES

1. Rees M., *Ap. J. Letters*, **153**, L1 (1968).
2. Caderni N., *et al.*, *Phys. Rev. D*, **17**, 1908 (1978).
3. Ceccarelli C., *et al.*, "The Birth of the Universe" – Proc. 17th Moriond Meeting, eds. J. Audouze and J. Tran Thanh Van, 191 (1982).
4. Nanos G.P., *Ap. J.*, **232**, 341 (1979).
5. Lubin P., and Smoot G., *Phys. Rev. Lett.*, **42**, 129 (1979).
6. Lubin P., and Smoot G., *Ap. J.*, **245**, 1 (1981).
7. Lubin P., Melese P., and Smoot G., *Ap. J. Letters*, **273**, L51 (1983).
8. Partridge R.B., Nowakowski J., and Martin H.M., *Nature*, **331**, 146 (1988).
9. Fomalont E.B., *et al.*, *Ap. J.*, **404**, 8 (1993).
10. Partridge R.B., *et al.*, *Ap. J.*, **483**, 38 (1997).
11. Wollack E.J., *et al.*, *Ap. J. Letters*, **419**, L49 (1993).
12. de Bernardis P., *et al.*, *Nature*, **404**, 955 (2000).
13. Hanany S., *et al.*, *Ap. J. Letters*, **545**, L5 (2000).
14. ALMA home page, http://www.mma.nrao.edu
15. De Petris M., *et al.*, *New Astronomy*, **1**, 121 (1996).
16. Hildebrand R.H., *Opt.Eng.*, **25**, 2 (1986).
17. Cudlip W., *et al.*, *Mon. Not. R. Soc.*, **200**, 1169 (1982).
18. Martin D.H., and Puplett E., *Infr.Phys.*, **10**, 105 (1969).
19. Lambert D.K., and Richards P.L., *Appl.Opt.*, **17**, n.10, 1595 (1978).
20. Pisano G.,"Progettazione di un Polarimetro per Misure di Fondo Cosmico nel Lontano Infrarosso", *Tesi di Laurea*, (1996).
21. Polavarapu P.L., in "Principles and applications of Polarization-Division Interferometry", (1998).
22. Arbuzov P., *et al.*, *Int. J. of Mod. Phys.*, **6**, 515 (1997).

The Milano polarimeter: an instrument to search for large scale polarization of the Cosmic Microwave Background

M. Zannoni[*], E. Battistelli[*], G. Boella[*], F. Cavaliere[†], M. Gervasi[*], A. Passerini[*] and G. Sironi[*]

[*]*Dipartimento di Fisica "G. Occhialini" - Università di Milano Bicocca - Piazza della Scienza 3, 20126 Milano, Italy*
[†]*Dipartimento di Fisica - Università di Milano - Via Celoria 16, 20133 Milano, Italy*

Abstract. The expected polarized component of the CMB signal is extremely faint. On degree angular scales and above it can be in the range of one part to 10^6 or 10^7. To detect such an elusive signal a dedicated instrument must be designed. In this contribution we present the new Milano polarimeter, a 33 GHz correlation system able to detect both linear and circular polarization. It has been designed to operate at large angular scale ($7° - 14°$) from the ground. In the light of the last observing campaign conducted from Antarctica we have partially redesigned the radiometer to improve its long term stability and sensitivity.

INTRODUCTION

Among the fine characteristics of the Cosmic Microwave Background (CMB), the polarization properties are still unknown. The residual polarization of the CMB was originated by the anisotropic scattering of photons with matter during the recombination era [1]. This is the reason why it is expected to have a linear polarization correlated with the spatial fluctuations of the CMB temperature. In particular the polarized radiation (ΔT_p) is expected to be a small fraction of the temperature anisotropy (ΔT_a): $\Delta T_p/\Delta T_a = \alpha$ with $\alpha \lesssim 0.1$. α is maximum at the horizon scale at recombination ($\sim 1\ deg$), but is rapidly decreasing at larger angular scales (for a review see [2]).

The positive detection of the CMB polarization is important to complete the picture of the universe as it is emerging from the existing measurements of the anisotropy. In fact, it will confirm the cosmological nature of the anisotropies discovered in the last decade [3], [4], [5]. A measurement of the polarization power spectrum will remove some degeneracy and help to distinguish among scalar, vector (vortices) and tensor modes of matter density perturbations. Besides the study of the CMB polarization at large angular scales ($\theta > 1\ deg$) is the most powerful instrument to investigate the partial reionization of the matter in the Universe after the recombination era.

Finally, deviations from a uniform and isotropic expansion or the existence of rotational modes in the expansion of the Universe, associated to primordial magnetic fields, produce a signature of polarized CMB. In this case circular polarization can be also detected [6]. No experimental evidence of CMB polarization has been obtained so far. In fact, only upper limits have been produced, mainly for linear polarization (for a review see the related papers in this volume). The best upper limits (few parts over 10^6) are still higher than the polarization expected, but the sensitivity of the experiments is now approaching the required one for a detection.

THE POLARIMETER'S CONCEPT AND PERFORMANCES

The instrument design

We have developed an instrument sensitive to the two polarized components of the incoming radiation. The polarimeter is a correlation receiver for ground based observations. It is sensitive to large angular scales, therefore it is able to investigate a possible reionization of the post-recombination universe [7]. The main characteristics of the

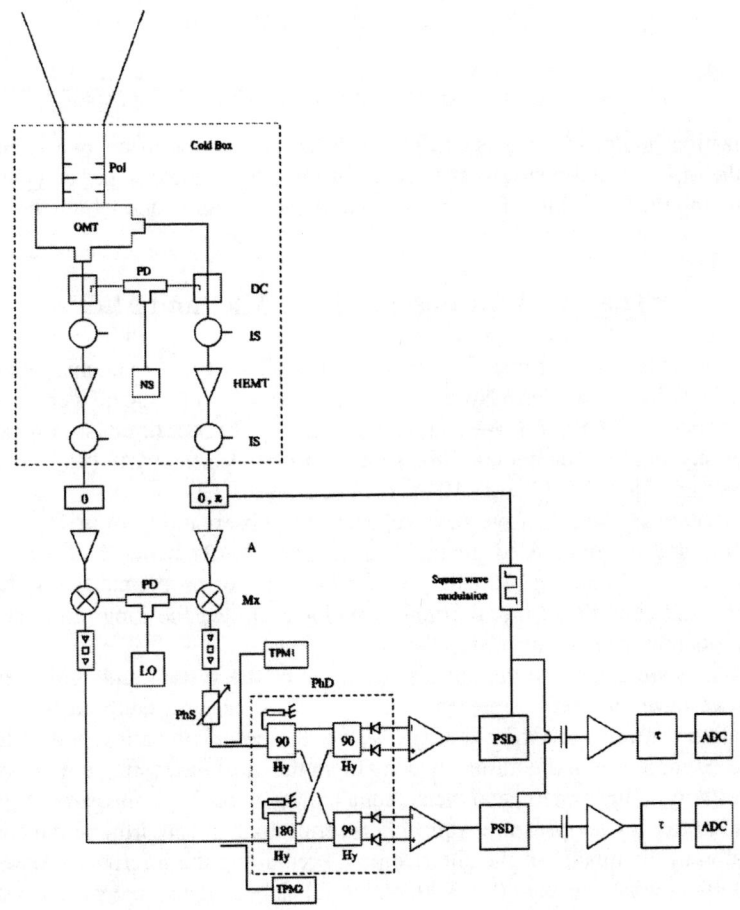

FIGURE 1. Schematic of the polarimeter. See the text for details.

polarimeter are summarized in Table 1 (for a deep discussion see [8]). In Figure 1 a schematic of the instrument, in the present setup (**Mk3**) is shown.

TABLE 1. Main characteristics of the polarimeter.

Detection technique	Etherodyne Correlation receiver
Observing frequency	33 GHz
Intermediate frequency	3 GHz
Frequency band	1.5 GHz
Beamwidth FWHM	$7° - 14°$
Polarization mode	Linear & Circular
Outputs	U, Q, V Stokes parameters

The angular resolution can be switched from 14° to 7° by adding an extension to the corrugated conical horn we use to observe the sky. The polarization mode can be switched from circular to linear by adding an iris polarizer (Pol) before the separation of the two components by means of an orthomode transducer (OMT). The front end is cooled down to $\sim 20\,K$ by a mechanical cryogenerator. The front-end amplifiers are low noise cryogenic HEMTs.

The RF ($\nu = 33\,GHz$) is downconverted (to $\nu_{if} = 3\,GHz$) using the same low phase noise oscillator (LO) for the two channels. A phase discriminator (PhD) represents the correlation unit. The outputs of the system give directly the Stokes parameters. Using the iris polarizer we will get the parameters U and Q describing the linear polarization:

$$O_1 = K_1 <E_A E_B> \cos(\delta+\phi) = K_1' (Q\cos(\phi) - U\sin(\phi)) \tag{1}$$

$$O_2 = K_2 <E_A E_B> \sin(\delta+\phi) = K_2' (Q\sin(\phi) + U\cos(\phi)) \tag{2}$$

Here δ is the polarization angle, while ϕ is the phase difference between the two channels introduced by the instrument. Adjusting the instrumental phase by the phase shifter (PhS) we can separate Q and U in the two outputs. On the other hand, removing the iris polarizer we get the parameters U and V describing the circular polarization.

The Antarctic observations: Mk1 and Mk2

The Milano polarimeter has been used for sky observation during two different campaign in Antarctica. In 1994-95 we have used the prototype (**Mk1**) from Terra Nova Bay, at see altitude. While the concept was the same as described, the front-end cooling was poor ($T \geq 100\,K$). We could operate the system continuously for two months observing the linear polarization of the sky, but the weather conditions have limited the sensitivity [9]. The results we have obtained are upper limits to both 7° and 14°: $\Delta T_p/T \sim 1 \cdot 10^{-4}$.

During the 1998-99 Antarctic campaign we have repeated the observations from Dome Concordia at 75°06′ S latitude and 3233 m altitude. We have used an updated version (**Mk2**) which had the front-end in the present configuration cooled down at $\sim 20\,K$. Unfortunately we had a series of problems during the observations, because of the extreme environmental conditions: first thermal drifts have limited the long term stability of the instrument; then a failure of the cryogenerator forced us to stop the observations.

In spite of these problems we could test the correct operation of the system and could verify the effectiveness of calibration techniques: we have an internal reference mark, an external grid calibrator, and a total power monitor. Combining the information coming from these sources we get the complete calibration of the system [10], [11]. In addition we have measured some useful quantities. Among them the most interesting are the system noise temperature (T_{sys}), related to the sensitivity of the system, and the antenna temperature (T_{ant}), measured by the total power monitor, and resulting by the sum of the sky temperature, atmospheric emission, and environmental contribution (see Table 2). T_{ant} gives us an estimation of the quality of the site (Dome C) regarding the microwave transparency. T_{sys}, as shown in Table 2, is a bit higher than what expected ($80 - 90\,K$), but it has been measured with a system poorly stable.

TABLE 2. The system and antenna temperature and calibrated gain factor at Dome C with the **Mk2** version of the polarimeter.

	T_{sys} (K)	T_{ant} (K)	Gain ($\mu K/ADU$)
Channel A	145 ± 2	7.5 ± 0.1	15.7 ± 0.9
Channel B	120 ± 2	8.5 ± 0.2	16.5 ± 0.9

Improvements: Mk3

In order to avoid the problems suffered during the Antarctic campaign in Dome C we decided to improve the polarimeter through the following actions:

1. We add a phase modulation $(0,\pi)$ with a synchronous detection (PSD), at a frequency of $781\,Hz$. This technique helps in preventing the $1/f$ noise and reducing the instrumental off-set. The improvement can be understand in terms of the ratio R between the fluctuations with (σ_l) and without (σ_{nl}) the phase modulation system [12]. We get:

$$R = \frac{\sigma_{nl}}{\sigma_l} \simeq \sqrt{2\ln\frac{T}{\tau}} \tag{3}$$

Here T is the total observing time (several days), while τ is the system response time (fraction of a second). It is evident the large improvement in few minutes of operation.

2. We insert the whole instrument inside a thermalized ambient, built using an insulated tent. Only the horn will stay outside. An active system of temperature control will complete the system. In this way we will protect the external electronics and the mechanic cryocooler from failure in case of large temperature variations. Besides we will help the thermalization of the sensitive electronics.

3. We replace the HEMT amplifiers with a novel twins provided by NRAO. The new HEMTs have a gain $\sim 5\ dB$ better and a noise temperature definitely lower ($T_N \sim 15\ K$ at $\sim 20\ K$ of operation temperature). Therefore we will have a system temperature $T_{sys} \sim 40 - 50\ K$, and a noise equivalent temperature $\Delta T_{min} \sim 1\ mK/\sqrt{Hz}$.

The current version of the polarimeter is more stable and allow to cancel out a large part of the instrumental off-set and to insert a larger gain. A better sensitivity ($\mu K/ADU$) is finally obtained.

OBSERVATION PROGRAMS

1. Observation from the **Testa Grigia Observatory**.
   (a) The polarimeter will start observation from the Testa Grigia site [13] on the Italian Alps, at 3480 m above see level. We will address first our investigation at large angular scales ($7° - 14°$). Observation will be carried out in transit mode, at a declination close to the North Celestial Pole.
   (b) Then we will place the receiver at the focal plane of the MITO telescope in order to observe with an angular resolution of $10 - 20\ arcmin$ [13], [14]. Attention has to be paid in order to avoid spurious polarization, even if the complete on-axis configuration of the Cassegrain optics helps [14].
2. Later the polarimeter will be operated from the **Antarctic Plateau** in order to take advantage of the long Antarctic night. In fact, long duration observations in a quasi static weather condition are possible, avoiding the solar radiation disturbances. Dome Concordia will become a permanent station in the next years and is a good place to operate the polarimeter. Natural targets are the South Celestial Pole (SCP) regions at $7°$ and $14°$. Observing at the zenith a circle around the pole, $15°$ apart, is described. Besides if we observe in direction of the SCP, in case of a linear polarization, we have a typical signature: $O_1 \propto \sin(4\pi t/T)$ and $O_2 \propto \cos(4\pi t/T)$ (t is the sidereal time, and T a sidereal day). Finally we can try to operate the receiver at the focal plane of some existing telescope, in order to explore the degree angular scale region.

ACKNOWLEDGMENTS

This research was carried out within the Concordia Project (supported by IFRTP and PNRA), and has been supported by PNRA and CSNA, by CNR, by MURST and Universities of Milano and Milano-Bicocca.

REFERENCES

1. Rees, M. J., *ApJ*, 153, L1, 1968.
2. Hu, W., and White, M., *New Astronomy*, 2, 323, 1997.
3. de Bernardis, P., et al., *Nature*, 404, 955, 2000.
4. de Bernardis, P., et al., *this volume*, 2001.
5. Hanany, S., et al., *ApJ*, 545, L5, 2000.
6. Kosowsky, A., *New Astronomy Rev.*, 43, 157, 1999.
7. Gervasi, M., et al., *AIP Conf. Proc.*, 476, 154, 1999.
8. Sironi, G., et al., *New Astronomy*, 3, 1, 1998.
9. Sironi, G., et al., *ASP Conf. Ser.*, 141, 116, 1998.
10. Zannoni, M., *PhD Thesys*, University of Milano, 1999.
11. Gervasi, M., et al., *SIF Conf. Proc.*, 68, 165, 2000.
12. Spiga, D., *Degree Thesys in Physics*, University of Milano, 2000.
13. De Petris, M., et al., *New Astronomy*, 1, 121, 1996.
14. De Petris, M., et al., *Applied Optics*, 28, 1785, 1989.

Scanning polarimeters for measurements of CMB polarization

S. Masi*, P. de Bernardis*, G. De Troia*, P. Natoli†, F. Piacentini* and G. Pisano*

*Università degli Studi di Roma La Sapienza - Roma - ITALY
†Università degli Studi di Roma Tor Vergata - Roma - ITALY

Abstract. The polarization status of the Cosmic Microwave Background (CMB) is still undetected. The BOOMERanG payload has been reconfigured as a scanning polarimeter, and its launch is planned for 2002. The same experimental approach has been selected for Planck-HFI and other experiments. Here we discuss some issues related to this methodology, such as the level of sensitivity required, the effects of instrumental cross-polarization and its measurement, and the best scan strategy.

INTRODUCTION

In the current cosmological scenario, CMB photons have been last scattered by free electrons at recombination ($z \sim 1100$). Thomson scattering produces linear polarization in the scattered light only if the incoming photon distribution has a non-zero quadrupole moment [1]. The small scalar (density) fluctuations which are believed to have originated the CMB anisotropy (and the large scale structure of the present universe) were present at recombination, and are a source of anisotropy in the distribution of incoming photons. The main terms of the anisotropy are monopole and dipole, while the quadrupole term is much smaller: for this reason the expected polarization is very low [4], [2]. The statistical properties of the polarization pattern we expect to see, due to this effect, can be computed together with the statistical properties of the anisotropy pattern. The latter is now starting to be constrained. It is thus possible to compute the power spectra of the polarization expected for the same cosmological scenario best fitting the current anisotropy data. This scalar component of the CMB polarization is expected to be curl-free, and is called the **E** component of the polarization. Moreover, in the inflationary scenario, tensor fluctuations (gravitational waves) are generated in the very early Universe, thus producing an additional (and different) pattern of polarization in the CMB. These fluctuations introduce both curl-free and curl (labeled **B**) components in the polarization pattern. There are thus several good reasons to search for such small polarization signature in the CMB sky. A cross correlation spectrum <ET> is expected as well (see e.g. [3]), with amplitude larger than the pure scalar and tensor polarization spectra. In fig.1 we plot the inflationary adiabatic power spectrum best fitting the BOOMERanG anisotropy data, and, for the same cosmological parameters, the cross spectrum anisotropy-E-polarization and the power spectrum of the **E** polarization. The **B** component is not plotted since it is expected to be very small. Changing the cosmological parameters within the ranges consistent with the BOOMERanG measurements of the anisotropy does not change much the resulting polarization. The big experimental problem is that the polarization signal is very small with respect to the CMB itself and even smaller with respect to many instrumental and astrophysical contaminants. Despite very intense and long lasting efforts [5], [6], [7], [8], [9], [10], [11], [13], [12], CMB polarization has not been detected yet. The current upper limits for the polarization power spectrum are reported in fig.1.

CMB POLARIMETRY WITH POLARIZATION SENSITIVE BOLOMETERS

Most polarimeters for CMB polarization studies use a modulating analyzer to extract, by means of synchronous demodulation techniques, the small polarized component embedded in a much higher unpolarized signal. Rotating wire grids, half wave plates, K-mirrors and Fresnel Rombs [14] have been used or proposed as polarization analyzers.

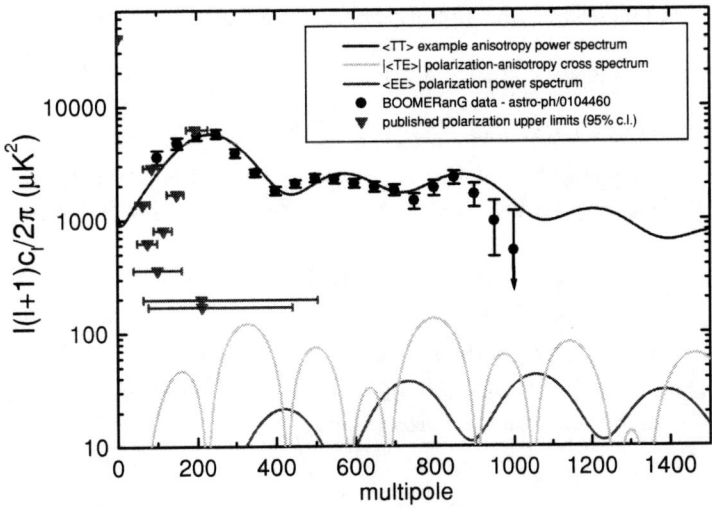

FIGURE 1. CMB anisotropy power spectrum, anisotropy-polarization cross spectrum, and polarization power spectrum. The upper limits come from the experiments cited in the text.

All these techniques suffer for the need of long integration time: one needs to point to the same sky pixel during many cycles of the analyzer. It is thus difficult to produce extended maps of the CMB polarization.

We have investigated the feasibility of mapping the two orthogonal components of the linear polarization with two separate bolometers, $B1$ and $B2$, each sensitive to one polarization, and combine the two signals to retreive the Stokes parameters Q and U. In some sense, this is similar to what is done with coherent detectors. Polarization sensitive bolometers of this kind have been prepared for the coming flight of BOOMERanG and for the High Frequency Instrument of the Planck satellite [15].

Signal modulation is obtained by scanning the sky at a constant rate v, so that the polarization signals are detected at a frequency f far from the 1/f knee of the noise and far from the effects of instrumental drifts. This sky scan approach is similar to the anisotropy measurements performed with BOOMERanG in its long duration flight in 1998 (B98) [16], [17]. In fact $f = v\ell/\pi$; with $\ell = 300 - 1000$ and $v = 1°/s$ we get $f = 1.7 - 5.5 Hz$, well inside the useful bandwidth of the 0.3 K bolometers used in the BOOMERanG instrument.

Combining signals from two different detectors instead of using always the same detector to measure the two polarization components is a significant concern (see e.g. [18]). One has to demonstrate that this procedure does not introduce significant systematics. We carried out simulations to investigate the problem in a quantitative way.

We define the Co-polar response of the polarimeter C as the response of the polarimeter to incoming radiation 100% polarized along the principal axis of the polarimeter. It is the product of the detector responsivity times the integral of the Co-polar beam response. The units will be V/W or V/K. The cross polar response X instead is the response of the polarimeter to incoming radiation 100% polarized and orthogonal to the principal axis of the polarimeter. It is the product of the detector responsivity times the integral of the cross-polar beam response. We will also define the polarization efficiency $\varepsilon = 1 - X/C$.

The signal detected from the two bolometers will be the sum of different components. We discuss now the frequency content of each of these components. A first order approximation for bolometer $B1$ will be:

$$S_1 = C_1(T_{CMB} + O + D + \Delta B/2 + \Delta T_U/2 + \Delta P_1) + N_1 + X_1(T_{CMB} + O + D + \Delta B/2 + \Delta T_U/2 + \Delta P_2)$$

where C_1 and X_1 are the co-polar and cross-polar response of bolometer 1; O and D are the instrumental offset and its slow drift; ΔB are the fluctuations of the background in the frequency range where CMB polarizations signals are expected; ΔT_U is the CMB anisotropy (unpolarized component); ΔP_1 is the CMB anisotropy component polarized along direction 1; N_1 is the noise of channel 1.

Looking to the frequency content of each of these components, we see that the dominant signals can be removed by means of a high-pass filter, removing the constant terms and slow drifts (T_{CMB}, O, D). The other components are all detected in the frequency range of interest. So the high-pass filtered signals can be written

$$S_1 = (C_1 + X_1)(\Delta B + \Delta T_U)/2 + C_1 \Delta P_1 + X_1 \Delta P_2 + N_1$$

$$S_2 = (C_2 + X_2)(\Delta B + \Delta T_U)/2 + C_2 \Delta P_2 + X_2 \Delta P_1 + N_2$$

Let's assume $C_1 C_2 \gg X_1 X_2$ (for example with a polarization efficiency = 90% we have already $0.81 \gg 0.01$) and do the math: we solve for ΔP_1 and ΔP_2 and compute the Stokes parameter. We find that the Stokes parameter $Q = \Delta P_1 - \Delta P_2$ is the sum of three terms:

- the difference signal $(S_1/C_1)(1 + X_2/C_2) - (S_2/C_2)(1 + X_1/C_1)$;
- the common mode term $(\Delta B + \Delta T_U)(X_1/C_1 - X_2/C_2)/2$;
- the noise term $(N_2/C_2)(1 + X_1/C_1) - (N_1/C_1)(1 + X_1/C_1)$.

The difference signal term is simply $S_1 - S_2$ with each of the two signals corrected for different co-polar and cross-polar responses, including different responsivities, efficiencies, etc. .

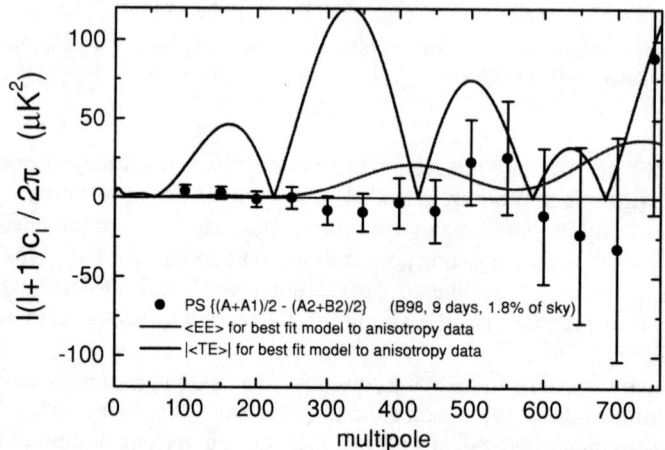

FIGURE 2. Power spectrum of the difference between independent channels in the B98 data. These channels were not sensitive to polarization, so the difference is expected to be 0. However, the size of the error bars gives an idea, based on real world data, of how sensitive a CMB polarization search based on comparing signals from different bolometers can be. The curves for $|<TE>|$ and $<EE>$ for the parameters best fitting the B98 anisotropy data are also plotted for comparison.

The common mode signal $(\Delta B + \Delta T_U)$ is reduced by a factor $(X_1/C_1 - X_2/C_2)/2$ with respect to the polarization signal. This means that it disappears if the two detectors have identical polarization efficiencies. Otherwise this term will become comparable to what we want to measure if $(X_1/C_1 - X_2/C_2)\Delta B/2 \sim \Delta P$ and $(X_1/C_1 - X_2/C_2)\Delta T_U/2 \sim \Delta P$. Notice that perfect responsivity matching of the two bolometers is not important for this term. If we are able to reduce the cross-polar response of the two bolometers below 5% of the co-polar one, then ΔB and ΔT_U can be up to 40 times larger than ΔP; if, in addition, the cross-polar and co-polar responses are matched within 5%, then ΔB and ΔT_U can be up to 8000 times larger than ΔP.

Possible sources of cross-polarization are errors in the relative calibration of the two bolometers being compared, asymmetries in the optical system, the filters, the feed horns, and the analyzers used in the detectors.

In the B98 experiment, it has been possible to estimate the relative calibration of two independent channels in the same band within 1% [20]. With the new generation of bolometers, featuring common feed optics and filters, we expect to do even better.

The best optical system is a completely azimuthally symmetric one: a Cassegrain telescope with a single detector in the center of the focal plane is perfect. However, this perfection degrades immediately if we consider off-axis locations of the detector in the focal plane. The same is true for off-axis telescopes, even if they satisfy the Mizuguchi-Dragone condition [21].

In BOOMERanG an off-axis telescope [17] [20] close to the Dragone condition has been used. We have analyzed in detail its polarization properties on a wide region of the focal plane. The telescope is an off-axis gregorian, with a

1.3m diameter aluminum primary enclosed in a low emissivity cavity. Secondary and tertiary are inside the cryostat, and reimage the primary focal plane inside the dewar. The cold tertiary acts as the Lyot stop of the system, and is surrounded by a 2K absorber, thus improving sidelobes rejection and significantly reducing the straylight on the detectors.

On the edge of the focal plane (0.75^o off-axis in azimuth, 0.25^o off-axis in elevation) we estimate $X_1/C_1 \sim 1 \cdot 10^{-3}$; $X_2/C_2 \sim 3 \cdot 10^{-3}$. Most of the cross polar response comes from the asymmetry of the primary illumination, due to the limited size of the tertiary. These predictions are currently being tested with extensive measurements.

The cross polarization intrinsic to the bolometers is lower than a few %, and its contribution to the common mode signal can be reduced even more by approaching perfect circular symmetry of the detectors feeds.

At this point it is needed an estimate of how large are the in-flight fluctuations in the background ΔB, in the frequency range of interest. In order to be really dangerous, these fluctuations should be synchronous with the sky. Otherwise, they just add noise. Instead of writing more or less optimistic estimates of ΔB, we have used real data from the B98 flight [22], where bolometers were not sensitive to the polarization, but were sensitive to all possible background fluctuations. In B98 six 150 GHz bolometers scanned the same sky pixel at different times. All the bolometers had independent readout channels, so we can have an idea from real world data of the effects of noise, drifts, responsivity changes one gets measuring the difference of maps obtained by independent bolometers. From column 5 in table 3 of [22] we see that, using 4 channels, the sensitivity would have been just enough to detect the first peak of the polarization spectrum at $\ell \sim 410$, and the first peak of the polarization-temperature cross-spectrum at $\ell \sim 330$ with high S/N. This result is evident from fig.2. We conclude that using polarization sensitive bolometers with the same performance as the unpolarized B98 ones, with the background conditions characteristic of the stratospheric LDB environment, it is definitely possible to start to constrain the polarization of the CMB. Next paragraph gives a few ways to improve from this starting point.

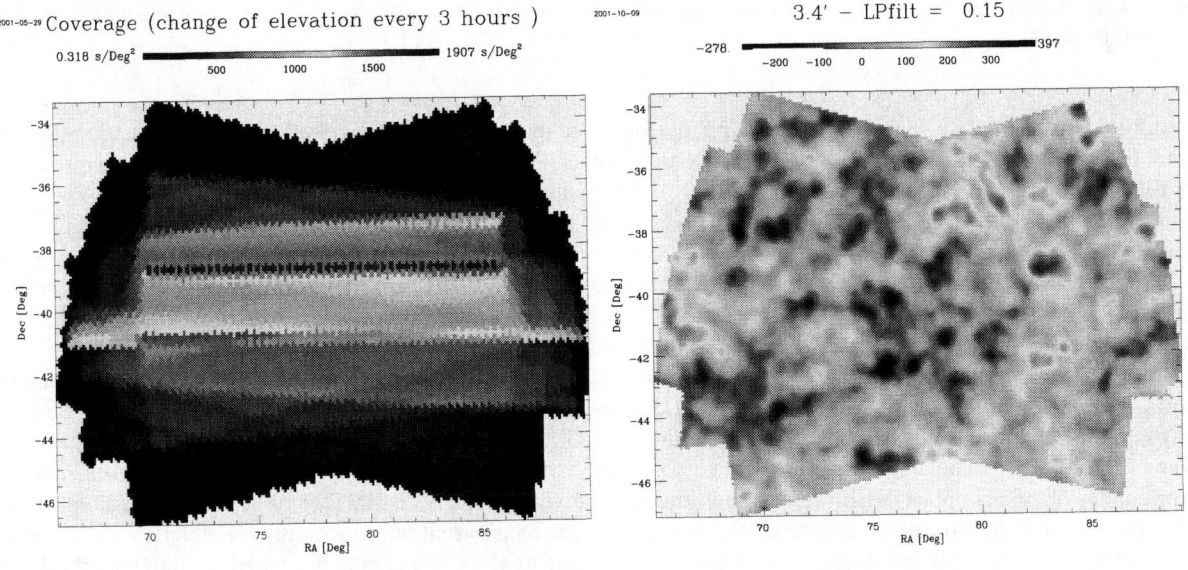

FIGURE 3. Left: Coverage map of a possible scan strategy for next flight of the BOOMERanG polarimeter. One day of observations from a location along a typical Antarctic LDB flight has been assumed. A simulation of temperature anisotropy in the same region is shown on the right. The CMB anisotropy power spectrum best fitting B98 has been used to compute this realization.

OPTIMIZATION OF THE SCANS AND OF THE FOCAL PLANE

There are many ways to optimize this experiment. The first one is to optimize the observation strategy. In the "discovery mode" we want to map a sky region much smaller than B98, accumulating much more integration time for each pixel. Since we are interested to features at $\ell > 300$, and the expected signal is so low, cosmic/sampling variance is a less important issue than instrumental noise.

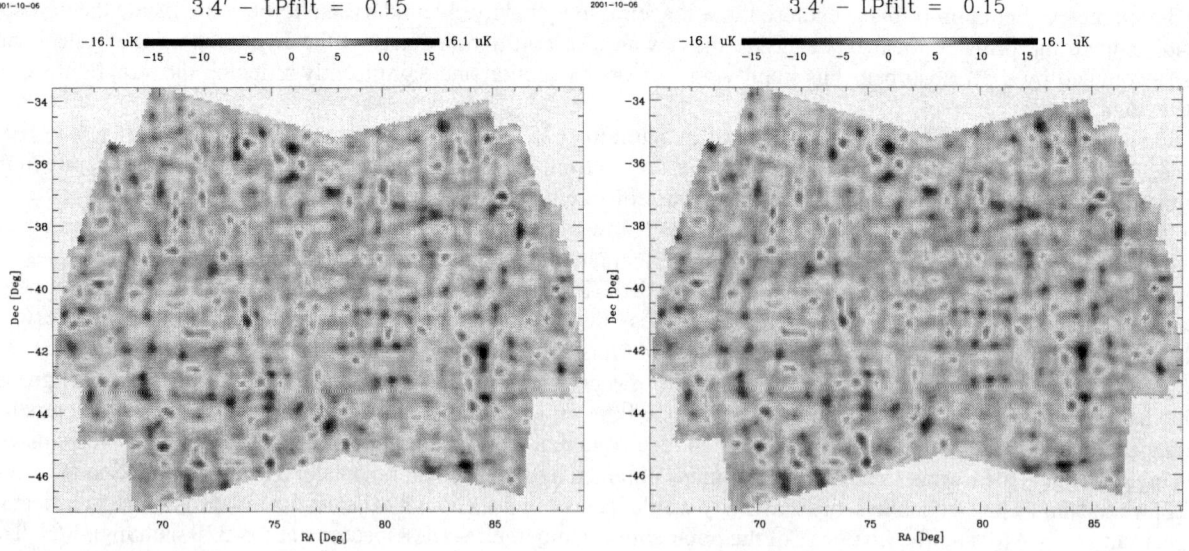

FIGURE 4. Simulation of Q polarization (left) in the target region, for the same seeds used for the anisotropy simulation of fig.3. In the right panel the Q map has been reconstructed from the data obtained by simulating a scanning polarimeter with 1% spurious polarization. System noise has been assumed to be negligible in order to make evident any small systematic effect induced by the spurious polarization.

In B98 we accumulated a few seconds per pixel, over more than 100000 (7'x7') pixels. In the polarization flight we want to get more than 100 seconds per pixel, observing some 3000 pixels. This will produce a factor ~ 5 improvement in S/N (sampling variance is not dominant). In fig.3 we plot a possible sky coverage map we can obtain in one day of the flight, with a scan strategy similar to the one used in B98. The payload scans the sky at constant elevation for 3 hours, with the center of the azimuth scans tracking the azimuth of the center of the target region. Elevation is re-adjusted every 3 hours, in order to limit the vertical size of the map. The azimuth scans are 10° p-p. Many hundreds of seconds per square degree can easily be accumulated with this strategy. The coverage can be concentrated on an even smaller region by reducing the size of the azimuth scans and by changing the elevation more often.

The performance of the scanning polarimeter is illustrated in fig.4 and fig.5. Here we have simulated temperature and CMB polarization map in the sky region of fig.3, and we have computed the resulting signals in the two detectors of the same PSB. The two signals have then been high-pass filtered as will be done to remove noise and drifts effects. The resulting detected polarization vectors are very similar to the input ones, anyway. The Q polarization map has been computed from the simulated PSB time-streams and is shown in fig.4 (left), in the case of an ideal the system. In fig.4 (right) we repeat the procedure assuming that the system acts as a partial polarizer with 1% polarization efficiency, i.e. with a spurious polarization of $\sim 1\%$. Note that, in order to show the results of systematic effects, in all these simulations we are assuming that detector noise is negligible. The resulting map is very similar to the original one. In fig.5 (left), instead, we assume the spurious polarization of the system is around 10%. It is evident that the common mode term, due to the unpolarized component of the CMB, is starting to affect significantly the measurements. In fig.5 (right) we use a 10% accurate measurement of the instrumental spurious polarization to correct the map of fig.5 (left). The resulting map is clean, and comparable to the original.

This demonstrates that if the CMB anisotropy is dominating the common mode signal ($\Delta B \ll \Delta T_U$), the instrumental polarization can be as high as a few % without affecting the measurement, and even higher if it can be measured in advance. The last resource, in case spurious polarization cannot be measured accurately and turns out to be important, is to remove from the contaminated polarization map the best fit anisotropy map. This can be obtained, as a first order approximation, by averaging the signals of the two bolometers. This procedure will also remove part of the polarization signal, but Monte-Carlo simulations can be used to correct for this effect.

The other thing to do is to optimize the focal plane for polarimetry. In BOOMERanG and Planck-HFI two bolometers have been mounted in the same photometer, thanks to the development in JPL/Caltech of the Polarization Sensitive Bolometers (PSB). In these devices two wire-grid-like absorbers with matched NTD thermistors are mounted, very

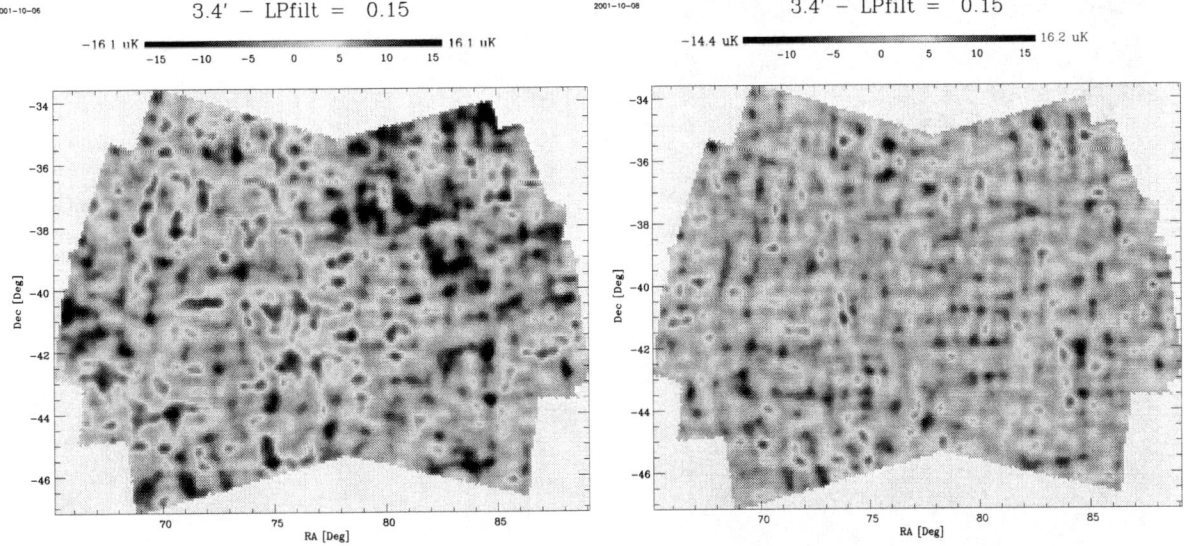

FIGURE 5. Left: Q map reconstructed from the data obtained by simulating a scanning polarimeter with 10% spurious polarization. Contamination from the common-mode CMB anisotropy signal is evident. Right: Q map reconstructed from the same data, corrected assuming a knowledge of the instrumental polarization with a relative error of 10%. System noise has been assumed to be negligible in order to make evident the systematic effects induced by the spurious polarization.

close to each other, in the same groove a circular corrugated waveguide. These devices are described in [15]. This approach improves significantly with respect to the performance estimated from the B98 difference spectra, because the two polarization components coming from the same sky direction are detected simultaneously. This avoids noise induced from pointing jitter. Moreover, it will be easier to have matched beam patterns and very good common mode interference rejection. It will also be easier to achieve temperature matching of the sensors. The filters stack will be in common, so that the two detectors will experience the same background.

POLARIMETER CALIBRATION

We have seen above how important it is to characterize the polarization properties of CMB polarimeters. The pre-flight calibration of a polarimeter is more difficult than the usual photometer calibration. In particular it is very important to study the co-polar and cross-polar response (beam and integral) of the polarimeter. We have developed a polarized, far field, sine-modulated source filling the beam of the instrument. This will allow us to carry out a through polarization characterization of all the detectors in the instrument. A 10' diameter beam is defined by a stop in the focus of a 1.3 m off-axis paraboloid (the spare of the BOOMERanG mirror). Placing a small size source in the center of the focal plane we limit the spurious polarization induced by the off-axis mirror to $\sim 10^{-4}$. Behind the stop there are two slant wire grid polarizers (P1 and P2), and a 77K black-body source. Rotating P2 at constant speed we modulate the signal, producing a sine wave pattern at twice the rotation frequency. Rotating P1 (in steps) we change the illuminator from co-polar to cross-polar (and all intermediate directions). The mechanics of the source is shown in fig.6. With this source we plan to measure C and X for each detector in the BOOMERanG focal plane with better than 0.1% accuracy.

ACKNOWLEDGMENTS

This activity is being supported by the Italian Agencies MURST, PNRA and ASI. We are deeply indebted to all the members of the BOOMERanG team for continuous constructive discussions and suggestions.

 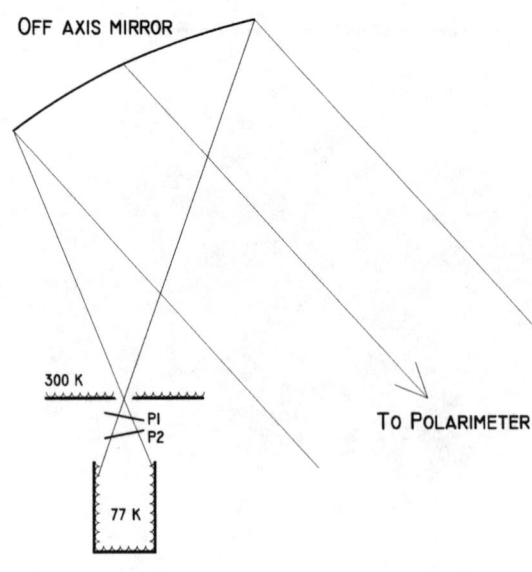

FIGURE 6. Beam filling, modulated polarized source used for the calibration of scanning polarimeters. The top mirror is a 1.3m diameter off-axis paraboloid (the spare of the BOOMERanG primary) producing a 5' beam filling the sensitive area of the BOOMERanG telescope. A set of two wire-grid rotating polarizers is used in front of a 77K cold load in the focus to produce sine-modulated 100% lineraly polarized radiation.

REFERENCES

1. Rees M., Ap.J., 153, L1, (1968)
2. Hu W., White M., New A. 2, 323, (1997)
3. Kaminokowsky M., Kosowsky A., Ann.Rev.Nucl.Part.Sci. 49, 77-123, (1999)
4. Kaiser N., MNRAS, 202, 1169, (1983)
5. Caderni N., Phys.Rev.D., 17, 1901, (1978); Phys.Rev.D., 17, 1908, (1978)
6. Nanos G.P., Ap.J., 232, 241, (1979)
7. Lubin P. and Smoot G., Ap.J., 245, 1, (1981)
8. Partridge B., et al., Nature, 331, 146, (1988)
9. Wollack et al., Ap.J., 476, 440, (1997)
10. Netterfield B., et al., Ap.J., 474, 47, (1997)
11. Hedman M.M. et al., Ap.J., 548, L111, (2001); astro-ph/0010592
12. Subrahmanyan R., et al., MNRAS, 315, 808 (2001).
13. Keating B., et al., 2001, "A Limit on the Large Angular Scale Polarization of the Cosmic Microwave Background", astro-ph/0107013
14. Pisano G., New Astron. Reviews, 43, 329, (1999)
15. Bock J. et al, this book.
16. Masi S., et al., in "3K cosmology", AIP Conf. Proc. 476, 237, (1999); astro-ph/9911520
17. Piacentini F., et al. 2001, "The BOOMERANG North America Instrument: a balloon-borne bolometric radiometer optimized for measurements of cosmic background radiation anisotropies from 0.3 to 4 degrees", astro-ph//0105148 (Ap.J. in press); see also his paper in this book.
18. Carretti E., et al., 2001, "Limits Due to Instrumental Polarisation in CMB Experiments at Microwave Wavelengths", astro-ph/0103318
19. Philhour B., et al., 2001, "The Polatron: A Millimeter-Wave Cosmic Microwave Background Polarimeter for the OVRO 5.5 m Telescope", astro-ph/0106543
20. Crill B.P., et al., in preparation. See also Crill B.P., Ph.D. Thesis, Caltech, 2000.
21. Dragone M., IEEE Trans. AP-30, 331, (1982); IEEE Trans.AP-22, 472, (1974)
22. Netterfield C.B., et al. (2001), " A measurement by BOOMERANG of multiple peaks in the angular power spectrum of the cosmic microwave background", astro-ph/0104460

POLAR: Instrument and Results

Brian G. Keating[*], Angelica de Oliveira-Costa[†], Christopher W. O'Dell[**], Lucio Piccirillo[‡], Nate C. Stebor[§], Max Tegmark[†] and Peter T. Timbie[**]

[*]*Division of Physics, Math, and Astronomy, California Institute of Technology, Pasadena, CA 91125*
[†]*Department of Physics, University of Pennsylvania, Philadelphia, PA 19104*
[**]*Department of Physics, University of Wisconsin – Madison, Madison, WI 53706*
[‡]*Department of Physics and Astronomy, University of Wales - Cardiff, Wales, UK CF24 3YB*
[§]*Department of Physics, University of California at Santa Barbara, Santa Barbara, CA 93106*

Abstract. We describe the design, performance, and results of a polarimeter used to make precision measurements of the 2.7 K cosmic microwave background. In the Spring of 2000 the instrument searched for polarized emission in three microwave frequency bands spanning 26-36 GHz. The instrument achieved high sensitivity and long-term stability, and has produced the most stringent limits to date on the amplitude of the large angular scale polarization of the cosmic microwave background radiation.

URL `http://cmb.physics.wisc.edu/polar`

INTRODUCTION

The 2.7K Cosmic Microwave Background (CMB) radiation is a vital probe of all modern cosmological theories. This radiation provides a "snapshot" of the epoch at which radiation and matter decoupled, approximately 300,000 years after the Big Bang, and carries the imprint of the ionization history of the universe. This information has been used to tightly constrain theories of cosmological structure formation, and has ushered in the era when "cosmological accuracy" is no longer a pejorative term.

Three properties are necessary to fully characterize the CMB: its spectrum, spatial isotropy, and polarization. The first two properties have been measured, whereas the polarization state of the CMB remains undetected. Detection of, or an improved upper limit on, the polarization of the CMB at large scales holds great promise for the determination of several fundamental properties of the standard cosmological model, such as the ionization history of the Universe and the contribution of gravitational waves to the spectrum of primordial perturbations. Most models predict that the magnitude of the polarization of the CMB at large angular scales is less than $1\mu K$. This is at least an order of magnitude below both the large scale anisotropy level of the CMB, as well as the best existing upper limits on its polarization.

Similar to the CMB anisotropy power spectrum, the polarization power spectrum contains information on all angular scales. Large angular scales (larger than $\simeq 1°$) correspond to regions on the last scattering surface which were larger than the causal horizon at $z \simeq 1000$. In the absence of reionization, these scales were affected *only* by the long wavelength modes of the primordial power spectrum. This region of the anisotropy power spectrum was measured by the *COBE* DMR, and establishes the normalization for models of large scale structure formation. Similarly, measurements of polarization at large angular scales will normalize the entire polarization power spectrum. Because the anticipated signal size is small at all angular scales, polarization measurements face more formidable challenges than anisotropy measurements.

The experiment described here measured polarization on large angular scales. While these signals may be weaker than signals on small scales, the design of a large angular scale measurement is comparatively simple and compact, with potentially lower susceptibility to sources of systematic error. This report describes the design of, and results from, Polarization Observations of Large Angular Regions (POLAR).

TABLE 1. POLAR K_a-Band Measured and Modeled FWHM Beam Widths.

Plane	ν [GHz]	$\theta_{fwhm} \pm 0.1°$
E	26	7.5°
E	29	7.0°
E	36	6.4°
H	29	7.1°
H	36	6.5°

INSTRUMENT

The POLAR radiometer is comprised of three main sections:

- Cold receiver components: optics, OMT, isolators, HEMT amplifiers.
- Room-temperature receiver components: warm RF amplifiers, heterodyne stage, warm IF amplifiers, band-defining filters, correlators.
- Post-detection components: pre-amplifiers, low frequency processing, and data acquisition.

POLAR is a superheterodyne correlation polarimeter that measures two orthogonal linear polarization states in three radiofrequency (RF) bands in the K_a-band between 26-36 GHz. In many respects, the signal processing techniques used in correlation polarimeters are similar to those used in polarization sensitive interferometers. The two polarization states $i \in \{x,y\}$, $E_i^{RF}(t,\nu,\phi_i) = E_i \cos(2\pi\nu t + \phi_i)$ enter a single-mode circular corrugated feedhorn and are separated by an orthomode transducer (OMT). The OMT's polarization isolation (−35 dB) and low cross-polarization (−30 dB) ensure low spurious polarization. The two polarizations are amplified by separate HEMT amplifiers [1] cooled to 25 K by a mechanical cryocooler. Downconversion from the RF band to an intermediate frequency (IF) band (2-12 GHz) is performed by Schottky diode mixers, driven by a Gunn diode local oscillator (LO) at 38 GHz. In the IF band the two polarization states are amplified producing $E_i^{IF} \propto E_i^{RF}$, and filtered into three separate IF bands, denoted J1, J2, and J3. The IF bands translate into RF bands: J1 (32-36 GHz), J2 (29-32 GHz), and J3 (26-29 GHz). Prior to filtering, two dedicated diode detectors measure the total power of each polarization state, which serve as atmospheric opacity monitors. The fields E_x^{IF} and E_y^{IF} are correlated by three Schottky Diode analog multiplier (correlator) circuits. The phase of the LO is switched between 0 and π at 1 KHz prior to mixing the E_y^{RF} waveform. The voltage produced by the correlators at this stage switches between $\kappa E_x^{RF} E_y^{RF}$ and $-\kappa E_x^{RF} E_y^{RF}$ at 1 KHz, where κ converts intensity to voltage. Phase-sensitive detection of this modulated signal reduces the effects of low frequency drifts in the LO output power and/or correlator sensitivity to negligible levels. After low-pass filtering, we record an audio-band signal with a DC component $I_{DC} = \kappa \langle E_x^{RF} E_y^{RF} \rangle$, where the brackets denote a time-average, and AC components proportional to the thermal noise from the radiometer, atmosphere, and celestial signals. These signals are referred to as the "science channels". A second lock-in amplifier for each correlator is referenced to the same 1 KHz waveform but delayed in phase by $\pi/2$ with respect to the phase switch. These signals, hereafter referred to as "quadrature phase channels" (QPC), contain only the noise terms of the RF band and no optical or celestial signals. The QPC are powerful probes of systematic effects produced solely by the radiometer and post-detection stages. The output from the QPC is proportional to the noise equivalent temperature (NET) of the instrument. For a correlation radiometer this is NET $= \sqrt{2/\Delta\nu}(T_{Rx} + T_{Ant})[K - s^{1/2}]$, where $\Delta\nu$ is the bandwidth of the radiometer, T_{Rx} is the receiver noise temperature, and T_{Ant} is the antenna temperature of observed optical sources, including diffuse sources such as the atmosphere and the CMB itself. A schematic outline of POLAR is presented in Figure 1.

POLAR uses two levels of ground screening. First an inner conical ground screen rotates with the instrument and is coated with Eccosorb. The Eccosorb panels absorb, rather than reflect the sidelobes to the sky. We estimate the antenna temperature of the inner shield to be < 1 K. Polarization generated by emission from the bare metal surface of the uncovered shield is believed to be much more troublesome than the slight increase in system temperature. POLAR'S inner ground screen co-rotates with the receiver, which ensures that if there is any residual polarized power produced by the inner screen, it would produce a constant polarized offset, which is subtracted during the analysis. The second level of shielding is of the more conventional reflective-scoop design, *e.g.* Wollack [2]. The scoop is mounted to the

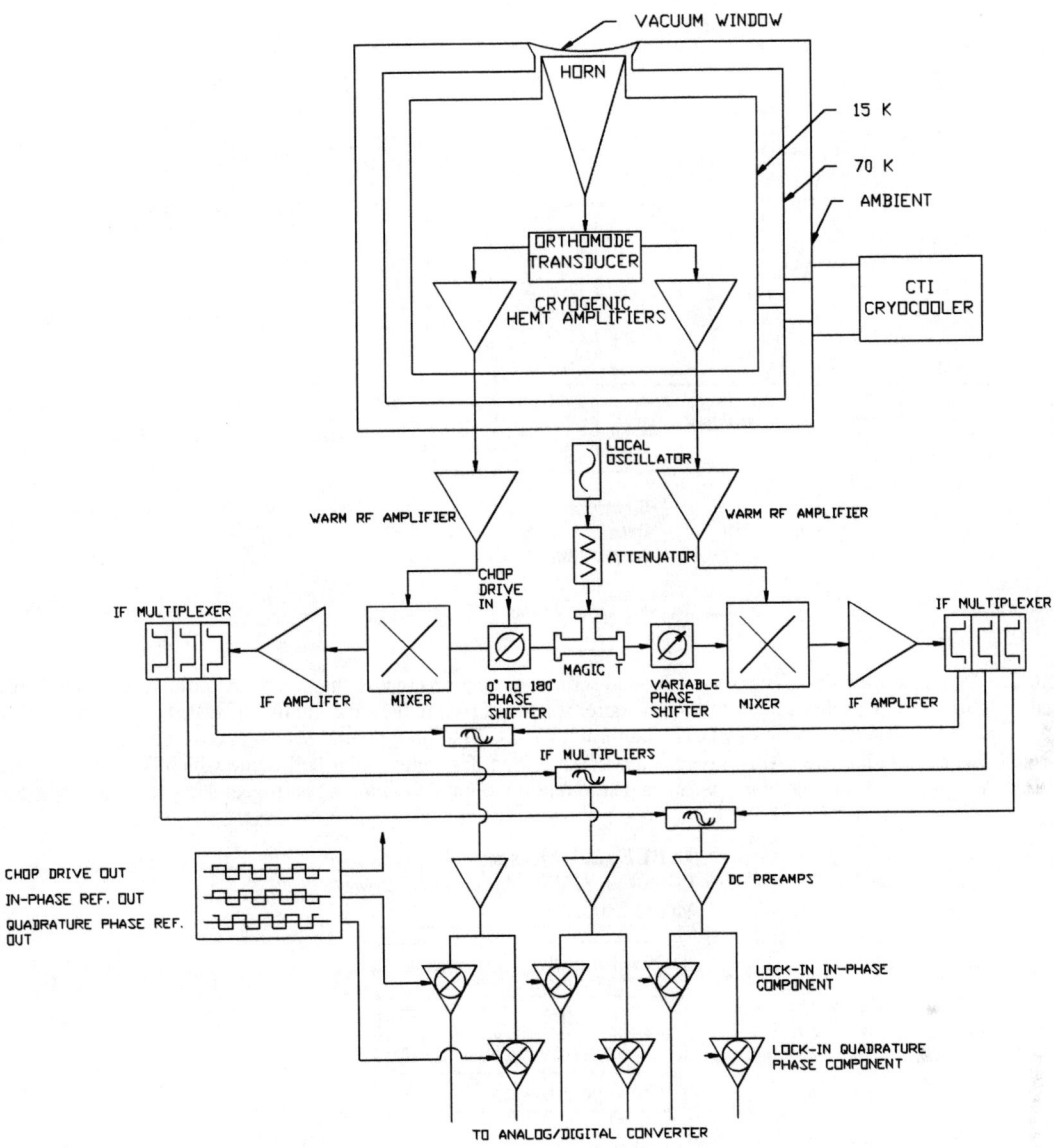

FIGURE 1. POLAR radiometer schematic.

side of the POLAR observatory, and is made of aluminum panels, 8' wide and 6' high. We estimate the level of sidelobe suppression to be ~ -40 dB.

SITE

POLAR's observations are conducted from a custom-built observatory located at the University of Wisconsin's Pine Bluff Observatory, (PBO). PBO is located at Longitude $+89°45'$, Latitude $+43°01'$, approximately 10 miles west of the campus and downtown Madison. It was determined that PBO was a relatively RF quiet region.

A motorized dome encloses the radiometer and rotating ground screen, keeping precipitation out, and maintaining a moderately thermally stabilized enclosure. The dome itself can be operated manually, or remotely via a WWW page in

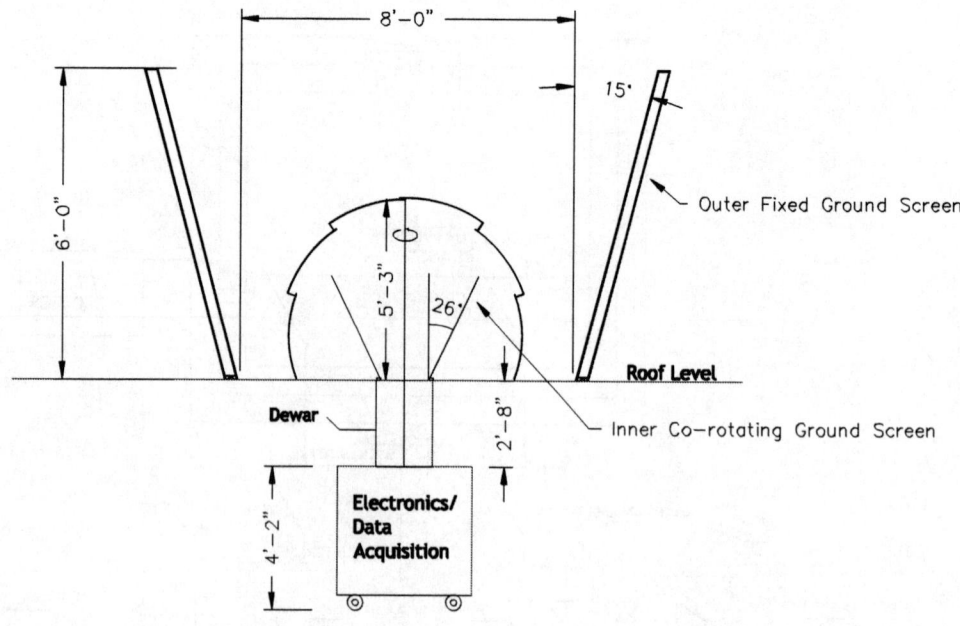

FIGURE 2. POLAR observatory. Two sets of ground screens are used to reduce the polarized spillover from the earth, as well as polarized emission from the shields themselves. The outer shield is fixed to the structure in which POLAR resides, and is composed of a lightweight steel skeleton covered by 0.05" aluminum sheets. The inner ground screen is covered with flat Eccosorb panels, and co-rotates with the radiometer. Also shown is the motor-driven, fiberglass clamshell-dome which is remotely operated via the World Wide Web in the event of inclement weather. The rotation mount, drive motor, bearing, and angular encoder are also shown.

TABLE 2. Measured Properties of POLAR'S OMT: Atlantic Microwave Model 2800.

Property	Value
Isolation	-35dB
VSWR	< 1.2
Cross-polarization	-30 dB

case of inclement weather developing while the experimentalist is elsewhere. The platform has a high-voltage power supply for operation of the CTI 8500 compressor, which requires 220 V at roughly 10 A. Also running to and from the pad is an Ethernet hub and cables which provide an intranet for data to be transferred from the rotating computer attached to the radiometer to a desktop workstation located $\sim 100'$ away in a separate building.

SYSTEMATIC EFFECTS

Because the anticipated polarization signal is a factor of ≤ 10 times smaller than the temperature anisotropy currently being detected, a thorough understanding of systematic errors is crucial. Polarimeters, such as POLAR have several advantages, however, that promise to make this effort possible. First, the atmosphere is known to be polarized only at a very low level; far below the expected level of CMB polarization [3]. Additionally, POLAR mapped polarization in a manner which de-correlated signals from neighboring pixels. POLAR compared the orthogonal polarization signals through the same airmass, and at the same time. In anisotropy observations, beam switching often adds noise and additional chop-dependent signals. Atmospheric effects had a smaller contribution to POLAR than to

TABLE 3. POLAR K_a Band Radiometer Components

Device	Manufacturer	Model
Circular-Square Transition	Custom Microwave	---
OMT	Atlantic Microwave	OM2800
HEMTs	NRAO	A29 & A30
Warm RF Amps	MITEQ	JS426004000-30-8P
Mixers	MITEQ	TB0440LW1
Gunn Oscillator	Millimeter Wave Oscillator Co.	---
Warm IF Amps	MITEQ	AFS6-00101200-40-10P-6
Triplexers	Reactel	---
Correlators	MITEQ	DBP112HA
Total Power Detectors	Hewlett Packard	HP 8474C
Lock-In Amplifiers	Analog Devices	AD630
K_a band Phase Switch	Pacific Millimeter Products	---
Dewar	Precision Cryogenic Systems	---
Cryocooler	CTI Cryogenics	8500 Compressor, 350 Cold Head

ground-based CMB anisotropy experiments, and allowed for longer observation times. Long-term observations were key to understanding and removing systematic effects [4]. Many spurious instrumental effects were isolated from astrophysical effects by long-term integration tests with the horn antenna replaced by a cold termination.

The most troubling aspect of these effects is that they may not be stable in time. For the correlator channels, the most pernicious contribution arises from the conversion efficiency and phase stability of the heterodyne stage. The conversion efficiency (loss) of the mixers is dependent on the Gunn Oscilator power. The oscillator's output power fluctuates (similar to an amplifier's gain), and therefore introduces mixer gain fluctuations, which can be misinterpreted as signals.

The relative phasing between the two arms of the radiometer is equally troublesome [5] since phase mismatch between the two RF paths reduces the effective bandwidth. There are several standard methods to improve the stability of the heterodyne stage of the receiver, including phase modulation at frequencies of ~ 1 KHz, and phase-locked loops to stabilize the LO. The latter is quite common in interferometers, though it is not incorporated in the POLAR K_a-band system. The former technique was performed for POLAR. In Table 4 we list some important systematic effects encountered in previous polarization measurements and summarize the solution adopted by POLAR.

TABLE 4. Expected Systematic Effects

Effect	Origin	Control Method
Mechanical Strain	Instrument Rotation	Zenith Drift Scan
Magnetic Coupling	Rotation in $\vec{B}_{earth}$	Minimal Ferrite Components (Isolators Only)
Microphonics	Mechanical Vibration	Vibration Isolation
EMI and RFI	Local Sources	Shield/Filter
Thermal Variations	Diurnal/Environment	Temp Control
Sidelobe Pickup	Sun/Moon/Earth	Low Sidelobe Antenna and Ground Screens

OBSERVATION STRATEGY AND SKY COVERAGE

Over a single night, POLAR swept out a $7° \times 360° \times \cos 43° = 1844°^2\,FWHM$ swath of the sky. Thirty-six $7°$ FWHM pixels comprise 5% of the sky. The data is binned into Stokes Q and U vs. RA, and multiple nights of data are co-added. The scan passes through the galaxy twice per day at RA$\sim 19h$ and again at RA$\sim 6h$.

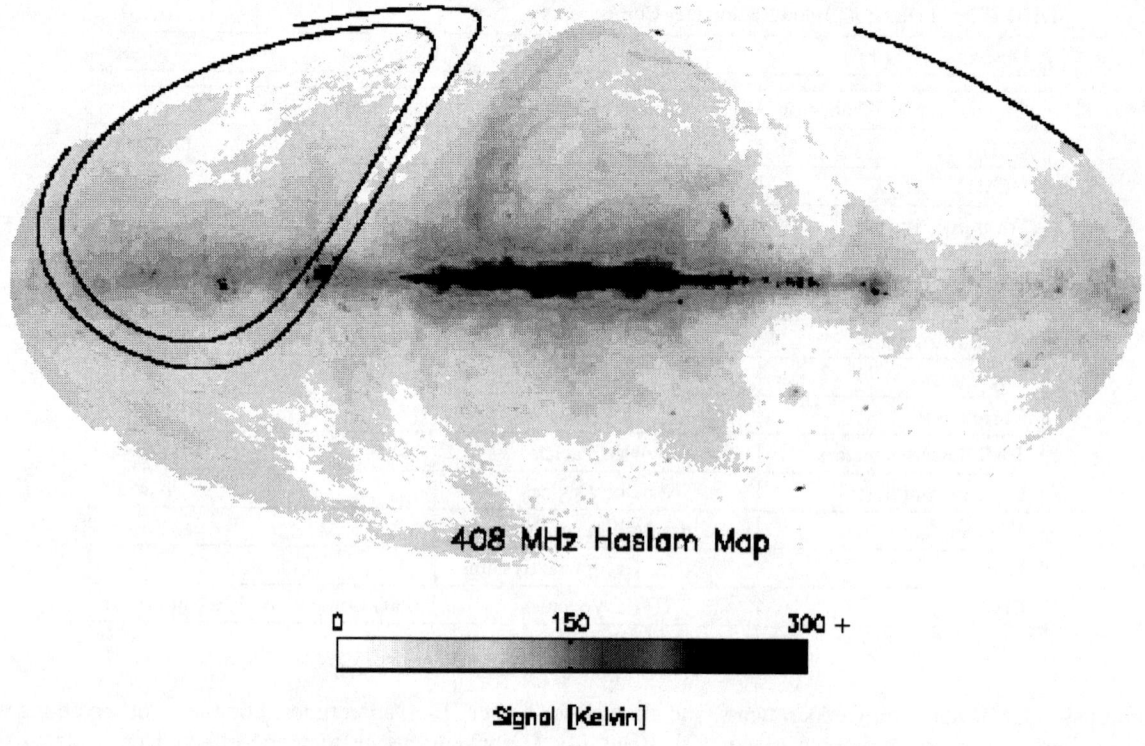

FIGURE 3. POLAR scan strategy in galactic coordinates, superimposed on the Haslam 408 MHz map

TABLE 5. POLAR Observing Parameters

Parameter	POLAR 2000
Declination of Drift Scan	43.03°
Beam width	$7° FWHM$
Fractional Sky coverage	$255° \times 7° FWHM \simeq 5\%$
Rotation Rate	0.033Hz
Point Source Sensitivity	$2.8\mu K$ Jy^{-1}
Post detection Bandwidth	5Hz

ATMOSPHERIC EFFECTS

Although the atmosphere is not expected to produce appreciable linearly polarized radiation, it produces a non-negligible contribution to the system temperature of the radiometer. Additionally, significant fluctuations of atmospheric loading increase the low-frequency noise spectrum of the receiver. We summarize the contribution to the antenna temperature seen by the radiometer in the K_a-band by computing the power spectrum of the atmosphere using a commercial code, AT[1]. To compute the antenna temperature AT requires as input the desired level of precipitable water vapor (PWV). With this specified, AT can compute the antenna temperature vs. frequency using a standard model of the earth's atmosphere. Figure 4 shows the atmospheric antenna temperature vs. frequency for various levels of PWV. Since the dominant contribution to the atmospheric temperature comes from the 22 GHz H_2O line rather than on the O_2 line at ~ 60 GHz, the dependence on PWV is quite noticeable.

---

[1] Airhead Software: Boulder, CO

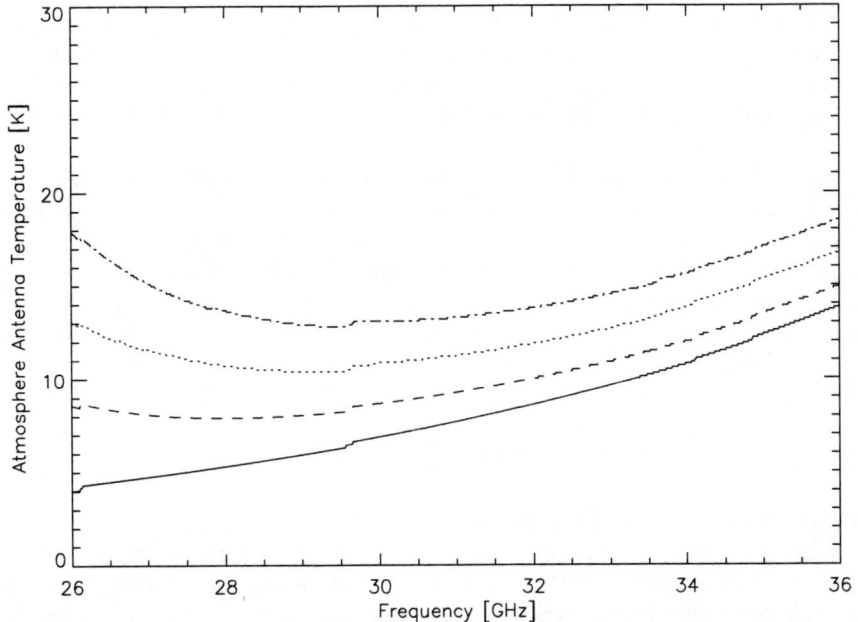

FIGURE 4. Calculated spectrum of Atmospheric Antenna Temperature in the K_a-band vs. Precipitable Water Vapor (PWV). From the top down, the four levels of PWV are: 30mm, 20mm, 10mm, 0mm. For 50% of the 2000 observing season, the atmospheric antenna temperature was < 16 K.

OBSERVATIONS AND ANALYSIS

The POLAR campaign collected a total of 746 hours of data between 2000 March 11 and 2000 May 29. We calibrated the polarimeter by rotating a 0.003 inch dielectric sheet (tilted at 45° with respect to the optical axis) around the optical axis [6]. This produced a ~ 500 mK polarized signal. In our coordinate system, defined with respect to the local geographic coordinates, we performed a χ^2 minimization of the angular binned data fit to the function

$$I(\theta_t) = I_o + C\cos\theta_t + S\sin\theta_t + Q\cos 2\theta_t + U\sin 2\theta_t, \qquad (1)$$

where $\theta_t = 2\pi ft$. In addition to the Stokes parameters Q and U, the terms C and S (which are synchronous with the rotation at frequency f) were monitored to determine our sensitivity to rotation-synchronous systematic effects, and to monitor atmospheric fluctuations. Three levels of diagnostics were used to detect and remove contaminated data:

- housekeeping and weather (dewpoint, cloud cover) cuts
- time ordered data (TOD) cuts
- rotation ordered data (ROD) cuts after constructing the Stokes parameters.

RESULTS

For each data segment that passed the data quality tests, the time-ordered data were converted to maps and covariance matrices via the minimum variance map-making procedure outlined in Tegmark [7]. The 25 total sections ranged in length from 4-24 hours. An overall offset was removed from each section, and the data covariance matrix was adjusted accordingly [7, 8]. All sections were combined into a final map and covariance matrix for each channel. Keating *et al.*(2001) [9] presents the maps of the Stokes parameters for all correlator channels versus RA.

To test the hypothesis that the maps were contaminated by foreground emission, we compared the likelihood for the null-hypothesis of no signal on and off the galactic plane. We tested a simple model of foreground contamination by a single fluctuating component [10]. Tests on these sections of the maps indicated that in the region $|b| > 25°$, no galactic contribution was detected. This indicates that our CMB upper limits were not contaminated by foreground emission, though any contamination would lead us to overestimate the upper limit rather than underestimate it.

Keating et al.(2001) presents a summary of the data analysis procedure and the results from the 2000 season. The limits on the large angular scale polarization of the CMB are $T_E < 10\mu K$ and $T_B < 10\mu K$ at 95% confidence for the multipole range $2 < \ell < 20$. Not included in these limits is the ($\sim 10\%$) calibration uncertainty. We also calculated the limits on T_E with $T_B \equiv 0\mu K$. Integrating the 1-D likelihood curve for T_E with $T_B \equiv 0\mu K$ we find $T_E < 8\mu K$ at 95% confidence.

We have also computed the limits that result from simply subtracting the offset from the 25 subsections of data and increasing the on-diagonal elements of the corresponding covariance matrix by a factor of $N_{pix}/(N_{pix} - 1)$, corresponding to the loss of one degree of freedom e.g., [10]. While limits obtained from this procedure do not correctly account for the Stokes parameter correlations induced by the offset removal, they do illustrate POLAR's raw sensitivity, and are useful for comparison with previous experiments. Following the notation of [9], the limits on CMB polarization from this procedure are $T_E < 8\mu K$ and $T_B < 8\mu K$ (both 95% confidence). Assuming $T_B \equiv 0\mu K$, we obtain $T_E = 6\mu K$ at 95% confidence.

CONCLUSION

With POLAR, we have demonstrated that the detection of the large angular polarization of the CMB is difficult but technologically feasible. Our limits are encouraging and suggest that future large angular scale, multi-pixel, correlation polarimeters (such as SPORT [11]) may have the capability to detect polarization of the CMB at the μK level. A detection would permit the discrimination between heretofore degenerate theoretical predictions. Polarization of the CMB has a unique signature in both real and Fourier space, as well as distinct spectral characteristics. CMB polarization, in conjunction with anisotropy measurements, is one of the most sensitive probes of the ionization history of the pre-galactic medium. This epoch of cosmic evolution is of great interest, and supplemental information from polarization detection will greatly advance our knowledge of the formation of structure in the early universe.

ACKNOWLEDGMENTS

This work has been supported by NSF grants AST 93-18727, AST 98-02851, and AST 00-71213, and NASA grant NAG5-9194. BK and CO were supported by NASA GSRP Fellowships. POLAR's HEMT amplifiers were provided by John Carlstrom.

REFERENCES

1. Pospieszalski, M. W., Lakatosh, W. J., Nguyen, L. D., Lui, M., Liu, T., Le, M., Thompson, M. A., and Delaney, M. J., "" Cryogenically-Cooled HFET Amplifiers and Receivers: State-of-the-Art and Future Trends"", in *1992 IEEE MTT-S Digest*, edited by L. Kirby, IEEE, Piscataway, NJ, 1992, p. 1369.
2. Wollack, E. J., Jarosik, N. C., Netterfeld, C. B., Page, L. A., and Wilkinson, D., *ApJL*, **419**, L49–+ (1993).
3. Keating, B., Timbie, P., Polnarev, A., and Steinberger, J., *ApJ*, **495**, 580+ (1998).
4. Wilkinson, D., "A Warning Label for Cosmic Microwave Background Anisotropy Experiments", in *Particle Physics and Cosmology, Proceedings of the Ninth Lake Louise Winter Institute*, edited by A. Astbury et al., World Scientific, Singapore, 1995, p. 110.
5. Thompson, A., et al., " *Interferometry and Synthesis in Radio Astronomy"*, Krieger Publishing Co., Malabar, 1998.
6. O'Dell, C. W. e. a., *IEEE-MTT* (2002).
7. Tegmark, M., *Phys. Rev. D.*, **56**, 4514–4529 (1997).
8. Bond, J. R., Jaffe, A. H., and Knox, L., *Phys. Rev. D.*, **57**, 2117–2137 (1998).
9. Keating, B. G., O'Dell, C. W., de Oliveira-Costa, A., Klawikowski, S., Stebor, N., Piccirillo, L., Tegmark, M., and Timbie, P. T., *ApJL*, **560**, L1–L4 (2001).
10. Wollack, E. J., Jarosik, N. C., Netterfeld, C. B., Page, L. A., and Wilkinson, D., *ApJL*, **419**, L49–+ (1993).
11. Cortiglioni, S., *These Proceedings* (2001).

COMPASS: a 2.6m telescope for CMBR polarization studies

L. Piccirillo[1], G. Dall'Oglio[2], P. Farese[3], J. Gundersen[4], B. Keating[5], S. Klawikowski[6], L. Knox[7], A. Levy[3], P. Lubin[3], C. O'Dell[6], P. Timbie[6], J. Ruhl[3]

[1]*University of Wales, Cardiff (UK)*
[2]*University of Rome III, (I)*
[3]*University of California, S. Barbara (USA)*
[4]*University of Miami (USA)*
[5]*California Institute of Technology, Pasadena (USA)*
[6]*University of California Davis (USA)*
[7]*University of Wisconsin, Madison (USA)*

Abstract. COMPASS (COsmic Microwave Polarization at Small Scale) is an experiment devoted to measuring the polarization of the CMBR. Its design and characteristics are presented.

INTRODUCTION

The 2.7K Cosmic Microwave Background Radiation (CMB) is one of the few tools we have for understanding the origin of the universe. This radiation provides a "snapshot" of the epoch at which radiation and matter decoupled, approximately 300,000 years after the Big Bang, and can tightly constrain theories of cosmological structure formation. There is no other known direct probe of the universe at such early times.

The three defining characteristics of this radiation are: its spectrum, spatial anisotropy, and polarization. The beginnings of structure formation through gravitational collapse should appear in the spatial distribution of the CMB; the COBE/*DMR* detected spatial anisotropy of the CMB on $10°$ scales of $\Delta T/T \approx 1.1 \times 10^{-5}$ (Bennett *et al.* 1996) and ground-based and balloon-borne experiments have also detected anisotropy at smaller scales (Netterfield et al., 2001, Halverson et al, 2001, Lee et al, 2001). Despite these new results, important questions remain. Does the observed CMB emanate directly from the decoupling era, or has it instead scattered from free electrons in an intervening reionization of the universe? What is the mechanism that produces the large-scale structure in the universe? What is the expansion rate of the universe? What causes the universal expansion and is it accelerating? What is the nature of the dark matter?

It is now clear, however, that the temperature anisotropy measurements alone will not answer all of these questions. Presently, the standard cosmological models require specification of more than 10 parameters. The anisotropy measurements must be combined with additional data sets to break the degeneracies in the models. Therefore we have undertaken a program to measure the third defining characteristic of the CMB, its *polarization*. The temperature anisotropy and the polarization of the CMB both depend on the power spectrum of fluctuations in the early universe as well as the ionization history of the universe, but they do so in different ways. The angular power spectrum of temperature anisotropy and polarization are both determined by factors such as: the source of the CMB anisotropy, the density parameter Ω, the baryon content of the universe Ω_B, and the Hubble constant, H_0. However, the CMB polarization is uniquely sensitive to photon rescattering after the decoupling era (*i.e.* it depends on the duration of recombination and the epoch of reionization), to the *velocity* of matter at the last scattering surface, and to the presence of gravitational waves in the early universe.

The predicted polarization amplitude is extremely small, less than one-part in 10^6 of the CMB intensity, or $\Delta T_{Pol}/T \leq 1\times 10^{-6}$. This signal is more than an order of magnitude below current upper limits; detecting it poses an extremely challenging task, requiring high-sensitivity detectors, careful attention to systematic effects, and knowledge that we do not yet possess about polarized foreground emission. Using an existing polarimeter from Madison, WI, we plan to detect this signal by integrating for a long period on a limited portion of the sky. However, these measurements will not be able to discriminate fully against foreground sources, and lack the sensitivity required to measure the angular power spectrum which encodes so much primordial information.

The most formidable unknown in these measurements is emission from foreground sources. COMPASS will complement other measurements of CMB polarization, including our own POLAR, by increasing the sensitivity and extending the frequency coverage beyond current observations. The latter is essential to discrimination against foreground emission from galactic and extragalactic sources.

COMPASS operates in the Ka-band (26-36 GHz) and has a beam size of 20' FWHM. In its first season of operation it searched for CMBR polarization in the North Celestial Pole region from Madison (WI-USA) from late winter through spring 2001. We give here few details of the instrument design and performance.

THE INSTRUMENT

COMPASS is a 2.6 meter diameter on-axis reflector with many novel features. The on-axis design was chosen to minimize the polarization offsets that can be induced by the primary mirror. The secondary mirror is held in position by a cone of transparent foam. We opted for this solution to minimise the scattering of radiation from any metal struts. The emissivity of the foam has been measured to be 0.5% ± 0.25%. The polarimeter at the focal plane is the POLAR receiver (Keating et al., 2001a and 2001b). The instrument uses two cooled HEMT amplifiers (at about 20K) in a correlation polarimeter configuration. The polarimeter achieved high sensitivity and stability over integration times of more than 100 hours. The RF band (Ka) is divided into 3 sub-bands spanning the 26-36 GHz frequency interval. We record three correlator outputs J1 (32-36 GHz), J2 (29-32 GHz) and J3 (26-29 GHz) corresponding to the three sub-bands, plus two total power outputs TP1 and TP2 integrating the full band of each HEMT amplifier. The receiver includes a corrugated scalar feedhorn (6 degrees FWHM) with a teflon lens to couple to the Cassegrain sub-reflector. The illumination on the sub-reflector and then on the primary mirror is respectively −25 and -16 dB. The resulting beam in the sky is about 20'.

Extensive noise measurements and tests have been performed before, during and after the 2001 winter/spring campaign. In Table 1 we report the measured sensitivity of COMPASS fully integrated and observing a blank sky region.

TABLE 1. Noise performances

	Noise mK $\sqrt{s}$ (thermodynamic)	RF band GHz
J1	1.1	32-36
J2	1.0	29-32
J3	0.8	26-29
TP1	≈5	26-36
TP2	≈5	26-36

The beam pattern has been measured by observing a remote GUNN oscillator positioned on a high tower in the far field. Repeated scans of the Gunn oscillator provided a very accurate measurement of the main beam. The beam has been checked also by observing known sky objects (Venus, Tau A and Cas A).

Calibration was performed by executing different scans of the polarized source Tau-A. This source is not resolved in our beam and is known to be polarized at a level of 6.6% (Flett & Henderson, 1979). The over-all preliminary calibration accuracy is about 20%.

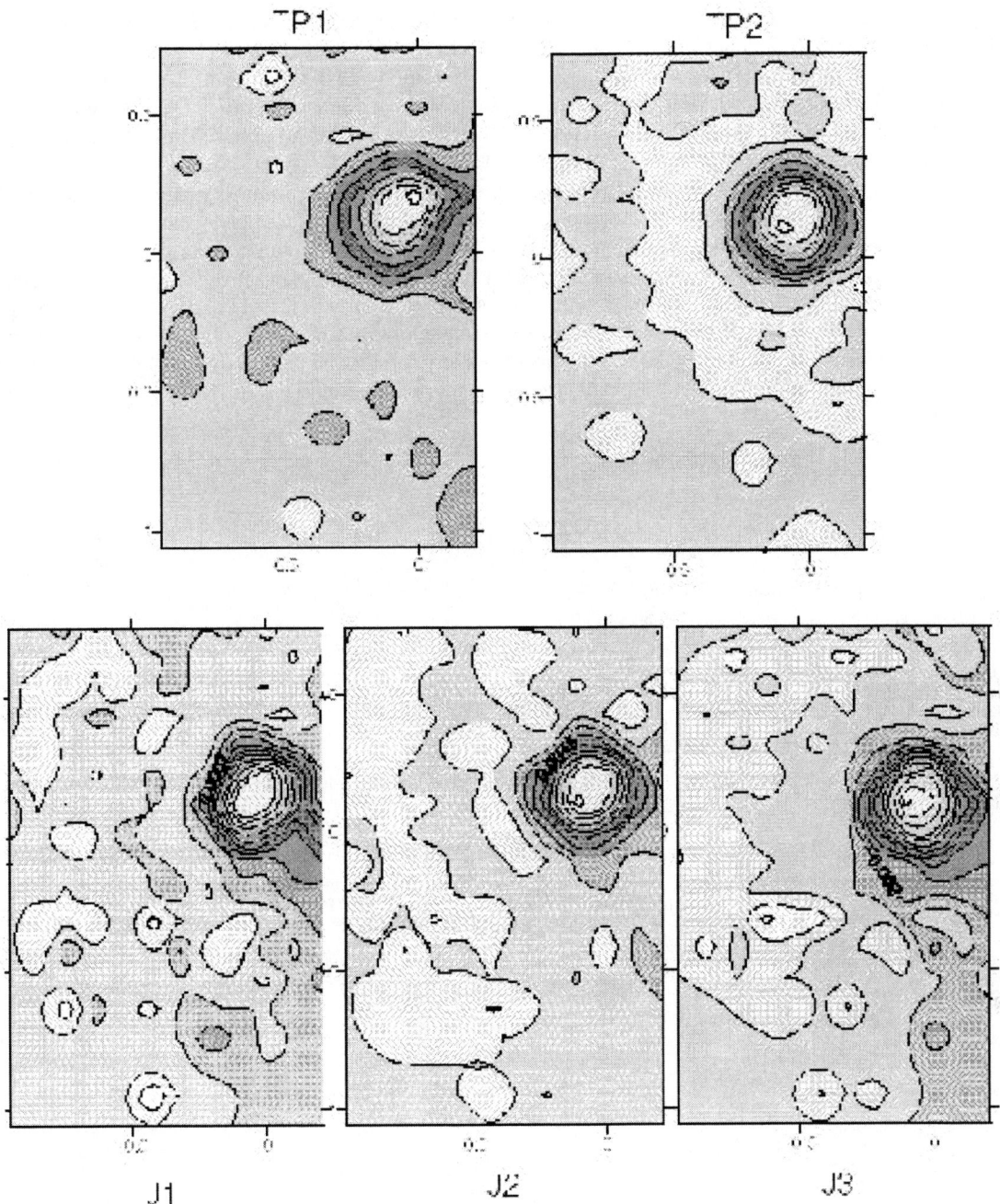

FIGURE 1. Maps of the polarized source Tau-A. TP1 and TP2 are the two total power channels. J1, J2 and J3 are the three correlator sub-bands.

The observing strategy was designed to achieve enough sensitivity per pixel in the sky and sky coverage given the noise performances and the amount of good observing time available. In Madison, WI, where the telescope is installed, we hope to have about 15% observing efficiency or roughly two full months per year. In our first observing campaign, from March to May 2001 we collected about 50 hours of usable data in Q mode. We scanned a cap (1.8 degrees diameter) around the North Celestail Pole (NCP) for a total of 30 pixels. This first campaign will produce a map with an expected noise of about 13 µK per pixel.

We are in the process of comparing our data set with other observations of the same sky region. In particular we will cross-correlate with the QMASK data looking for potential foregrounds. A companion survey with the Effelsberg 100 meter telescope at 32 GHz is under way looking for point sources contamination.

We plan to use COMPASS to observe the same region of the sky again starting in fall 2001. The polarimeter observes only one Stokes parameter at a time, but by rotating the polarimeter with respect to the telescope in steps of 45 degrees, we can measure U and Q on the sky. We are also developing a 90 GHz polarimeter that we hope to mount on the COMPASS telescope in early 2002 to help sort out any foreground signals that might be present in the 30 GHz data. In this configuration we expect a beam size of 7' and plan to map the same NCP region again in both Stokes parameters.

REFERENCES

1. Bennett, C. L., *et al* Ap. J. Lett. **464**. L1. 1996.
2. Flett, A. M. and Hendrson, C., MNRAS, 189, 867 (1979)
3. Halverson, N. W., *et al.*, astro-ph/0104489 (2001)
4. Keating B., *et al.*, Ap. J. Letters, 560, L1 (2001)
5. Keating B., *et al.*, Ap. J., 495, 580 (1998)
6. Lee, A. T., *et al.*, astro-ph/0104459 (2001)
7. Netterfield, C. B., *et al.*, astro-ph/0104460 (2001)

PolKa: A Tunable Polarimeter For Mm/Submm Wavelengths

G.Siringo, L.A.Reichertz and E.Kreysa

Max-Planck-Institut für Radioastronomie
Auf dem Hügel 69
D-53121 Bonn, Germany

Abstract. We present a new polarimeter for mm/submm wavelengths. Very low insertion losses and the possibility to tune its operating wavelength in a wide range make it a versatile instrument. It will be used with MPIfR bolometer arrays at different wavelengths to produce high-resolution maps of polarization.

INTRODUCTION

Measuring the linear polarization of the radiation emitted by interstellar dust is a very efficient way to get information about the magnetic fields present in the observed region. Dust grains tend to be aligned along the lines of the magnetic fields and for that reason the radiation emitted is partially polarized. They also behave as a polarizing filter modifying the state of polarization of the radiation transmitted through the interstellar medium [1, 2, 3].

The recent development of arrays of bolometers has given a strong improvement in photometric observations of mm/submm sources. Our goal is to realize a polarimeter that can be coupled to any array of bolometers as a permanent facility. Therefore, the polarimeter should not affect photometric observations and must be possible to switch in a short time from one observational mode to the other.

INSTRUMENT OVERVIEW

The typical configuration of a polarimeter, to measure linear polarization, has mainly two parts:
- a polarization modulation device, normally a rotating half-wave-plate (HWP). The rotation of the retarder produces a rotation of the plane of the linear polarization leaving unaffected the unpolarized radiation. Retarders are normally made of birefringent materials and so they are strongly affected by selective absorption. The phase shift is a function of the wavelength.
- an analyzer, normally a linear polarizer. In the mm/submm range a linear polarizer is normally made of a grid of thin parallel free standing conducting wires (called wire-grid). It reflects the component polarized parallel to the wires and transmits the orthogonal one.

Our polarimeter PolKa (Polarimeter für Bolometer-Kameras) uses a special type of HWP, based on metallic reflection. The reflection-type-HWP (RHWP) is made of two parts: a wire-grid polarizer and a mirror. Tuning the distance in between the two parts (fig.1), according to the working wavelength, it's possible to produce a 180 degrees phase shift in between the two components of polarization, because one is reflected by the wires, the other one by the mirror following a longer path. The phase shift is given by the simple relation

$$\text{phase shift} = 2\pi \frac{t}{\lambda \cos\alpha} \qquad (1)$$

where t is the distance in between the wire-grid and the mirror, α is the angle of incidence of the incoming radiation and λ is the operating wavelength.

It is very important to stress out that this device uses just metallic reflections and absorption is therefore negligible.

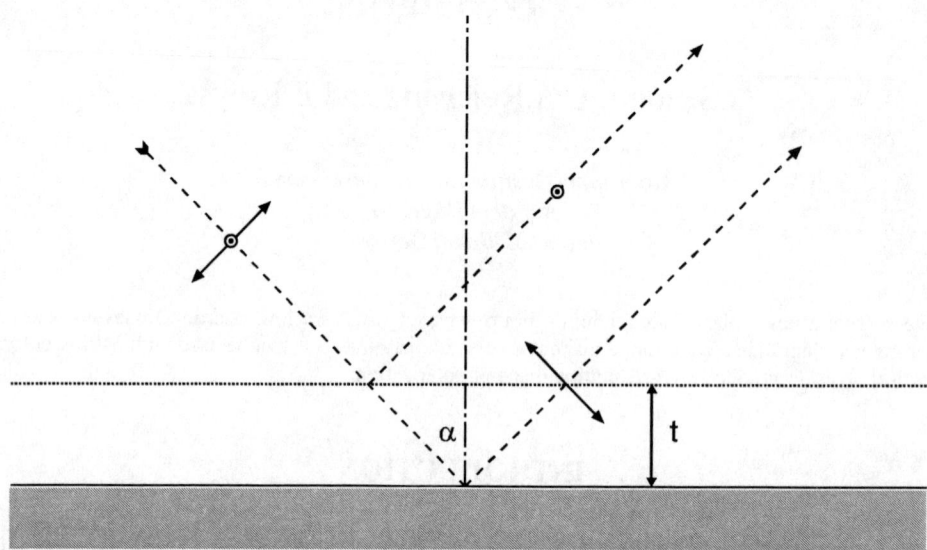

FIGURE 1. The production of a phase shift in the two components of polarization due to the reflection from the wire-grid (dotted line) and the subreflector. t is the tickness of the gap and α is the angle of incidence.

Given the possibility to tune the operating wavelength, PolKa can be used with different receivers and with different telescopes and it will be used together with any of the MPIfR bolometer arrays [4].

FIRST TEST

A PolKa prototype has already been tested at the Heinrich Hertz Telescope with a 19 channel array at .87 mm wavelength. Its RHWP has a wire-grid with a clear aperture of 146 mm, wires of 25 micron of diameter and a spacing of 100 micron. The RHWP is mounted on a motorized air bearing. The resulting high stability of the rotation axis will minimize spurious modulation of the total power. The measurements are done in a continuous rotation mode, at a modulation frequency of 16 Hz, that is four times the mechanical rotation frequency. The data are acquired using an ADC at 512 samples/sec and demodulated via software. The amplitude of the signal gives information about the degree of polarization of the radiation. The phase gives information about the angle of the polarization vectors in the sky (fig.2).

We made some calibrations using a 100% polarized source and using planets (Venus, Mars, Jupiter and Uranus were observable) as unpolarized sources. Observations of some galactic clouds and of some extragalactic objects have been done too and the analysis of the data is in progress.

An enhanced version, with a larger HWP, is planned to be used at the IRAM 30 m telescope with the MAMBO arrays at 1.3 mm and also with the HUMBA array at 2 mm.

CONCLUSIONS

The test at the telescope confirmed very low insertion losses from the RHWP, that behaves as a plane mirror once the analyzer is removed. This means that is possible to switch easily from polarimetry to photometry. By tuning the gap between grid and mirror is also possible to switch in between receivers operating at different wavelength. These special features make PolKa suitable to become a permanent facility.

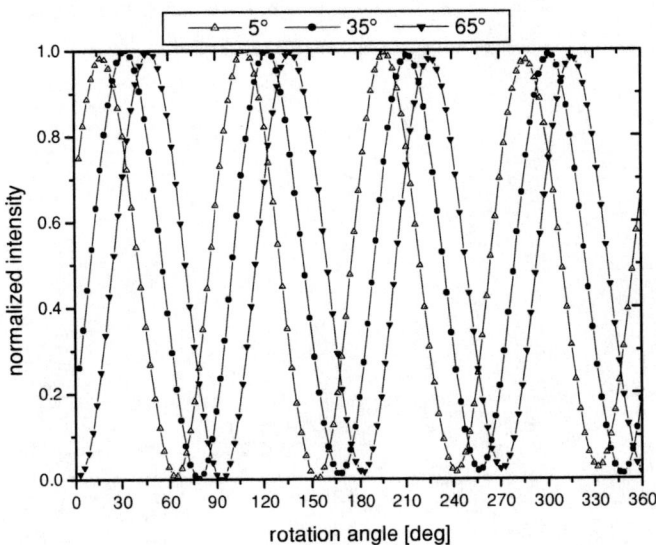

FIGURE 2. Example of calibration curves for different elevation angles of the telescope measured at the HHT using a 100% polarized source in front of the secondary mirror.

REFERENCES

1. Draine, B.T. and Lee, H.M., *The Astrophysical Journal* **285**, 89 (1984).
2. Lee, H.M. and Draine, B.T., *The Astrophysical Journal* **290**, 211 (1985).
3. Hildebrand, R.H., *Quarterly Journal of the Royal Astronomical Society* **29**, 327 (1988).
4. Kreysa, E. et al., *Infrared Physics and Technology* **40**, 191 (1999).

2K1BC Workshop
Experimental Cosmology at millimetre wavelengths

SATELLITE EXPERIMENTS

Planck Low Frequency Instrument

N. Mandolesi[1], M. Bersanelli[2], R.C. Butler[1], C. Burigana[1], D. Maino[3],
A. Mennella[4], G. Morgante[1], L. Valenziano[1], F. Villa[1]

On behalf of the LFI Consortium

[1]*TESRE-CNR – Bologna, Italy*
[2]*Università di Milano, Dip. Di Fisica, Milano, Italy*
[3]*Osservatorio Astronomico, Trieste, Italy*
[4]*IFC – CNR, Milano, Italy*

Abstract. The Low Frequency Instrument (LFI) is one of the two instrument onboard the ESA Planck Mission. LFI will image the Cosmic Microwave Background anisotropies in four different bands, from 30 GHz to 100 GHz, with an array of 54 radiometers. The instrument characteristics and expected performances are presented, with particular attention to the control of systematic effects.

INTRODUCTION

Planck will measure and accurately characterize small fluctuations in the Cosmic Microwave Background (CMB) temperature, which are only about one part in 10^5 of the CMB temperature [1,2,3]. In addition the LFI detectors are sensitive to the linear polarization of the received radiation [2], which for the CMB is expected at the 10^{-6} level. These measurements should reveal the primordial structure that grew to form galaxies and will test ideas about the properties and evolution of our Universe and the structures within it. By mapping CMB anisotropies precisely, the Planck satellite will provide decisive and extraordinarily exciting measurements of the geometry, constitution, and early history of the Universe.

The Planck/LFI will play a key role in the Planck mission for a number of reasons [2].

Its 70 and 100 GHz channels are optimal to get the cleanest possible view of primordial CMB fluctuations: Galactic emission, dominating the astrophysical foreground noise on scale $\geq 30'$, is minimum around 60 GHz; the confusion noise due to extragalactic sources, dominating on smaller angular scales, is minimum in the range 100-200 GHz, where it is primarily due to radio sources. A combination of data from all LFI channels will allow an accurate subtraction of contaminating foreground signals; the removal of Galactic dust emission will be further improved using HFI maps [3].

At 100 GHz the LFI reaches an angular resolution of 10', allowing to determine the power spectrum of CMB anisotropies up to multipoles $l \approx 1000 - 1500$. In the standard Cold Dark Matter model, anisotropies on smaller scales are quasi-exponentially erased by photon diffusion: little cosmological information remains there.

Polarization measurements, which should be possible with an accuracy of few μK towards the ecliptic caps, will provide independent estimates of cosmological parameters and will help in breaking any degeneracies in the determination of cosmological parameters.

The LFI data are essential to accurately subtract the radio source contribution to fluctuations [4], which are the main astrophysical limitation for CMB anisotropy measurements also in the HFI 143 and 217 GHz channels.

The high sensitivity, accurately calibrated maps of the Galactic synchrotron and free-free will constitute a unique tool for investigating the distribution of: i) the interstellar magnetic field; ii) relativistic electrons; iii) the ionized components of the interstellar medium.

The multifrequency LFI all-sky surveys will be unique in providing complete samples of bright extragalactic sources with strongly inverted spectra.

INSTRUMENT DESCRIPTION

A definitive CMB anisotropy measurement requires both extremely good sensitivity and rejection of systematic effects. The design of the experiment is driven by these two requirements. The detectors are cooled to low temperatures to achieve the best possible sensitivity. Avoidance and removal of systematic effects influence every aspect of the mission. An observation scheme that controls and tests systematic effects is imperative. Each pixel on the sky will be measured repeatedly by several different detectors at each of several frequencies on many different time scales. This process allows full characterization and subtraction of any instrumental offsets and drifts and provides for separation of astrophysical foregrounds.

The LFI is a multi-frequency radiometer array, located in the focal surface of the Planck telescope [5], designed to cover the 30-100 GHz spectral range, and will operate at four frequencies: 30 GHz, 44 GHz, 70 GHz and 100 GHz [2]. LFI takes full advantage of the dramatic progress of transistor amplifier technology achieved over the last decade, particularly for low noise performance and reliability of cryogenically cooled indium phosphide (InP) high-electron-mobility transistors (HEMTs). These devices exhibit the best noise performance in the LFI frequency range. In addition, the overall sensitivity requirement pushes the LFI design towards an array of 54 channels, compatible with power dissipation and available focal area. Transistor amplifiers based on InP, as opposed to the more traditional GaAs devices, ensure not only the minimum noise contribution from the receiver but also minimum heat dissipation. The LFI design includes an array of 27 feeds, each coupled to two channels exploiting the two linear polarizations, distributed in the four selected frequencies to obtain adequate sensitivity in each band.

To further minimize the power dissipation in the focal area the LFI radiometers are split into two sub-assemblies (front-end unit and back-end unit) connected by a set of waveguides, one located at the telescope focus, the other on the 300K portion of the spacecraft. For all frequencies, about 35 dB of front-end gain is sufficient to guarantee that the overall noise is dominated by the low-noise front-end amplifiers. In this way, the power dissipation for the LFI front-end array can be kept below 0.6 W, a sufficiently low value to enable active cooling of the HEMTs to 20 K. This is achieved with a vibration-less hydrogen sorption cooler, which also provides the 18~K pre-cooling stage to the HFI. Active cooling of the LFI front end reduces the noise temperature substantially, roughly a factor of 3 compared to an optimized passive cooling design.

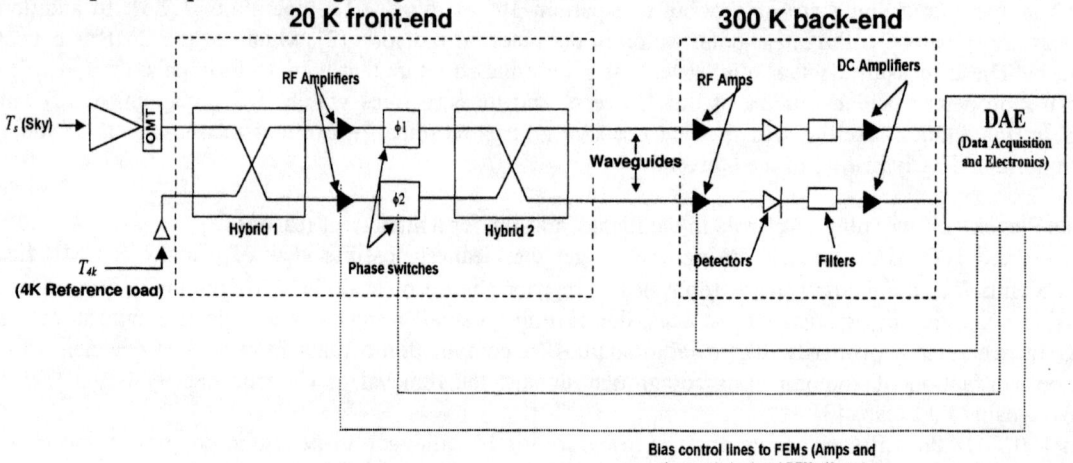

FIGURE 1. LFI radiometer block diagram.

The low-noise amplifiers use InP HEMTs in cascaded gain stages. The amplifiers at 30 and 44 GHz will use discrete InP HEMTs incorporated into a microwave integrated circuit (MIC). At these frequencies, cryogenic MIC amplifiers have demonstrated noise temperatures approaching 10 K, with 20 % bandwidth. At 70 and 100 GHz, where as many as 44 channels are present, LFI will use MMICs (Monolithic Microwave Integrated Circuits), which incorporate all circuit elements and the HEMT transistors on a single InP chip. Cryogenic MMIC amplifiers have been demonstrated at 75-115 GHz which exhibit ≤ 50 K noise from 90-105 GHz. The LFI will fully exploit both MIC and MMIC technologies at their best.

To reach the goal sensitivity the radiometric performance of the front-end passive components (corrugated feed horns, orthomode transducers – OMTs, hybrid couplers) must be outstanding over the entire 20% baseline bandwidth. Prototyping work shows that electroforming techniques are capable of producing high quality components at frequencies well above 100 GHz. The compact focal array requires miniaturized designs with no sacrifice in performance. The solution of extracting the two orthogonal linearly polarized signals form each feed maximizes the number for channels per available focal plane area, and, at the same time, adds polarization capability to the instrument.

The final LFI maps are expected to reach a sensitivity of approximately 6 μK in each reference pixel of 30' size. Coupled with the LFI nominal angular resolution [6] at the various frequencies (30', 24', 14' and 10' at 30, 44, 70 and 100 GHz respectively) this means a ΔT/T sensitivity per resolution element of 3 to 7 x 10^{-6}.

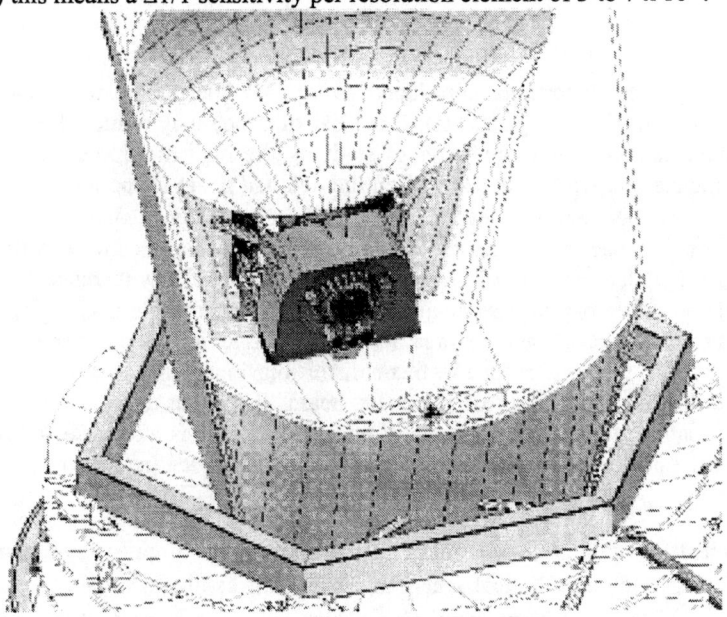

FIGURE 2. The LFI inside the Planck payload.

SYSTEMATIC EFFECTS

In order to reach the LFI scientific objectives, it is crucial to limit at μK level the residual instrumental systematic errors [7], particularly those that are synchronous with the spacecraft spin period. Spin-synchronous signals, in fact, are virtually indistinguishable from any *true* sky signal and, therefore, will affect permanently the final LFI maps. The choice of the Planck orbit (Lagrangian Point L2 of the Earth-Sun system) and basic scanning strategy (spin axis in anti-solar direction, and the field of view will scan near-great circles in the sky at 85 degrees from the spin axis) are based on the need to provide a stable environment to minimize thermal and radioactive systematic effects. In general we can classify systematic errors into five broad categories:

Main Beam distortion. Here we are concerned by errors caused by deviations of the real main beams from pure symmetric Gaussian shapes because of optical distortions and by differences between beam shapes at the same frequency [6].

Instrument Intrinsic. These are systematic effects that arise from non-idealities of the radiometers themselves, and are characterized by short- to medium-time scale fluctuations. Two typical examples are gain and noise temperature oscillations in the amplifiers, which display a $(f_k/f)^\alpha$ spectrum, where $\alpha \approx 1$ and f_k is the so-called knee frequency.

External Straylight. These systematic errors are generated by signals coming from celestial sources entering the instrument through the antenna side-lobes. Because of the difficulty to accurately measure the beam pattern at very low response level, it may be hard to properly remove straylight effects from data stream a posteriori. It is then necessary to control them in hardware to strongly reduce their impact.

Thermal Effects. This kind of effects is caused by thermal oscillations on time-scales ranging from less than a minute to more than an hour. A typical example in Planck is represented by thermal instabilities caused by the Sorption Cooler (the 20 K cryostat on-board the Planck satellite, that will cause periodic oscillations in the focal plane temperature with peak-to-peak amplitude ≤ 50 mK and with a wide frequency spectrum). See also [8] for further discussion.

Pointing. The pointing of the spacecraft will be known with some degree of uncertainty, which depends on both the limitations of the satellite attitude control subsystem and on the ability to reconstruct the orientation of the beam using star sensor and instrument data.

We will now focus on some of the above topics.

MAIN BEAM DISTORTION EFFECTS

The impact of main beam distortions introduced by optical aberrations on Planck measurements has been carefully studied in several works (e.g. [9,10,11]). It is clear that they may reduce the nominal angular resolution of the optical system, then degrading our capability of studying the high multipole range of CMB fluctuations, and make the measured antenna temperature dependent on the detailed beam shape and orientation, then introducing an additional systematic effect in the data at a level of some μK in terms of rms value.

As shown in [9], at high galactic latitudes the combined effect of main beam distortions and of Galaxy emission fluctuations increases the added error at 30 GHz by about a factor of 3 with respect to the case of a pure CMB fluctuation, whereas it produces only a very small additional effect in the cosmological channels. In addition, the combined effect of beam distortions and extragalactic source fluctuations is found to be very small at all LFI frequencies [12] compared to the noise induce by beam distortions in the case of a pure CMB sky.

We focus here further on the impact of the main beam distortions on the determination of angular power spectrum of CMB fluctuations to identify the range of multipoles which is most affected by this effect. We report in Figure 3 and 4 the angular power spectrum C_l as function of the multipole l in terms of

$$\Delta T_\ell = \sqrt{\ell(2\ell+1)C_\ell / 4\pi}$$

We have computed here a full year simulation both for a pure symmetric gaussian beam with $FWHM = \sqrt{8\ln 2}\,\sigma_b = 30'$ and for an elliptical gaussian beam with axial ratio r = 1.3 and with the same effective resolution ($\sqrt{\sigma_x \sigma_y} = \sigma_b$) of the considered symmetric beam (r=1). We computed the difference between the maps obtained with the elliptical and the symmetric beam by coadding the corresponding data streams and calculate the angular power spectrum of this difference map (we use the Healpix scheme [13]). This provides an estimate of the error introduced by main beam distortions on the determination of CMB angular power spectrum, in absence of appropriate deconvolution techniques able to take accurately into account the main beam detailed shape, owing for example to uncertainties in its reconstruction. As shown in Figure 3, this effect is particularly relevant at quite large multipoles, close to the CMB peak; the magnitude is clearly related to the value of r.

1/f NOISE

The effect of $1/f$ noise, in particular, has been minimized using the "pseudo-correlation" receiver concept for the LFI radiometers (see Figure 1). In fact for the Planck spin rate of 1 rpm (f = 0.017 Hz) and the knee frequencies measured to date for HEMT amplifier, a total power radiometer design would have unacceptably high $1/f$ noise, resulting in strong striping effects in the sky images. On the other hand, a much faster spin rate would damage the sensitivity of the HFI bolometers through the effect of their time response.

This receiver is a modified version of a class of radiometer configurations widely used in radio and microwave astronomy since the late 50's and at frequencies ranging from 408 MHz to 97 GHz. In the LFI scheme, the signals from the sky $T_x \approx 3K$, and from a reference load T_y, are combined by a hybrid coupler, amplified in two independent amplifier chains, and separated out by a second hybrid. The sky and the reference load temperatures are then measured and differenced. Since the reference signal has been subject to the same gain variations in the two amplifier chains as the sky signal, the true sky power can be recovered. Similar insensitivity to fluctuations in the detectors themselves is realized by switching in and out ± 90 deg phase shifts synchronously in each amplifier chain. Two square law diodes convert the signals to DC voltages, each carrying the power level of the sky T_x or reference load T_y, depending on the state of the phase shifter. The outputs of each of the two diodes will be first processed to take differences between successive sky and load measurements, and then the two diodes signals will be differenced.

Our early simulation work showed that the *1/f* noise of the radiometers with reference loads at 20 K (the front-end environment temperature) would result in substantial striping effects in the product maps. Software analysis indicate that signal processing algorithms can be constructed which are able to reduce the effect of stripes [14, 15]. These algorithms work better if the effect of 1/f noise is previously reduced "in hardware", by designing the instrument with reference loads operating at a temperature $T_y \approx T_x$. The design of the Planck focal plane unit provides a natural way to achieve this, i.e., by mounting the reference loads on the HFI front-end unit, where they have a temperature of about 4 K [16].

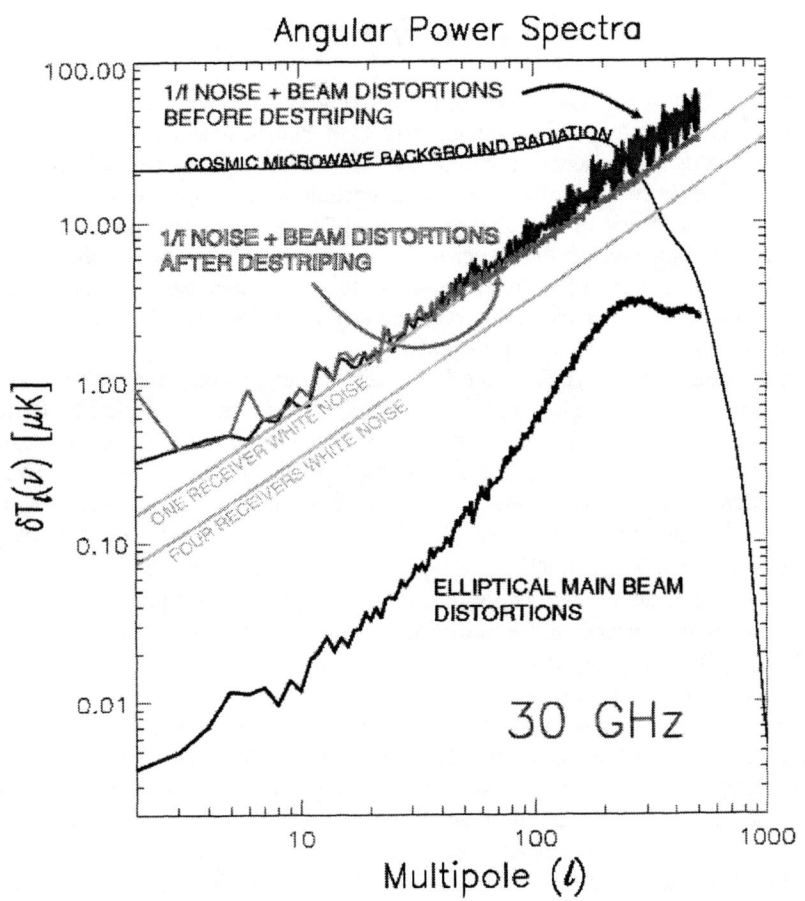

FIGURE 3. Comparison between the power spectra of some classes of systematic effects and the CMB power spectrum.

We performed detailed simulations to quantify the impact of *1/f* noise in Planck observations. In a flight simulation code which includes all the main properties of the Planck survey we have included the computation of simulated noise stream. By coadding the data streams we can obtain simulated observed maps of instrumental noise that can be quantitatively analyzed in a standard way by evaluating the angular power spectrum. As reported by [17] the effect of *1/f* noise can be seen as one or more additive levels, different for each scan circle. To remove these additive levels and "clean" the "observed" map we have then implemented a destriping code [18] based on the global minimization of the differences between the temperatures measured at the same sky positions in different scan circles [1]. This allows us to quantify the impact of *1/f* noise after applying our destriping algorithm for different scanning strategies, beam locations, and magnitude of the 1/f effect.

Our flight simulator and destriping code as well as the main results for the *1/f* noise and their implications are extensively presented in [18] and [15]; the efficiency of the destriping technique has been there quantified by using estimators such as χ^2 or the rms temperature of the stripes, and, in particular, by studying the impact of the residual *1/f* noise on the CMB power spectrum reconstruction.

In Figure 3 we show the combined effect of *1/f* noise and main beam distortions in terms of angular power spectrum before and after applying our destriping code. We find that the efficiency of the destriping algorithm is not significantly affected by the additional uncertainty introduced by the ``systematic'' differences among the observed temperatures resulting from different orientations of the main beam at the crossing points of different scan circles [19]. As evident, the excess of noise with respect to the case of pure white noise is significantly reduced by this kind of algorithm.

GALAXY STRAYLIGHT CONTAMINATION

The straylight contamination due to Galaxy emission entering at large angles from antenna pointing direction may be one of the most relevant sources of systematic effects in Planck observations. Since the antenna response features at large angular scales from the beam center (far sidelobes) are determined largely by diffraction and scattering from the edges of the mirrors and from nearby supporting structures, they can be reduced by reducing the illumination of the edge of the primary, or in the jargon of antenna design, increasing the edge taper, defined as the ratio of the power per unit area incident on the center of the mirror to that incident on the edge: a lower edge taper leads to a lower sidelobe level. On the other hand, lowering the edge taper has a negative impact on the angular resolution (for a detailed study on the Planck telescope angular resolution see [20]). A trade off between angular resolution and straylight contamination has to be performed.

The main astrophysical source of straylight at the LFI channels derives from the Galactic emission and depends on the observed sky region, on the frequency and on the shielding efficiency. At the LFI frequencies, the Galaxy straylight is expected to be particularly crucial at the lowest frequencies, due to the increasing of synchrotron emission and anisotropies with the wavelength, so our simulations focus on the 30 GHz case. We have considered as a reference case the old Planck Carrier configuration studied in [21]. The simulations include the convolution of the full sky with the full 30 GHz antenna pattern [19, 22]. We compute separately the absolute signals from three reference pattern regions: the main beam (at $\theta \leq 1.2°$), the intermediate beam (at $1.2° \leq \theta \leq 5°$) and far pattern (at $\theta \geq 5°$). For the 30 GHz channel, the pattern region outside $\approx 1.2°$ corresponds to antenna responses lower than ≈ -40 dB with respect to the peak response, where the beam response is difficult to measure in flight through planets [2]; $\theta \approx 5°$ approximately divides pattern regions where significant response variations occur on angular scales much less than $1°$ from those where they occur on $\sim$ degree or larger scales.

By co-adding the data streams obtained with the flight simulator we obtain the corresponding maps which are then analyzed in terms of angular power spectrum (see Figure 4) [19]. We show separately the contribution from intermediate (solid light gray) and far (solid dark gray) pattern regions. The most important contamination in terms of angular power spectrum derives from the intermediate pattern when all the sky is considered; on the contrary, considering only the regions at $|b| \geq 30°$ (dashed lines: light gray for the intermediate pattern, dark gray for the far pattern) the galactic straylight power spectrum is dominated by the far sidelobes (dark gray dashed line).

In general, this effect is relevant at low multipoles and becomes less than 20 times smaller than white noise at $l \approx 50$, due to the decreasing of Galaxy fluctuation spectrum at high multipoles. Our simulations show that galactic straylight peaks at ≈ 15 μK, a value comparable with the sensitivity per pixel, owing to the signal entering at few degrees from the beam center; this relatively large effect occurs in the regions close to the galactic plane where in any case the "direct" (i.e. observed by the main beam) contamination from the Galaxy prevents an accurate determination of CMB fluctuations, being not trivial the subtraction of Galaxy emission at levels less than few % accuracy. On the other hand, the large straylight contamination values, although critical for CMB anisotropy measurements, are not crucial for the determination of Galaxy emission, which is several order of magnitude larger.

In general, the typical value of galactic straylight is less than the 50 % of the white noise sensitivity. The most crucial contamination for the Planck goal is that introduced by far pattern features, the corresponding signal peaking at ≈ 6 μK; this contamination, although smaller than that from intermediate pattern regions in terms of absolute values, dominates at high galactic latitudes where we can extract the maximum cosmological information. Here, the effect is comparable to the noise sensitivity. This issue may be particular crucial for polarization measurements that take maximum advantage from the highest sensitivity regions.

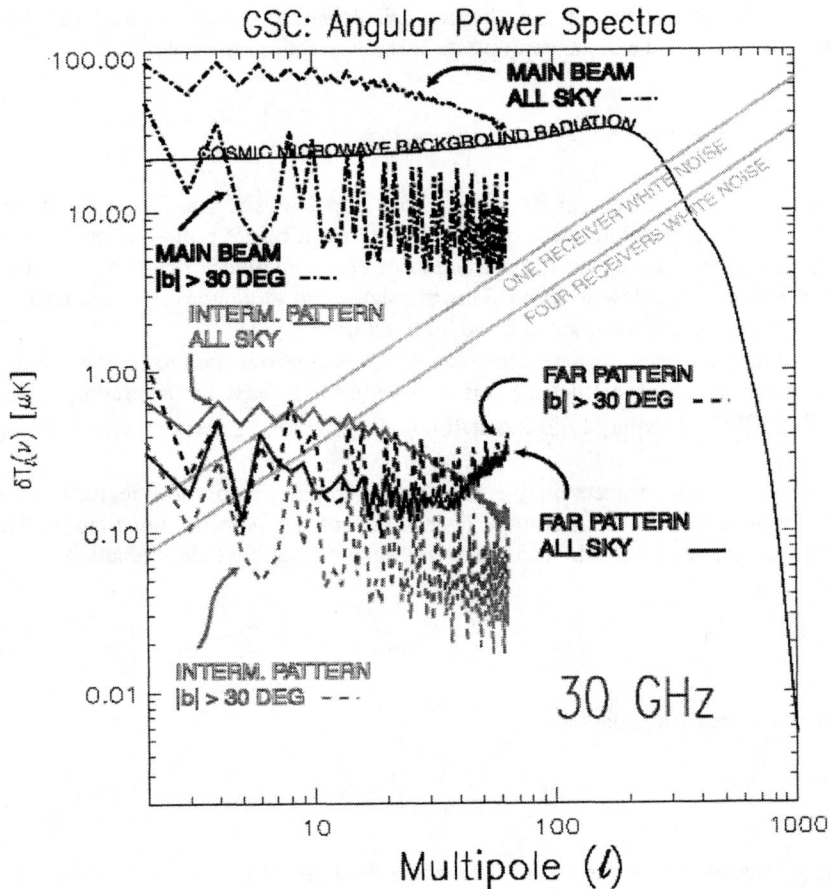

FIGURE 4. Galactic Straylight Contamination (GSC): Impact on the Power Spectrum reconstruction.

THERMAL EFFECTS

To meet the highly stringent cryogenic requirements of the Planck instruments a dedicated chain of cryo-coolers is being developed. Its implementation in the Planck satellite implies the development of a very complex thermal environment, that will have to be carefully controlled in order to avoid thermal fluctuations generating unwanted systematic effects.

A first category of thermal instabilities is represented sources that are external to the Planck satellite i.e. bright celestial objects (Sun, Moon, Earth). In particular the Moon and the Earth may potentially induce temperature variations over very long time-scales (greater than 1 day) because of their varying distance from the satellite during the mission, while the Sun is a potential source of spin-synchronous fluctuations, that may be caused by deviations from the nominal anti-sun position of the spin axis (e.g. because of planned precession movements or because of oscillations in the satellite attitude after repointing). Although these effects are not expected to be a major source of systematic errors, their effect will be assessed as much quantitatively as possible.

A second and most important category of thermal fluctuations is represented by "internal" instabilities, i.e. oscillations in the temperature at the various thermal boundaries between the spacecraft and the instrument. LFI, in particular, will be sensitive to thermal fluctuations caused by the Sorption Cooler that propagate through the satellite and affect virtually any thermal interface. An example showing the effect of temperature fluctuations of the 20 K stage induced by the Sorption Cooler is detailed in [8]. LFI will be also sensitive to the behavior of the HFI 4K cooler, which provides LFI with the reference load signal. A high degree of stability in this reference signal is of crucial importance for LFI, because any oscillation in the reference load will be transferred directly to the final measurement.

The impact of such systematic effects on the ultimate LFI scientific results is being carefully assessed so that potentially critical areas are identified and the appropriate corrective actions are taken.

POINTING

The pointing uncertainty introduces two main effects on CMB map analysis. The first is due to the error induced in the in-flight recovery of the beam resolution (i.e. its FWHM) through bright point sources, mainly external planets as Jupiter and Saturn. As a consequence, our reconstruction of the "convolved" sky angular power spectrum suffers from the uncertainty of the beam window function. This translates into an increase of the uncertainty of the angular power spectrum estimation, most relevant at middle and high multipoles [23].

In fact, the error induced on the recovery of the deconvolved C_l is approximated to the first order in D, by $2l^2\sigma_b^2 D$, where D is the relative error on the reconstructed FWHM. As a result, to have an uncertainty on C_l less than that due to the white noise at $\ell \cong 1200$ we require $D \leq 2\%$ at 100 GHz. This can be assured with a pointing accuracy better than 1' at 2σ level.

The second effect due to pointing uncertainty is the "direct" angular resolution degradation, resulting from the convolution of the beam pattern with the statistical distribution of the pointing uncertainty. This affects the sky angular power spectrum at high multipoles. In the case of a symmetric beam window function

$$e^{-l^2\left(\frac{\sigma_b}{rad}\right)^2}$$

this effect introduces a decrease by a factor

$$e^{-\left[\left(\frac{\sigma_e}{\sigma_b}\right)^2-1\right]l^2\left(\frac{\sigma_b}{rad}\right)^2}.$$

This factor is the ratio between the power of the fluctuation field and the power of the instrumental noise at the multipole l, where

$$\sigma_e = \left(\sigma_b^2 + \sigma_n^2 + \sigma_u^2\right)^{\frac{1}{2}}$$

is the global effective resolution [24] due to the additional contributions from the pointing (1σ) error of the telescope direction associated to the spin axis nutation, σ_n, and the pointing (1σ) error of the telescope direction associated to the pointing evaluation uncertainty, σ_u [25]. This pointing effect, less important for the lower frequency channels, represents a non negligible effect at 100 GHz. The signal to noise ratio decreases by about the 20-30 % at $\ell \approx 1200$ if $\sigma_n=\sigma_u=1$'; it means that the intersection between the power spectra of the unsubtractable white noise and the CMB anisotropy occurs at a multipole smaller of about 30 – 100 (the proper value depending on the underlying CMB model) than that derived with no pointing uncertainty [23]. Clearly, this effect is much more important for an accurate determination of the CMB angular power spectrum at very high multipoles with the HFI 217 GHz channel which has a nominal FWHM of about 5'. Therefore, from this point of view, our requirement of a pointing accuracy better than 1' at 2σ level, as discussed above, is strongly desirable also for LFI in view of the sinergy between LFI and HFI.

CONCLUSIONS

The Planck Low Frequency Instrument is conceived to obtain high resolution, high sensitivity, full-sky imaging of the Cosmic Microwave Background. Its design is driven by the need to couple the required low noise performances with very effective suppression of all potential sources of systematic effects. In fact, several aspects of the entire Planck mission (payload design, orbit, scanning strategy) are driven by the need to control unwanted spurious signals at µK level. We have presented some aspects of the on-going extensive work carried out by the LFI team to understand in detail the instrument properties and their potential effects on the product maps. This effort is crucial in order to fully extract the enormous scientific information expected from the Planck mission.

ACKNOWLEDGMENTS

The HEALPix package use is acknowledged. We wish to warmly thank the Planck Low Frequency Instrument Data Processing Center for the support to the simulation work.

REFERENCES

1. Bersanelli, M., Bouchet, F.R., Efstathiou, G., Griffin, M., Lamarre, J.M., Mandolesi, N., Norgaard-Nielsen, H.U., Pace, O., Polny, J., Puget, J-L., Tauber, J., Vittorio, N., and Volonté, S., "ESA, COBRAS/SAMBA Report on the Phase A Study", **D/SCI(96)3** (1996)
2. Mandolesi, N. et al., "The Low Frequency Instrument, a Proposal Submitted to the ESA in response to the Announcement of Opportunity for the FIRST/Planck mission" (1998)
3. Puget, J-L., et al., "The High Frequency Instrument, a Proposal Submitted to the ESA in response to the Announcement of Opportunity for the FIRST/Planck mission" (1998)
4. Toffolatti, L., Argueso Gomez, F., De Zotti, G., Mazzei, P., Franceschini, A., Danese, L., and Burigana, C., *MNRAS*, **297**, 117-127 (1998)
5. Villa, F., Bersanelli, M., Burigana, C., Butler, R.C., Mandolesi, N., Mennella, A., Morgante, G., Sandri, M., Terenzi, L., and Valenziano, L., "The Planck telescope", this Conference proceedings
6. Sandri, M., Bersanelli, M., Burigana, C., Butler, R.C., Malaspina, M., Mandolesi, N., Mennella, A., Morgante, G., Terenzi, L., Valenziano, L., and Villa, F., "Planck Low Frequency Instrument: Beam Patterns", this Conference proceedings
7. Bersanelli M., et al., *Astroph. Lett and Comm.*, **31**, 171 (2000)
8. Mennella, A., Bersanelli, M., Burigana, C., Maino, D., Ferretti, R., Morgante, G., Prina, M., Mandolesi, N., Butler, R.C., Valenziano, L., and Villa, F., this Conference proceedings.
9. Burigana, C., Maino, D., Mandolesi, N., Pierpaoli, E., Bersanelli, M., Danese, L., and Attolini, M.R., *Astron. Astroph. Suppl.*, **130**, 551-560 (1998)
10. Mandolesi, N., Bersanelli, M., Burigana, C., Gorski, K.M., Guzzi, P., Hivon, E., Maino, D., Malaspina, M., Valenziano, L., and Villa, F., "On the Planck effective angular resolution", ", *Int. Rep. TeSRE/CNR.*, **199/1997**, (1997)
11. Mandolesi, N., Bersanelli, M., Burigana, C., Gorski, K.M., Hivon, E., Maino, D., Valenziano, L., Villa, F., and White, M., *Astron. Astroph. Suppl.*, **145**, 323-340 (2000)
12. Burigana, C., Maino, D., Mandolesi, N., Villa, F., Valenziano, L., Bersanelli, M., Danese, L., Toffolatti, L., and Argueso Gomez, F., *Astro. Lett. Comm.*, **37**, 253-258 (2000), astro-ph/9903137
13. Gorski K.M., Hivon E., Wandelt B.D., "Analysis Issues for Large CMB Data Sets", in *Proceedings of the MPA/ESO Conference on Evolution of Large-Scale Structure: from Recombination to Garching*, edited by A.J. Banday, 1998, p. 37, astro-ph/9812350
14. Delabruille, J., *Astron. Astroph. Suppl.*, **127**, 555 (1998)
15. Maino, D., Burigana, C., Maltoni, M., Wandelt, B.D., Gorski, K.M., Malaspina, M., Bersanelli, M., Mandolesi, N., Banday, A.J., and Hivon, E., *Astron. Astroph Suppl*, **140**, 383-391 (1999)
16. Valenziano, L., Bersanelli, Butler, R.C., Cuttaia, F., Mandolesi, N., Mennella, A., Morigi, G., Morgante, G., Sandri, M., Terenzi, L., and Villa, F., "The 4K reference load for the Planck Low Frequency Instrument", this Conference proceedings
17. Janssen, M., Scott, D., White, M., astro-ph/9602009 (1996)
18. Burigana, C., Malaspina, M., Mandolesi, Danese, L., Maino, D., Bersanelli, M., and Maltoni, M.," A preliminary study on destriping techniques on Planck-LFI measurements vs. observational strategy", *Int. Rep. TeSRE/CNR.*, **198/1997**, (1997), astro-ph/9906360
19. Burigana, C., Maino, D., Gorski, K.M., Mandolesi, N., Bersanelli, M., Villa, F., Valenziano, L., Wandelt, B.D., Maltoni, M., and Hivon, E.,*Astron. Astroph.*, **373**, 345-358 (2001)
20. Mandolesi, N., Bersanelli, M., Burigana, C., and Villa, F., *Astroph. Lett. and Comm.*,**31** (2000), astro-ph/9904135
21. De Maagt, P., Polegre, A.M., and Crone, G., "Planck - Straylight Evaluation of the Carrier Configuration", *Technical Report ESA, PT-TN-05967, 1/0* (1998)
22. Wandelt, B.D., and Gorski, K.M., *Phys. Rev. D*, **63**, 123 (2001), astro-ph/0008227
23. Burigana, C., Butler, R.C., and Mandolesi, N., "Planck-LFI pointing accuracy requirements", *PL-LFI-PST-TN-023* (2001)
24. Puget, J-L., La marre, J.M., Sygnet, J.F., and Hivon, E., Document No. PL-HFI-IAS-TN-POINT002 (2001)
25. ALCATEL Space, Document No. ASPI-DCMI-00-APC-279, (2000)

ODIN preliminary results, HERSHEL and other stories

Pierre Encrenaz

Paris Observatory, 75014 – Paris - France

Abstract. The first observations of the Odin Satellite are presented. Their implications for the Herschel satellite are mentionned. The necessity of large teams to build these sofisticated instruments leaves room for small teams in very specialized areas. Historical examples concerning the Cosmic Microwave Background are reminded.

FROM GROUND TO SPACE MILLIMETER ASTRONOMY

The early millimeter antennas (5m at Fort Davies, 36' at Kitt Peak) have been designed and built for continuum observations, in particular to observe the moon and planets after the first satellite launches. After phase locked heterodyne receiver could be installed at the prime focus of these antennas, observations of interstellar and planetary molecules started. It was then easy to get weeks – even months – of observing time on the first generation of millimeter telescopes. It has been soon realized that the observatories at sea level (Bordeaux POMI, BTL 7m antenna) were severely limited by the amount of pricipitable water of these sites (Fig.1). Higher altitude sites were sought for (IRAM 30m and interferometer, Cerro Tololo mini, La Silla SEST, Hawaï CSO). The size of the teams needed to operate these telescopes, increased from a few in 1970 to hundred in 1990 (Hawaï JCMT, IRAM) and will continue to grow for ALMA in chilean desert. Millimeter astronomy on the ground reaches then its golden age. However two limitations remain: transitions from telluric molecules (H_2O, O_2 mostly) severely limit ground observations, and observations of large frequency bands in continuum are severely hindered. Suborbital platforms on which small teams developed the most sensitie receivers to be used to overcome ground limitations: kas, balloons paved the way to satellites like COB, ISO, SWAS, ODIN, ROSETTA (MIRO), HERSCHEL AND PLANCK. The last two involve the wordwide community at large, and imply hundreds of scientists to prepare the instruments and model the analysis of the data to be acquierd during their limited lifetime.

HETERODYNE IN SPACE AND THE ODIN SATELLITE

In order for your instrument to be selected on a satellite, you need to pass a very severe peer review panel and convince them that the technology presented is ripe, affordable and will bring the most important science. For historical reasons (poor communication between the radioastronomical and astrophysical communities, technical difficulties of the heterodyne techniques) it took more than 15 years to convince space agencies to accept these radio receivers on board spacecraft. The success of MLS (Microwave Limb Sounding) [1] on UARS changed the situation and small satellite began to be fonded (SWAS, ODIN). Mostly devoted to O_2 and H_2O in the interstellar medium and the earth atmosphere, ODIN built by the Swedish Space Corporation in collaboration with Canada, France and Finland has been launched in February 2001. Its characteristis are given Table 1, a spectrum obtained recently is shown in Figure 2. The major goal is the detection of interstellar O_2 whose abundance remains a challenge for the models of interstellar chemistry.

The success of these two missions make possible the construction of a larger satellite (HERSCHEL) for a launch in 2007 (with Planck on Ariane V). It will include technologies developed for ISO. Its main features are given [2].

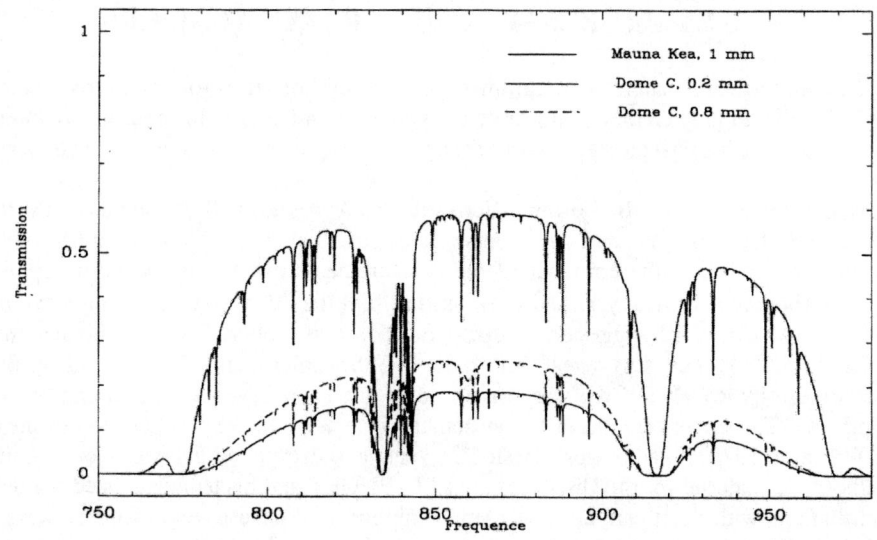

FIGURE 1. The ATM model developed by Juan Pardo Carrion predicts to 0.5 % the true absorption of the atmosphere as a function of altitude.

TABLE 1.

Diameter	Observed Frequencies	System Noise
1.1m	119 GHz	500 K
	487 – 572 GHz	2500 K
	UV (OSIRIS)	

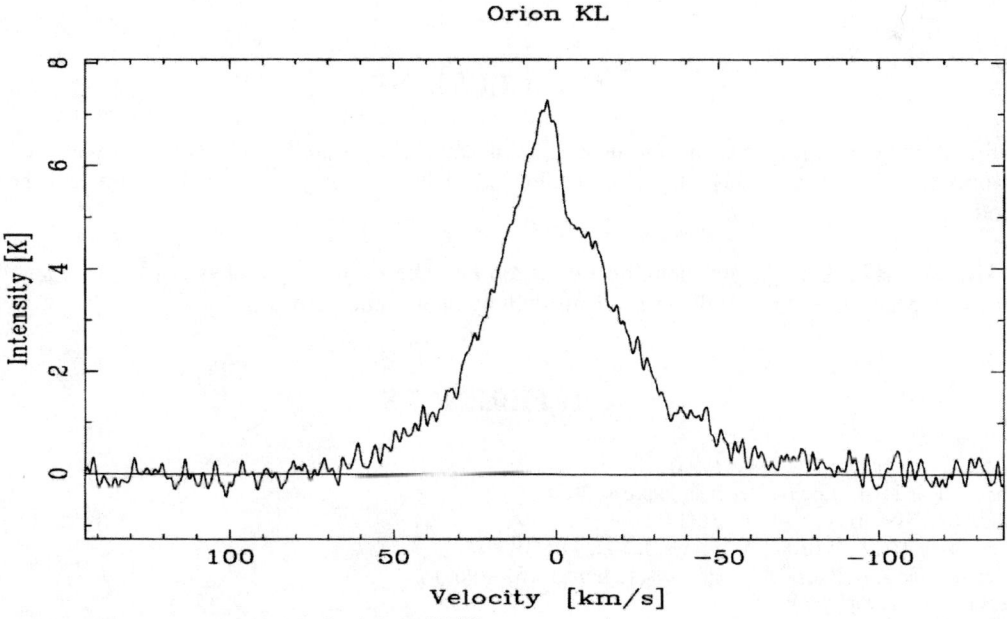

FIGURE 2. ORION KL spectra recently obtained from ODIN.

KEY PROBLEMS AND PARTIAL ANSWERS

When a field of science matures like millimeter and submillimeter astronomy, it may be difficult for a small team to find a niche. How to give credit and promote the young scientists who have designed and built an instrument while more than 2000 astrophysicist are ager to use it (High Energy Physics faces the same challenges).

Two examples related to CMB (Cosmic Microwave Background) will be used as an illustration. In 1961 Ed Ohm, physicist at Bell Labs, found an excess noise at 5 GHz using the Holmdel horn antenna build for the TELSTAR satellites. He enlarged the error bar of the measurements made with the most sophiticated equipement available at the time (hundreds of man year had been used to bint the 4095 MHz traveling wave maser) to include ϕ in the error bar [3]. A. Crawford suggested to Arno Penzias and Robert Wilson who had just gratuated from Columbia and Cal Tech Universities to repeat and reanalyse the experiment. Upon a visit by B. BURKE at BTL, The Princeton team lead by Robert DICKE was convinced of the discovery of the CMB and two consecutive papers published in Ap.J. [4]. The understanding of the systematic effects of the experiment led to the Nobel Prize to A. Penzias and R. Wilson in 1978. It had been suggested 30 years ago that young galaxies would emit in the far infrared [5]. COBE produced the material for this discovery, but J.L. PUGET and his team analyzed the data in retrieving the foreground contributions with great care and discovered this infrared excess. Even so they were not on the COBE team, their profound understanding of systematic effects made them authors of this important discovery [6].

SUGGESTIONS

If you have in your small team unique technical or scientific capabilities, your contribution can be decisive.

* The submillimeter segmented telescope for the PRONAOS gondola [6] lead to the selection of the SiC technology for the 3,5m HERSCHEL telescope.

* The ATM atmospheric model of Juan Pardo taking into account very far away wings of telluric lines is now used world wide [7].

* Major projects may never be funded : the NRAO 25m millimeter telescope rated 1 in the FIELD was never built. The US Large Haadron Collider has been cancelled while the tunnel was already digged out. The european astroplane has never been built.

CONCLUSIONS

65 Millions ago the largest creatures on the earth, dinosaurs disappeared suddenly (brontosaurus, archeopteryx, tyrannosaurus rex). Small mammals adapted to ecological niches not only survived but modelled the earth as we know it today.

The same is happening with submillimeter astronomy: the young bright physicist we have heard during this conference will make the best use of the instruments built by their predecessors.

REFERENCES

1. J.W. Waters, et al Nature 362, 597, (1993).
2. G. Pilbratt, Toledo Meeting on Hershell. in press, 2001
3. E.A. Ohm Bell Syst. Tech. J. 40. 1065, (1961).
4. A.A. Penzias and R.W. Wilson : Astrophys. J. 142, 419 (1965)
5. P.J. Encrenaz and R.B. Partridge : Astrophys. Letters 3, 161 (1969)
6. J.L. Puget et al : Astrophys. J. (1996)
7. J. Pardo-Carrion, unpublished PhD from Paris VI University, (1995)

Submillimeter and millimeter wave sky mapping in the space project Submillimetron

Vladimir D. Gromov*[¶], Nikolay S. Kardashev*, Leonid S. Kuzmin[¶]

*Astro Space Center of Lebedev Physical Institute, Profsoyuznaya 84/32, 117810 Moscow, Russia
[¶] Chalmers University of Technology, Microtechnology Centre, 41296 Göteborg, Sweden

Abstract. Submillimetron is an international project of the space telescope for astronomical studies at submillimeter and millimeter wavelengths using free-flying spacecraft and facilities of the Russian segment of the International Space Station. The payload is the 60-cm telescope cooled to liquid helium temperature with arrays of a novel type detector, Normal-metal Hot-Electron Bolometer (NHEB). The angular resolution is 1-10 arcmin, field of view - about 1 degree, detectors sensitivity - about 10^{-18} W/Hz$^{1/2}$, spectral region - 0.2 - 2 mm. Parameters of the instrument and complementarity to other experiments including CMB measurements are discussed.

INTRODUCTION

A general aim of modern astrophysics and fundamental physics is to understand the processes by which the universe evolved from initial simplicity to complexity visible now in all scales. This evolutionary approach to astronomy was clear enough in early sixties along as understanding of importance of submillimeter waves. At that time our activity was initiated in this region for ground-based [1-3] and for air and space [4-6] observation. In 70's – 80's the perspective concepts of submillimeter space telescopes were determined: fully cryogenically cooled one similar to [7] and [6], radiatively-cooled like [8], and deployable reflector [9]. Last concept is more adapted for large mirrors but at cost of larger complexity. It is under development now in Millimetron project [10]. Radiative cooling was proposed for project POIROT and later for project EDISON [8] in combination with machine-cooling, lightweight mirror and very distant orbit (L2). The concept was finally accepted by ESA in 1993 for realization in 2007 in large projects Hershel (previously FIRST) [11, 12] and Planck (previously COBRAS/SAMBA) [13, 14]. As noted in review [15], the principles of space-suitable cryocoolers were defined in 60's [16, 17] and Planck already included close-loop refrigerators. Hershel uses a large ISO-like cryostat. Both systems maintain low-level temperature only for focal devices. Radiatively cooled mirrors have thermal emission greater than extraterrestrial background in submillimeter spectral region. A goal of background-limited sensitivity may be achieved by use of a cryogenically cooled telescope in project Submillimetron [18-21].

CONCEPT OF SUBMILLIMETRON PROJECT

Sensitivity of sensors is a main factor defining possibilities to observe faint distant objects. Fundamental limit of direct detector sensitivity δP depends on background power P_{bg} as $\delta P^2 = n(n+1)m(h\nu)^2 \Delta\nu$, where $\Delta\nu$ is spectral bandwith, $h\nu$ - energy of photons, $m = g(A\Omega)/\lambda^2$, $g=2$ (for one polarization $g=1$), $(A\Omega)$ – geometrical factor, λ - wavelength. A value of n is determined by $P_{bg} = nm(h\nu)\Delta\nu$. Here is an important difference with receivers using front-end mixers or amplifies, which noise limit is proportional to $(n+1/2)$ and can't be very small even at low background. $P_{bg} = \varepsilon(e^{h\nu/kT}-1)^{-1}m(h\nu)\Delta\nu$, where k is Boltzmann constant, T - temperature, ε - emissivity. Background with surface brightness νI_ν produces $P_{bg} = (\nu I_\nu)(A\Omega)(\Delta\nu/\nu)$. It was noted in [22] that previously used simplified evaluation $\delta P = \varepsilon^{1/2} \delta P_{\varepsilon=1}$ overestimated δP in submillimeter and mm regions where usually $n_{\varepsilon=1} > 1$.

Extraterrestrial background was measured by instruments on COBE satellite [23]. Submillimeter brightness spectra was determined in [24] and shown in Fig. 1 (left). Instrumental emission spectra are shown in Fig. 1 (right).

Even for low emissivity ($\varepsilon=1\%$) of radiatively cooled mirrors (T > 10-40 K) their emission sufficiently exceeds that of the background. A cryogenically cooled telescope can reach background-limited sensitivity with detector "noise equivalent power" (*NEP*) of 10^{-19}-10^{-18} W/Hz$^{1/2}$ comparable with δP_{bg}. Principles of such devices were proposed in early 90's [25, 26] along as principles of electron cooling [27, 28]. In both cases, superconductor-insulator-normal metal (SIN) tunnel junctions were used. Novel detector technologies for extremely low astronomical backgrounds permit to reach unique observational sensitivity even for moderate size telescope.

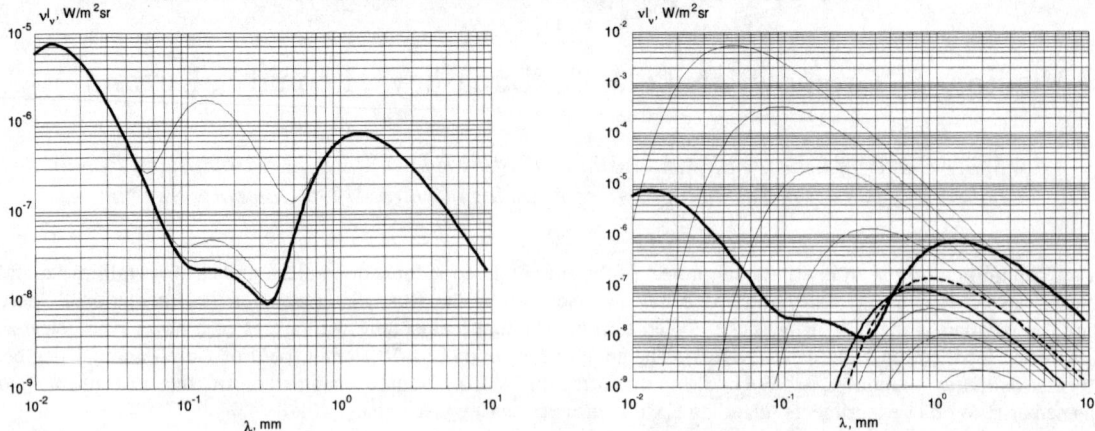

FIGURE 1. Spectra of background brightness νI_ν, W/m^{2}sr. <u>Left panel</u>: extraterrestrial background [24] in regions with galactic coordinates $b>60°$, $30°<b<60°$, $10°<b<30°$, and $b=0, l=180°$ corresponds to curves from bottom to top. <u>Right panel</u>: instrumental background for temperatures of telescope mirrors T = 80, 40, 20, 10, 5, 4, 3, and 2 K corresponding to thin curves from top to bottom, and emissivity $\varepsilon=0.01$; dashed curve - $\varepsilon=0.04$, T= 5 K. Thick solid curve shows extraterrestrial background for comparison.

International project Submillimetron [18-21] has been initiated by Astro Space Center of Lebedev Physical Institute. It uses a spacecraft with cryogenically cooled submillimeter telescope [20] flying separately from the International Space Station (ISS). The concept of a free-flying module with periodic docking to the ISS gives possibility to combine a low cost with reliability, refilling, repairment and maintenance. Main parameters of the instrument are given in Tables 1 for telescope of diameter 0.6 meters and with *NEP* of NHEB of 10^{-18} W/Hz$^{-1/2}$ [29-30]. The NHEB uses SIN tunnel junction as a temperature sensor. SQUID readout gives additional flexibility for combining the detectors in arrays providing opportunity for low noise multiplexing. Additional advantage of "built-in" electron cooling is compensation of background power load for bolometers [31]. As shown on Fig. 1, intensity of the background varies more than an order of magnitude depending on Galactic coordinates.

TABLE 1. Parameters of the submillimeter photometric instrument

λ_L, mm	0.2	0.3	0.4	0.5	0.6	0.8	1.0	1.5	2.0
λ_S, mm	0.15	0.2	0.3	0.4	0.5	0.6	0.8	1.0	1.5
θ, arcmin	1	1.4	2	2.6	3.2	4	5.2	7.2	10
P_{bg}, fW	0.09	0.2	0.54	1.7	4.4	24	54	244	300
δP_{bg}, W/Hz$^{-1/2}$	$3\cdot10^{-19}$	$4\cdot10^{-19}$	$5.5\cdot10^{-19}$	$8.7\cdot10^{-19}$	$1.3\cdot10^{-18}$	$2.6\cdot10^{-18}$	$3.4\cdot10^{-18}$	$6.3\cdot10^{-18}$	$6\cdot10^{-18}$
δP_{tot}, W/Hz$^{-1/2}$	$1\cdot10^{-18}$	$1.1\cdot10^{-18}$	$1.1\cdot10^{-18}$	$1.3\cdot10^{-18}$	$1.6\cdot10^{-18}$	$2.8\cdot10^{-18}$	$3.6\cdot10^{-18}$	$6.4\cdot10^{-18}$	$6.1\cdot10^{-18}$
NEFD, mJy/Hz$^{1/2}$	0.76	0.8	1.7	3.2	5.7	8.1	17	23	44
$NE\nu I_\nu$ pW/m^{2}srHz$^{1/2}$	120	43	33	29	29	20	20	10	6.9

Temperature of mirrors T = 5 K and total emissivity ε = 1% were used for estimations of P_{bg} and δP_{bg} in Table 1. Rise of ε to 4% leads to rise of δP_{tot} not greater than 35% in the worst case. The NHEB detector as an antenna-coupled microbolometer is optimal for high-resolution observations (diffraction limited, polarization sensitive) corresponding to m=1, g=1. The δP_{tot} includes noise of background and detector. For NHEB temperature of T = 0.1 K the used *NEP* is a "pessimistic" value. The wavelengths λ_L and λ_S are long-wave and short-wave boundaries of spectral bandwidths. Angular resolution $\theta=\lambda/D$. NEFD is noise equivalent of flux density for observations of point sources, $NE\nu I_\nu$ is noise equivalent of brightness for measurements of extended emission.

An orbit of Submillimetron is that of the ISS with period 90 min, very low eccentricity, 51.6 ° inclination and altitude up to 425 km. During observation a large enough distance from the ISS is possible. Due to precession of the orbit ~5°/day, any sky point is observable several times on neighborhood turns and ~10 times a year during repetition of sky survey. Multiple observations give tools for separation of moving bodies of Solar system and for investigation of flux variability. Each observation continues 0.2-2.5 s and total integration time increases as number of paths.

SCIENTIFIC OBJECTIVES AND COMPLEMENTARITY TO OTHER PROJECTS

Confusion-limited full-sky survey is a main goal of the Submillimetron experiment. It should reveal more than 10^6 individual sources in submillimeter spectral region. Information for these wavelengths has foremost importance for cosmology and problems of galaxies and stars formation. General interest should represent unbiased data on submillimeter spectral distributions and their variability in scale between an hour and a year. It shouldn't be underestimated the significance of massive data-sets on clusters of galaxies (SZ effect), active galactic nuclei (AGN), old and young stars (AGB, envelopes, protoplanetary disks). Catalog (database) of sources even with spectra and variability is relatively small extract from experimental data. Most of them should be in form of maps of brightness distribution for all spectral bands. These data contain information on relic background (CMB anisotropy), extragalactic IR background, unresolved faint sources, dust distribution in Galaxy and Solar system.

Figures 2, 3 shows sensitivity of Submillimetron (Table 1) in comparison with other space projects. Corresponding data are from the following publications: IRAS [32], ISO [33], IRIS (Astro-F) [34], SIRTF [35], Hershel [36, 37], Planck [13, 14]. Noise equivalent in units per $Hz^{1/2}$ corresponds to statistical error of 1σ for integration time of about 1 s.

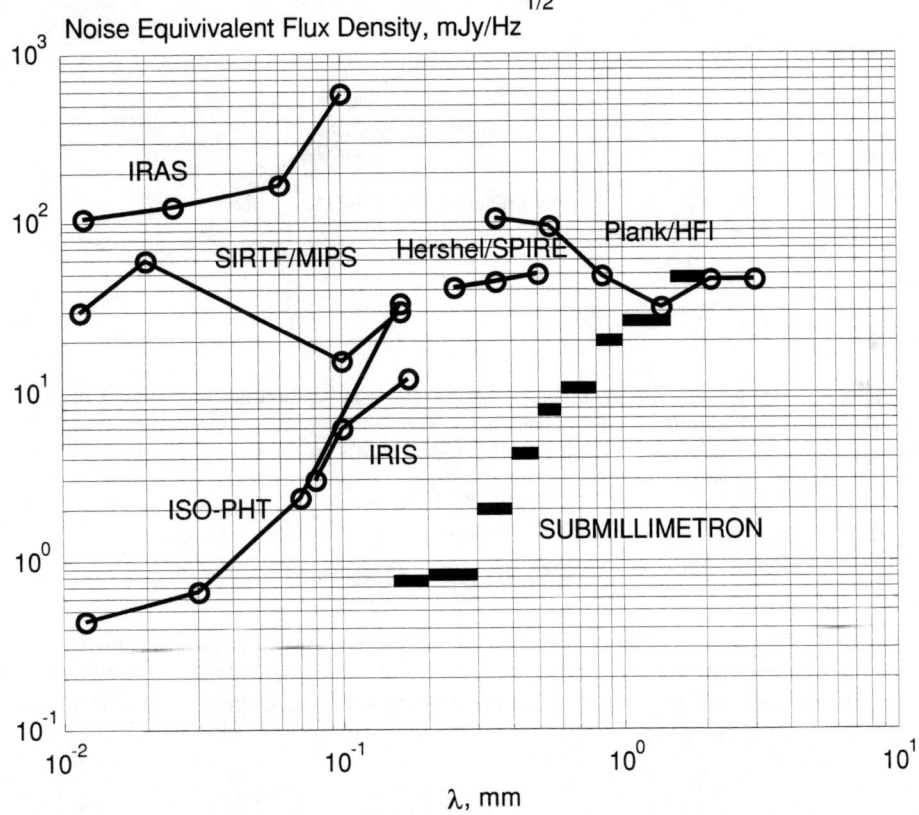

FIGURE 2. Sensitivity for point sources observation. Filled rectangles – Submillimetron project, open circles connected with lines – data for other projects (designations are given near corresponding lines).

Last decade of submillimeter (mainly ground-based) observations gives a lot of impressive results. They show particularly that up to 50-80 % energy of electromagnetic radiation from distant ($z \geq 1$) objects in Universe came in submillimeter region. Current decade is time of a number of infrared and submillimeter projects [13-14, 34-37]. Concurrent analysis of a totality of scientific data produces effect of synergy when final result is greater then sum of complementary component. It increases a value of each experiment. Due to high sensitivity, the Submillimetron should detect most targets of other space telescopes and has excellent complementarity to all these projects. Submillimeter spectra should expand infrared measurements (SIRTF and IRIS). Full sky survey in spectral regions of Hershel permits to generalize its investigation of individual objects and selected fields. Submillimetron data complement Planck experiments by data on foreground sources. Better angular resolution and sensitivity (in submillimeter channels) permit to reveal sources unresolvable in Planck maps. Their subtraction from millimeter-wave data should increase accuracy of measurements of CMB anisotropy.

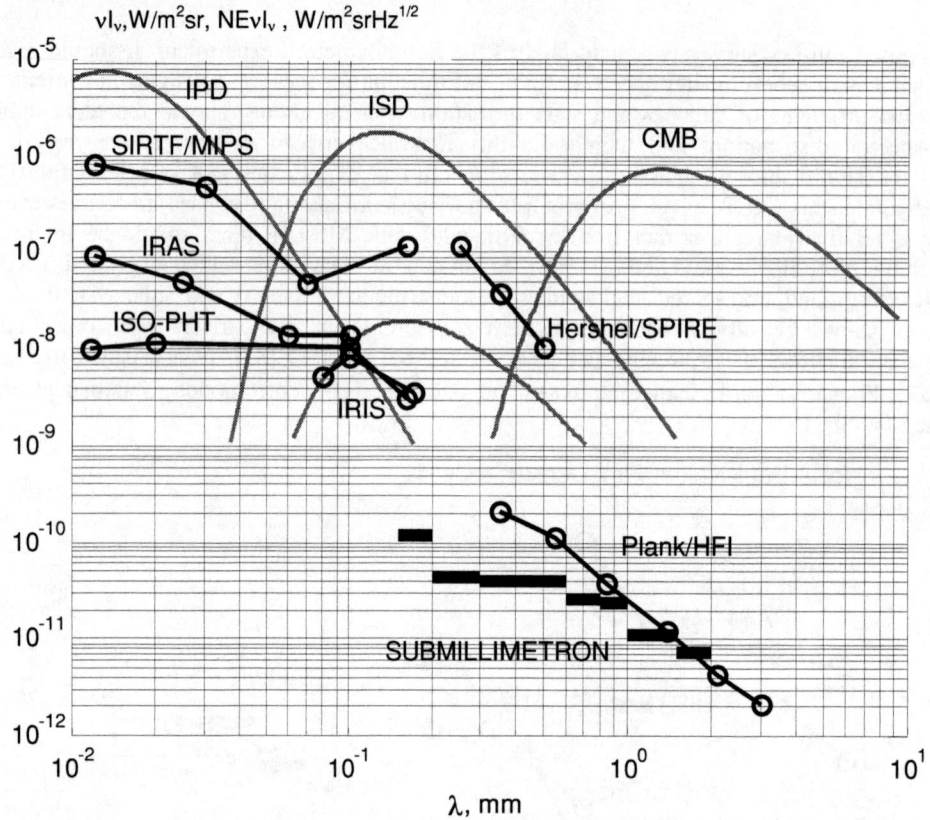

FIGURE 3. Sensitivity for extended emission observation. Filled rectangles – Submillimetron project, open circles connected by lines – data for other projects (designations are given near corresponding lines). Gray curves represent components of extraterrestrial background: left – interplanetary dust emission, right – CMB spectrum, upper curve in center - galactic emission near galactic plane, bottom one – near galactic poles.

CONCLUSIONS

Concept of the Submillimeron project is based on cryogenically cooled telescope and supersensitve antenna-coupled microbolomers. It permits to reach the high sensitivity in survey for cataloging submillimeter sources and mapping a sky in sub-mm and mm-wave bands. This experiment offers an optimum solution for submillimeter full-sky survey and provides a good complementarity to other space projects.

ACKNOWLEDGMENTS

It is a pleasure to acknowledge a large number of individuals for their contributions and interesting discussion on the project. The Submillimetron initiative group includes Alexander Andreev, Michael Tarasov, Arttu Luukanen, Vyacheslav Slysh, Kees van't Klooster, Lionel Duband, Alain Ravex, Harald Merkel, Alexey Ustinov. Important contributions were made by Anatoliy Trubnikov (Astro Space Center), Leonid Gorshkov, Sergey Stoiko, Andrey Adov (Rocket Space Corporation Energia), Alexander Vystavkin (Inst. Radioengineering and Electronics RAN).

This research was supported in part by grants INTAS 97.731, ISTC 1239 and Wenner Gren foundation.

REFERENCES

1. Vasil'chenko, N. V., Kardashev, N. S., Moroz, V. I., et al, " The Submillimeter and Infrared Transmittance of the Atmosphere at the North Pole of Cold" (in Russian) *Soviet Astronomy* **11**, 720 (1968).
2. Burova, L.P., Gromov V.D., et al., *Sov. Astron. Lett. (ed. Am. I. of Phys.)* **15**, 339-341 (1986).
3. Maslov, I.A., Gromov, V.D., et al., *Pisma v Astronomicheskii Journal* **15**, 667-672 (1989).
4. Sholomitski, G.B., Gromov, V.D., et al., "Spectral and polarimetric instrumentation for the aircraft astrophysical investigations in the range 50-500mkm" in *Galactic and Extragalactic Infrared and Submillimeter Astron.*, 19th COSPAR Symp. Proceedings, Astrophysics and Space Science Library, D. Reidel Publishing Co., Dordrecht, 1976.
5. Salomonovich et al., " A submillimeter telescope for the orbital piloted station Salyut-6" *Radiotekhnika* **34**, 33 -40 (1979).
6. Balebanov, V.M., Gromov V.D., et al., "Submillimeter cooled telescope" *Astronomicheskii Journal* **63**, 867-873 (1986).
7. Neugebauer, G., Habing, H. J., et al., *Astrophys. J.* **278**, L1-L6 (1984).
8. Thronson H.A., Hawarden, T.G., Penny, A.J., Sholomitski, G.B., Matsumoto, I., "The Edison infrared space observatory" in *Hamburg COSPAR Workshop on the Cool Universe*, Proceedings, 1995.
Thronson H.A., et al., *Adv. Space Res* **18**, 171-183 (1996). http://ast.star.rl.ac.uk/ajp/ajp_edison.html
9. Swanson, P. N., et al., "Plans for a large deployable reflector for submillimeter and infrared astronomy from space" in *International Conf. on Advanced Technology Optical Telescopes*, Proceedings, SPIE, Bellingham, USA, 1982, pp. 151-155.
10. Kardashev, N., et al., "Project Millimetron " (in Russian), *Trudy FIAN* **228**, 112 (2001).
11. Douglas, N. G., Frisk, U.O., "A proposed cornerstone mission: FIRST" in *Space-Borne Sub-Millimetre Astronomy Mission*, Proceedings of an ESA Workshop, ESA SP-260, 1986, p. 219.
12. Pilbratt, G.L., et al., "The Herschel Mission, Scientific Objectives, and this Meeting" in *The Promise of the Herschel Space Observatory*, Proceedings of symposium, ESA SP-460, 2001, pp. 13-20.
13. Bersanelli, F.R., et al., *Report on the Phase A Study of COBRAS/SAMBA*, ESA, 1996.
14. Tauber, J. A., "The Planck Mission" in *IAU Symposium 204*, Proceedings, M. Harwit and M. Hauser, (eds), 2000.
15. Troitsky V.F., "Modern status of space cryogenics" (in Russian), Lebedev Phiscal Institute, Moscow, 2001.
16. Schulte C.A., et al., *Adv.in Cryo. Engineer.* **10**, 477-485 (1965).
17. Breckenridge R.W., *Adv.in Cryo. Engineer.* **14**, 387-393 (1969).
18. Gromov V., et al., "Project Submillimetron" (in Russian), *Trudy FIAN* **228**, 143 (2001).
19. Kuzmin L., Kardashev N., Gromov, V., et al., "Submillimeter Telescope for the Russian Segment of the ISS: Submillimetron Project" in *The 2nd European Symposium on the Utilisation of the International Space Station - November 1998*, Proceedings, ESA SP-433, ESTEC, Noordwijk, The Netherlands, 1999, pp. 127-134.
20. Gromov, V., et al., "Submillimeter Telescope for the Russian Segment of the ISS: Submillimetron Project" in *The Promise of FIRST -12-15 December 2000*, edited by G.L. Pilbratt et al., Symposium Proceedings ESA SP-460, 2001 (in press), http://arxiv.org/abs/astro-ph/0108491.
21. Gromov, V., et al., "The space cryogenic submillimeter telescope based on hot-electron microbolometers (Project Submillimetron)" in *Space Astrophysics Detectors and Detector Technologies - June 2000*, Conference Proceedings, STScI, Baltimore, USA, 2001. http://www.stsci.edu/stsci/meetings/space_detectors/pdf/Gronovkuzmin-e.pdf
22. Gromov V.D., Quantum limit of radiation detectors at nonisothermal background (in Russian), Space Research Institute, USSR Academy of Sciences, Moscow, 1983.
23. Boggess, N.W., et al., *Astrophys. J.* **397**, 420-429 (1992).
24. Reach, W. T., et al., *Astrophys. J.* **451**, 188-199 (1995).
25. Nahum, M., Martinis, J.M., *Appl. Phys. Letters* **63**, 3075-3077 (1993).
26. Nahum, M., Richards, P. and Mears, C., *IEEE Trans. on Appl. Superc.* **3**, 2124-2127 (1993).
27. Nahum, M., Eiles, T. M., and Martinis, J. M., *Appl. Phys. Letters* **65**, 3123-3125 (1994).
28. Leivo, M., Pekola, J.P, and Averin, D.V., *Appl. Phys. Letters* **68**, 1996-1998 (1996).
29. Kuzmin, L. et al., *Proc. SPIE* **3465**, 193-199 (1998).
30. Vystavkin A., et al., "Terahertz Andreev Reflection Based Normal Metal Hot-Electron Bolometer for the Cryogenic Telescope of the International Space Station" in *The 10th International Symposium on Space Terahertz Technology*, T.W. Crowe and R.M. Weikle (eds.), Proceedings, Univ. of Virginia, Charlottesville USA, 1999, pp. 372 - 389.
31. Golubev, D., Kuzmin L., *J. of Appl. Phys.* **89**, 6464-6472 (2001).
32. Beichman, C. A., et al., *IRAS. Explanatory supplement*, Jet Propulsion Lab., California Inst. of Tech., Pasadena, 1985.
33. Lemke, D., et al., *ISOPHOT observer's manual*, ed. U. Klaas, H. Krüger, I. Heinrichsen; A. Heske, R. Laureijs, ESA, 1994.
34. Kawada, M., *FIS - Far-Infrared Surveyor onboard the ASTRO-F*, Report SP No. 14, ISAS, 2000, pp. 273-280.
35. Werner, M. W., "SIRTF Surveys" in *Sky Surveys. Protostars to Protogalaxies*, edited by B.T. Soifer, Conference Proceedings, v. 43, Astronomical Society of the Pacific, San Francisco, California, 1993.
36. Griffin, M., "The Design of a Bolometer Instrument for FIRST" in *The Far InfraRed and Submillimetre Universe*, Proceedings of symposium, ESA SP-401, 1997, pp. 31-35.
37. Griffin, M.J., et al., "The SPIRE Instrument for Herschel" in *The Promise of the Herschel Space Observatory*, edited by G.L. Pilbratt et al., Proceedings of symposium, ESA SP-460, 2001, pp. 37-44.

Herschel Space Observatory

V. Natale

CAISMI – CNR, L.go E. Fermi 5, 50129 Firenze, Italy

Abstract. The ESA Cornerstone mission Herschel Space Observatory [1], devoted to photometric and spectroscopic observations in the far-infrared and submillimetric spectral region, is described. Key scientific objectives are summarized. A concise description of the three focal plane instruments, proposed by International Consortia, is given. These instruments will allow to observe the Universe from high-z epochs down to very nearby objects such as planets and comets.

INTRODUCTION

Herschel Space Observatory is the fourth cornerstone mission in the "Horizon 2000" long term ESA science program plan. *Herschel* is a multi user observatory type mission intended to allow observations in all the submillimetre and far-infrared part of the electromagnetic spectrum (60 - 700 µm), which is mostly inaccessible to the observational astronomy from ground sites and even from balloon platforms.

The scientific objectives and mission requirements of this cornerstone were firstly discussed in 1986 in a workshop in Segovia [1]. In 1987 a Science Advisory Group (SAG) was established by ESA, to trade off between scientific objectives and technical complexity and cost constraints. *Herschel* was selected as Cornerstone number four by SPC in November 1993.

The baseline concept for *Herschel* includes a 3.5 meter, f/9.8, Cassegrain telescope radiatively cooled to an operational temperature of about 80 K in the L2 orbit. The mirror will be made of aluminum, polished to guarantee the diffraction limited operation at 60 µm. A superfluid helium cryostat will provide cooling power for three focal plane instruments designed for the photometry and the high resolution spectroscopy in the wavelength range 60 – 700 µm. The predicted lifetime for the liquid helium has been computed to be about 4.5 years. The advantages of *Herschel* can be summarized as follow:
- the low emissivity 80 K telescope will provide an extremely low thermal background, much lower than could even be achieved with ground based, airborne or balloon facilities;
- the high spatial resolution will lower the "confusion" noise due to the presence of several sources in the antenna beam;
- the complete absence of atmospheric absorption;
- the very large sky coverage;
- the large amount of observing time;
- the low radiation environment with the consequent reduction of "spikes" on the detectors.

In February 1998, three proposals have been submitted to ESA in response to the Announcement of Opportunity for participation in the *Herschel* Mission. Because of the technological difficulties and the related costs of the instruments, these proposals have been prepared by International Consortia which include leading expertise in the far infrared - submillimetre astronomy and instrumentation.

---

[1] formerly called **FIRST** The Far Infrared and Submillimetre Telescope. More information on the mission can be found at: http://astro.estec.esa.nl

The satellite is foreseen to be launched in the early 2007. In Sec.2 the scientific objectives are highlighted while in Sec.3 and Sec. 4, the instruments and the Science Operations are respectively described.

SCIENTIFIC OBJECTIVES

Herschel will open the last major part of the electromagnetic spectrum mainly inaccessible from ground-based and airborne facilities. It will perform photometry, medium and high resolution spectroscopy of selected objects, as well as deep surveys. Black bodies with temperatures in the range 5 - 50 K peak in the wavelength interval covered by *Herschel*. In the same wavelength interval, gases with temperature between few K and few hundred K emit their brightest molecular and atomic emission lines.

The science objectives were deeply discussed in a number of dedicated Symposia including those which taken place in Segovia (1986) [1], Liege (1990) [2] and Grenoble (1997) [3]. Submillimetre and far infrared observations are crucial for a deeper understanding in a variety of key areas; as a consequence, most of the observing time will be devoted to the following, among the others, Key Project:

- deep broadband photometric surveys in the band 150 – 500 µm. This program will allows a) the study of star formation in galaxies up to a redshift of about 5, b) the search of a population of high-z dusty star forming galaxies, c) the study of the formation and the early evolution of AGN and quasars and d) the study of the large scale structure in the high redshift universe.
- follow-up spectroscopy of selected objects. The observation of the brightest cooling lines of the interstellar gas will give very important information on the physical processes and energy production in galaxies. Optical and radio observations cannot give detailed information of the bolometric luminosity since most of the energy is emitted in the far-infrared.
- physics and chemistry of the ISM in our Galaxy and in redshifted galaxies. High resolution spectroscopy will probe the physical conditions and chemical processes of the interstellar medium as well the star forming regions in all their evolutive stages. The importance of *Herschel* with respect to other facilities is the possibility to have a complete spectral coverage over a large frequency band and hence will provide a complete inventory of the atoms, molecules in the various phases in the interstellar medium and determine the cooling rates of the gas. It will be possible to observe, for instance, all the submillimetric H_2O and O_2 transitions.
- high resolution spectroscopy in comets and outer planets.

THE INSTRUMENTS

The proposed instruments[2], optimized with respect to the identified key science areas, are:
- PACS (A Photoconductor Array Camera and Spectrometer) covering the wavelength range 60-210 µm;
- SPIRE (Spectral and Photometric Imaging REceiver) covering the wavelengt region 200-670 µm;
- HIFI (Heterodyne Instruments) including five receivers continuously covering the frequency interval 480 - 1250 GHz (625 - 240 µm) plus one receiver for the band 1.41 - 1.91 THz (213 - 157 µm).

Only a concise description of the instrument is given here; an updated and deeper description together with the expected performances may be found in the Web sites of the leading Institutions.

PACS (Photoconductor/Bolometer Array Camera & Spectrometer) uses two photoconductor arrays and two bolometer arrays. It will perform imaging line spectroscopy and imaging photometry in the 60 - 210 µm wavelength band. In photometry mode, it will simultaneously image two bands, 60 - 90 µm or 90 - 130 µm and 130 - 210 µm, over fields of view of 1.75 x 3.5 arcmin. In spectroscopy mode, it will image a field of about 50 x 50 arcsec with an instantaneous spectral coverage of ~1500 km/s and a spectral resolution of ~175 km/s.

---

[2] The leading Institutions of the three international Consortia, one for each instrument, are respectively :
PACS: Max-Planck Institute fur Extraterrestrische Physik, Garching (**http://pacs.ster.kuleuven.ac.be**)
SPIRE: Queen Mary and Westfield College, London; (**http://www.ssd.rl.ac.uk/SPIRE**)
HIHI : SRON, Groningen (**http://saturn.sron.nl/hifi/index.html**)

SPIRE (Spectral and Photometric Imaging REceiver) has three photometric bands with separate bolometer arrays and employs an imaging Fourier transform spectrometer (FTS) to provide variable resolution spectroscopy from 200 to 670 μm. The photometric channels simultaneously image a field of view of 4 x 4 arcmin onto three bolometer arrays centered respectively at 250, 350 and 500 μm The spectroscopic channel uses a Martin Puplett polarizing interferometer which images a separate field of view of 2 x 2 arcmin into two more bolometer arrays, optimized to cover the bands 200 - 300 μm and 300 - 670 μm. The spectral resolution can be varied from about 0.04 and 20 cm^{-1}.

HIFI (Heterodyne Instrument for FIRST) will provide very high resolution spectroscopy. HIFI will combine the spectral resolving power capability of the radio heterodyne technique with quantum noise limited detectors like SIS or HEB. It will be possible to have a continuous frequency coverage in the range 0.48 - 1.25 THz in five bands and in both polarizations using SIS as mixers. A sixth band will provide coverage of 1.41 - 1.91 THz frequency interval in single polarization and will use the newly developed HEB (Hot Electron Bolometers) mixers. This high frequency band have been added for the observation of the peculiar lines of the FeH (1.411 THz) and CII (1.898 THz). Each receiver will have an instantaneous bandwidth of 4 GHz which will be analyzed in parallel by a pair of wideband (WBS) and a pair of high resolution (HRS) spectrometers. The WBS will use acousto-optic technology and will provide a frequency resolution of 1 MHz over a bandwidth of 4 GHz for each of the two polarization. The HRS will be based on a digital autocorrelator and will be able of a spectral resolution of 200 KHz over a bandwidth of 1 GHz.

SCIENTIFIC OPERATION

Each Consortium must provide, one for each instrument, the Instrument Control Center (ICC) which has the responsibility in the areas of data reduction, instrument calibration and scientific analysis. This is a novel concept on the conduction of Scientific Operation envisaged from ESA for this mission. As matter of fact, ICC is part of the Ground Segment and will provide, during observatory operations, a) the monitoring of the operations and status of the instrument and its subsystems, b) the planning, execution and analysis of calibration observations and c) the update of the data processing software.

CONCLUSIONS

The Herschel Mission together with some scientific objectives and the proposed instruments have been presented. The performances of the payload are expected to be by far superior to any other existing instruments for submillimetre and far infrared bands allowing to address key topics in astrophysics and cosmology.

ACKNOWLEDGMENTS

The italian participation to Herschel is supported by ASI, the Italian Space Agency.

REFERENCES

1. Proc. of Space-Borne Sub-Millimetre Astronomy Mission, ESA SP-260, Segovia, August 1986
2. Proc. of From Ground-Based to Space-Borne Sub-Millimetre Astronomy, ESA SP-314, Liege, December 1990
3. Proc of The Far Inra-Red and Submillimetre Universe, ESA SP-401, Grenoble, April 1997

The High Frequency Instrument of Planck: Requirements and Design

J.M. Lamarre[1], B. Maffei[2], P.A.R. Ade[2], M. Piat[1], J. Bock[3], J.L. Puget[1], P. de Bernardis[4], M. Giard[5], A. Lange[6], A. Murphy[7], J.P. Torre[8], A. Benoit[9], R. Bhatia[6], F.R. Bouchet[10], R. Sudiwala[2], V. Yurchenko[7]

1 Laboratoire d'Etude du rayonnement et de la Matière en Astrophysique, Observatoire de Paris, 61 Bd de l'Observatoire, 75014 Paris. Institut d'Astrophysique Spatiale, Orsay, France, email: lamarre@ias.fr, puget@ias.fr, piat@ias.fr
2 Cardiff University, Wales, UK, email: P.A.R.Ade@qmw.ac.uk, bruno.maffei@qmw.ac.uk
3 Jet Propulsion Laboratory, Pasadena, Ca, USA, email: james.Bock@jpl.nasa.gov
4 Universita La Sapienza, Roma, Italy, email: debernardis@roma1.infn.it
5 CESR, Toulouse, France, email: martin.giard@cesr.fr
6 Caltech, Pasadena, Ca, USA, email: ael@astro.caltech.edu, rsb@astro.caltech.edu
7 University of Maynooth, Ireland, email: amurphy@may.ie, v.yurchenko@may.ie,
8 Service d'Aéronomie, Verrières le Buisson, France, email: Jean-Pierre.Torre@aerov.jussieu.fr
9 CRTBT, Grenoble, France, email: benoit@labs.polycnrs-gre.fr,
10 Institut d'Astrophysique de Paris, France, email: bouchet@iap.fr

Abstract. The Planck satellite is a project of the European Space Agency based on a wide international collaboration, including United States and Canadian laboratories. It is dedicated to the measurement of the anisotropy of the Cosmic Microwave Background (CMB) with unprecedented sensitivity and angular resolution. The detectors of its High frequency Instrument (HFI) are bolometers cooled down to 100mK. Their sensitivity will be limited by the photon noise of the CMB itself at low frequencies, and of the instrument background at high frequencies. The requirements on the measurement chain are directly related to the strategy of observation used for the satellite. Due to the scanning on the sky, time features of the measurement chain are directly transformed into angular features in the sky maps. This impacts the bolometer design as well as other elements: For example, the cooling system must present outstanding temperature stability, and the amplification chain must show, down to very low frequencies, a flat noise spectrum.

INTRODUCTION

The most distant, and therefore the most ancient source of radiation that can be directly observed from Earth is the Cosmic Microwave Background (CMB). The satellite COBE has measured its submillimeter emission, which is that of a nearly perfect blackbody at 2.73K. This emission is attributed to the primordial universe when it was about 300 000 years old and warm enough (3000K) to ionize the hydrogen gas that constitutes most of its mass. Due to the expansion of the universe, this radiation was red-shifted by Doppler effect by a factor of about 1000, and thanks to the cooling due to the expansion, it could travel and reach us through the very transparent neutral hydrogen. The discovery of the CMB and its refined observation by COBE are pillars that support the big bang theory. The CMB is isotropic over the sky down to a level of 10^{-5} at small scales. The tiny deviations from uniformity give us unique information on the physics of the primeval universe, on the cosmological parameters describing the geometry of the universe, and on the history of matter and radiation since the Big Bang and up to our times.

New results from the balloon-borne experiments BOOMERanG [1] and MAXIMA [2] and from ground-based experiments using radio detectors and interferometry were recently published [3], [4]. They gave a first view of the small scale anisotropy of the CMB, unveiling the predicted peaks in the power spectrum of its angular distribution.

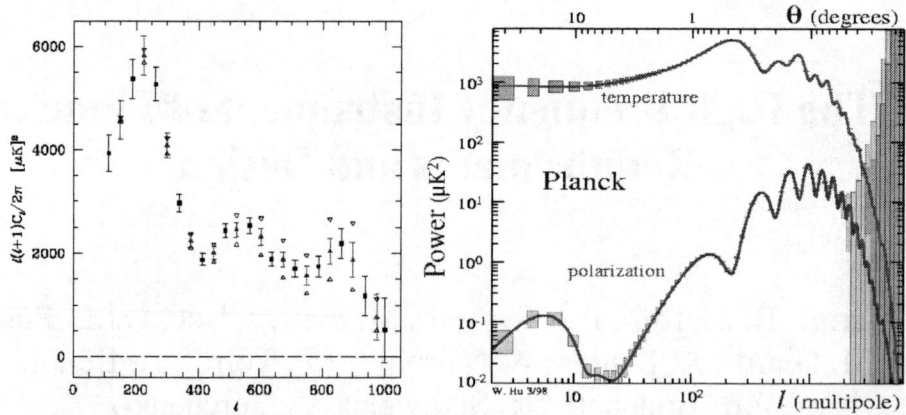

FIGURE 1. Left (a): The power spectrum of CMB anisotropies as seen by BOOMEranG. The first peak corresponds to an angular scale of about 1 degree as predicted for a "flat Universe" (i.e., a Universe where the spatial part of the metric is Euclidean). (b): Simulation of the Planck capability in the domain of angular frequencies. Smaller details will be measured with a much better sensitivity (wrt BOOMERanG) on the whole sky. The position of the first peak informs us on the curvature of the universe, while the whole spectrum give much more about other cosmological parameters.

Following COBE, NASA launched in the summer of 2001 the MAP satellite which will be a second generation CMB space experiment. The detectors are uncooled radio-type receivers with a sensitivity comparable to the balloon-borne bolometer experiments but with significantly better capabilities for large scale measurements and control of systematics.

The 'Planck' project of the European Space Agency, to be launched in 2007, is intended to be the third generation of CMB space experiments, pushing to its limits the knowledge that will be retrieved from the CMB observation with unprecedented angular resolution and sensitivity.

HFI PERFORMANCES

The high frequency instrument (HFI) covers the frequency range between 100 and 1000GHz with an angular resolution of about 5 arcmin in for of its six spectral bands. Its sensitivity will be limited, in the CMB channels, by the statistical fluctuations [5] of the CMB itself [6], which makes it a kind of ultimate experiment. It will also measure the polarization of the CMB in three channels, which will give independent and unique information on the CMB anisotropy [7]. The I, Q and U Stokes parameters will be measured thanks to four independent bolometers sensitive to polarizations separated by $45°$ from each other.

The accuracy on the CMB can be achieved only by removing the various foregrounds formed by the evolving universe situated between us and the warm primordial universe emitting the CMB. Among these, we see the emission of dust and gas in our own galaxy and from other galaxies. Clusters of galaxies, that contain high temperature gas detected in the X-rays, distort the CMB by inverse Compton scattering. This is the Sunyaev-Zeldovich Effect (SZE), that makes clusters of galaxies good tracers of the dynamics of the universe at large scales. Six bands in the HFI and four more in the LFI are needed to separate these various components thanks to their spectral and spatial signature. An additional benefit of the increase in complexity resulting from this approach is that all these astrophysical sources will be known much better, which is in many cases of major interest for astronomy and an important motivation for the HFI science team.

TABLE 1. Planck HFI performances.

Central Frequency	(GHz)	100	143	217	353	545	857
Beam Full width Half Maximum	arcmin	9.2	7.1	5.0	5.0	5.0	5.0
Number of unpolarized detectors		4	4	4	4	4	4
Number of polarized detectors		-	8	8	8	-	-
Total sensitivity (•T/T)	µK/K	2.2	2.4	3.8	15	80	8000
U and Q sensitivity (•T/T)	µK/K	-	4.8	7.6	30	-	-
Flux sensitivity	mJy	9.0	12.6	9.4	20	46	52
YSZ per FOV (x10^6)		1.2	2.1	440	6.4	32	730

Planck has to be considered not only as the third generation of CMB satellites, but also as the first sub-Terahertz sky survey of modern astronomy. Several thousands of galaxies, of young stellar objects, of clusters of galaxies will be observed in a new way. Nearly every field of astronomy will benefit from its results, from the study of the solar system (Trans-Neptunian objects for example) to the large scale structure of the universe (SZE results), as well as new insight on the cold components of galaxies, a possible candidate for dark matter. The Planck project is committed to deliver a set of well-defined products to the scientific community at large and will be one of the inputs of the future virtual observatories of the future.

DESIGN OF THE HFI SYSTEM

The Planck orbit will be a "halo" orbit around the L2 Lagrangian point of the Sun-Earth system, at about 1.5 million km of the Earth (figure 2). The satellite will rotate at 1RPM around an axis nearly anti-solar, allowing its field of view to scan large circles in the sky. Every 60min, the axis of rotation will be shifted by 2.5 arcmin, in order to follow the movement of the Earth around the Sun. In six months the successive circles will cover the full sky with unequal integration times.

The useful data from the telescope is therefore a quasi-periodic signal, with a period of one minute. The signal from the detection chain must therefore be stable from one observation of a source to the next one [8]. Lower frequencies can be filtered out in the data reduction process. This stability is a major requirement for the cryogenics and the electronics of HFI. The scan rate is 6deg/s. With 5 arcmin beams, the response time of the bolometers must be less than a few milliseconds. This is the driving requirement for the choice of the bolometers. In particular, it was understood from the beginning of the project that only very low temperature (100mK or less) would allow to reach this speed for the large bolometers needed in the millimeter range. This choice is consistent with the requirement that the sensitivity must be limited by the photon noise in the HFI wavelength range.

TABLE 2. Required bolometer performances.

Frequ. (GHz)	Optical Load (pW)	Maximum NEP 10^{-17}WHz$^{-1/2}$	Goal time constant (ms)	Maximum time constant (ms)
100	1.0	1.2	3.9	7.8
143	1.1	1.5	2.9	5.7
217	1.1	1.8	2.2	4.4
353	1.0	2.2	2.2	4.4
545	5.0	6.0	2.2	4.4
857	16.0	13.5	2.2	4.4
143P	0.57	1.1	3.0	5.7
217P	0.54	1.3	2.2	4.4
545P	2.50	4.3	2.2	4.4

The absorber of the bolometers is of the Spider web type [9] for unpolarized detectors. For polarized channels; the bolometers are sensitive to polarization, thanks to an absorber made with parallel wires (J. Bock). NTD Ge thermometers with impedance of about 10MΩ are the sensitive elements. The required performances are listed in table 2.

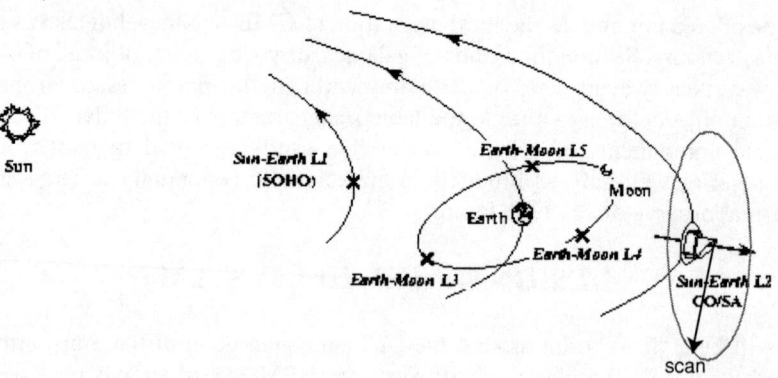

FIGURE 2. Schematic representation of the Planck orbit and scanning strategy. The readout electronics must not be the limiting factor of the detection chain. A specially developed amplifier [10] using tunable AC bias and cold J-FET preamplifier shows a 5nVHz$^{-1/2}$ noise down to 10^{-2} Hz (fig.3).

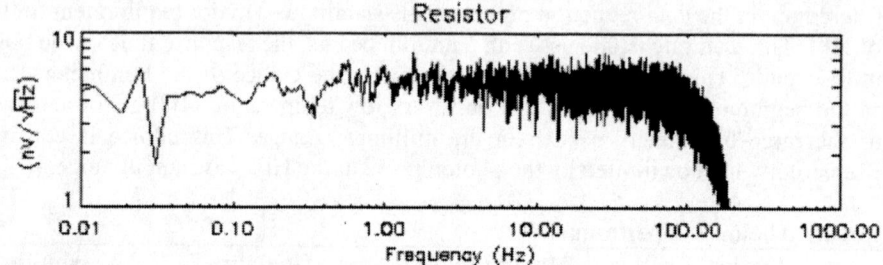

FIGURE 3. Noise spectrum of the readout electronics

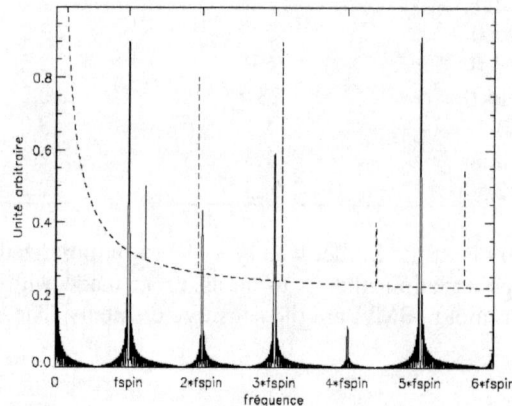

FIGURE 4. The scientific signal is periodic at the frequency of the spin rate of the satellite (1RPM). The first 5000 harmonics contain useful information, which determines the working frequency range of the bolometers: from 0.016Hz to about 100Hz.

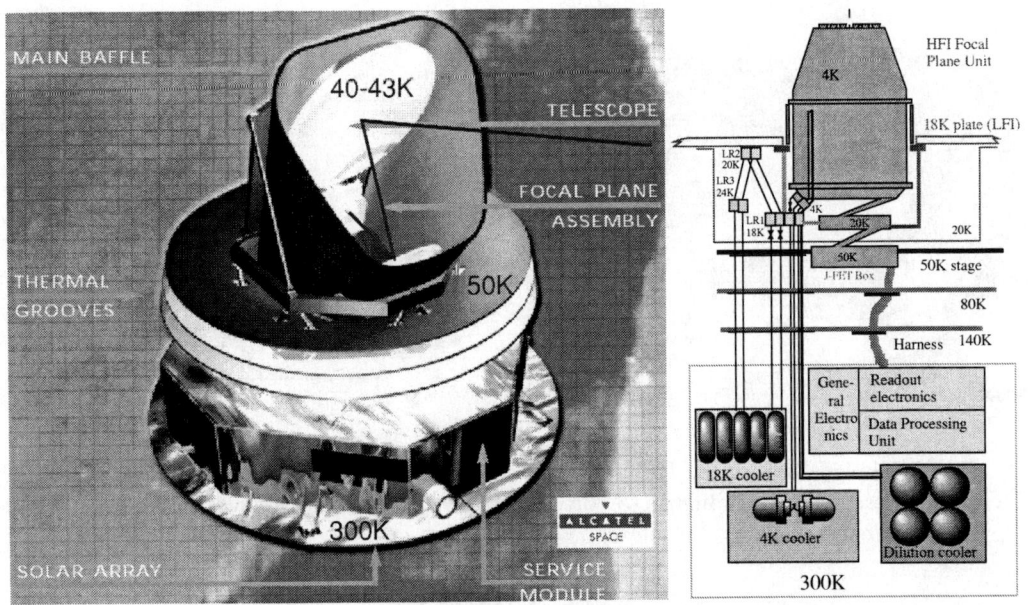

FIGURE 5. Left: The solar panels cover the bottom part of the satellite. The thermal architecture allows to passively cool the telescope down to 40K. Right: Three different active coolers are needed to cool the bolometers down to 0.1K.

Thermal requirements have been the driver for the design of the Planck Satellite. Thanks to its stable orientation with respect to the sun, the moon and the earth, it was possible to obtain an efficient passive cooling of the science payload down to 50K or less (see figure 5).

Coolers based on J-T expansion of hydrogen and sorption pumps cool both LFI and HFI down to 20K. The J-T valve delivers a mixture of liquid and gas at about 17.5K. A first high efficiency heat exchanger pre-cools the fluids for the coolers at lower temperature. Mechanical compressors provide helium for the 4K J-T compressor. This is used to cool the whole focal plane unit of the HFI. Such a low temperature was needed to keep to low values the background radiation reaching the bolometers. The open loop helium3/helium4 dilution cooler provides both the cooling of the bolometers down to 100mK and cools an intermediate stage at 1.6K by J-T expansion of the helium mixture.

The internal architecture of the HFI focal plane unit is shown on figure 6. This architecture, based on a classical Russian dolls design, is essential to the proper operation of HFI. As shown on the right section of figure 6, radiation from the various stages loads the bolometers. This radiation must be very stable, because any fluctuation in the range of useful frequencies (0.016 to 100Hz) could be taken as coming from the sky. This implies very good temperature stability of the various stages of the HFI-focal plane unit: $20nkHz^{-1/2}$ for the 100mK stage (directly connected to the bolometers), $28\mu KHz^{-1/2}$ for the 1.6K stage, and $10\mu KHz^{-1/2}$ for the 4K stage.

The amount of consumables of the 0.1K cooling system is the limiting parameter for the duration of the mission of Planck-HFI, which is nominally of 14 months.

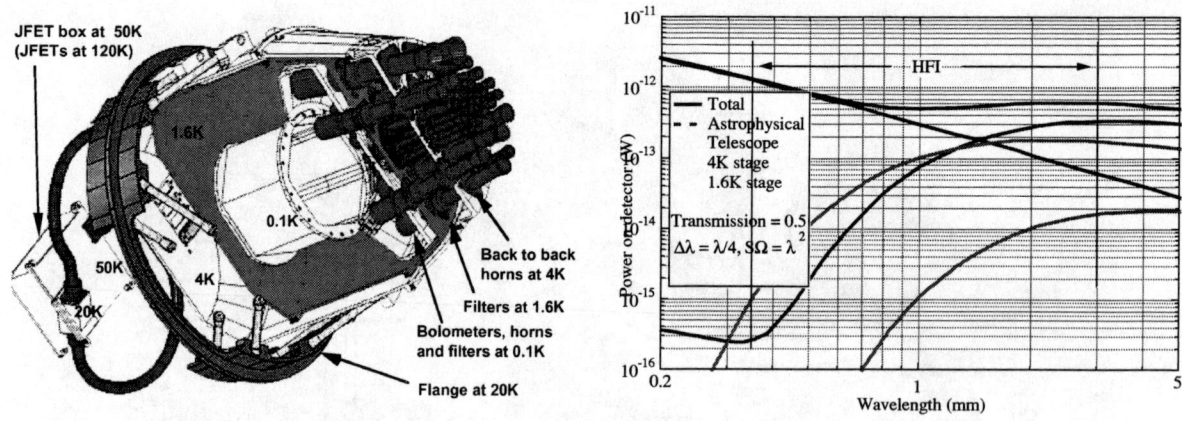

FIGURE 6. The focal plane unit of the HFI consists of three stages at 4K, 1.6k, and 0.1K. The coupling of the bolometer with the telescope is made thanks to corrugated horns at 4K. These stages radiate on the detectors and contribute to the background as well as the telescope (right).

CONCLUSIONS

Photon noise of the CMB radiation itself limits the sensitivity in the three most sensitive channels. Fundamental fluctuations of the flux reaching the detectors are therefore the main parameter limiting the instrument sensitivity. The only way to increase the sensitivity of such an instrument is to significantly increase the number of detectors. The scanning strategy plays a major part in the derivation of the instrument specification, since it gives, for signal and noises, the relation between the time domain and the domain of angular scales. A study of the bolometer requirements in the Fourier domain proved to be an efficient tool for several aspects of the instrument design.

ACKNOWLEDGEMENTS

Planck HFI is now under construction, for a launch early 2007. We thank all the technical and scientific persons who contributed to the design of this instrument.

REFERENCES

1. P. de Bernardis et al., *A flat Universe from high resolution maps of the CMBR*, Nature, 404, 955, 2000.
2. A.T. Lee et al., *A high spatial resolution analysis of the MAXIMA-1 cosmic microwave background anisotropy data*, astro-ph/0104459.
3. A.D. Miller et al., *A measurement of the angular power spectrum of the CMB from l=100 to 400*, ApJL, in press, astro-ph/9906421.
4. N.W. Averson et al., *DASI first results: A measurement of the cosmic microwave background angular power spectrum*, astro-ph/0104489.
5. J.M. Lamarre, *Photon noise in photometric instruments at far infrared and submillimeter wavelengths*, Appl.Opt., 25, 870, 1986.
6. J.M. Lamarre et al., *The High Frequency Instrument of PLANCK: Design and performances*, Astro. Lett. And Comm., 37, 161, 2000.
7. F.R. Bouchet, J.L. Puget, J.M. Lamarre, *The cosmic microwave background: from detector signals to constraints on the early universe physics*, in *The Primordial Universe*, Binétruy et al. Editors, EDP sciences, Les Ulis, Paris, 103, 2000.
8. M. Piat et al., Proceedings of *LTD8*, Dalfsen, the Netherlands, 1999, NIMA, 444, 419, 2000.
9. A.D. Turner et al., *Si_3N_4 micromesh bolometer array for sub-millimeter astrophysics*, Appl.Opt., in press, 2001.
10. S. Gaertner et al., *A new readout system for bolometers with improved low frequency stability*, Astron. and Astrophys. Suppl., 126, 151, 1997.

The 4K Reference Load for the Planck Low Frequency Instrument

L. Valenziano[1], M. Bersanelli[2], R.C. Butler[1], F. Cuttaia[1], N. Mandolesi[1],
A. Mennella[3], G. Morigi[1], G. Morgante[1], M. Sandri[1], L. Terenzi[1], F. Villa[1]

on behalf of the LFI Consortium

[1]*TESRE-CNR – Bologna, Italy*
[2]*Università di Milano, Dip. Di Fisica, Milano, Italy*
[3]*IFC – CNR, Milano, Italy*

Abstract. The Low Frequency Instrument (LFI) on-board the Planck satellite, is composed by an array of 54 receivers. They follow a pseudo-correlation scheme, continuously observing the sky and a stable reference load, at a temperature T~4K. The so-called 4K Reference Load is made of 54 small absorbing targets, mechanically and thermally connected to the High Frequency Instrument (HFI) cryostat. The 4K Reference Load requirements pose severe limits to its mass, dimensions, power dissipation and performance.

INTRODUCTION

The Low Frequency Instrument [1,2] is one of the two instrument on-board the ESA Planck mission.

LFI will use a pseudo-correlation receiver concept, based on High Electron Mobility (HEMT) amplifiers. This radiometer concept is chosen to maximize the stability of the instrument by reducing the effect of non-white noise generated in the radiometer itself [3].

In this scheme, the difference between the inputs to each chains (the signal from the telescope and the signal from the reference black body respectively - the 4K Reference Load) is continuously being observed. To minimize the 1/f noise of the radiometers, the reference blackbody temperature should be as close as possible to the sky temperature (~ 3K).

Differing DC gains are applied after the detector diodes to compensate this temperature difference. Adjusting the ratio r of DC gains makes it possible to null the output signal, minimize the sensitivity to RF gain fluctuation and to minimize the white noise at the output.

The purpose of the 4 K reference load is twofold. First, the load provides the radiometer with a lower input offset (the radiometric temperature difference between the sky and the reference load). Reducing the input offset reduces the minimum achievable radiometer knee frequency for a given amplifier fluctuation spectrum. This minimum achievable knee frequency assumes perfect phase and gain matching in the two "legs" of the radiometer and assumes other ideal characteristics in radiometer components. An *ideal* reference load temperature would match the sky temperature (approximately 2.7 K), but there is no convenient spacecraft source of 2.7 K with sufficient cooling power, and routing reference horns through the instrument to use another part of the sky as the reference appears impractical.

The second purpose of the 4 K reference load is to maximize the probability of achieving something close to the ideal performance of the radiometer. Careful attention is paid to the fluctuations of reference load signals, which can result in systematic effects in the final maps. Main source of fluctuations are: load temperature fluctuations, radiometer fluctuations reflected by the loads, external radiation leaking in the target-horn gap (see next paragraph for details).

4K REFERENCE LOAD UNIT DESCRIPTION

The 4K Reference Load is composed by small absorbing targets, made of ECCOSORB CR series, assembled in a mounting structure. Each target is facing a small pyramidal horn. This result in one target for each radiometer polarization, that is two target for each Front-End Module (FEM). 4K Reference Load is sketched in Figure 1.

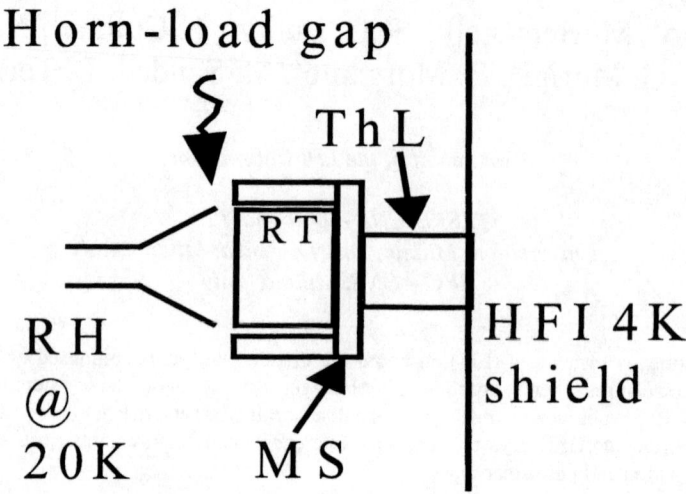

Figure 1: Sketch of the different parts which constitute the 4K reference Load unit.

Target (RT) have different design for each of the LFI frequency: their dimension increase with reducing frequency. They are located in front of each reference horn (RH), one for each polarization of a single radiometer (that is two for each FEM). Reference target design is optimized for each LFI frequency. This results in at least four target design, which show different mechanical and thermal properties.

Reference horns provide input of the reference signal from the reference blackbody to the radiometer *reference* arm. Two reference horn, one for each polarization, are connected to each Front End Module (FEM). High frequency radiometers (100 and 70 GHz) RHs are integrated in FEMs. Reference horns for low frequency radiometers (44 and 30GHz), which are placed in an external sector of the LFI Radiometer Array Assembly, are connected via waveguides.

Targets are assembled in a mounting structure (MS), which provides mechanical interface to the HFI [4]. Separated holding structures are provided for the high frequency RTs (100 and 70 GHz), located in the upper section of the HFI, and for the low frequency RTs (44 and 30 GHz), located in the lower section of the HFI (see Figure 2). The HFI will provide the reference temperature to the 4KRL unit. The mounting structure, holding the reference targets, is connected to the HFI outer radiation shield, whose temperature is approximately 4K. Requirement on temperature fluctuations of the 4K stage are very stringent ($\Delta T \leq 10 \mu K/\sqrt{Hz}$)

Temperature monitoring of the 4KRL is provided by temperature sensors inside the HFI outer radiation shield. Housekeeping data from this sensors, acquired by the HFI data acquisition system, will be provided to LFI during the Planck flight activities and during ground tests.

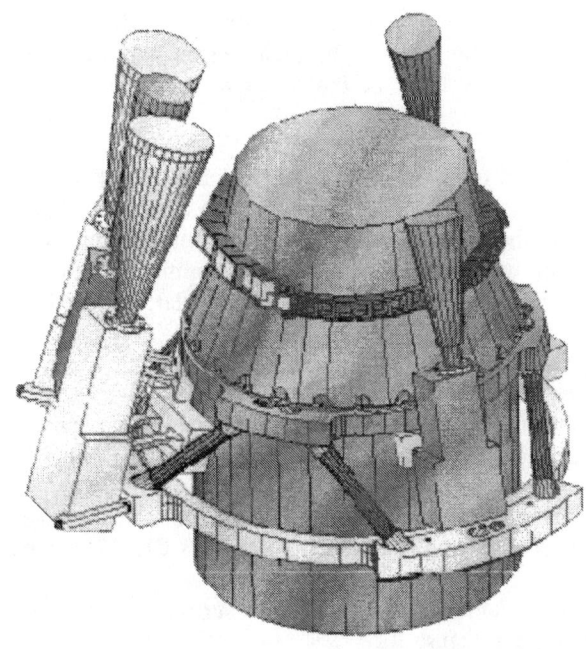

FIGURE 2. The 4K Reference Load unit mounted on the HFI cryostat. Lower frequency channel only are shown. Mounting structure design is preliminary.

Reference Horn design

In the Reference Horn design, some considerations have been followed:

- Horn dimensions must be small, to allow it to be included in the limited dimensions of the FEMs.
- The spacing between the horn and the target must be small to minimize leakage.
- The leakage itself must be minimized
- The horn Return Loss must not limit the 4K Reference Load performance.

The above mentioned constraints led to a pyramidal design for RH. Moreover, horn aperture dimensions and flare length is very small. The aperture dimensions range from approximately 20 x 10 mm at 30 GHz to 7 x 1.3 mm at 100 GHz. The maximum flare length is one wavelength. The horn-target gap is less than 2 mm.

A critical design driver is the leakage requirement: any signal entering the reference arm of the radiometer is indistinguishable from a true sky signal. This is most relevant for Spin Synchronous signals. To improve the coupling between RHs and RTs, grooves have been designed around the horn aperture. This led to a significant reduction of the leakage.

Reference Target design

Target design philosophy is driven by some main constraints:

- Targets must behave as a blackbody at a level of 1% at all the LFI frequencies
- Targets must fit inside the envelope between the LFI and the HFI
- Targets mass must be minimized

The *ideal* material for targets should have minimum reflectivity at LFI frequencies. ECCOSORB CR110 could have been a good choice. However, due to its low attenuation [5], the volume needed to reach the required Return Loss would have exceeded the maximum allowed envelope.

In order to reach low reflectivity at microwave frequencies, the refractive index *n* of the medium must be as close as possible to 1 or its variation must be smooth. This is the main reason why many high performance absorbing materials are commonly shaped as *bed of points*. It can be shown that *n* vs. the height of the pyramid has a parabolic shape.

A second design consideration is derived from the electric field shape at RH aperture. In the horn very near field it is reasonably to assume that the electric field is not much different from what is inside a WG.

The RT refractive index variation in the field propagation direction must match as close as possible this shape, that is it must be as smooth as possible where the electric field is maximum, at the centre of the horn aperture. Moreover, the RH has a shape of a WG flare, where aperture is rectangular, with the wider side orthogonal to the electric field.

These considerations lead to a cross design in front of the reference horn. Horn aperture dimensions are equal to the cross arm. A pyramid is placed in the centre of the cross, where the field is maximum and *n* must vary smoothly. All this elements are made of low reflection material, that is ECCOSORB CR110. Four CR110 block can be placed at the sides of the pyramid, to further improve Return Loss.

To get the required RL in the allowed volume, a composite target it has been chosen. The back part (the one closer to the HFI shield) is made of high absorbing material, ECCOSORB CR117, assembled to the front part with ECCOSORB specific cement.

To minimize radiation leaks at the lateral sides of the target, this is surrounded by a metal box.

The actual design for targets is sketched in Figure 3.

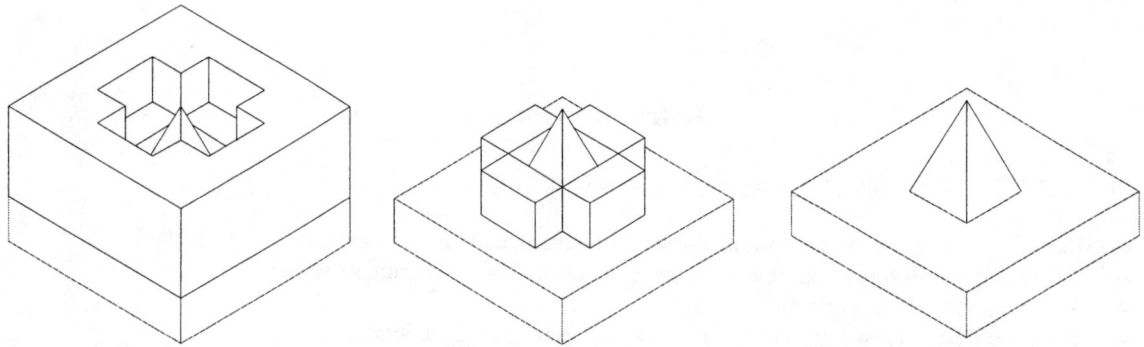

Figure 3: Reference target design. Side dimensions range from ~ 10 mm at 100 GHz to ~ 30 mm at 30 GHz.

LABORATORY TEST

An extensive test activity is being carried out by the 4K Reference Load developing Team to fully characterize the unit performances.

Test on absorbing materials

Reflectivity and transmission properties of CR and MF series (type 110 and 117) samples are being measured at 44GHz at room temperature, 77K and 7 K (actual minimum temperature of the Cryo facility).

Samples are cast by the 4KRL Team in dimensions that fit inside a WG. They are cut at different length, following specifications in [5]. Material and samples are numbered and all the relevant information (kind if material, date of deliver, expiring date, process date, process procedures, operators, overall quality, etc.) is included in a database.

The cryo facility is composed by a small dewar vessel. Reference temperature is given by liquid Nitrogen (~77 K) and liquid Helium (~4.2 K) at ambient pressure. Temperature is monitored by a dedicated cryo sensor. Test results show a general agreement with data in the literature [6,7,8]. A detailed analysis is in progress.

Test on targets

Target and reference horns have been tested at 30, 44 and 100 GHz for Return Loss and Leakage. Results are quite good, showing a RL ≤ 20 dB in the four LFI bands. Further test on leakage are in progress; preliminary results indicate that power leaking in the horn-target gap is less than −35 dB at 44 GHz.

CONCLUSIONS

The 4K Reference Load is a critical part of the Low Frequency Instrument. A detailed analysis of all the systematic effects is being carried out by the developing teams to minimize their effect in hardware. An extensive set of measurements has been set up to accurately characterize the unit. The actual design showed extremely good performances and further test are foreseen to asses the compliance with mission requirements.

ACKNOWLEDGMENTS

The authors wish to thank L. Pagan, S. Mariotti and S. Levin for participating in this activity.

REFERENCES

1. Bersanelli, M., Bouchet, F.R., Efstathiou, G., Griffin, M., Lamarre, J.M., Mandolesi, N., Norgaard-Nielsen, H.U., Pace, O., Polny, J., Puget, J-L., Tauber, J., Vittorio, N., and Volonté, S., "ESA, COBRAS/SAMBA Report on the Phase A Study", **D/SCI(96)3** (1996)
2. Mandolesi, N. et al., "The Low Frequency Instrument, a Proposal Submitted to the ESA in response to the Announcement of Opportunity for the FIRST/Planck mission" (1998)
3. Mandolesi, N., Bersanelli, M., Butler, R.C., Buriana, C., Maino, D., Mennella, A., Morgante, G., Valenziano, L., and Villa, F., this Conference proceedings
4. Puget, J-L., et al., "The High Frequency Instrument, a Proposal Submitted to the ESA in response to the Announcement of Opportunity for the FIRST/Planck mission" (1998)
5. Emerson & Cuming Data Sheet, 1-20
6. Peterson, J.B., and Richards, P.L., *Int. J. Infr. Mill. Waves*, **5**, 1507-1515 (1981)
7. Hemmati, H., Mather, J.C., and Eichhorn, W.L., *Appl. Opt.*, 24, 4489-4492 (1985)
8. Halpern, M., Gush, H.P., Wishnow, E., and De Cosmo, V., *Appl. Opt.*, **25**, 565-570 (1986)

The Planck Telescope

F.Villa[1], M.Bersanelli[2], C.Burigana[1], R.C.Butler[1], N.Mandolesi[1], A.Mennella[3], G.Morgante[1], M.Sandri[1], L.Terenzi[1], L.Valenziano[1]

[1] *Istituto TESRE/CNR – Bologna – Italy*
[2] *Università di Milano – Milano – Italy*
[3] *Istituto di Fisica Cosmica / CNR – Italy*
On behalf of the Planck Collaboration

Abstract. In this paper we present an overview of the Telescope designed for ESA's mission dedicated to map the Cosmic Microwave Background Anisotropies and Polarization. Two instrument, LFI and HFI, operate in an overall frequency range between 25 and 900 GHz and share the focal region of the 1.5 meter optimized aplanatic telescope. The optimization techniques adopted for the optical design and the telescope characteristic are reported and discussed.

INTRODUCTION

Planck is a mission dedicated to imaging the anisotropies of the Cosmic Background Radiation with a typical angular resolution of 10 arcmin and an average sensitivity per pixel of $2 - 4$ μK/K. Two instrument, the radiometric Low Frequency Instrument (LFI) [1] and the bolometric High Frequency Instrument (HFI) [2], are coupled to an optimized dual reflector off–axis telescope with a 1.5 meter of projected aperture. Both instrument, cryogenically cooled respectively at 20 K and 100 mK, will measure the radiation coming from the sky and scattered by the telescope between 30 GHz and 100 GHz for LFI and between 100 GHz and 857 GHz for HFI. The telescope is passively cooled at about 50 K by a thermal design of the payload based on a set of three dedicated V–grooved shields. The electromagnetic design of the telescope, carried out after approximately three years of intensive studies and simulations, has been consolidated during the Planck Payload Architect industrial activity, kicked off on Jan 99 and completed on Dec 99. The study was performed by ALCATEL, under the control of ESA, with a strong scientific and technical support of the instrument teams, and of the Telescope Provider.

An accurate study of the telescope performances plays a fundamental role in the understanding of the systematic effects in Planck. Any non ideality in the telescope performance will contaminate the measurements to some extent. As a consequence the arising effects must be analyzed and understood in detail. The non–ideal behavior can be divided in two main categories: the aberration of the main beam which degrade the angular resolution [3] [4] and the near and far side lobes which contribute to the Straylight Induced Noise, or briefly SIN [5]. In this work a description of the Planck telescope is given, emphasizing the evolution of the design from the first layout to the present baseline. The electromagnetic performances are briefly discussed.

THE PLANCK TELESCOPE

The Telescope is based on a two–mirror off–axis scheme (see Figure 1) which offers the advantage to accommodate large focal plane instruments with an unblocked aperture an thus maintaining the diffraction by the secondary mirror and struts at very low levels. The field of view is as large as +/– 5° and centered on the line of sight (LOS) which is tilted at about 3.7 degrees with respect to the main reflector axis. The LOS is 85° away from the spin axis which points in the anti – sun direction.

The Telescope will work in the frequency range between 25 GHz to 1 THz. In this range the total emissivity of the telescope will be less than 1 % at the beginning and less than 5% at the end of the mission.

Both the primary and the secondary mirrors have an ellipsoidal shape as in the case of the Gregorian aplanatic design. The sub reflector revolution axis is tilted with respect to the main reflector revolution axis. The primary mirror physical dimensions are about 1.9 x 1.5 meters, allowing a projected circular aperture of 1.5 meter of diameter. The secondary reflector has been oversized up to approximately 1 meter of diameter to avoid any additional under illumination of the primary. LFI and HFI are located in the focal region which is approximately a 8° tilted plane with respect to the plane perpendicular to the Z_{RDP}. The typical roughness of the mirror surface is required to be 1 μm at any scale up to 0.8 mm and the average mechanical surface error will be 10 μm with respect to the best fit surfaces (2 μm of amplitude maximum for periodic structures).

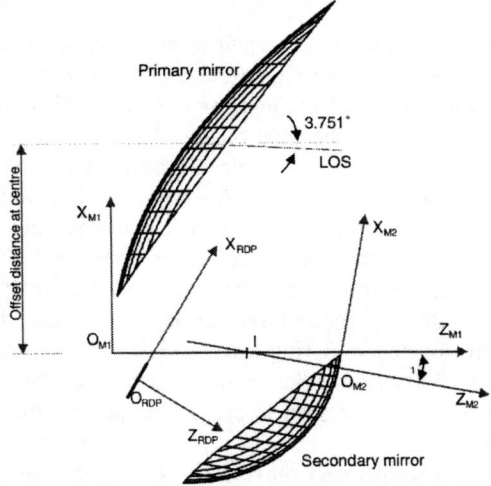

FIGURE 1. Layout of the Planck Telescope. The Line of Sight is tilted at 3.751 degree with respect to the primary mirror axis. The secondary mirror axis is tilted at 10.1 degrees with respect to the primary mirror axis. The main coordinate systems are indicated: the primary reflector frame (M1), the secondary reflector frame (M2) and the Reference Detector Plane (RDP) which represents the center of the focal plane. The Line of Sight (LOS) is also showed. Z_{M1} and Z_{M2} are the revolution axes of the primary and secondary ellipsoids, respectively.

The Planck Telescope as a complete satellite sub–unit (see Figure 2) includes the primary reflector, the primary mirror supporting structure, the secondary reflector, the secondary reflector fixation struts, the telescope frame (holding the two mirror support structures, the focal plane array and the interfaces with the payload module struts), the payload module straylight baffle (interfaced with the coldest V–groove shield at 50 K), the telescope inner baffle between the focal plane array and the secondary mirror, the primary reflector extension baffle, the instrumentation for the telescope hardware, and the hardware to interface with the instruments.

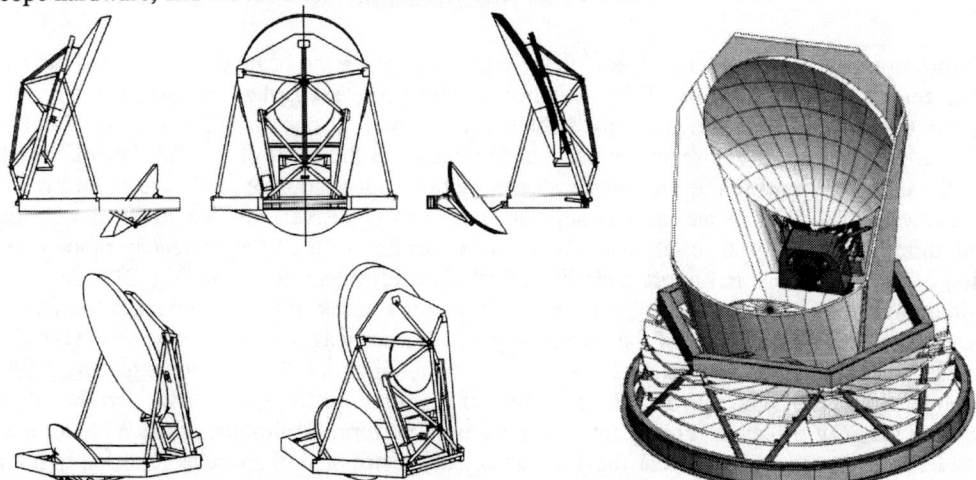

FIGURE 2. Left: the Planck Telescope as a single unit to be integrated in the payload module. Right: the Planck payload and the telescope. The V-grooves are seen as well as the focal plane array accommodating the two instruments. The hexagonal structure, visible on the top of the third V-grooved shields, is the telescope frame.

The primary and secondary mirrors will be fabricated using Carbon Fiber (CFRP) technology. The baseline consists of an all-CFRP rounded triangular tubes sandwich array arranged in a honeycomb–like structure. This kind of structure has been chosen to satisfy the requirements of low mass (< 120 Kg including struts and supports), high stiffness, high dimensional accuracy, and low thermal expansion coefficient. The sandwich concept consists of a thick (4-10 cm) honeycomb-like core with the desired shape, and to which are bonded two thin (1-1.5 mm) reflecting skins.

The History

The current Planck Telescope Configuration is the result of an evolution of several designs started from the one reported in the COBRAS proposal in 1993 [6]. The COBRAS telescope was a *"clear – aperture, 0.6 m focal length Gregorian with primary parabolic reflector of 1.5 m diameter (1.0 meter illuminated) and an elliptical secondary of 0.57 m diameter"*. The design was modified at the time of the Phase–A proposal in 1996 [7] when COBRAS and SAMBA were joined in a single mission. A 1.3 meter projected aperture off-axis telescope satisfying the Dragone–Mizuguchi (DM) condition [8] was chosen for COBRAS/SAMBA. The DM configuration exhibits an ideal response only in the center of the focal region. In this point the telescope is equivalent to a blockage free on–axis configuration. Unfortunately, as the LFI optical working group demonstrated, the aberrations (especially the coma) increase significantly for the feeds located outside the center of the focal surface [9]. The degradations of the telescope performances (mainly the effective angular resolution) in the main beam region affected the scientific capabilities of LFI, whose feeds are located in a ring around HFI with a typical beam location approximately 3 degrees away from the telescope optical axis. After the announcement of opportunity for the FIRST/Planck program, in February 1998, the telescope design was increased in projected aperture from 1.3 to 1.5 meters of diameter. To avoid any change of the focal plane design only the main reflector was changed improving the stray light rejection by adding a ring around the top edge of the primary mirror. The so called *"Carrier Configuration"* was in fact a 1.3 – like telescope with a more under illuminated 1.5 meter primary mirror. Several optical configurations were studied and designed by the LFI team and DSRI with the purpose of investigating the best 1.5 meter telescope configuration for Planck.

An aplanatic design was obtained starting from a Gregorian standard configuration by modifying the conical constants and curvature radius according to the analytical *"aplanatic"* condition [10]. The aplanatic choice of the mirror shapes (both ellipsoidal) minimizes the spherical and comatic aberrations even in the case of the off–axis scheme, as readily seen in the left panels of Figure 3 and extensively reported in [11]. Several aplanatic designs were considered and proposed as alternative solutions for the telescope. The improvements on the main beam shape due to the aplanatic configuration are evident also for the off–axis feeds. During the Planck Payload Architect industrial activity on 1999, a telescope optimization process was carried out.

The Optimization

The optimization [12] has been addressed with the aim to minimize the beam distortions and the straylight noise. The spillover energy and the ellipticity of the main beams have been assumed as the quality evaluation parameters. Moreover, one of the additional parameter used for selecting the best configuration was the flatness of the focal surface in order to minimize the obscuration between the feeds on the focal plane unit. Operatively the study was mainly devoted to the optimization of the mirrors shapes by the minimization of the Wave Front Error (WFE), using the optical software ®CODEV. Although this software has been conceived for systems in the optical wavelength range, a minimization of the WFE or alternatively the maximization of the Strehl Intensity ratio corresponds to the minimization of the aberrations in a more general context. The advantage is that in ®CODEV (or similar software) an automatic optimization of the WFE can be possible. However a check of the optimization results was required by the LFI team and a set of simulations of the telescope performances was run using Physical Optics based software like GRASP8. The following parameters have been tuned to optimize the telescope starting from an aplanatic design: the conic constant of both mirrors, the curvature of the sub reflector, the distance and the tilting between the two mirrors, and the FOV direction. The optimization has been performed minimizing the WFE for a total of sixteen (eight for LFI and eight for HFI) different focal points equally distributed in space as reported in the right panel of Figure 3. The so–called CASE1 configuration has been selected as baseline among several optimized designs and it has been shown to be the best compromise between the optical performances and the available room on the payload module.

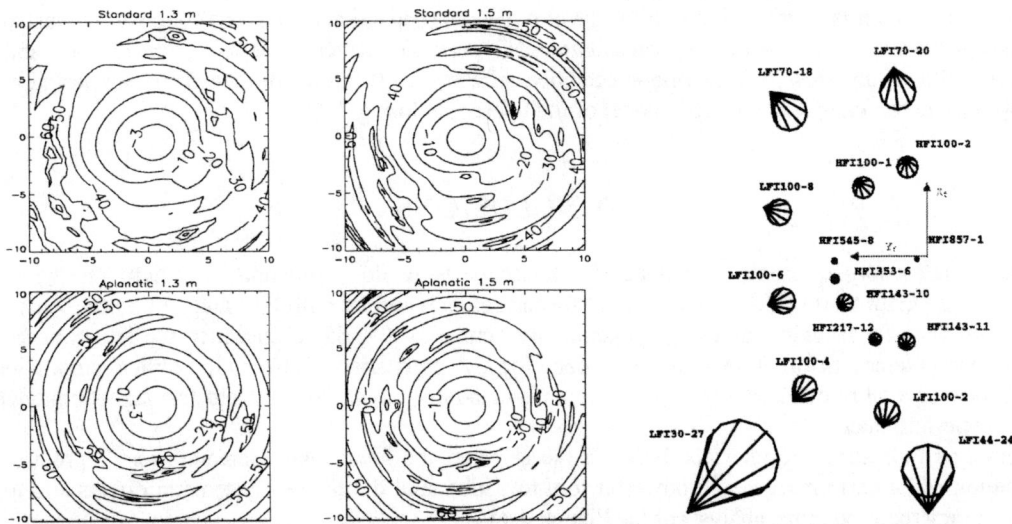

FIGURE 3. The four contour plots on the left show the improvement of the beam symmetry due to the aplanatic design proposed by the LFI team. The first row of contour plots reports the main beam response of two Dragone–Mizuguchi different configurations with 1.3 m and 1.5 m of diameter from left to right respectively (Standard 1.3 m and Standard 1.5 m). The second row reports from left to right the response of the aplanatic configurations with 1.3 m and 1.5 m of aperture respectively. These two last configurations were proposed by the LFI team as alternative solutions before starting the optimization of the telescope. All the calculations have been done at TESRE and at 100 GHz with a typical LFI off–axis feed. The right panel shows the feed locations have been used for the optimization.

The Performances

The CASE1 configuration shows a significant improvement of the telescope performances for all the LFI and HFI detectors concerning the beam shape and the far side lobes, with respect to the Carrier configuration. The optical quality of the optimized telescope has been carefully investigated by the LFI team because of the highly off–axis location (3 degrees on average) of the LFI horns, particularly for the most important cosmological frequency channel at 100 GHz. The left plot of Figure 4 shows the main beam shape on the sky of all of the LFI detectors. It is readily seen that the coma and spherical aberrations are kept at very low levels. The simulated beams show elliptical shape up to the 20 dB contour approximately, which means a really good improvement with respect to the carrier configuration (see the contour plots of Figure 3 and the right panel of Figure 4) even for the most offset beams. As an example of the response of the telescope at the HFI frequencies, the simulated beam shape of one of the 217 GHz HFI channel is reported in the center plot of Figure 4.

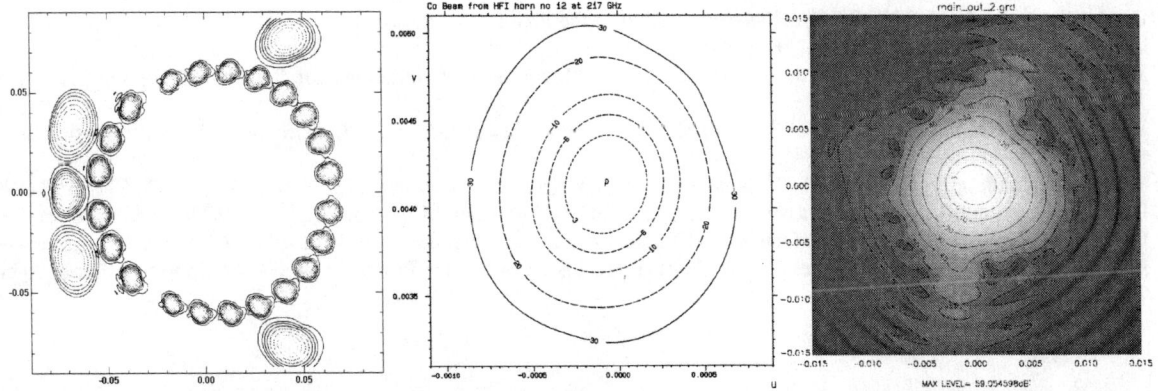

FIGURE 4. The performances of the Planck Telescope are summarized in these plots. The contour plots are calculated on the (U,V) plane. Left: the finger print on the sky of the LFI detectors. Center: one of the HFI 217 GHz detector response on the sky. Right: One of the LFI 100 GHz beam. Note that obviously the scale is different for each plot and that the beam symmetry is acceptable even for the most off–axis beam position.

The beam calculation is only the first step for the complete modelization of the telescope which must also include the analysis of roughness, tolerances, misalignments, periodic structure effects, molecular and particles contamination effects, cooldown and thermo–mechanical effects. All the simulations will be supported and verified with an appropriate test campaign on a RF–model of the telescope during 2002.

CONCLUSIONS

The Planck telescope represents a challenge for telescope technology and optical design. The wide frequency coverage (from 25 GHz to 1000 GHz), the high performances required by both HFI and LFI instruments sharing the 400 x 400 mm wide focal region, and the cryogenic environment (40 – 65 K) in which the telescope will operate, have never been obtained before in experimental cosmology. A comparison between the optical simulations and the measurements at several frequencies and feed locations is mandatory to validate the electromagnetic models used for predicting the performances.

The telescope represents the interface between the sky and the focal plane detectors and a precise study and characterization of its performances is a powerful tool to understand the related systematic effects and to reach the scientific accuracy required in the analysis of the Planck data.

ACKNOWLEDGMENTS

We wish to thank the European Space Agency (ESA), TICRA engineering consultant, the HFI consortium, the LFI consortium, and the TP consortium.

REFERENCES

1. Mandolesi, N., et al., *The Low Frequency Instrument*, response to the Announcement of Opportunity for the FIRST/Planck Programme, Feb. 1998 .
2. Puget, J-L., et al., *The High Frequency Instrument*, response to the Announcement of Opportunity for the FIRST/Planck Programme, Feb. 1998.
3. Mandolesi, N., Bersanelli, M., Burigana, C., Gorski, K.M., Hivon, E., Maino, D., Valenziano, L., Villa, F., and White, M., *Astron. Astroph. Suppl.*, **145**, 323-340 (2000)
4. Mandolesi, N., Bersanelli, M., Burigana, C., Gorski, K.M., Guzzi, P., Hivon, E., Maino, D., Malaspina, M., Valenziano, L., and Villa, F., "On the Planck effective angular resolution", *Int. Rep. TeSRE/CNR.*, **199/1997**, (1997)
5. Burigana C., Maino D., Gorski K.M., Mandolesi, N., Bersanelli, M., Villa, F., Valenziano, L., Wandelt, B.D., Maltoni, M., Hivon, E., *Astron. & Astrophys.* **373**, 345-358 (2001).
6. Mandolesi, N., Smoot, G.F., *COBRAS – Cosmic Background Radiation Anisotropy Satellite*, A proposal submitted to ESA M3 Call for Mission Ideas, 1993.
7. Bersanelli, et al., *The COBRAS/SAMBA mission*, Report of the Phase – A, 1996.
8. Dragone, C., *The B.S.T.J.*, **57**, No. 7,2263 (1978).
9. Villa, F., Valenziano, L., Mandolesi, N., Bersanelli, M., "Main Beam Simulations for the Planck Telescope", *Int. Rep. TeSRE/CNR.*, **231/1998**, (1998).
10. Wilson, R. N., "Aberration Theory of Telescopes," in *Reflecting Telescope Optics I*, Astronomy and Astrophysics Library, Springer, Germany, 1996, pp. 90-93.
11. Villa, F., Mandolesi, N., Burigana, C., "A Note on the Planck Aplanatic Telescope", *Int. Rep. TeSRE/CNR.*, **221/1998** (1998).
12. Dubruel, D., Cornut, M. , Fargant, Passvogel, T., De Maagt, P., Anderegg, M., Tauber, J., "Very Wide Band Antenna Design for Planck Telescope Project Using Optical and Radio Frequency Techniques" in *Millennium Conference on Antennas & Propagation - 2000*, edited by Danesy, D. & Sawaya, H., ESA Conference Proceedings SP-444, European Space Agency, 2000, CD-ROM.

Analysis of thermally-induced effects in Planck Low Frequency Instrument

A. Mennella[1], M. Bersanelli[2], C. Burigana[3], D. Maino[4], R. Ferretti[5], G. Morgante[3,6], M. Prina[6], N. Mandolesi[3], C. Butler[3], L. Valenziano[3], F. Villa[3]

On behalf of the LFI Consortium

[1] *IFC-CNR, Milan, Italy*
[2] *Università di Milano, Dip. Di Fisica, Milano, Italy*
[3] *TESRE-CNR, Bologna, Italy*
[4] *Osservatorio Astronomico di Trieste, Trieste, Italy*
[5] *LABEN S.p.A., Vimodrone, Milan, Italy*
[6] *Jet Propulsion Laboratory, Pasadena, USA*

Abstract. The Planck mission will provide full-sky maps of the Cosmic Microwave Background with unprecedented angular resolution (~ 10') and sensitivity ($\Delta T / T \sim 10^{-6}$). This requires cryogenically cooled, high sensitivity detectors as well as an extremely accurate control of systematic errors, which must be kept at µK level. In this work we focus on systematic effects arising from thermal instabilities in the Low Frequency Instrument. operating in the 30-100 GHz range. Our results show that it is of crucial importance to assure "in hardware" a high degree of stability. In addition, we provide an estimate of the level at which it is possible to reduce the contamination level in the observed maps by proper analysis of the Time Ordered Data.

INTRODUCTION

Planck is an European Space Agency (ESA) mission to map spatial anisotropy in the Cosmic Microwave Background (CMB) over a wide range of frequencies with an unprecedented combination of sensitivity, angular resolution, and sky coverage. It consists of a Low Frequency Instrument (LFI) and a High Frequency Instrument (HFI) observing the sky through a common telescope [1, 2].

The LFI is constituted by an array of 54 radiometers actively cooled at 20 K that will collect the microwave radiation in four well defined frequency bands, centered at 30, 44, 70 and 100 GHz, while the HFI will image the sky in six frequency channels between 100 and 857 GHz with its array of 48 bolometric detectors cooled at 0.1 K. High thermal stability is required to avoid spurious systematic effects that need to be kept at µK level in order to retain the scientific value of the measured data.

To meet the cryogenic requirements a dedicated chain of cryo-coolers will be implemented on-board the Planck satellite. The 20 K stage, in common between the two instruments, will be provided by a Sorption Cooler [3], a vibration-less cryostat with a cooling capability of >1 W at 20 K. The heart of the cooler is composed by six hydride compressor beds in which hydrogen is alternatively absorbed and released as the temperature of the beds is modulated, thus creating a constant gas flow. The activity of such compressors generates second-order temperature oscillations that will propagate through the satellite and may affect the thermal boundaries between the spacecraft and the Planck instruments.

In this work we present the results of a preliminary assessment of the impact of temperature fluctuations of the 20 K stage of the LFI instrument. In our study we have estimated the maximum systematic error generated by such fluctuations of the LFI maps, taking into account the ability to reduce the level of these effect by applying destriping algorithms to the Time Ordered Data. Our results show that although the impact of such temperature instabilities can be reduced "in software" it is crucially important to guarantee "in hardware" a temperature stable at the mK level in the 20K stage.

PLANCK THERMAL ENVIRONMENT

The instruments on board the Planck satellite will measure the CMB with a very high sensitivity, of the order of few µK per pixel at the end of a 14 month mission; in order to maintain the instrumental noise at very low levels a specifically designed chain of cryo-coolers is being currently developed. In Fig. 1 (a and b) we show a sketch representing the Planck thermal environment. The satellite, in particular, presents two main temperature stages (50 K and 300 K) thermally decoupled by three thermal shields ("V-grooves") at different temperatures which radiate heat into deep space. The instruments in the focal plane present different temperature stages ranging from 0.1 K (HFI bolometers) to 20 K (LFI radiometers).

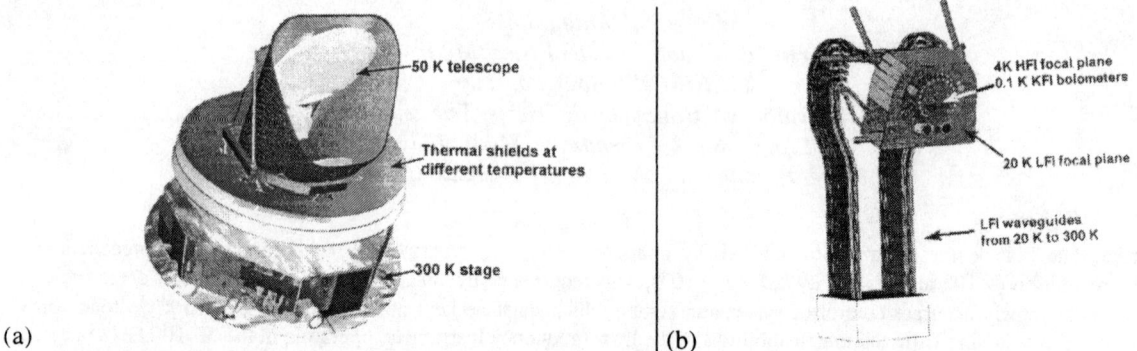

FIGURE 1. (a) Temperature environment in the Planck satellite. The three shields (at 140 K, 100 K and 50 K) thermally decouple the warm (300 K) and cold (50 K) satellite stages. (b) Temperatures of the Planck instruments.

Such a complex thermal environment clearly requires a very careful control of the temperature stability at the thermal boundaries between the spacecraft and the instruments, because any temperature variation will leave its signature in the measured data thus causing spurious systematic effects. The main source of temperature instability in Planck is represented by the Sorption Cooler (located in the 300 K Service Module) which drives the entire Planck cryo-chain and provides the 20 K environment to the LFI.

The temperature provided by the Sorption Cooler will not be perfectly stable, but will display oscillations caused mainly by pressure fluctuations in the compressor beds. These instabilities will impact on the behavior of many temperature-sensitive components of the LFI radiometers. As shown schematically in Fig. 2, each LFI radiometer is composed by a 20 K front-end stage, and a 300 K back-end. Each radiometer compares the signal from the sky to a stable 4 K reference load. In this work we consider the impact of temperature instabilities on the most sensitive 20 K front-end components, i.e. feed horns and orthomode transducers, the 4 K reference horn antenna and the front end RF amplifiers.

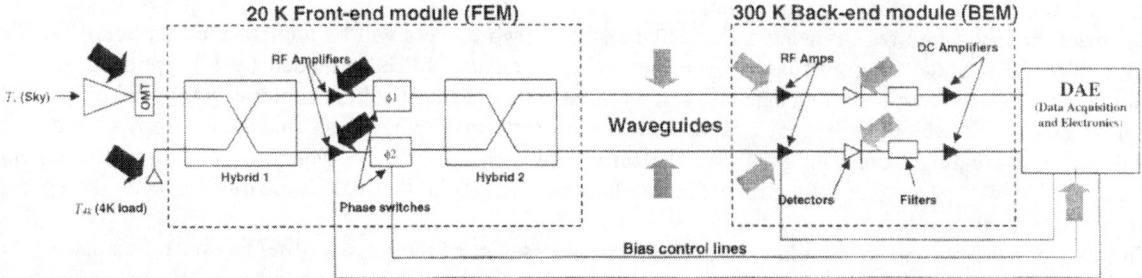

FIGURE 2. Schematic of Planck-LFI radiometers that highlights the components that are sensitive to temperature fluctuations. In this work we focus on the effect temperature instabilities on the most sensitive 20 K front-end components (black arrows).

Instability in the radiometer physical temperature will cause oscillations in the measured signal and, therefore, spurious anisotropies in the final maps. The impact of signal oscillations with a period that is not synchronous with the spacecraft spin (1 rpm) can be mitigated by applying so-called "destriping" algorithms to the Time Ordered Data, while spin-synchronous signal fluctuations will permanently affect the final maps. The high level of systematic error rejection required for Planck-LFI imposes strict limits on the maximum acceptable peak-to-peak error per pixel[1] caused by thermal effects (± 0.8 µK for spin synchronous effects and ± 1.1 µK for other periodic effects).

EFFECT OF FRONT END TEMPERATURE INSTABILITY

Temperature fluctuations at the Sorption Cooler cold end (20 K)

Although experimental data concerning the shape of the physical temperature fluctuations caused by the Sorption Cooler are not available yet, recent simulations have provided a first insight into the effect of non-idealities in the compressor assembly behavior on the temperature stability of the 20 K cold end. In Figs. 3 and 4 we show two examples of the cold end temperature oscillation in two different scenarios: an ideally behaving cooler (Fig. 3) and a cooler with a non-homogeneous hydride distribution in one of the compressor beds (Fig. 4). The main result of these simulations is that Sorption Cooler-induced temperature fluctuations are characterized by a high number of harmonics of the compressor bed period (which currently set to 667 s). Furthermore a non ideal behavior in the compressor assembly results in an increased impact of higher and lower harmonics; in particular Fig. 4 shows a tail of higher harmonics in which it is visible (see Fig. 4b) a quasi-spin synchronous harmonic close to 60 s with an amplitude of the order of 1 mK.

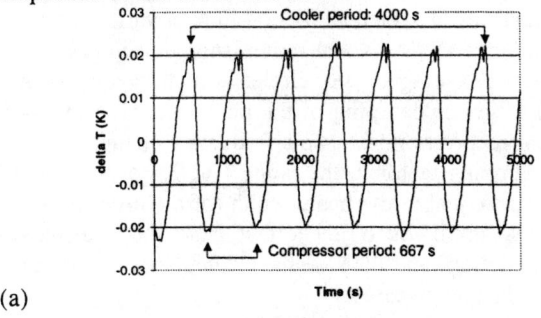 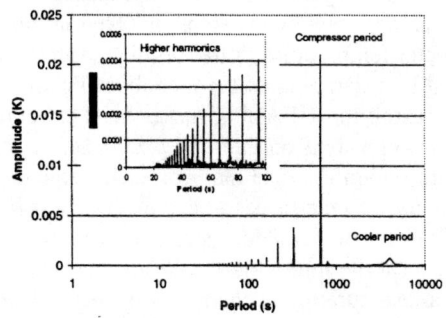

(a) (b)

FIGURE 3. Simulation of temperature fluctuation at the sorption cooler cold end for an ideally-behaving set of compressor beds (6 equal compressors with homogeneous beds). (a) Oscillation in time domain. (b) Fourier transform. The inset is a zoom of the high frequency tail of the spectrum.

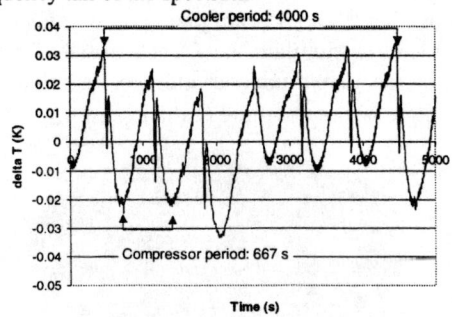 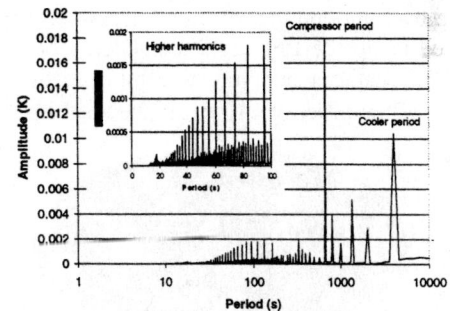

(a) (b)

FIGURE 4. Same as in Fig. 3 for a Sorption Cooler characterized by 5 equal compressors with homogeneous beds and 1 compressor with non-homogeneous hydride distribution.

Damping by LFI mechanical structure

In the previous section we have discussed the behavior of the physical temperature at the Sorption Cooler 20 K cold end. Now we want to evaluate how these fluctuations are transferred through the LFI mechanical structure to

---

[1] Here we refer to a pixel size equal to the physical beam size, θ_{FWHM}. Note the LFI maps will be generated with a lower pixel size, of the order of $\theta_{FWHM}/3$, which corresponds to the data sampling resolution.

the LFI radiometers. The schematic in Fig. 5a shows the damping effect caused by the thermal mass and thermal resistance of the instrument, which acts as a low-pass filter for thermal oscillations. The net effect of this filter is that the actual temperature variation at the front-end radiometers display a lower peak-to-peak amplitude and a much more limited content of high frequency harmonics. In Fig. 5b we report results obtained using our current LFI thermal model which quantify the frequency dependence of the damping factor for the four LFI frequency channels.

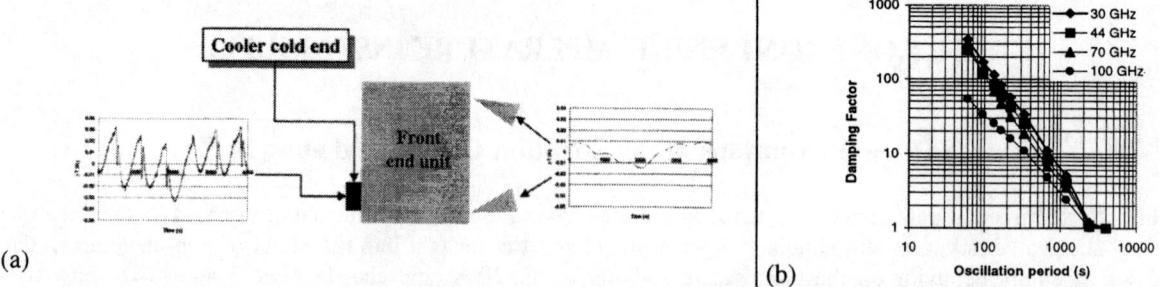

FIGURE 5. Damping of thermal by LFI mechanical structure. (a) The sketch shows that the LFI mechanical structure is a *low-pass* filter for thermal fluctuations. (b) Damping factor vs. fluctuation period for the four LFI frequency channels.

Effect on LFI maps

The next step is to evaluate the effect of physical temperature fluctuations on the measured maps. In this paper we limit our analysis to the 30 GHz channel which appears also to be the most sensitive to temperature fluctuations.

Let us now review the steps involved in this process: (1) the instrument transfer function links a variation in the radiometer temperature to the oscillation in the measured signal; (2) the measurement redundancy during each scan yields a first damping of the signal oscillation; (3) the data stream of "averaged" sky circles is then "mapped" into the sky using the HEALPix hierarchical structure [4]; this yields a further damping of the peak-to-peak oscillation which is dependent on the map pixel-size; (4) maps from different receivers in the same frequency channel are co-added to obtain a single map for each frequency, with a further damping factor in the range 1 to 2.5 depending on the channel; (5) the peak-to-peak amplitude is rescaled on a pixel size equal to the beam width (of the order of 36' at 30 GHz). To perform this step we have first evaluated the r.m.s. amplitude of the effect from the map power spectrum (at the multipole l corresponding to the beam angular resolution) and then we have calculated the peak-to-peak value assuming no change in the map distribution at the various angular scales.

Here we neglect the map co-adding step so that each map is relative to one particular LFI receiver, which has no impact on the results al lower frequencies (30 and 44 GHz).

In Fig. 6 we show 30 GHz maps (pixel size: 13.7') of the systematic error (in thermodynamic temperature) induced by front-end temperature fluctuations for the cases shown in Figs. 3 and 4. Map pixel coordinates have been produced using the LFI flight simulator, a code specifically developed to simulate the Planck mission. The figure shows that without applying destriping to the Time Ordered Data the maximum variation on the final maps is relevant, of the order of ±16 µK in the ideal case and about ten times greater in the more realistic case.

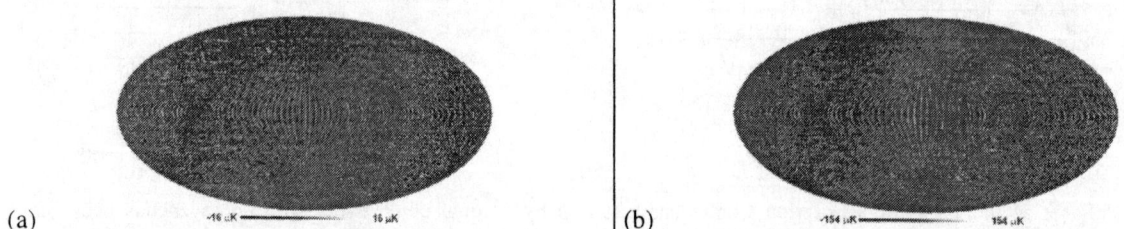

FIGURE 6. Map of the final systematic error (thermodynamic temperatures) at 30 GHz (13.7' pixel size) for the two Sorption Cooler scenarios considered in this note: (a) ideal Sorption Cooler, (b) Cooler with one non homogeneous bed.

Let us now evaluate the ability of reducing such systematic effects by applying a destriping algorithm to the Time Ordered Data[2]. The algorithm makes use of intersections in the sky scans (determined by the scanning strategy) to reduce the impact of periodic spurious signals (see [5] for further details). In general the destriping efficiency increases with the period of the spurious signal, while spin synchronous signals will not be altered by this

---

[2] Note that, in general, the destriping code is applied to data streams that contain also the instrumental noise. Some tests have shown that the presence of a superimposed noise (white and white+1/f) has a negligible impact of the code ability to reduce the effect of periodic fluctuations.

procedure. This is shown clearly in Fig. 7, that shows the various damping factors (measurement redundancy, mapping and destriping) for signals having different periods.

In Table 1 we present a summary of the maximum peak-to-peak systematic error per pixel after destriping the maps shown in Fig. 6. The values summarized in the table indicate that even in the most favorable case the residual systematic is of the same order of the maximum error accepted from all thermal effects. Although this analysis is still preliminary and it is based on simulations of the Sorption Cooler behavior, it shows that a temperature stability at the 20 K cold end of order 10 mK would be needed to maintain the induced systematic errors at a negligible level.

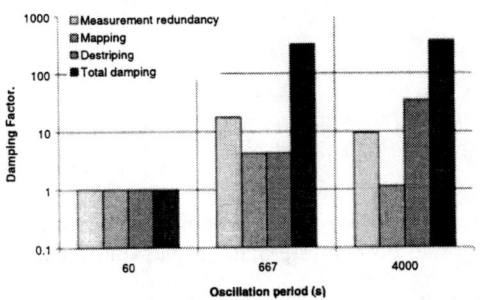

FIGURE 7. Typical damping factors for an (multiple pixel measurements and destriping) vs. fluctuation period. Spin synchronous fluctuations are not removed (damping = 1) while "longer" oscillations are reduced (after destriping) by a total factor of about 300.

TABLE 1. Residual systematic error on final maps (on 36' pixel) caused by front-end temperature fluctuations

Shape of temperature fluctuation	Peak-to-peak amplitude of systematic error in map (µK)	
	Non spin synch.	Spin synch
⟨∧∧∧∧∧⟩	1.0	0.3
⟨∧∧∧∧∧⟩	3.1	1.5

CONCLUSIONS

In this paper we have presented a preliminary study of the impact of Sorption Cooler temperature fluctuations on the LFI measurements. Our results show that several factors contribute to reduce the impact of such fluctuations on the measured maps: the LFI mechanical structure that very efficiently reduces the amplitude of fast temperature oscillations (i.e. with a period less than 500s) propagating from the Sorption Cooler Cold end to the radiometers, the measurement redundancy during each scan circle and the application of destriping algorithms to the Time Ordered Data. Nevertheless, assuming non-homogeneous beds in the cooler, the expected maximum peak-to-peak systematic error per 36' pixel in the final maps at 30 GHz is at the 2-4 µK level: although apparently small, this is significantly more than the systematic error allocated to this effect. Future work will be focused at refining the analysis (including the use of experimental data from the Sorption Cooler breadboard, when available) and at identifying means to improve the temperature stability of the LFI front-end .

ACKNOWLEDGMENTS

The HEALPix package use is acknowledged (see HEALPix home page at http://www.eso.org/science/healpix/). We also wish to thank the Planck LFI Data Processing Center for the support to the simulation work.

REFERENCES

1. Mandolesi, N., et al, "Planck Low Frequency Instrument," in *Proceedings of 2K1BC Workshop on Experimental Cosmology @ mm-waves*, AIP Conference Proceedings, 2001 (this issue).
2. M. Bersanelli, N.Mandolesi, *Astrophisical Letter and Communication*. **37**, 171-180 (2000).
3. Bhandari, P., et al, *Astrophysics Letters and Communications* **37**, 227-237 (2000).
4. Gorski, K.M., Hivon, E., Wandelt, B.D., "Analysis Issues for Large CMB Data Sets", *Proceedings of the MPA/ESO Conference on Evolution of Large-Scale Structure: from Recombination to Garching*," edited by Banday, A.J., Sheth, R.K., Da Costa, L., 37-42, 1998.
5. Maino, D., et al, Astronomy and Astrophysics Supplement Series **140**, 383-391 (1999).

Measuring CMB Polarization with ESA PLANCK SubMM-Wave Telescope

Vladimir Yurchenko[*][†]

[*] *Experimental Physics Department, National University of Ireland, Maynooth, Co. Kildare, Ireland*
[†] *Institute of Radiophysics and Electronics, National Acad. Sci., 12 Proskura St., Kharkov, 61085, Ukraine*

Abstract. We analyze the polarization properties of the tilted off-axis dual-reflector submillimeter-wave telescope on the ESA PLANCK Surveyor designed for measuring the temperature anisotropies and polarization characteristics of the cosmic microwave background.

INTRODUCTION

The dual-reflector submillimeter-wave telescope on the ESA PLANCK Surveyor is being designed for measuring the temperature anisotropies and polarization characteristics of the cosmic microwave background (CMB).

Our research is concerned with one of its focal plane instruments, Far-IR High Frequency Instrument (HFI) [1], which will cover six frequency bands centered at 100, 143, 217, 353, 545 and $857 GHz$ providing the sensitivity of $\Delta T/T \sim 10^{-6}$ and the angular resolution down to 5 arcminutes. The HFI consists of an array of 36 horn antenna structures (Fig. 1, a) feeding the bolometric detectors which will be cryogenically cooled to a temperature of $100 mK$.

One objective of the research is the optimization of the HFI optical design and the computation of the HFI beam patterns. The challenge of the problem is that the telescope is electrically large ($D/\lambda = 4300$ at $\lambda = 350 \mu m$) and consists of two essentially defocused ellipsoidal reflectors providing a very large field of view at the focal plane.

Another objective is the characterization of the polarization properties of the multi-beam telescope system. The CMB polarization is expected to be at a level of only 10% of the temperature anisotropy quadrupole, and the success of the measurements will depend crucially on the precise knowledge of the polarization properties of the telescope.

FORMULATION OF THE PROBLEM

While the performance of the antenna structures can be thoroughly tested in terrestrial conditions, the coupling of the HFI with the telescope is, basically, optimized through the computer simulations.

Among various simulation techniques, physical optics (PO) is the most adequate one for the given purpose. However, conventional implementations of the technique [2, 3] do not fit the size of the problem. Commercially available packages are also very limited in their capacity to rigorously answer this sort of questions. For example, even the best commercial software requires many hours of the full-scale physical optics computation of the main beam of the telescope at the relatively low frequency of $143 GHz$, while all conventional physical optics codes collapse at the highest frequencies of $545 - 857 GHz$.

To solve the problem, we developed a special PO code [1] that allowed us to overcome the limitations of a generic approach for large multi-reflector systems and perform typical simulations of the telescope in the order of minutes. Generally, it requires only 1 minute for a polarized Gaussian beam at the frequency of $143 GHz$ and about 30 minutes for the beam of 30 modes at $857 GHz$ using a PC Pentium III ($500 MHz$) under the Linux operating system.

We simulated the beams of twelve linearly polarized Gaussian horns operating in the transmitting mode at the nominal frequencies 143, 217, and $353 GHz$. Each horn is designed for the simultaneous measurement of two orthogonal linear polarizations, 'a' and 'b', of the incoming radiation. The polarization is characterized by the tilt ψ of the electric field at the beam axis in the sky with respect to the local vertical as defined below. The tilt is $\psi_a = +45°$ and $\psi_b = -45°$

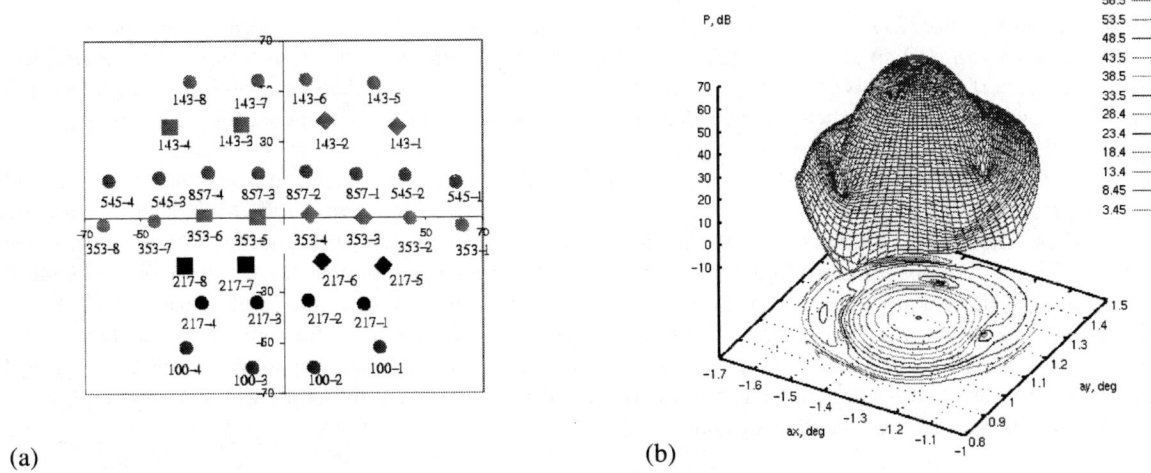

FIGURE 1. (a) Horn positions on the focal plane as seen from the telescope. (b) Total power of the telescope beam 143-1a computed for the clipped Gaussian feed model F3 with $a = 3.9mm$, $w = 2.8mm$, and $L = 120mm$ ($f = 143GHz$).

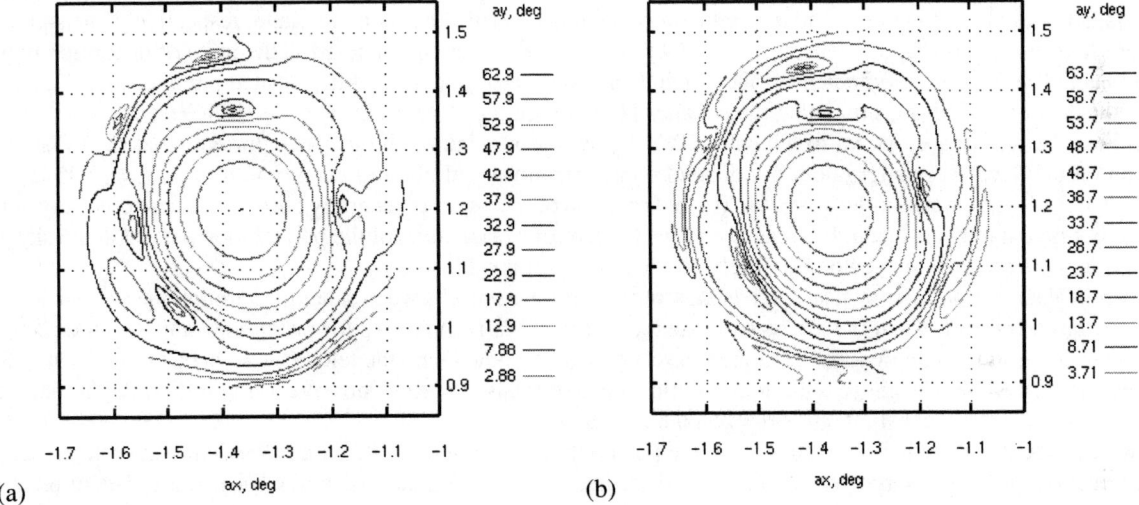

FIGURE 2. Power pattern of the telescope beam from the enhanced horn HFI-143-1a ($\psi_a = +45°$, feed model F1) at the frequency (a) $121GHz$ and (b) $166GHz$.

for the horns 143-1, 143-2, 217-5, 217-6, 353-3, and 353-4 (see Fig. 1, a), while $\psi_a = 0°$ and $\psi_b = 90°$ for the other six polarized horns (the angle ψ is measured clockwise from the upward vertical direction as seen from the telescope).

We compared three different models of the horn feed, F1, F2 and F3, when the source field is specified in different manner. The feed model F1 is represented by the far-field power and phase patterns of the actual corrugated horn in $0°/45°/90°$ cuts which are used in a somewhat approximate manner for computing the incident complex-vector electromagnetic field on the secondary mirror. The feed F2 is specified by the 'clipped Gaussian' model distribution of the horn aperture electric field which is accurately propagated to the secondary mirror and further through the system. Finally, the feed F3 is represented by the far-field patterns similar to the model F1 which, however, are computed precisely from the clipped Gaussian aperture field of the model F2 and then used in an approximate manner identical to the feed model F1.

The electric field at the aperture of the Gaussian horn in the feed models F2 and F3 is specified as

$$\vec{E}(r) = E_0 \exp(-r^2/w_a^2) \exp(-iqr^2)\, \vec{e}, \qquad 0 \leq r \leq a \qquad (1)$$

where r is the radial coordinate, w_a is the beam radius at the horn aperture, a is the horn aperture radius, $q = \pi/\lambda L$,

L is the curvature radius of the wavefront at the aperture (approximately, L is the horn slant length), and $\vec{e} = \vec{e}_{a,b}$ is the unit polarization vector ($\vec{E}(r) = 0$ if $r > a$). The aperture model of this kind is quite accurate in representing the field of real Gaussian horns. In particular, both the power and phase patterns of such a horn computed at the frequency $100GHz$ coincide perfectly well with the experimental data available for the model Gaussian horn designed for this frequency. However, some discrepancies in the sidelobes at the level of about $-30dB$ appear for the horns specifically optimized for the higher angular resolution.

The comparison of the feed models F2 and F3 shows that the minor differences can only be observed in the polarization patterns at the periphery of the telescope beam while the power patterns of the main beam are identical for both models. The beams computed with the feed models F1 and F3 are also very similar when the parameters of the model F3 are properly adjusted. In this case, also, it is only the polarization at the periphery of the beam that differs slightly for different models while the power patterns are very similar even at the level below $-30dB$. Finally, the orientation of the polarization vector on the beam axis in the sky field is precisely the same for all the feed models, and the polarization pattern, in general, is rather independent of both the horn pattern and of the fine features of the propagation model used in the simulations. It proves that the far-field model F1 can safely be used for the simulation of the telescope beams from the real horns despite the approximations used in the model.

POLARIZATION OF THE GAUSSIAN BEAMS

Fig. 1, b, shows the power pattern of the telescope beam H-143-1 as projected on the plane normal to the telescope line-of-sight at the $(0,0)$ point (a_x and a_y are the horizontal and vertical axes on the plane, respectively, measured in degrees). The beam axis is at the point $a_x = -1.3645°$, $a_y = 1.1975°$ which is defined as the point of maximum power of the beam. The pattern is computed for the clipped Gaussian feed model F3 with adjusted parameters $a = 3.9mm$, $w = 2.8mm$, and $L = 120mm$, simulating the enhanced horn operating at the frequency $f = 143GHz$.

The beam is well shaped down to almost $-30dB$ below the maximum and can be approximated by a Gaussian function (at this level, the pattern does not depend on polarization) with the full beam width of $W_{min} = 7.05$ arcmin and $W_{max} = 7.50$ arcmin measured at $-3dB$. The polarization of the beam is generally elliptical except precisely at the beam axis where it remains linear. In order to achieve the required orientation of the polarization pattern in the sky, we should orient the polarization vector $\vec{e}$ properly on the horn aperture [1].

For immediate comparison of polarizations measured by different horns when scanning through the sky, we should use easily aligned directions in the sky as equivalent reference polarization axes for different beams. Such directions are the meridians in the spherical frame of the telescope, with the pole being the telescope spin axis tilted by $\eta = 85°$ with respect to the line-of-sight (the meridians define local verticals at various observation points while the parallels are local horizontals that constitute the orthogonal directions).

Now, we should define the reference axis for the polarization vector $\vec{e}$ of a tilted horn. We define the reference axis as the direction of $\vec{e}$ in the horn aperture plane that is projected on the vertical axis of the focal plane. The orientation of $\vec{e}$ is specified by the angle ϕ in the horn aperture plane measured from this reference in a clockwise direction when looking from the horn to the secondary mirror.

Using these definitions, we have found that the beam of the H-143-1 horn is polarized at its axis at the required angles of (a) $\psi_a = +45°$ and (b) $\psi_b = -45°$ if the horn polarization vector $\vec{e}$ is specified by the angle $\phi_a = +42.99°$ and $\phi_b = -47.01°$, respectively.

When the polarization vector $\vec{e}$ is properly oriented, the cross-polarized component of the far field measured along the respective orthogonal direction in the sky is minimized. In such a case, the power pattern of the cross-polarized component is, basically, determined by the power pattern of the semi-minor axis of the polarization ellipse at each observation point of the beam. This pattern is very much the same for any orientation of the polarization vector and can be approximated as follows

$$P_{cr} = P_{cr0} \left[(\theta/\theta_0) \sin(\varphi - \varphi_0) \right]^p \exp[(\theta/\theta_0)^q] \qquad (2)$$

where P_{cr0}, dB, is the maximum power of the minor axis component achieved at the points specified by the polar angle θ_0 and the azimuthal angles $\varphi_0 \pm 90°$ measured from the center of the pattern which is located at the beam axis (the values of the parameters in this approximation depend on the position of the horn considered but the magnitude of P_{cr0} is, typically, more than $30dB$ below the maximum value of the total power of the beam).

Since the actual frequency bands of the horns are rather wide ($121 - 166$, $182 - 252$, and $295 - 411 GHz$), the far-field patterns of real horns differ essentially at the lower and upper frequencies of each band. It results in somewhat

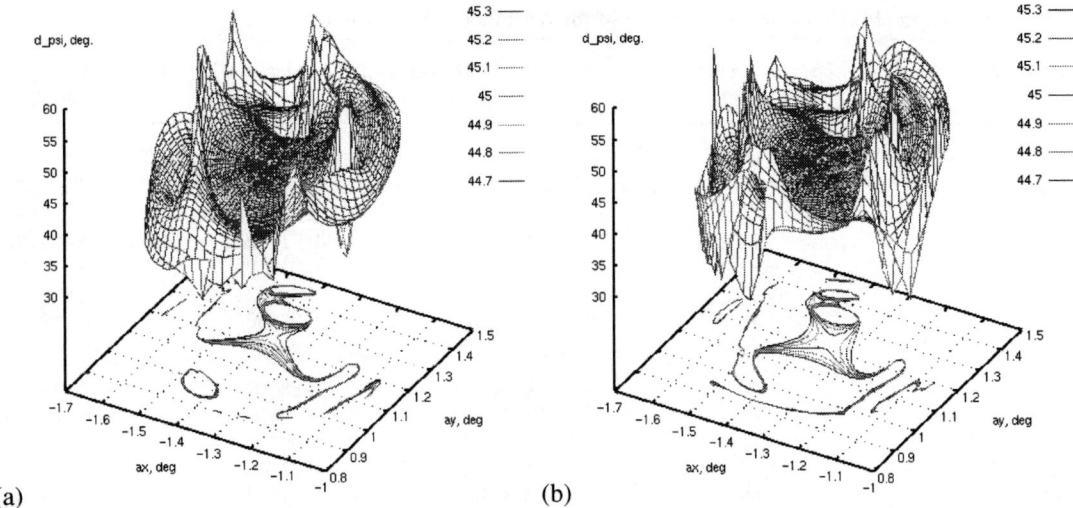

FIGURE 3. Deviation of the major axis of polarization ellipse from local vertical in the beam from the enhanced horn HFI-143-1a ($\psi_a = +45°$, feed model F1) at the frequency (a) $121 GHz$ and (b) $166 GHz$.

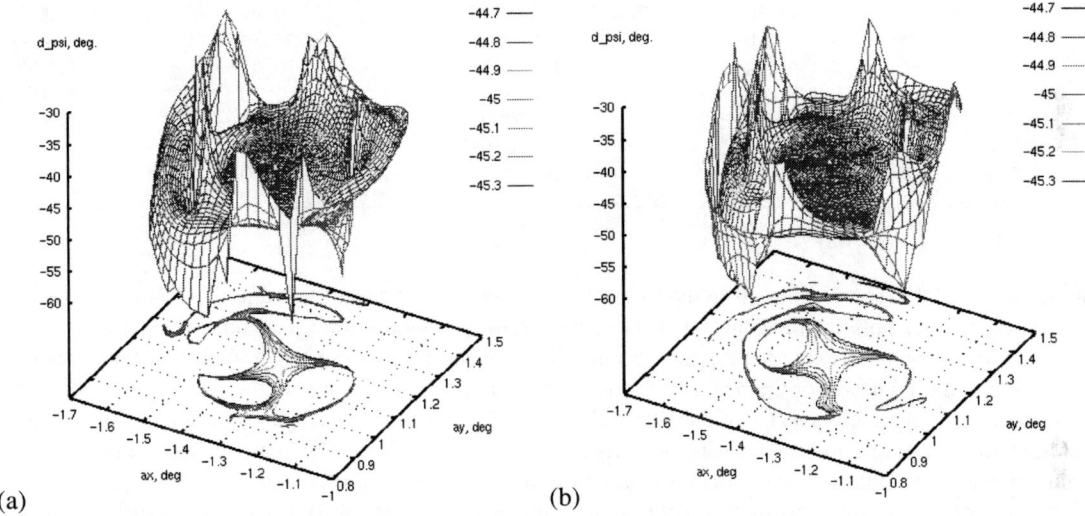

FIGURE 4. Deviation of the major axis of polarization ellipse from local vertical in the beam from the enhanced horn HFI-143-1b ($\psi_b = -45°$, feed model F1) at the frequency (a) $121 GHz$ and (b) $166 GHz$

different power and polarization patterns of the telescope beam as shown in Figures 2 - 4. Nevertheless, despite these differences, the polarization angles at the beam axis are the same irrelevant of the horn patterns and of the operating frequency.

The values of the aperture polarization angle φ computed for all the HFI horns as well as the direction cosines (the Cartesian components of the unit vector) of the aperture electric field with respect to the coordinate frame $M1$ of the primary mirror are summarized in the Table 1.

CONCLUSIONS

A fast physical optics code has been developed for the analysis of the dual-reflector submillimeter-wave telescope on the ESA PLANCK Surveyor. The code overcomes the limitations of a generic approach for large multi-reflector quasi-optical systems and can perform typical simulations of the telescope in the order of minutes.

TABLE 1. Orientation of the Polarization Vector $\vec{e}$ of the Aperture Electric Field

HFI horn	Aperture Angle ϕ [degrees]	Unit Vector of the Aperture Electric Field		
		$e_{X_{M1}}$	$e_{Y_{M1}}$	$e_{Z_{M1}}$
143-1-a	42.99	0.629015	0.679549	0.377562
143-1-b	-47.01	0.521115	-0.728982	0.443874
143-2-a	44.17	0.597142	0.696388	0.398076
143-2-b	-45.83	0.552563	-0.716862	0.425186
143-3-a	0.82	0.813844	0.014303	0.580907
143-3-b	-89.18	0.031357	-0.999321	-0.019324
143-4-a	2.10	0.812137	0.036506	0.582323
143-4-b	-87.90	0.080095	-0.995568	-0.049291
217-5-a	42.99	0.661010	0.680077	0.317114
217-5-b	-47.01	0.567397	-0.729548	0.381863
217-6-a	44.11	0.634880	0.695677	0.336097
217-6-b	-45.89	0.593249	-0.717633	0.364772
217-7-a	0.89	0.869188	0.015525	0.494238
217-7-b	-89.11	0.029397	-0.999362	-0.020308
217-8-a	2.01	0.867847	0.034982	0.495598
217-8-b	-87.99	0.066276	-0.996754	-0.045699
353-3-a	43.70	0.634926	0.690048	0.347422
353-3-b	-46.30	0.571370	-0.722094	0.390020
353-4-a	44.70	0.610581	0.703339	0.364012
353-4-b	-45.30	0.594981	-0.710743	0.375289
353-5-a	0.58	0.853206	0.010120	0.521476
353-5-b	-89.42	0.020092	-0.999707	-0.013472
353-6-a	1.51	0.852069	0.026308	0.522767
353-6-b	-88.49	0.052455	-0.998000	-0.035275

Simulations of the telescope beams from the linearly polarized horns have shown that the far-field of the telescope is, generally, elliptically polarized except precisely at the beam axis where the linear polarization is preserved. The magnitude of the minor semi-axis of the polarization ellipse in the telescope beam remains at the level of $-30dB$ below the maximum total power of the beam even for the most tilted horns located at the edge of the HFI horn array.

When rotating the polarization vector of the horn field about the horn axis, the major axes of the polarization ellipses at the central part of the telescope beam rotate virtually by the same angle about the beam axis while the power patterns of both the co- and cross-polarized components of the far field remain basically unchanged.

Thus, the orthogonal polarization directions at the beam axis in the sky are transformed into the orthogonal directions in the focal plane of the telescope. Some deviations from this basic rule which occur mainly at the periphery of the beam are of minor importance since they virtually do not contribute to the measured total power of the polarized component.

ACKNOWLEDGMENTS

The author is grateful to Bruno Maffei for the power and phase patterns of enhanced horns and to Yuying Longval for providing the updated positions and aiming angles of the HFI horns along with the map of the focal plane. The author would like to acknowledge the support of Enterprise Ireland.

REFERENCES

1. Yurchenko, V., Murphy, J. A., and Lamarre, J. M., *Int. J. Infrared and Millimeter Waves*, **22**, 173-184 (2001).
2. Diaz, L., and Milligan, T., *Antenna Engineering Using Physical Optics: Practical CAD Techniques and Software*, Artech House, London, 1996.
3. Scott, C., *Modern Methods of Reflector Antenna Analysis and Design*, Artech House, London, 1990.

The Infrared Telescope for Submillimetron Mission

Viktor I.Bujakas, Vasiliy N.Leonov, Vladimir F.Troitsky

P.N.Lebedev Physical Institute, Russian Academy of Sciences
Profsoiuznaya st.84/32, Moscow, 117810, Russia.
E-mail bujakas@ asc.rssi.ru

Abstract. We propose to modernise the far infrared telescope with metal mirrors, early developed for balloon experiments, and use it as a main instrument of the Submillimetron Mission.

1. The Submillimetron Project for space astronomy is now under consideration in the Astro Space Center of P.N.Lebedev Physical Institute (Russia) and in the Chalmers University of Technology (Sweden) [1]. Within this Project the submm-band space telescope will be put in orbit and operate in 0.3 – 1.5mm wavelengths. The main characteristics of the Mission are described in the paper of Dr. V. Gromov.

2. We propose the far infrared telescope with metal mirrors [2] (Fig.1), early developed for balloon experiments, be considered as a possible instrument for Submillimetron Mission [3]. The optical diameter of the telescope is 1-meter, it contains primary, secondary and turning metal mirrors made of special aluminium alloy. The optical scheme of the telescope is presented in Fig.2. A similar telescope, however a smaller one (optical diameter 0.85 m), was early used for balloon infrared studies (Fig.3.).

FIGURE 1 The 1- meter far infrared telescope with metal mirrors.

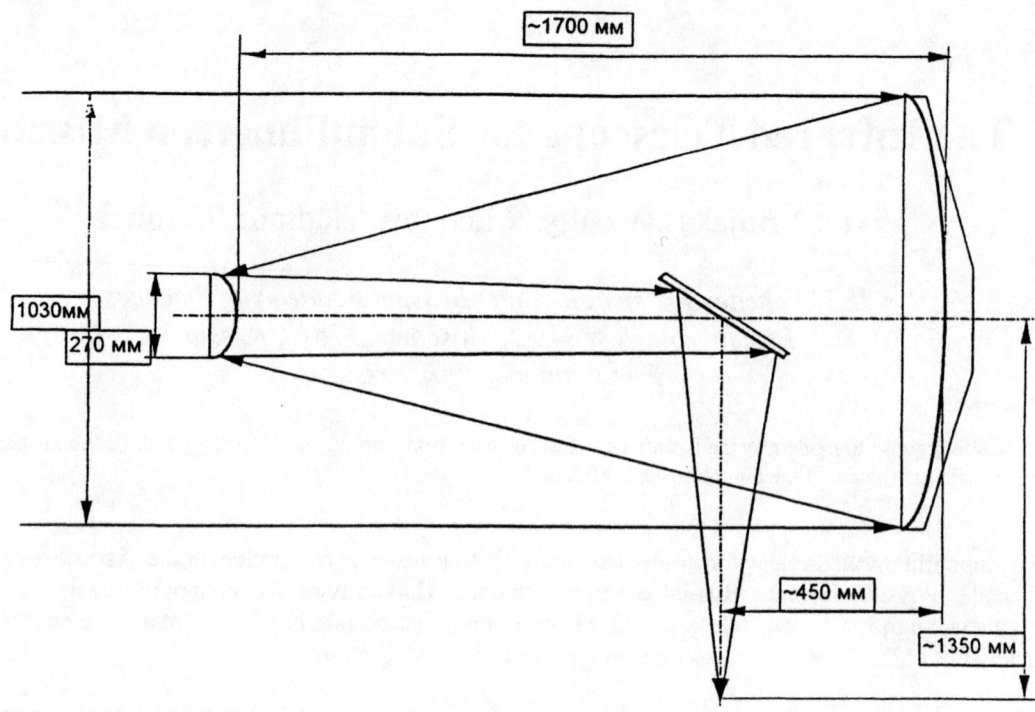

FIGURE 2. The optical scheme of the telescope

3. The balloon telescopes were equipped with sufficiently large integral detectors and the technical requirements to admissible aberration of the instruments were not severe. Therefore spherical primary and secondary mirrors were used. To meet technical requirements of the Submillimetron Project the substitution of the spherical secondary mirror by an aspherical one is planned. Preliminary estimates confirm the possibility of the modernisation.

4. Traditional cooling systems for infrared space telescope mirrors are consumption systems and use liquid helium from a space tank. To increase time of operation and reduce requirements to transport facilities, a closed-loop system to cool the mirrors of the Submillimetron telescope is proposed [3].

5. To check whether it is possible to use deeply cooled metal mirrors, for space telescopes the mirror deformation measurements at low temperatures are needed. The experience of the Astro Space Center in ground based cryogenic tests of space equipment [4] was used and the experiment to check the thermal deformation of the main mirror at 80 K ° has been developed.

References

1. Gromov V. The Submillimetron mission: astronomical tasks; comparision with other space projects for infrared and submillimeter astronomy *Proceedings of 4-th International Workshop on the Project Submillimetron,* Gothenburg, 2000, pp.23-53

2. Lapshin V.I., Salomonovich A.E., Leonov. V.N. et. all. About the possibility to fabricate large light metal mirrors for far infrared telescopes, *Preprint of P.N.Lebedev Physical Institute.* Moscow, 1983, n 291 (in Russian).

3. Kardashev N.S. et. all. The Millimetron project. *Proceedings of P.N.Lebedev Physical Institute.* Moscow, 2000, v. 228, pp.112 – 129. (in Russian).

4. Bujakas V.I., Troitsky V.F. et.all. The cooling of low noise amplifiers of space radiotelescope Radioastron in the space and on the ground. *Proceedings of the 6-th European Symposium on Space Environmental Control Systems,* Noordwijk, The Netherlands, 1977, pp 549 – 555.

FIGURE 3. The 0.85 m far infrared telescope for balloon studies.

Planck Low Frequency Instrument: Beam Patterns

M.Sandri[*], M.Bersanelli[†], C.Burigana[*], R.C.Butler[*], M.Malaspina[*], N.Mandolesi[*], A.Mennella[**], G.Morgante[*], L.Terenzi[*], L.Valenziano[*] and F.Villa[*]

[*]*TeSRE – CNR, 40122 Bologna, Italy*
[†]*Università di Milano, 20133 Milano, Italy*
[**]*IFCTR – CNR, 20133 Milano, Italy*

Abstract. The Low Frequency Instrument on board the Planck satellite is coupled to the Planck 1.5 meter off-axis dual reflector telescope by an array of 27 corrugated feed horns operating at 30, 44, 70, and 100 GHz. We briefly present here a detailed study of the optical interface devoted to optimize the angular resolution (10 arcmin at 100 GHz as a goal) and at the same time to minimize all the systematics coming from the sidelobes of the radiation pattern. Through optical simulations, we provide shapes, locations on the sky, angular resolutions, and polarization properties of each beam. *(On behalf of LFI Collaboration)*

INTRODUCTION

The Planck Telescope is designed as an off-axis tilted system offering the advantage of an unblocked aperture. The baseline configuration has been selected among thirty different optical designs. The current configuration has been obtained by optimizing the telescope performance for a set of equally distributed in frequency (from 30 to 857 GHz) and space (within the Focal Plane Box) representative feed horns. Both mirrors have an ellipsoidal shape (Aplanatic Configuration). The conical constants, the focal lenght, the tilting, and the decenter of the mirrors have been combined to reduce the main beam aberrations, the curvature of the focal surface, and the spillover as well.

The Low Frequency Instrument (LFI) is one of the two instruments onboard the Planck satellite [1] that shares the focal region of the telescope. It is an array of 54 HEMT-based pseudo-correlation receivers coupled to the Telescope by 27 dual profiled corrugated feed horns. The baseline layout of the LFI Focal Plane Unit (FPU) foresees the 27 feed horns located around the High Frequency Instrument (HFI): 16 feed horns at 100 GHz, 6 at 70 GHz, 3 at 44 GHz and 2 at 30 GHz. The LFI team has carried out an exhaustive study of the LFI optical interface devoted to optimize the angular resolution (12 arcmin at 100 GHz as a requirement, 10 arcmin at 100 GHz as a goal) and at the same time to minimize all the systematics coming from the side lobes of the radiation pattern.

OPTICAL SIMULATIONS

Different techniques can be applied to predict the radiation pattern: Geometrical Optics (GO), Geometrical Theory of Diffraction (GTD), Physical Optics (PO) and Physical Theory of Diffraction (PTD). The simulations have been performed considering each feed (Gaussian, X– axis polarized [2]) as a source and computing the pattern scattered by both reflectors on the far field. The GO/GTD methods have been used to model the sub reflector, while the main reflector has been modelized using the PO.

Angular Resolution vs Edge Taper

The Angular Resolution (expressed here in term of Full Width Half Maximum, FWHM) of the beam on the sky depends on the illumination of the primary mirror. The illumination can be represented by the Edge Taper, defined as the ratio of the power per unit area incident on the centre of the mirror to that incident on the edge. Decreasing the edge taper has a positive impact on the angular resolution but lower is the edge taper and higher is the side lobes level since they are largely due by diffraction and scattering from the reflector edges. The trade-off between the Angular Resolution (which impacts the ability to reconstruct the anisotropy power spectrum of the Cosmic Microwave Background Radiation at high multipoles [3]) and the Edge Taper (which impacts the systematics of the detected signal from receivers [4]) has been carried out for some LFI feed horns [5]. The dependence of the angular resolution improvement on the edge taper degradation is almost linear until a certain edge taper is reached, when increasing the illumination on the primary mirror doesn't involve a further angular resolution betterment. The latter because a strong illumination of the mirrors increases the aberrations on the main beam. Obviously, the amount of the improvement depends on the feed horn location, since the primary mirror is illuminated in a different way.

Field Distribution on Primary Mirror

The amplitude field incident on the main reflector has been computed for each LFI feed horn. The model of the feed we used is a X− axis polarized gaussian horn with an edge taper of 30 dB at 22° of angle [2]. The X− axis (Y− axis) of the contour plot lies on the symmetry (asymmetry) plane of the telescope. The +Z direction corresponds to the main beam peak direction and the (X,Y,Z) is a standard cartesian frame. As a consequence, the top edge of the main reflector is at $X \simeq 750$ mm and $Y \simeq 0$ mm on the contour plot coordinate system. We used the Geometrical Optics (GO) and the Geometrical Theory of Diffraction (GTD) on the sub reflector to calculate the total amplitude of the field incident on the surface of the primary mirror, in the reference system of the main beam.

The contour plots show that, as expected, the illumination of the primary mirror is roughly elliptical. As a consequence the field amplitude on the primary mirror rim is not constant (see Figure 1, left panel). The amplitudes of the field on the main reflector contour have been used to set the requirements on the edge taper values for all the LFI feed horn illuminations. The field amplitude on the mirror contour is a function of the φ angle ($E = E(\varphi)$), defined in the reference system of the main beam ($\varphi = 0$ in the direction of the top edge of the main reflector and has a counterclockwise direction).

The edge taper of each feed, at a reference angle (22° or 24°), has been chosen comparing the field amplitude, $E(\varphi)$, with that corresponding to a *worst reference* case, $\tilde{E}(\varphi)$, for which a full straylight analysis has been performed, showing acceptable contamination levels from the galactic emission [4]. The edge taper correction of each feed horn, in order to assure a straylight rejection analogous to the reference case, was calculated by computing the lowest difference between the edge taper curve of each feed and the reference edge taper curve ($min|E(\varphi) - \tilde{E}(\varphi)|$). We decrease by this difference the horn's edge taper at the reference angle and the results are reported in Table 1. In this way, for each LFI feed horn, no single point on the main reflector rim has an edge taper value lower than that of the reference case [6].

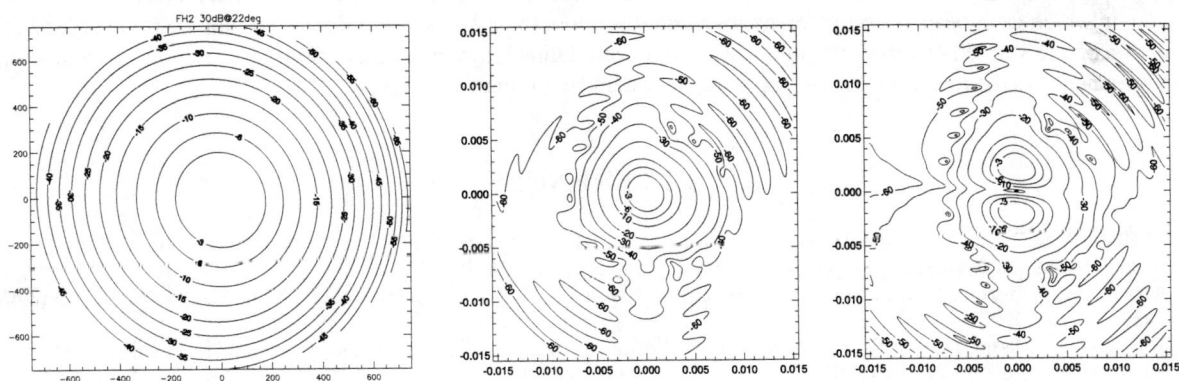

FIGURE 1. Left: field distribution on the main reflector for the feed horn 2 at 100 GHz (X− and Y− axis are in mm). Centre and right: UV plot of the copolar and crosspolar components of the feed horn 5 at 100 GHz after the polarization alignment.

Main Beam Response

The main beam power pattern has been simulated for each feed horn, with the edge taper value found out by the Edge Taper optimization procedure [6]. Edge Taper, Angular Resolution, Ellipticity and Directivity of each beam are shown in Table 1. These results (adopted as requirements for the current LFI baseline) can be considered as conservative. In fact, although the reported simulations have been done using a Gaussian model for each feed horn, preliminary studies show that dual profiled horns allow to reach a further improvement on the angular resolution. Moreover, a dedicated straylight study for each feed horn will be performed with the aim to relax the edge taper and subsequently to improve the angular resolution more.

TABLE 1. Main Beams characteristics.

Feed	Frequency (GHz)	ET (dB @ °)	FWHM (Min, arcmin)	FWHM (Max, arcmin)	FWHM (Ave, arcmin)	Ellipticity	Directivity (dBi)
27 – 28	30	30.00 - 22.00	29.16	40.68	34.92	1.40	50.47
25 – 26	44	30.00 - 22.00	25.32	31.68	28.50	1.25	52.08
24	44	30.00 - 22.00	20.04	27.84	23.94	1.39	53.77
23 – 18	70	20.50 - 22.00	12.36	15.72	14.04	1.27	58.37
22 – 19	70	23.10 - 22.00	12.60	16.08	14.34	1.28	58.23
21 – 20	70	25.00 - 22.00	12.60	16.44	14.52	1.30	58.07
9 – 10	100	25.50 - 24.00	9.48	11.88	10.68	1.25	60.59
8 – 11	100	26.80 - 24.00	10.08	12.36	11.22	1.23	60.16
7 – 12	100	27.60 - 24.00	10.56	12.84	11.70	1.22	59.79
6 – 13	100	27.70 - 24.00	11.16	13.20	12.18	1.18	59.52
5 – 14	100	27.40 - 24.00	11.40	13.20	12.30	1.16	59.34
4 – 15	100	28.10 - 24.00	11.88	13.68	12.78	1.15	59.16
3 – 16	100	28.30 - 24.00	11.88	13.80	12.84	1.16	59.07
2 – 17	100	28.10 - 24.00	12.12	13.80	12.96	1.14	59.05

Polarization properties of the beams are mainly determined by the feed position and orientation in the Focal Plane Unit. The LFI polarization properties have been optimized rotating each feed horn about its axis, in order to obtain the right orientation of the polarization direction of each beam. The polarized radiation in the far field is fully described by giving the projection of the electric field vector in two mutually orthogonal directions. Far field amplitude radiation patterns reported here are given in the Co and Cross polar basis according to the Ludwig's 3rd definition. The reference system in which each beam is computed has been rotated about their Z– axis in order to find out the main polarization direction on the sky. When this condition is reached, a well defined minimum appears in the crosspolar component, in corrispondence to the maximum of the copolar component (see Figure 1).

REFERENCES

1. PLANCK Telescope Design Specification, Tech. Rep. SCI-PT-RS-07024, ESA, ESTEC, Netherlands (2000).
2. Villa, F., Gaussian Beam for Corrugated Feed Horns, Tech. rep., ITESRE 286 (2000).
3. Mandolesi, N., Bersanelli, M., Burigana, C., Gorski, K., Hivon, E., Maino, D., Valenziano, L., Villa, F., and White, M., *A&ASS*, **145**, 323–340 (2000).
4. Burigana, C., Maino, D., Gorski, K., Mandolesi, N., Bersanelli, M., Villa, F., Valenziano, L., Wandelt, B., Maltoni, M., and Hivon, E., *A&A*, **373**, 345–358 (2001).
5. Sandri, M., Villa, F., Bersanelli, M., and Mandolesi, N., Planck/LFI Optical Simulations: on the trade-off between angular resolution and edge taper, Tech. rep., ITESRE 308 (2001).
6. Sandri, M., Villa, F., Bersanelli, M., and Mandolesi, N., Planck/LFI Optical Simulations: edge taper evaluation, Tech. rep., ITESRE 309 (2001).

Sources Variability With Planck LFI

L.Terenzi[1], M.Bersanelli[2], C.Burigana[1], R.C.Butler[1], G.De Zotti[3], N.Mandolesi[1],
D.Mennella[4], G.Morgante[1], M.Sandri[1], L.Valenziano[1], F.Villa[1]

[1]*Istituto Te.S.R.E./ CNR – Bologna – Italy,*
[2]*Università di Milano – Milano – Italy,*
[3]*Osservatorio Astronomico di Padova – Padova – Italy*
[4]*Istituto di Fisica Cosmica/ CNR – Italy*
On behalf of the Planck Collaboration

Abstract. Planck LFI (Low Frequency Instrument) will produce a complete survey of the sky at millimeter wavelenghts. Data stream analysis will provide the possibility to reveal unexpected millimeter sources and to study their flux evolution in time at different frequencies. We describe here the main implications and discuss data analysis methods. Planck sensitivities typical for this kind of detection are taken into account. We present also preliminary results of our simulation activity.

INTRODUCTION

The LFI (Low Frequency Instrument) [1] on board of the Planck satellite is designed and optimized to measure primary anisotropies in the CMB (Cosmic Microwave Background); the images that it will produce at 30, 44, 70 and 100 GHz will have an unprecedented combination of sky coverage, calibration accuracy, freedom from systematic errors, stability and sensitivity. The LFI data will represent also a good opportunity to study the astrophysics of extragalactic radiosources. In particular, LFI is expected to be efficient in the detection of extragalactic radiosources in active phases characterized by high emission levels (BL Lac, blazar). We collected informations on typical flux intensities, spectral variabilities and light curves of these objects (see, e.g., [2]) to properly exploit the experiment observation strategy, based upon a full sky coverage in periods of about six months at four frequency channels.

PLANCK LFI SENSITIVITY

Starting from the LFI sensitivities at different frequencies (Planck Low Frequency Instrument, Instrument Science Verification Review, October 1999, private reference), considering the main properties of the Planck scanning strategy [3] and the LFI beam positions on the telescope field of view [4], we are able to evaluate the averaged instrumental sensitivities of the LFI array for the study of variable sources on different time-scales and the number of relevant observations with the quoted sensitivities (Table 1; see [5] for further details). The global rms noise in the Planck data streams is the sum in quadrature of all the relevant contributions, assumed independent. The CMB, Galaxy and extragalactic source confusion noises per beam ($FWHM^2$) vary respectively from about 250, 100 and 60mJy to 190, 7 and 20mJy, when the frequency goes from 30 to 100 GHz.

Therefore, in the final recovery of radiosource flux variability we can take advantage from the knowledge of diffuse component fluctuations at the highest and middle HFI frequency channels (at high resolution) since the CMB dominates over the astrophysical confusion noise. The global noise, relevant in this context, is then dominated by the LFI data stream receiver noise.

The Planck sensitivity depends also on the sky position, mainly on the ecliptic colatitude θ_e; the baseline scanning strategy implies a sensitivity, expressed in terms of sensitivity averaged on the sky, approximated by the law $(\sin\theta_e/\sin 50°)^{1/2}$; sources at high ecliptic latitudes are observed much longer than those at low latitudes.

TABLE 1. Instrumental Performances And Number Of Measurement For Typical Time-Scales.

Period	>14 Months (Aux. Data)	14 Months	1-6 days/sin50°	Few-12 hours/sin50°
σ_{noise} (mJy) at 30 GHz	13.4	19.0	--	--
N_{meas} at 30 GHz	1	2 (3)	--	--
σ_{noise} (mJy) at 44 GHz	20.5	29.0	35.7 - 50.0	--
N_{meas} at 44 GHz	1	2 (3)	2	--
σ_{noise} (mJy) at 70 GHz	28.0	39.6	--	68.6
N_{meas} at 70 GHz	1	2 (3)	--	3
σ_{noise} (mJy) at 100 GHz	32.2	45.5	71.0 – 132.9	108.5 – 132.9
N_{meas} at 100 GHz	1	2 (3)	6 - 7	2 - 3

SIMULATIONS

We have simulated the LFI observations of a number of sources located at different positions in the sky, in order to evaluate the impact of different source observation durations.

We have considered a representative set of beam positions among those recently simulated for the current baseline [4] corresponding to the LFI feed horns 27 (at 30 GHz), 25 (at 44 GHz), 21 (at 70 GHz), and to the feed horns 2, 6 and 9 (at 100GHz).

Time ordered data (TODs), expressed in term of simulated antenna temperature, have been generated (see Fig. 1) for sources with different fluxes (1, 3 and 5 Jy). The relationship between the source flux and the observed antenna temperature is used to translate the simulated antenna temperature TODs into source flux evaluations.

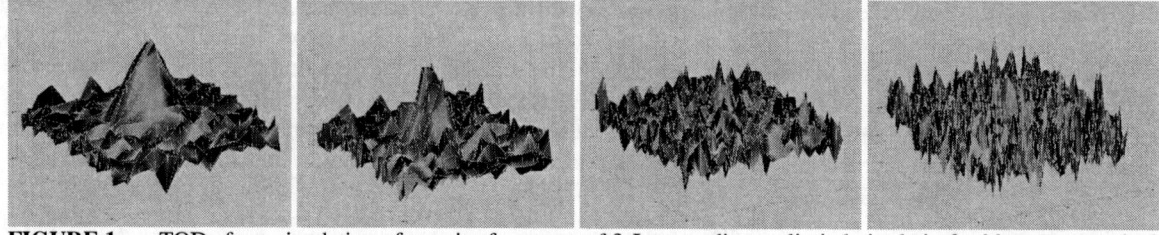

FIGURE 1. TODs from simulation of transit of a source of 3 Jy at medium ecliptic latitude in feed horns respectively at 30, 44, 70 and 100 GHz.

Next, we analyze the instrument efficiency for a preliminary source flux reconstruction. As evident, the 100 GHz receivers are too noisy to allow a satisfactory flux reconstruction by using a single beam, so in subsequent analyses we have considered the combination of six beams (12 receivers).

Of course, starting from TODs, this calculation brings out to ugly flux fluctuations when the beam center points to regions relatively far from the sources. Therefore, in order to efficiently reconstruct the source flux we have to optimize the choice of the region extent around the source in a way more or less strict according to the source luminosity and the channel sensitivity.

We firstly assume a good knowledge of the beam pattern and neglect possible pointing errors.

Estimates of the optimized number of samples to properly reconstruct the source flux at low and high ecliptic latitudes are shown in Tables 2 and 3 for fluxes of 1 and 5 Jy. Along the scan circle direction, we find that it is advantageous to work with three samples around the source; in that direction the sampling interval is equivalent to 1/3 of the FWHM, so we find advantageous to use an angular scale of about one FWHM. $\Delta\phi_{eff}$ reported in the table is the optimal interval considered along the spin axis re-pointing direction multiplied by $\sin\theta_e$; it ranges between one and two times the beam FWHM. The accuracy in the flux recovery is also reported in the tables.

We have further analyzed the effect of the pointing uncertainty upon flux reconstruction, assuming as a reference a pointing error of 1arcmin at 1σ level. The impact on source flux reconstruction is shown in columns 2-5 in the table 4, where, in analogy with previous panels, we report relative errors in flux at low and high ecliptic latitudes (LEL and HEL). The present error estimates are based on 30 simulations.

The pointing error implies also a systematic error in the main beam in-flight recovery during the mission. Given the current estimates on the relative uncertainty in beam resolution evaluation (ΔFWHM/FWHM of about 0.007 at 30 GHz and 0.022 at 100 GHz [6,7]) introduced by such a level of pointing uncertainty, we find a relative error, $\Delta F/F$, on source flux recovery from a TOD sample of about 0.007 (0.028) for samples at $\theta \cong 1\sigma_{beam}$ ($2\sigma_{beam}$) from the source for the 30 GHz channel, and 0.022 (0.089) at $1\sigma_{beam}$ ($2\sigma_{beam}$) for the 100 GHz channel. Note as, in general, these effects induced by pointing uncertainty are relevant for the highest frequency channels because of their better resolution and become comparable to the noise sensitivity in the case of bright sources. Therefore, a pointing accuracy significantly better than $\cong$ 1arcmin at 1σ level, at least by a factor 2 as in the current LFI requirements [6], is extremely important not only for the LFI cosmological aim but also for accurate radiosource variability studies.

TABLE 2. Sampling Width and Accuracy in Flux Recovery of a Source at Low Ecliptic Latitude.

Frequency (GHz)	$\frac{\Delta F}{F}$ (1 Jy)	$\frac{\Delta \phi_{eff}}{FWHM}$	$\frac{\Delta F}{F}$ (5 Jy)	$\frac{\Delta \phi_{eff}}{FWHM}$
30	0.085	0.7	0.016	0.6 - 1.5
44	0.120	1.8	0.026	1.7
70	0.239	1.6	0.056	1.6
100	0.41	1	0.092	1
100 (6 beams)	0.131	1.2	0.033	2

TABLE 3. Sampling Width and Accuracy in Flux Recovery of a Source at high Ecliptic Latitude.

Frequency (GHz)	$\frac{\Delta F}{F}$ (1 Jy)	$\frac{\Delta \phi_{eff}}{FWHM}$	$\frac{\Delta F}{F}$ (5 Jy)	$\frac{\Delta \phi_{eff}}{FWHM}$
30	0.051	1.5 - 1.8	0.011	1.6
44	0.083	1	0.017	0.9 – 1.3
70	0.193	1.1	0.035	1.2
100	0.243	1	0.058	1
100 (6 beams)	0.090	0.8	0.021	1.8

TABLE 4. Effect of Pointing uncertainty on the Accuracy in Flux Recovery.

Frequency (GHz)	$\frac{\Delta F}{F}$ Rms (LEL)	$\frac{\Delta F}{F}$ Half Disp. (LEL)	$\frac{\Delta F}{F}$ Rms (HEL)	$\frac{\Delta F}{F}$ Half Disp. (HEL)
30	0.0082	0.016	0.0046	0.0087
44	0.0146	0.0237	0.0095	0.0161
70	0.0246	0.0501	0.0198	0.0419
100	0.050	0.1073	0.0267	0.0537
100 (6 beams)	0.0176	0.0417	0.0117	0.02407

REFERENCES

1. Mandolesi, N., et al., Planck LFI, A Proposal Submitted to the ESA, (1998)
2. Terasranta, H., Tornikoski, M., Mujunen, A., et al. *Astron. Astrophys. Suppl. Ser.* **132**, 305-331 (1998)
3. Bersanelli, M., et al., ESA, COBRAS/SAMBA Report on the Phase A Study, D/SCI(96)3 (1996)
4. Sandri, M., Bersanelli, M, Burigana, C., Butler, R. C., Malaspina, M., Mandolesi, N., Mennella, A., Morgante, G., Terenzi, L., Valenziano, L., Villa, F., these proceedings
5. Burigana, C., Planck/LFI: Sensitivity Estimates for the Detection and Study of Variable and Moving Sources, Int. Rep. TeSRE/CNR 298/2000, October (2000)
6. Burigana, C., Butler, R.C. and Mandolesi, N., Planck LFI Pointing Accuracy Requirements, PL-LFI-PST-TN-023 (2001)
7. Burigana, C., Natoli, P., Vittorio, N., Mandolesi, N., and Bersanelli, M., astro-ph / 0012273

2K1BC Workshop
Experimental Cosmology at millimetre wavelengths

DETECTORS, OPTICS AND CRYOGENICS

Bolometers for Millimeter-wave Cosmology

James J. Bock

Jet Propulsion Laboratory
M/S 169-327
4800 Oak Grove Dr.
Pasadena, CA 91107

Abstract. Bolometers offer high sensitivity for observations of the cosmic microwave background, Sunyaev-Zel'Dovich effect in clusters, and far-infrared galaxies. Near background-limited performance may be realized even under the low background conditions available from a space-borne platform. We discuss the achieved performance of silicon nitride micromesh ('spider web') bolometers readout by NTD Ge thermistors. We are developing arrays of such bolometers coupled to single-mode feedhorns. CMB polarization may be studies using a new absorber geometry allowing simultaneous detection of both linear polarizations in a single feedhorn with two individual detectors. Finally we discuss a new bolometer architecture consisting of an array of slot antennae coupled to filters and bolometers via superconducting microstrip.

INTRODUCTION

The decade of the spectrum from 100 GHz to 1 THz is rich in scientific content, yet remains one of the least explored. This region contains the bulk of the energy in the Cosmic Microwave Background (CMB), much of the Cosmic Infrared Background (CIB), believed to represent the integrated emission from ultra-luminous infrared galaxies at high redshifts, and virtually all of the observable emission from cool (T<15K) clouds in our own galaxy. It is in this range of wavelengths that significant advances in space astrophysics will best be realized: making definitive maps of the temperature and polarization anisotropy of the CMB, understanding the spectrum, spatial distribution and origins of the CIB, and unveiling the epoch in which the first stars and galaxies in the universe formed.

The sensitivity of current bolometers allows for background-limited performance for photometry at millimeter- to far-infrared wavelengths. At millimeter-wavelengths, the brightness of the cosmic microwave background dominates the sky brightness and sets a photon noise level at the detector of $\sim 1 \times 10^{-17}$ W/$\sqrt{\text{Hz}}$ for broadband photometry, a sensitivity now achieved routinely with bolometers operating at temperatures $T \leq 300$ mK. Instruments with a moderate number of background-limited bolometers are intended for the next generation of millimeter- and sub-millimeter space-borne observatories (Planck/HFI with 48 bolometers, and Herschel/SPIRE with 326 bolometers). Future technical challenges are developing large-format detector arrays for imaging applications, and improving detector sensitivity for spectroscopy.

Advances in detector technology are likely to remain vital for cosmology in the post-Planck/Herschel epoch. For example, it has recently been determined [1] that the careful study of CMB structure, in particular degree angular scale *polarized* sky structure, can be used to detect not just potential perturbations but also gravitational waves. Unlike photons, these gravitational waves travel freely through the ionized early universe, so the study of CMB polarization may allow us to look much further back in time - and at much higher particle energies - than has been possible before. However, the sensitivity needed for these observations exceeds that attainable with the Planck satellite, and will require a significant advance in focal plane format to 10^3 to 10^4 background-limited pixels. If a new instrument can be built with sufficient sensitivity to polarization, the resulting observations have the potential to allow the study of the physics that dominated the universe $\sim 10^{-33}$ seconds after the Big Bang!

MICROMESH BOLOMETERS

Bolometers consist of a thermally isolated radiation absorber coupled to a sensitive thermometer. The sensitivity of these detectors are limited by thermal fluctuations in the thermal link to the cold plate, NEP = $\sqrt{4kT^2G}$, where G is the thermal conductance of the link. In practice, the thermal conductance must be tailored to optical power absorbed by the bolometer. The speed of response, set by the heat capacity C and steepness of the thermistor $\alpha = d\ln(R)/d\ln(T)$, must be fast enough to respond to the fastest signal set by the observing scheme of the instrument.

We have developed single-element bolometers made from a micromachined, free-standing silicon nitride membrane. This structure is ideal for detection of millimeter-wave radiation as it minimizes the large absorber volume. The membrane consists of an inner absorber region metalized with a thin Ti/Au layer providing optimal coupling to electro-magnetic radiation, and radial support legs providing mechanical support and thermal isolation. The thermal rise of the absorber from radiation is detected by a semi-conducting Neutron Transmutation Doped (NTD) Ge thermistor attached to the absorber via In bump bonds and readout with lithographed electrical leads. The thermal conductance of the device is tailored by adjusting the thickness of the leads metalization.

TABLE 1. Properties of Micromesh Bolometers

T [K]	G [pW/K]	G(min) [pW/K]	C [pW/K]	1/f v_c [Hz]	NEP [1e-17 W/$\sqrt{Hz}$]	NEP $\sqrt{\tau}$ [1e-18 J]
0.1	4	0.06	0.2	0.015	1.3	0.5
0.3	10	2	0.7	0.03	0.2	2

Notes: 1. G, C, 1/f knee frequency, NEP, and NEP $\sqrt{\tau}$ refer to minimum achieved values to date.
2. G(min) is estimated minimum G allowed by silicon nitride support beams.

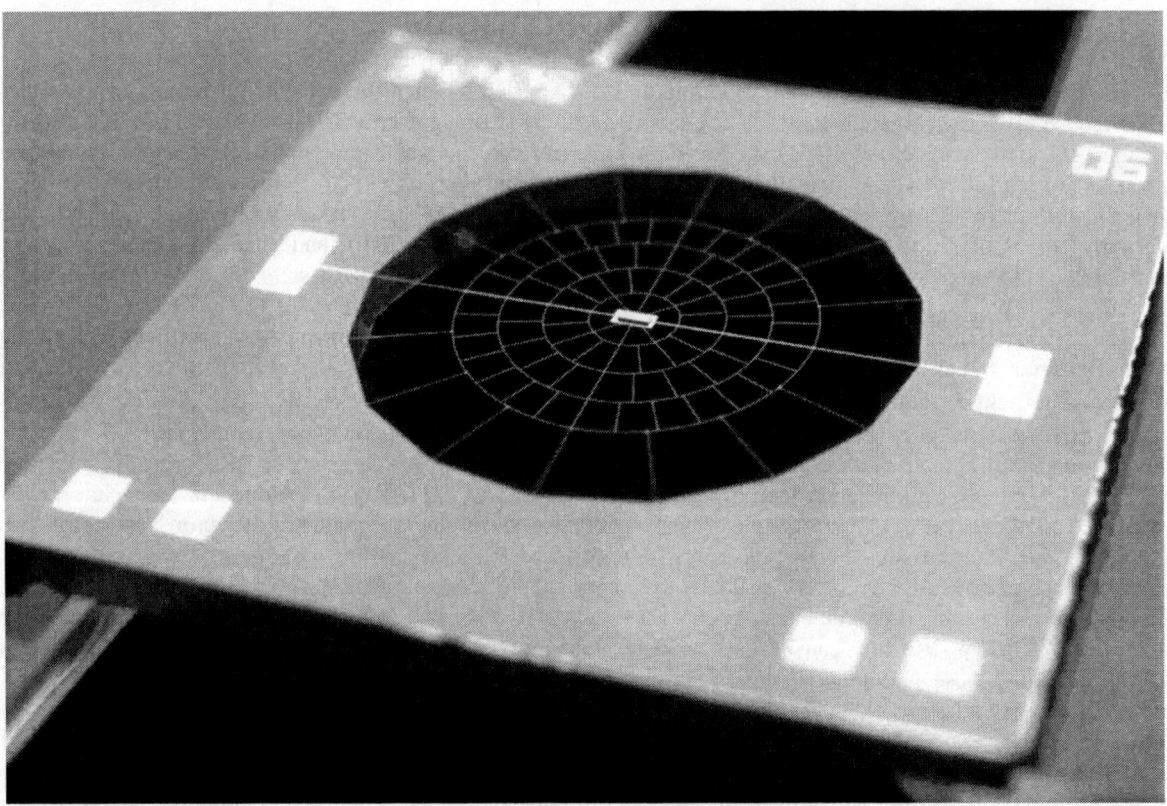

FIGURE 1. Silicon nitride micromesh bolometer developed for the 143 GHz channel of the Planck High Frequency Instrument (HFI). This device has a 3.4 mm diameter absorber (the inner region patterned as grid) suspended by 16 radial support legs 1 mm long. The NTD Ge thermistor, located in the middle of the absorber, is readout by two lithographed electrical leads connected to the large contact pads at the left and right.

The best device parameters achieved to date with In bump-bonded NTD Ge thermistors are described in Table 1. These bolometers have been applied to millimeter-wave cosmology from balloon-borne (BOOMERANG, MAXIMA, ARCHEOPS, see these proceedings) and ground-based (ACBAR) experiments. NTD Ge thermistors offer extreme 1/f stability often desirable in slow-scanned observations. The achieved NEP is sufficient for background-limited, single-mode broad-band photometry set by the astrophysical foreground. The minimum achievable NEP, determined by the thermal conductance of the silicon nitride support beams, allows for sensitivities approaching 1e-19 W/$\sqrt{Hz}$ at 100 mK. Although such ultra-sensitive devices are not practical for photometry, they can be applied to space-borne sub-millimeter spectroscopy. For exemple, high sensitivity bolometers coupled to a dispersive spectrometer and a cooled telescope promise orders of magnitude improvement over the planned line survey capabilities of Herschel. Unlike heterodyne receivers, direct detectors can achieve background-limited sensitivity even under these extremely low background conditions.

BOLOMETER ARRAYS

Further improvement in background-limited photometers can only be realized by implementing arrays of detectors. We are developing bolometer arrays consisting of a planar detector array of micromesh bolometers coupled to conical feedhorns, similar to arrays developed for SCUBA and MAMBO, for ground-based (BOLOCAM), balloon-borne (BLAST), and space-borne (SPIRE) cameras (see these proceedings). Feedhorn coupling maximizes the sensitivity per bolometer. In the case of background-limited detectors, a feedhorn-coupled array approaches the theoretical mapping speed achievable with Nyquist-sampled bare arrays that use 16 times more pixels and require multiplexing. Feedhorn coupling controls the spatial response of the detector, and allows for high end-to-end optical efficiency.

TABLE 2. Parameters of Micromesh Bolometer Array

Quantity	@ Q = 0 pW	@ Q = 2.4 pW	Units
Yield	0.9	-	
R_0	180 ± 44	-	Ω
Δ	42 ± 0.8	-	K
G_0 (300 mK)	55 ± 8	-	pW/K
β	1.85 ± 0.16	-	
τ	-	11.3 ± 2.4	ms
C (390 mK)	-	1.3 ± 0.3	pJ/K
S_e	5.0 ± 0.3	3.6 ± 0.3	10^8 V/W
V_n (calc)	11.7 ± 0.6	21.0 ± 1.2	nV/$\sqrt{Hz}$
V_n (meas)	14.1 ± 1.0	21.2 ± 1.2	nV/$\sqrt{Hz}$
NEP	2.9 ± 0.3	5.8 ± 0.2	W/$\sqrt{Hz}$
NEP_{tot} / NEP_{photon}	-	1.21 ± 0.03	
η	-	0.45 – 0.65	

Notes: 1. Standard deviations over the array quoted for all parameters.
2. S_e, V_n, and NEP determined at 22 mV bias (near peak responsivity).
3. V_n, NEP include 3.5 nV/$\sqrt{Hz}$ amplifier noise.

We have developed a prototype bolometer array intended to demonstrate technological readiness for future space-borne applications (e.g. the Herschel Space Observatory) [2]. The bolometer array, designed for operation at λ = 350 μm, demonstrates a dark NEP = 2.9 × 10^{-17} W/$\sqrt{Hz}$ and mean heat capacity of 1.3 pJ/K at 390 mK. The bolometer array demonstrates theoretical noise performance arising from photon, phonon and Johnson noise, with photon noise dominant under the design background conditions. We measure the ratio of total noise to photon noise to be 1.21 under an absorbed optical power of 2.4 pW. Excess noise is negligible for audio frequencies as low as 30 mHz. The optical efficiency of the feedhorn and bolometer combination (η = 0.45 – 0.65) is somewhat lower than anticipated, perhaps due machining errors in the feedhorns or due to thermal losses in the bolometer. The optical efficiency achieved in millimeter-wave feedhorn-coupled bolometers [3] is consistent with a bolometer and feedhorn efficiency greater than 0.8.

FIGURE 2. Array of silicon nitride micromesh bolometers developed for the SPIRE instrument on Herschel. The bolometer array is designed to be operated at $\lambda = 350$ μm coupled to an array of conical feedhorns.

POLARIZATION-SELECTIVE BOLOMETERS

We are developing polarization-selective bolometers for measurement of cosmic microwave background polarization anisotropy. Two detectors, each sensitive to linear polarization, are housed in corrugated waveguide at the end of a scalar feedhorn (see Fig. 3). A single Stokes parameter (Q or U) may be obtained from the difference in signal between the bolometers. The bolometers are located a distance $\lambda/4$ from the back of the waveguide for maximal optical coupling, and are separated by a small distance to minimize cross-polarization. This compact arrangement results in well-matched beams, since both detectors observe through the same feedhorn optics, maximizing the rejection of common-mode optical signals. The close proximity of the bolometers also helps reject common-mode environmental signals from the instrument (temperature drifts, microphonics, etc).

Simulation of the cavity, feed, and waveguide with HFSS indicates that the optical efficiency of the feedhorn and detector system alone may be high (> 95 %) with low-cross polarization (< 1 %). To date, we have achieved a maximum end-to-end optical efficiency, including millimeter-wave filters and an additional back-to-back feedhorn, of 20 % with a minimum cross-polarization < 5 % (W.C. Jones, California Institute of Technology, private communication). The effective sheet impedance of bolometers developed for these tests was significantly lower than optimal, resulting in high reflection, and further improvement in the optical efficiency can thus be expected. Polarization-selective bolometers are planned QUEST and Planck/HFI (see these proceedings).

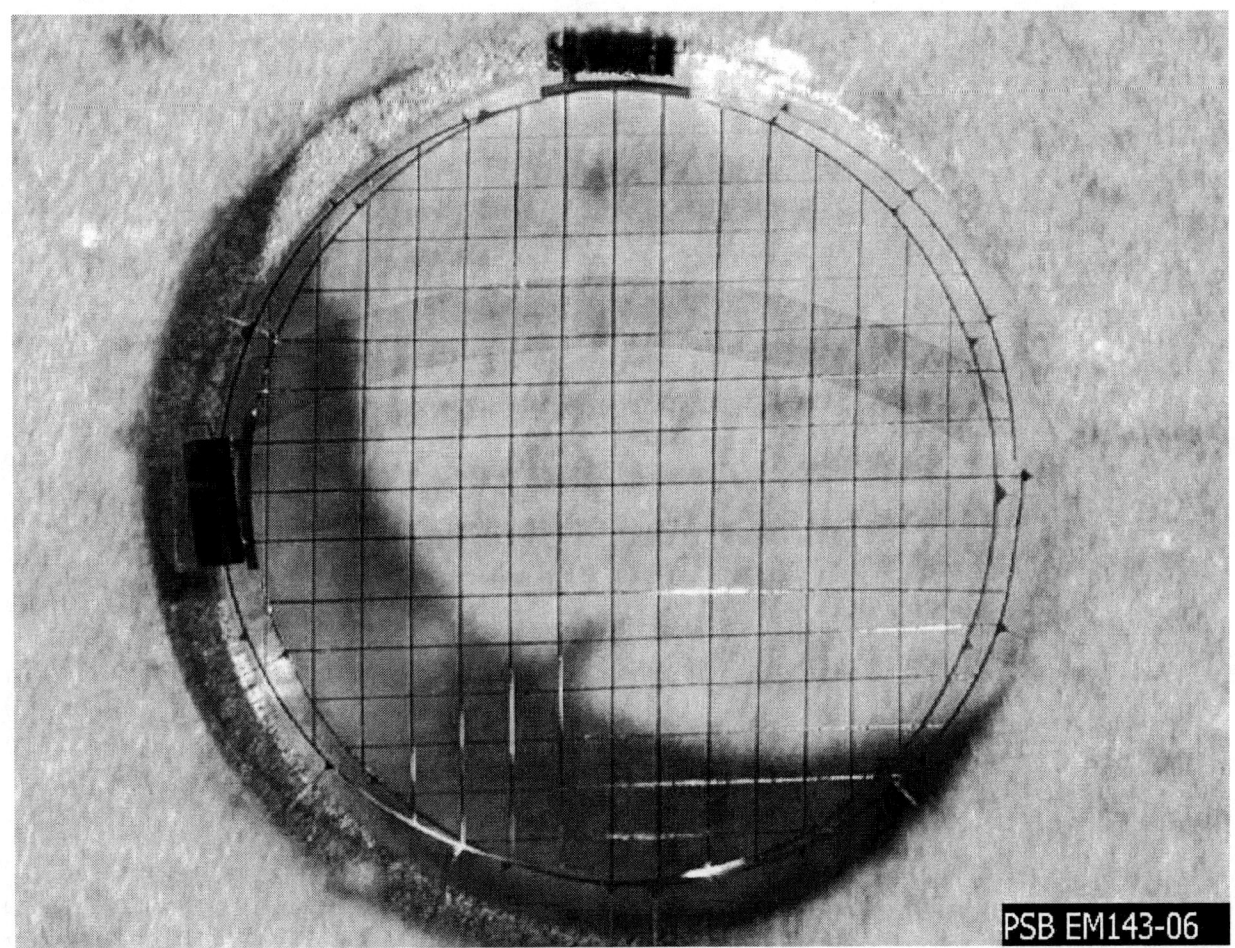

FIGURE 3. Pair of polarization-selective bolometers placed in corrugated waveguide, coupled to a single-mode corrugated feed horn. Each bolometer has a linear absorbing grid designed to couple to a linear polarization. The thermistors are located at the end of the absorber. The radial mechanical support legs and silicon frame are obscured by the waveguide.

ANTENNA-COUPLED BOLOMETER ARRAYS

Further advances in the sensitivity of millimeter-wave focal planes must come from increases in the array format. At millimeter-wavelengths, large format bolometer arrays not only require multiplexing at the sub-K stage, but a significant advance in focal plane architecture. In practical applications, millimeter-wave bolometers require beam-collimating optics to control their illumination on warm optics. Cryogenic re-imaging optical systems can be used to define the illumination pattern of infrared detectors. However, in the millimeter it is difficult to cool the necessarily large optics below 1-2 K. Since the brightness of the cold optics are similar to the 2.75 K sky brightness, only a modest fraction of the detector throughput can be allowed to couple to black 2 K surfaces. Thus at millimeter wavelengths, bare absorbing pixels cannot be used in low background applications without filtering and collimating optics at $T < 1$ K. While feedhorns provide collimation in current instruments, they can only used with modest numbers of pixels as feedhorns create a significant mass penalty at $0.1 - 0.3$ K, preventing practical implementation of large-format focal plane arrays. In the case of Planck, the HFI focal plane fills much of the available useful field at $\nu > 100$ GHz and approaches the maximum practical mass at 100 mK for a space application. A significant improvement in sensitivity over Planck will be needed to map the polarization of cosmic microwave background anisotropy, yet clearly the HFI focal plane cannot be scaled up to 10^3 to 10^4 pixels. A post-Planck mission requires not only multiplexing, but a compact, low-mass focal plane that provides beam collimation without feedhorn optics.

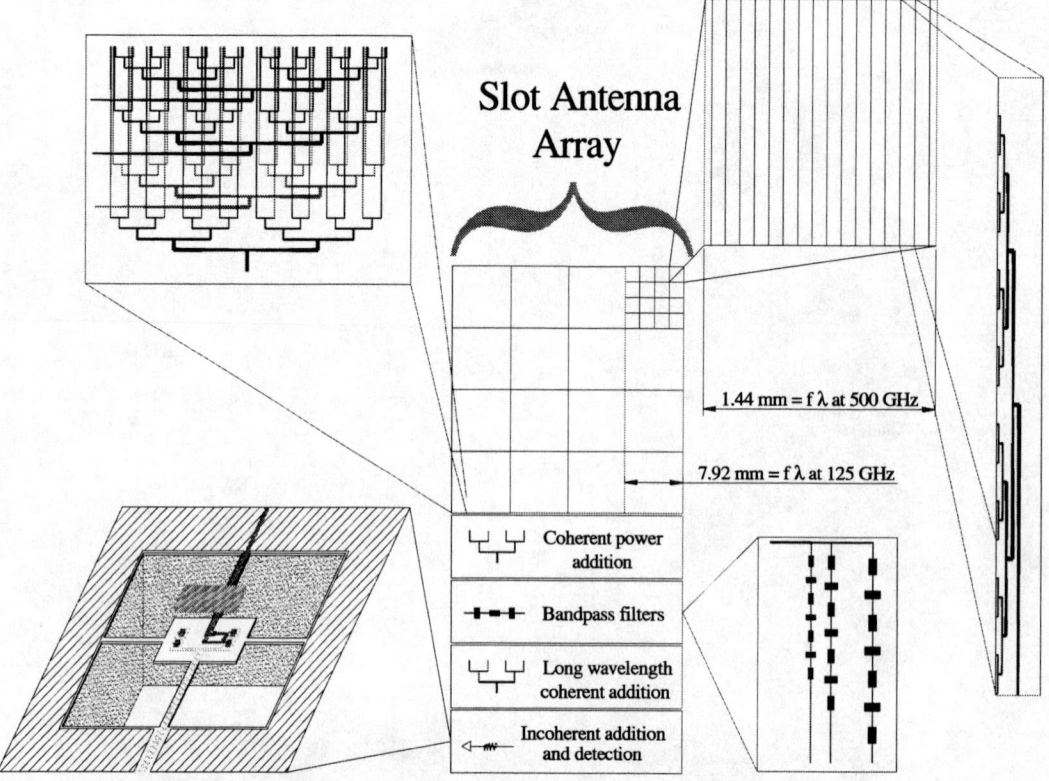

FIGURE 4. Focal plane architecture designed for polarimetry of the Cosmic Microwave Background (CMB) at the focus of an f/3.3 optical system. The focal plane contains 3 overlapping, close-packed arrays of synthesized, diffraction-limited pixels: a 4x4 array at 125 GHz, an 8x8 array at 250 GHz and a 16x16 array at 500 GHz. There are thus a total of 336 detectors coupled to an active area of 32x32 mm. A densely packed (100 μm slot spacing) array of slot antennas in Nb ground plane deposited on a silicon wafer gives an antenna impedance of ~20 Ω. Microstrip taps cross the slot antennas at 100 micron intervals and are coherently summed, first along columns (inset, upper right), and then along rows (inset, upper left) to a single transmission line per diffraction-limited 500 GHz pixel. The microstrip is adiabatically tapered after each summation to maintain an impedance of 20 Ω. The microstrip lines pass through a bank of stripline filters, are coherently summed for the low frequency pixels, and terminate in a detector (inset, lower left).

We propose a new architecture for millimeter-wave focal plane arrays of bolometers coupled to antennae and filters via low-loss superconducting Nb microstrip. We are developing a concept for a polarization-sensitive bolometer array, shown in Fig. 1, that uses bolometers coupled to slot antenna array with superconducting Nb microstrip.

Microstrip-Coupled Bolometres

Microstrip-coupling provides several important advantages over coupling via a distributed absorber. Antennas may be coupled to a bolometer by means of superconducting microstrip terminated in a resistor. Unlike a distributed absorber, which requires an active area at least as large as λ^2 in order to couple efficiently, the microstrip termination resistor may be as small as lithographic techniques permit. The termination resistor does not couple easily to stray radiation. Future space-borne applications in which a cooled aperture and/or narrow spectral bandwidth will reduce the background loads to fWs, stray radiation from 2 K alone (~1 pW/mm^2) makes use of bolometers with radiation absorbers intractable.

Antenna-coupled bolometers have a field-of-view defined by the antenna. In contrast, bolometers with radiation absorbers require single- or multi-mode feeds to define the field of view, or are subject to an unrestricted view of stray light from 2π sr. Feedhorns typically dominate the suspended mass and volume of the focal plane. Antenna-coupled bolometers may take advantage of lithographed stripline filters. Stripline filters provide high transmission and excellent out-of-band rejection, and are a well-developed technology. They replace the much larger and more massive metal-mesh filters required by bolometers with radiation absorbers. The high frequency blocking requirements of mm-wave bolometers are quite severe. Antennas and microstrip do not efficiently propagate high frequency radiation.

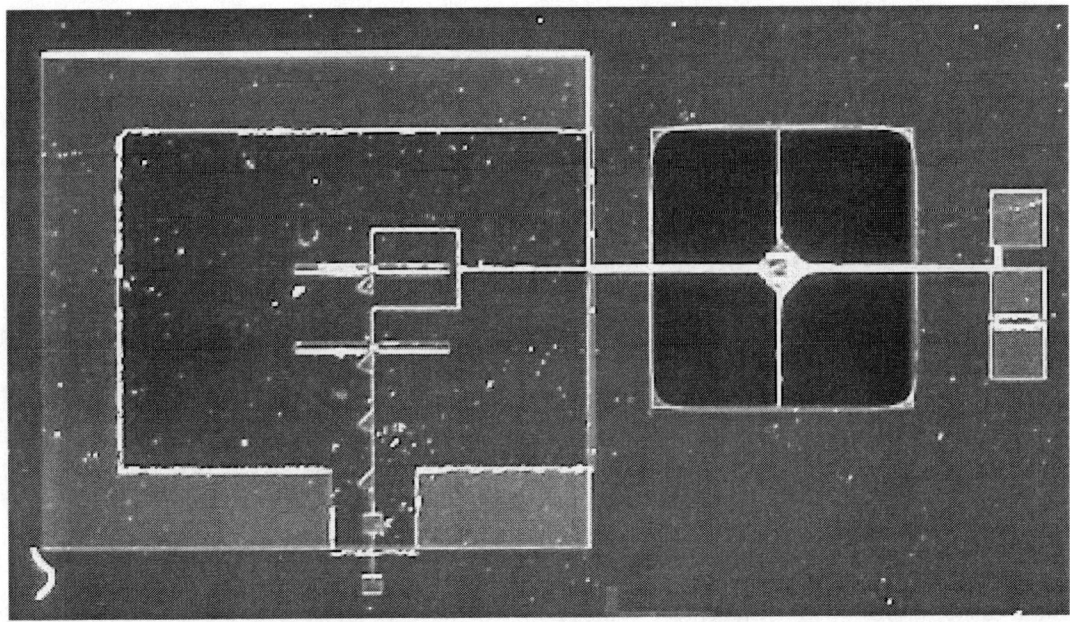

FIGURE 5. Single-element antenna-coupled TES bolometer, currently under development, consists of a dual slot antenna (at left) coupled to superconducting Nb microstrip. The microstrip passes over a suspended beam of silicon nitride to a termination resistor located on the thermally isolated diamond-shaped region. An Al/Ti/Au transition-edge superconductor, located near the termination resistor, is readout via superconducting Nb leads to contact pads located on the right-hand side.

Antenna-coupled bolometers have active areas that can be orders of magnitude smaller still than a micromesh bolometer. Our initial tests of stripline-coupled bolometers will use a normal metal resistor to terminate the stripline, and a separate trilayer TES with Nb leads for readout (see Fig. 5). This simple architecture, ideal for testing, can realize the sensitivity required for background-limited operation at 300 mK. However, further improvement in detector sensitivity, possible due to the small active areas allowed by stripline coupling, could later be realized by any of a number of microstrip-coupled detectors (e.g. SQPC, HEB, kinetic inductance, e-phonon decoupled TES, etc).

Microstrip Filters and Antennas

We are developing single antennae and filters, sub-elements in the larger architecture shown in Fig. 4. We have designed a two antenna arrangements for testing with tunnel junction detectors. The first architecture consists of long slots with periodic taps to Nb microstrip. The microstrip are summed in phase with adiabatic tapers onto a single transmission line in a binary tree. Because the microstrip are added in phase, and the slots are continuous, the intrinsic bandwidth of the antenna is large. The common transmission line can then be split with a diplexer to allow simultaneous imaging in several broad photometric bands. The second geometry uses an array of modified dual-polarization, dual-slot antennae. Each polarization is summed onto a common transmission line. Because the slots are necessarily discontinuous due to the desire to receive both polarizations simultaneously, this antenna can only be used in a single photometric band. These unit antennae can later be combined into a large-format bolometer array with directed beams and high density.

We are developing broad-band filters in Nb microstrip. The filter design we have developed is a lumped, 3-pole filter that provides high out-of-band impedance. This provides high attenuation of out-of-band signals. The high

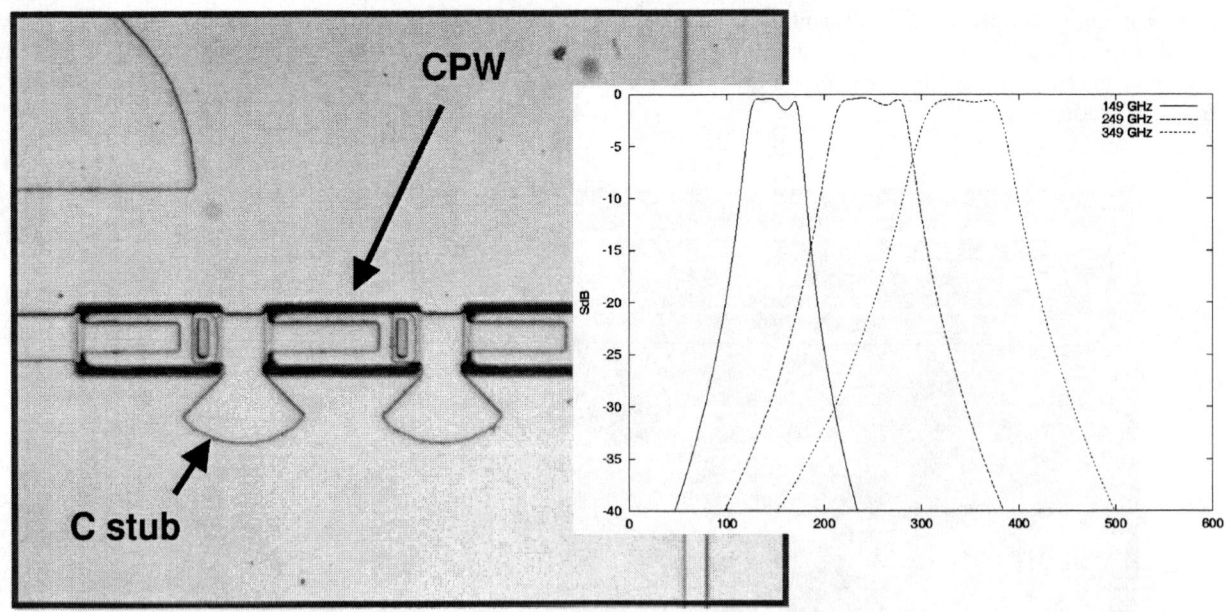

FIGURE 6. A three-pole capacitive inductive filter fabricated in Nb microstrip for operation at 150 GHz. The calculated transmittance of three filters designed to work in a diplexer is shown in the right-hand panel.

out-band-impedance also allows the filters to be combined into a diplexer which splits a single transmission line into several spectral bands. More information on the design of the filters and antennae may be found in Goldin *et al*. [4].

The key to the successful implementation of microstrip-coupled bolometers is the ability to transport signals across macroscopic distances via superconducting microstrip transmission line. Loss measurements on niobium superconducting films with an SiO dielectric layer [5] indicate the Nb microstrip has sufficiently low loss for the focal plane architecture in Fig. 4.

CONCLUSIONS

Silicon nitride micromesh bolometers demonstrate sensitivities required for background-limited imaging of the csomic microwave background in ongoing instrumentation efforts from balloon-borne and ground-based receivers. Arrays of micromesh bolometers are planned in the next generation of millimeter- and sub-millimeter space-borne observatories. Further improvements require large-format detector arrays with integral filtering and beam direction. Microstrip-coupled focal plane structures greatly reduce the size, mass, cooling requirements, risk, and cost of mm- and sub-mm focal planes for future space-borne astrophysics.

ACKNOWLEDGMENTS

The author would like to acknowledge support from NASA grant NAG5-10317 and from the JPL DRDF.

REFERENCES

1. Caldwell, R. R. *et al*., 1998, Phys. Rev. D 5902, 7101.
2. Turner, A. D. *et al*., 2001, *Applied Optics*, in press.
3. S.E. Church *et al*., "A compact high-efficiency feed structure for cosmic microwave background astronomy at millimeter wavelengths," Proc. 30th ESLAB Symp., 'Submillimetre and Far-Infrared Space Instrumentation,' ESA SP-388 (1996).
4. Goldin, A. *et al*., 2001, LTD-9 conference proceedings, in press.
5. Vayonakis, A. *et al*., 2001, LTD-9 conference proceedings, in press.

New Technologies for the Detection of Millimeter and Submillimeter Waves

P.L. Richards,[1,2,4] J. Clarke,[1,2] J.M. Gildemeister,[1,2] T. Lanting,[4] A.T. Lee,[1,3]
M.J. Myers,[1,4] D. Schwan,[4] J.T. Skidmore,[1,4] H.G. Spieler,[3]
and Jongsoo Yoon[1,2]

[1]*Physics Department, University of California, Berkeley, CA 94720-7300, U.S.A.*
[2]*Materials Sciences Division, Lawrence Berkeley National Laboratory, Berkeley, CA 94720, U.S.A.*
[3]*Physics Division, Lawrence Berkeley National Laboratory, Berkeley, CA 94720, U.S.A.*
[4]*Space Sciences Laboratory, University of California, Berkeley, CA 94720-7450, U.S.A.*

Abstract. Voltage-biased superconducting bolometers have many operational advantages over conventional bolometer technology including sensitivity, linearity, speed, and immunity from environmental disturbance. A review is given of the Berkeley program for developing this new technology. Developments include fully lithographed individual bolometers in the spiderweb configuration, arrays of 1024 close-packed absorber-coupled bolometers, antenna-coupled bolometers, and a frequency-domain SQUID readout multiplexer.

INTRODUCTION

The current generation of CMB experiments are done with two very different types of receiver. High electron mobility transistor (HEMT) amplifiers are phase-conserving linear photon amplifiers whose sensitivity is limited by quantum noise. Bolometers, by contrast, are square-law detectors which do not preserve phase so have no such sensitivity limit. In principle, bolometers can reach the sensitivity limit set by photon statistics. The best photometers using current spider web bolometers, with germanium thermistors and JFET amplifiers operate close to this limit for the photon rates observed near the peak of the CMB spectrum. In spite of their limited sensitivity, however, HEMT receivers are used for many current CMB temperature anisotropy experiments including the interferometers CBI, DASI, and VSI, and the MAP and PLANCK-LFI orbital missions. Part of the reasons for this preference is that the operating characteristics of HEMT amplifiers are generally superior to those of conventional bolometers. These characteristics include dynamic range, linearity, dependence of the responsivity on cryostat temperature and infrared power loading, speed, sensitivity to rf interference, required operating temperature, reproducibility, etc. These operating limitations make bolometric systems very difficult to optimize and often limit the performance achieved. Both HEMT and bolometer technologies are currently limited by complexity to relatively small arrays. The CMB interferometers have up to 14 receivers, bolometric systems use 16-100 bolometers.

A number of cosmological experiments now under consideration will require much deeper or more rapid mapping of the sky than is now possible. These include proposed measurements of the E and B-mode polarization anisotropy of the CMB, statistical studies of the Sunyaev-Zel'dovich effect and studies of the sources of the far infrared background. For these experiments we need large arrays of 10^3-10^4 bolometers with improved operating characteristics.

THE VOLTAGE-BIASED SUPERCONDUCTING BOLOMETER

A new bolometric technology is now being developed which promises to supply the required performance in large format arrays. The voltage-biased superconducting bolometer (VSB) [1,2] uses a superconducting thin film transition-edge temperature sensor (TES). The voltage bias creates a strong negative electrothermal feedback that keeps the bolometer operating temperature constant despite changes in cryostat temperature or infrared power loading. The VSB has a number of advantages over conventional bolometer technology. The feedback increases the bolometer speed and the linearity, and reduces the sensitivity to environmental factors.

The Johnson noise is suppressed. The bolometers operate at low impedance, where microphonic effects are reduced, and use superconducting quantum interference (SQUID) amplifiers. These amplifiers dissipate negligible power, operate at low temperatures and have very low noise. Perhaps most important, the VSB is produced entirely by thin film deposition and optical lithography, and so can be fabricated in large format arrays. Also, the large noise margin of the SQUID amplifiers makes it possible to build multiplexers which can read out many bolometers through a single amplifier.

Status of VSB Development

Because of the attractive properties of the VSB, a number of groups are actively pursuing development. Much of the work in the open literature comes from the Berkeley group which will be featured here. More detailed descriptions of the Berkeley devices including illustrations can be found at http://bolo.berkeley.edu/. Leg-isolated bolometers with round (spiderweb) mesh absorbers have been successfully produced by thin film deposition and optical lithography [3]. As with conventional spiderweb bolometers, these devices will be used with undersampled horn-coupled arrays. Sources of excess noise have been explored and techniques to eliminate them have been found [4]. Excellent low frequency (1/f) noise performance has been demonstrated without ac bias, square format mesh absorber bolometers with folded legs suitable for compact arrays have been fabricated and tested [5]. Mechanical structures (legs and mesh absorbers) have been fabricated with high yield in a 1024 bolometer close-packed array format [5]. These arrays will be used to obtain Nyquist sampling for maximum mapping speed. A primitive hot electron (electron-phonon isolated) VSB has been successfully tested [1]. The Berkeley group (and others) are investigating antenna coupled VSB's. The antenna output can be coupled to the bolometer with a superconducting transmission line which can have very low loss at CMB frequencies and bolometer operating temperatures. The advantages of this approach are that the antennas can be polarization sensitive and that the antenna output can be split into a number of bands before detection by transmission line RF multiplexers and bandpass filters. This is particularly convenient for CMB polarization measurements where simultaneous measurements at several frequencies and two polarizations are required.

Output Multiplexers

Large format arrays of 10^3 or more bolometers will be enabled by output multiplexing. Present technology requires one amplifier in the cryostat for each bolometer. Each amplifier requires ~5 wires. An array of 10^3 bolometers would then require 5×10^3 wires entering the cryostat. The technologies used to detect electromagnetic waves at infrared, optical and higher frequencies rely on output multiplexing schemes such as the CCD. The great scientific potential of the millimeter/submillimeter band will not be realized without an analogous output multiplexer scheme.

The group at the National Institute of Standards and Technology (NIST) was the first to address this critical need. The NIST group chose an approach in which SQUID's are used to switch the output of a row of detectors sequentially to a single amplifier [6]. The number of bolometers which can be read out through a single amplifier without increase in noise is given by the square of the ratio of the bolometer noise to the SQUID noise. The Berkeley group is exploring a frequency-domain multiplexer which ac biases each detector in a row with a different frequency and separates the signals with lock-in amplifiers at the SQUID output [7]. In this case, the limitation depends on the ratio of the slew rate of the SQUID to the flux noise [8]. A test chip has been fabricated with Nb thin-film lithography and signals from eight simulated bolometers have been successfully multiplexed and de-multiplexed with low crosstalk and noise. A design optimization has been completed and a more capable chip fabricated for tests with real bolometers. It is not clear at this stage whether time-domain or frequency-domain multiplexing will be best. Both appear to be capable of providing the output multiplexing that is absolutely essential for successful large format bolometer arrays.

ACKNOWLEDGMENTS

The authors are indebted to X. Meng for technical support in device fabrication. All devices were fabricated in the Berkeley Microfabrication Laboratory.

This research was supported by the National Science Foundation, NASA, and by the Director, Office of Science, Office of Basic Energy Sciences of the U. S. Department of Energy under Contract No. DE-AC03-76SF00098.

REFERENCES

1. Lee, A.T. et al., A Superconducting Bolometer With Strong Electrothermal Feedback, Appl. Phys. Lett. **69**, 1801 (1996).
2. Lee, Shih-Fu et al., A Voltage-Biased Superconducting Transition Edge Bolometer with Strong Electro Thermal Feedback Operated at 370 mK, Applied Optics **37**, 3391 (1998).
3. Gildemeister, J.M., Lee, A.T., and Richards, P.L., A Fully Lithographed Voltage-Biased Superconducting Spiderweb Bolometer, Appl. Phys. Lett. **74**, 868 (1999).
4. Gildemeister, J.M., Lee, A.T., and Richards, P.L., A Model for Excess Noise in Voltage-Biased Superconducting Bolometers, Appl. Optics Lett. (in press).
5. Gildemeister, J.M., Lee, A.T., and Richards, P.L., Monolithic Arrays of Absorber-Coupled Voltage-Biased Superconducting Bolometers, Appl. Phys. Lett. **77**, 4040 (2000).
6. Chervenak, J.A., et al., Superconducting Multiplexer for Arrays of Transition Edge Sensors, Appl. Phys. Lett. **74**, 4043-4045 (1999).
7. Yoon, J., et al, Single Superconducting Quantum Interference Device Multiplexer for Arrays of Low-Temperature Sensors, Appl. Phys. Lett. **78**, 371 (2001).
8. Yoon, J., et al, Single SQUID Multiplexer for Arrays of Voltage-Biased Superconducting Bolometers, presented at 9th International Workshop on Low Temperature Detectors, LTD-9, Madison Wisconsin, 23-37 July 2001.

Bolometer Arrays For Mm/Submm Astronomy

E. Kreysa, H.-P. Gemünd, A. Raccanelli, L.A. Reichertz, and G. Siringo

Max-Planck-Institut für Radioastronomie, Bonn, Germany

Abstract. Arrays consisting of large numbers of sensitive bolometers have become powerful tools for Mm/Submm Astronomy. On large ground-based telescopes for example they were essential in the discovery of a population of faint, highly redshifted point sources which provide important clues to the star-formation history of the universe. The Bolometer group at the Max-Planck-Institut für Radioastronomie has been developing bolometer arrays since 1980. This paper is meant to give an overview of the state and future of this effort.

INTRODUCTION

The Bolometer group at The Max-Planck-Institut für Radioastronomie (MPfIR) in Bonn has an active program of developing arrays of bolometers for ground-based Mm/Submm Astronomy. The purpose of this contribution is to present those arrays that are currently in operation at different telescopes and look at future prospects, especially also at new facilities. Due to space limitations, only especially interesting features of the arrays will be discussed here, as more complete descriptions will be the subject of future publications.

MAMBO

MAMBO stands for Max-Planck Millimeter Bolometer Array, and refers an array of 37 bolometers operating at 300 mK with an effective wavelength of 1.2 mm. It has operated since 1997 at the 30 m Millimetre Radio Telescope (MRT) of IRAM (Institut für Radioastronomie im Millimeterbereich), situated on Pico Veleta, Sierra Nevada, in Spain. MAMBO has been used by a large user community with good success. Its sensitivity at the 30m MRT is strongly weather dependent, but can reach 20 mJy Hz-1/2, which is within a factor of two of the thermal background limit.

For our group at MPIfR, a particularly interesting astronomical result of MAMBO has been the discovery of highly redshifted point sources in surveys of "empty" fields. These so called MAMBO sources provide important clues concerning the formation of stars in the early universe. The rate of discovery of these sources would increase dramatically if one could map large areas with high efficiency. Because this requires large format arrays it provides enough motivation to go on using the present, proven technology and to explore the limits of array size.

Horn-Coupled Arrays

The architecture of MAMBO is typical for the present MPIfR bolometer arrays: the thermal, electrical and mechanical structure of the bolometers is produced microlithographically with high precision on a Silicon Wafer. Millimeter radiation is coupled to the bolometers through an array of conical horns, the antenna properties of which are well known. In front of the horn array there are filters at different temperatures for defining the bandpass, and for rejecting high-frequency thermal background. The transformation of the focal ratio of the horns to that of the telescope is done with a room temperature optical system of lenses and mirrors. A small cryostat is therefore sufficient to accommodate even fairly large arrays.

Bolometer Design

The MAMBO bolometers are of composite type, which allows the separate optimisation of critical bolometer parameters, like thermal conductivity, heat capacity and absorptivity. Freestanding membranes of Silicon Nitride provide low thermal conductivity for the bolometers by virtue of their amorphous structure. The membranes are manufactured by standard LPCVD techniques and, with a thickness of about 1 micron and a size of 3.4 mm square, are very strong. Therefore the only manual manufacturing step, namely the attachment of the Neutron Transmutation Doped (NTD) Germanium thermistors, can safely be done on a freestanding membrane. Utilising the wide bandwidth available in the 1 mm atmospheric window, the thermal background on a ground-based telescope is so high that the thermal conductivity of the unstructured membrane is sufficiently low already [1]. The phonons created by absorption of photons are collected by a "split ring" of Gold, with a thickness of 200 nm, in the center of the membrane, achieving spatially uniform absorption within the ring (Figure 2.). NTD Germanium thermistors with electrical contacts on one side ("flatpacks") are then Indium-soldered across the one gap of the Gold ring. Electrical connection with negligible thermal conductivity is provided by 80 nm thick sputtered Niobium wires. If lower thermal conductivity should be required, the membrane can be structured as indicated in figure 2. The fact that the time constant is about 4 ms shows that the contribution made by the Gold and the thermistor to the heat capacity is not excessive.

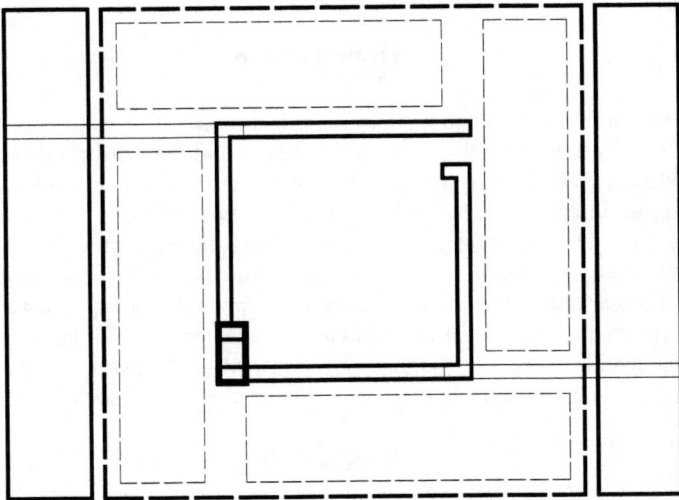

FIGURE 1. Bolometer layout. The large square, marked by a thick dashed line is the Silicon Nitride membrane, 3.4 x 3.4 mm in size. The NTD Germanium thermistor is shown across one of the gaps of the split Gold ring. Gold and Niobium layers are drawn in thick and thin continuous outlines respectively. The thin dashed lines mark openings in the membrane that would lower the thermal conductivity if this should be necessary.

Electromagnetic Modelling

The properties of bolometer absorbers consisting of one layer of dielectric with a metal film on one or both sides, have been described in the literature in the approximation of planar incident waves and layers of infinite lateral extent [2]. These approximations are no longer valid for small absorbers placed close to the exit of the circular waveguide. Three-dimensional numerical simulations on the basis of a finite difference program package were carried out in order to maximise the wideband absorption. The first result was that absorbers based on a dielectric with high refractive index were just as effective behind the waveguide as in free space. For example, a thin Silicon layer, a quarter wave thick (90 μm @ 1.2 mm wavelength) with a metal film on the back side, with maximum absorption at 1.2 mm, was entirely satisfactory and was therefore used in some early arrays. However, the fabrication would be much easier if one could do without the extra Silicon layer of about 90 μm thickness. Simulations with the metal film just on the Silicon Nitride membrane and a quarter wave reflector behind it led to the conclusion that satisfactory absorption could only be achieved if the waveguide was flared into a small horn. Further simulations showed that the metal ring around the absorber had no adverse effect.

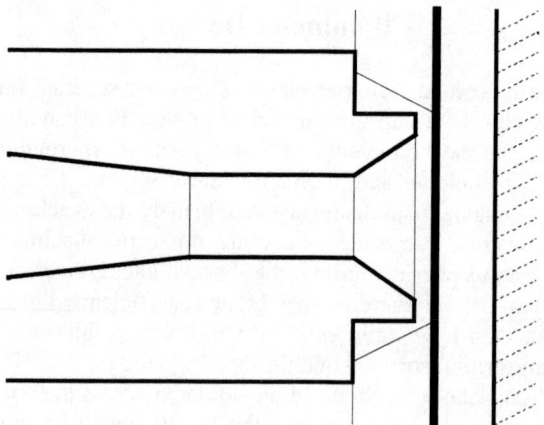

FIGURE 2. Detail of the horn bolometer interface. Shown are from left to right: the vertex of the conical horn with the circular waveguide and the small flared horn, the Silicon Nitride membrane (thick vertical line) on the Silicon wafer and the quarter wave reflector.

Horn Design

Nearly all MPIfR bolometers are designed to detect a single mode of the radiation field in the focal plane. The full spatial resolution of the telescope will therefore be preserved and the sensitivity to point sources maximized. In this case the same considerations with respect to aperture and beam efficiency apply as for coherent receivers. For example, for general purpose observations, one is led to the same goal of about -13dB edge-taper of the illumination on the telescope primary. Each bolometer or array is designed for one frequency and can therefore be optimized for that frequency without any compromise. The feedhorns are corrugated or smooth-walled conical horns. For arrays they are combined in a closepacked hexagonal grid. Each horn feeds into a circular waveguide, which is about two diameters long. The waveguide acts as a mode filter and at the same time as a highpass filter, taking advantage of the steep cutoff of the fundamental waveguide mode (H11). An additional lowpass filter in front of the horn array restricts the bandwidth to that of the fundamental mode of the circular waveguide of about 27%.

Rf Shielding

Ever since the time of an incident of strong radar interference on Pico Veleta, all MPIfR bolometer arrays are equipped with two layers of RF shielding. The first shield is the outer shell of the cryostat. All wires entering the cryostat are filtered, and the entrance window is covered on the inside with a specially designed inductive mesh. The bolometer-mount with the horn array presents the second shield. All signal and bias wires enter the bolometer cavity via feedthrough capacitors with a chip resistor in series.

Cryostat and Preamplifier

All the MPIfR arrays that are cooled to 300 mK, fit comfortably into a small, commercial He-4 cryostat with a 4" diameter cold work surface. MAMBO is equipped with one of the very compact He-3 sorption coolers, developed in France [3], with high pressure internal storage of the He-3 and designed for side-looking configurations. For the large diameter windows and filters of MAMBO2 a bottom-looking configuration is more convenient and a homemade He-3 stage with low pressure external storage was fitted in the same size He-4 cryostat. During operation the He-4 is pumped on continuously, as this will decrease significantly the thermal load on the He-3 stage. The hold time between cooling cycles is typically two days.

The preamplifiers are based on junction field-effect-transistors (FETs) with low noise at low frequencies, mounted in an RF tight preamplifier box at 300K on top of the cryostat. The transistor noise at 300 K is not significantly higher than at the optimum temperature around 100 K and microphonics can be reduced to insignificant levels by careful wiring inside the cryostat. Only in the cryogenically more complex HUMBA array was it necessary to put the transistors on a heated stage in the cryostat.

Laboratory Tests

In bolometer receivers where the spectral response can be affected by several components, a simple multiplication of the transmission curves of each individual component does not lead necessarily lead to a correct spectral system response. There could be interactions between components, and the response of the horn bolometer assembly is difficult to calculate or measure. Before a system goes on a telescope we try to characterise the system response in our lab with a Martin-Puplett interferometer. In this measurement a blackbody is used as the source and the complete array cryostat as the detector. Assuming a flat response of the Martin Puplett interferometer, the resulting spectrum should be that of the system multiplied by that of the blackbody.

The angular response of MAMBO was checked recently in the feed-pattern measurement facility of the MPIfR. The facility has absorbing walls such that low level sidelobes can be detected. The source was a 230 GHz coherent source and in this setup no significant sidelobes were seen above those expected theoretically.

MAMBO2

After the commissioning of MAMBO it became clear that a larger array (MAMBO2) would greatly improve the efficiency of the search for faint cosmological sources, the study of which is one of the main scientific activities of the Millimeter and Submillimeter Astronomy group of the MPIfR. If the area to be mapped is much larger than the array itself, then the time for mapping to the same depth depends as 1/n, if n is the number of array elements. As a compromise between technical risk and speed of development the number of array elements was set to around 120. MAMBO2 was planned to be a copy of the successful MAMBO, similar except for the higher number of bolometers.

Wafer Layout

FIGURE 3. View of MAMBO2 from the side opposite to the horns. The dark squares are the Silicon Nitride membranes, carrying the (square) Gold rings and Niobium wires. The NTD-Germanium chips are barely visible across one corner of each Gold ring. The octagonal Silicon wafer is fixed mechanically and electrically to thermal shunts on the surrounding mounting ring via ultrasonically bonded Gold wires. From the thermal shunts, wires are bonded to the center conductor of the feedthrough capacitors, which are visible on the mounting ring around the wafer. Note that two membranes are broken.

The layout of MAMBO2 still fits comfortably on a standard 4" silicon wafer (Figure 3.). With horn diameters of 5.5 mm and 3.4 x 3.4 mm membranes there is plenty of room for wiring between the rows of membranes. There is one common ground line in each row. The wafer is glued with wax, wiring side down, to a Sapphire wafer of the same size, then etched with KOH from the back in order to release the membranes, and diced to the desired shape.

Bolometer Mount

The bolometer wafer is held by 50 micron diameter Gold bonding-wires in the center of a Gold-coated Copper ring containing the feedthrough resistors (Figure 3.). The wires end on thermal shunts on the copper ring, and in this way provide robust electrical, thermal and mechanical connections between the ring and the wafer. Further bonds connect the thermal shunts to the center conductors of the feedthrough capacitors. The Copper ring with the wafer is sealed on the wiring side by the reflector plate and on the opposite side by the horn array. While this whole assembly is at 300 mK, the bias resistors are on the He-4 surface at 1.5 K. The parts at 300 mK are enclosed by a radiation shield at He-4 temperature (1.5K), which also serves as mounting surface for the filters and as attachment for Kevlar strings between the He-3 and He-4 cooled parts (Figure 4.).

FIGURE 4. View of the array of horns in MAMBO2. The scale is indicated by the array grid constant of 5.5 mm. Connectors for the bolometer signals are visible on two sides of the horn array. The radiation shield at 1.5K carries the wide band filters and serves as attachment for Kevlar strings that fix the components at 0.3K relative to those at 1.5K.

Optical Design

In the Gaussian beam approximation, a beam can be transformed to a similar one with different beam parameters by a Gaussian beam telescope (GBT). A GBT is a combination of two lenses (or mirrors) with a common focus between them. A beamwaist at the front focus of the first lens is transformed into another one at the back focus of the second lens. It can also be shown that this transformation is broadband and that the waist radii will be in the ratio of the focal lengths. These results are easily derived in the thin lens approximation for beams on axis. For a large field, this condition is no longer valid, but one can start with a design that satisfies the GBT condition on axis. By ray tracing, one can then optimise the image quality across the image plane within the boundary conditions of vertical incidence of off-axis bundles and a low curvature of the focal plane. For the large field of MAMBO2, a solution with a spherical mirror and an aspherical lens (made of high density polyethelene – HDPE), fitting within the space available in the receiver cabin of the 30 m MRT, illustrated in Figure 5. MAMBO uses two asperical HDPE lenses, which were optimised in the same procedure, by ray tracing.

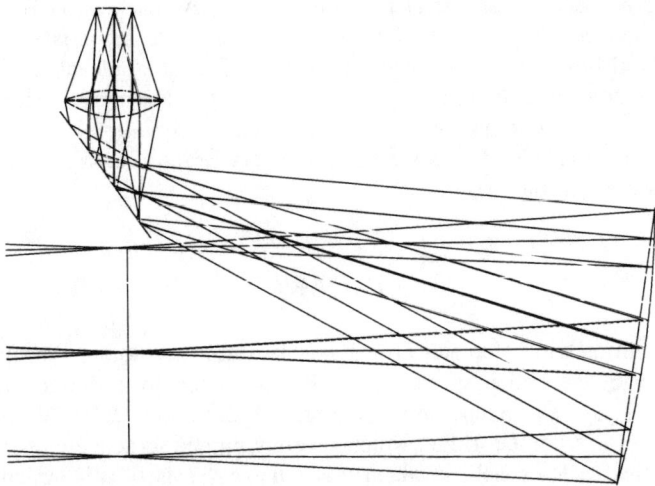

FIGURE 5. Optics for matching MAMBO2 to the IRAM 30 m MRT. The radiation from the secondary is entering from the left. The spherical mirror has a diameter of 450 mm. Because of the Nasmyth focus of the 30 m MRT, the bottom-looking cryostat of MAMBO2 will stay vertical in the final focus behind the HDPE lens.

MAMBO2 at the IRAM 30m MRT

FIGURE 6. MAMBO2 in the receiver cabin of the IRAM 30 m MRT. The spherical mirror with a diameter of 450 mm is visible in the lower right corner, while the cryostat, HDPE-lens and flat folding mirror are in the upper left. The whole assembly is mounted on a vibration isolated optical table.

In February 2001, MAMBO2 was briefly installed for the first time in the receiver cabin of the 30 m MRT. The field diameter in the Nasmyth focus of the 30 m MRT is limited by the size of the Nasmyth mirrors, and the field of MAMBO2 is already close to that limit. Major mechanical changes were necessary to fit MAMBO2 with its optics into the limited space reserved for bolometers; at the same time a new backend (ABBA) for the large number of detectors had to commissioned. Alignment of the beams of an array on to secondary mirror is a slow process in a Nasmyth focus. For the central beam to hit the center of the secondary at all elevations it is necessary for this beam to propagate along the elevation axis, before being reflected by the Nasmyth mirror. After only a preliminary alignment, a beammap on Saturn was obtained as a first light observation. This observation was useful for debugging the system. Problems that were found were not of a very serious nature, so that it is likely that MAMBO2 will be online for the next winter season.

HUMBA

HUMBA, the "hundred millikelvin bolometer array", is a 19-element bolometer array for 2 mm wavelength and designed primarily for Sunyaev-Zeldovich studies. HUMBA is cooled by a dilution refridgerator, which can be adjusted to operate continously to any temperature between 50 and about 200 mK. Until recently HUMBA was subject to exess noise originating from the fridge, which severely limited its sensitivity. The solution of this problem is the subject of the paper by A. Raccanelli et al. in this volume. Tests of this system at the IRAM 30m are in progress.

POLARIMETRY

Polarimetry with arrays could be very exiting. The paper by G.Siringo et al. in this volume describes a retardation device that can be tuned to different wavelengths. The insertion of an additional polariser in front of the cryostat will transform any bolometer array into a polarimeter. Although the power in other polarisation is lost, one does not have to provide a second identical array; this represents a substantial saving of cost and effort. This is work in progress at the MPIfR.

SIMBA

SIMBA, the "SEST imaging bolometer array" is a joint project of the European Southern Observatory (ESO), Onsala Space Observatory, Bochum University and the MPIfR. The array and the cryostat are copies of MAMBO. In the Cassegrain focus of the 15m SEST telescope, SIMBA has to operate over the whole elevation range of 90 degrees. The same type of coupling optics as for MAMBO2, but now with a 45 degrees angle between the incident and final beams, allows the coverage of the whole elevation range, within an inclination range of the cryostat of +/- 45 degrees. Efficient mapping is the main purpose of using arrays and this is usually performed by scanning slowly in the chopping direction. SEST does not have a chopping secondary, therefore a new mapping mode, called "fast scanning", had to be developed. This observing mode is explained in the paper by L.A. Reichertz et al. in this volume. SIMBA was commissioned succesfully in June 2001.

APEX

With ESO and Onsala Space Observatory as partners, MPIfR is going to build a submm telescope of 12 m diameter, to be placed on the ALMA site in Chile. The location is at 5000 m altitude in a high region of the Chilean Atacama desert. APEX (Atacama pathfinder experiment), as the telescope is called, is a copy of an ALMA prototype telescope. It will offer unique opportunities for Submm astronomy in the southern hemisphere, making use of the exellent atmospheric conditions there. While the optical and mechanical characteristics of the main mirror will be identical to those of the ALMA telescopes, care will be taken in the design of secondary optics, to allow very large bolometer arrays take advantage of the full field of view. First light is foreseen for 2003.

CONCLUSIONS

It seems clear that the near future will see very large bolometer arrays, with a tendency to fill the available telescope focal plane. New submm telescopes will be designed for the maximum field; this is a new development in Radioastronomy. Arrays are often compared by the numbers of their elements, which can be misleading if used as the sole characteristic of their performance. The area covered by two arrays each with the same number of elements can differ by a factor of 16 depending on whether the elements are sized for instantaneous Nyquist sampling or full efficiency with respect to point sources. A more significant figure for comparison would be the array throughput $A\Omega$, where A is the effective array area. Arrays with several hundred elements can hardly be envisaged without cold multiplexers. Besides the advantage of fully lithographic fabrication, the attractiveness of superconducting bolometers [4] lies in the promise of multiplexed SQUID readout [5]. These are exiting times in this field, and the bolometer group of the MPIfR is planning to participate in these developments.

ACKNOWLEDGMENTS

The succes of the bolometer development at MPIfR owes a lot to the skill and dedication of our engineers and technicians, W. Esch, G. Lundershausen and B. Ufer.

E.E. Haller and J. Beeman of LBNL Berkeley have always been a reliable source of excellent NTD-Ge thermistors.

The group of V. Hansen at the University of Wuppertal performed the electromagnetic field simulations of the bolometer absorber structures, and developed software for calculating mesh filters.

During micromachining campaigns in the microlab of UC Berkeley, E.K. enjoyed the friendly atmosphere and the helpfulness of the staff and many fellow labmembers. Special thanks go to X. Meng for the deposition and etching of the Niobium layers.

The Millimeter and Submillimeter Group of the MPIfR, under the direction of K. Menten, can always be trusted to spur on the technical effort by proposing observations which make new demands.

REFERENCES

1. Holmes, W., Gildemeister, J.M., Richards, P.L. and Kotsubo, V., Appl. Phys. Letters, 72, 2250-2252, (1998).
2. Carli, B.and Iorio-Fili, D., Journal Opt. Soc. Am., 71, 1020-1025, (1981).
3. Torre, J. P. and Chanin, G., Review Sci. Inst., 56, 318-320, (1985).
4. Gildemeister, J.M., Lee, and A.T., Richards, P.L., Appl. Phys. Letters, 77, 4040-4042, (2000).
5. Yoon, J., Clarke, J., Gildemeister, J.M., Lee, A.T., Myers, M.J., Richards, P.L., and Skidmore, J.T., Appl. Phys. Letters, 78, 371-373, (2001).

A Filled Bolometer Array Camera For Ground-Based Observations

Vincent Reveret[a], Patrick Agnese[b], Philippe Andre[a], Eric Doumayrou[a],
Rene Gastaud[a], Jean Le Pennec[a], Louis Rodriguez[a],

(a) Service d Astrophysique, CEA Saclay, Orme des Merisiers
91191 Gif Sur Yvette, FRANCE
(b) LETI / LIR, CEA-Grenoble, 38054 Grenoble, FRANCE

Abstract. This paper describes the design of a bolometer camera using a filled array, for submillimeter ground astronomy. The architecture of the array is presented, as well as the calculated performances of a (16 x 16) pixels array, mounted on a 3 m diameter telescope, such as KOSMA.

INTRODUCTION

In 1996, following the successful achievement of ISOCAM [1], CEA has started the development of a new type of bolometer array for the detection of submillimeter radiation in astronomy. The first bolometer array [2] made of 256 pixels has been tested in the lab in 1999. Since, this technology has been chosen for the PACS instrument on the Herschel Space Observatory for the 60 — 210 µm band. At the same time, we have decided to develop an instrument for ground observations using this kind of bolometer array, for the submillimeter atmospheric bands. We describe here the principle of the CEA bolometer array, and we present the design of a camera suitable for ground applications.

THE BOLOMETER ARRAY

The main purpose in the early development of this new type of detector was to realize a CCD-like filled array, with a large number of bolometric pixels. Previous space missions like ISO or CASSINI/CIRS have given the CEA / LETI an expertise on some technological processes. ISOCAM s multiplexing system has permitted to reach the goal of having a large number of bolometric pixels without individual readout circuit. The array is realized in a collective process using silicon micromachining conception. See Fig. 1. CEA/LETI arrays are made of 256 individual bolometers working at 300 mK. Each pixel has a 5 µm thick silicon mesh with deposited absorber (TiN or TaN) and a silicon implanted thermometer. The mesh alone could absorb 50% of the incoming radiation (horizontal resonance). To increase the absorption, this mesh forms a quarter-wave cavity with a gold reflector (vertical resonance). The height of the cavity is adjusted by indium bumps.

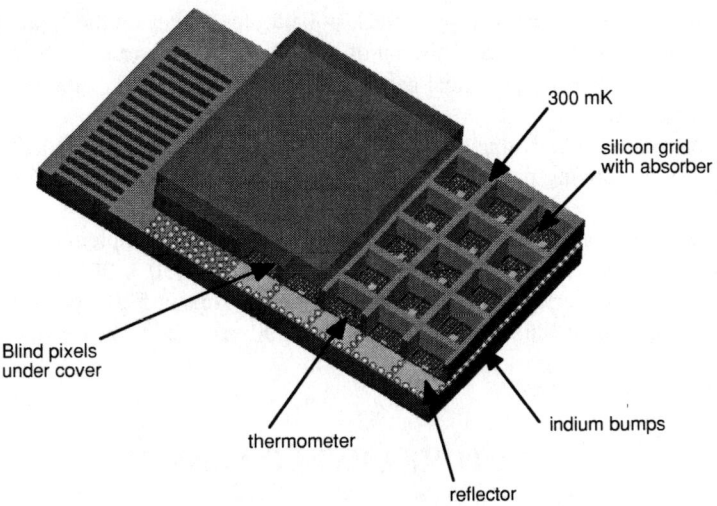

FIGURE 1. Schematic representation of the CEA filled array (15 optical pixels are seen).

The energy absorbed by the metal is transmitted via the silicon mesh, to the implanted thermal sensor. The thermometer is silicon doped with phosphorus (boron compensated). At very low temperatures, electrical conduction is realized by electrons hopping between phosphorus sites. The Efros model [3] predicts an exponential dependence of the resistance with temperature:

$$R = R_0 \exp\left(\sqrt{\frac{T_0}{T}}\right) \exp\left(-\frac{qL(T)E}{kT}\right)$$

where, T_0 and R_0 are parameters related to thermometer, E is the applied electric field, k is Boltzmann's constant and L(T) is the hopping length. The parameters of the Efros model have been measured experimentally and verify the theory over a wide range of temperature and electric field.

The use of monocrystalline silicon meshes gives a very low heat capacity to the system, at these temperatures. Microscopic silicon beams (5 μm x 2 μm cross section, 420 μm long) evacuate the heat from the mesh to the heat sink.

The output signal is measured at the middle point of a resistive bridge composed of the bolometer thermometer and a reference resistance maintained at 300 mK. MOS transistors are used to adapt the very high impedance of the detection stage (300 mK) to the next amplification stage (2 K). The multiplexing stage then converts 16 outputs to 1. These electrical circuits are disposed under each pixel of the array.

In order to be immune from any outside perturbation, two rows of blind pixels are disposed on a side of the array, with the same characteristics than the detecting ones. Heaters are implanted on these pixels to produce the same thermal load than that received by the optical pixels. A fully differential system permits to remove the correlated noise of the signal.

At the moment, the detector noise is dominated by the noise of the MOS transistors, which is lower than 1 μV / Hz$^{1/2}$ at 1Hz. The responsivity at $\lambda = 350$ μm has been measured up to 2×10^{10} V/W, giving a total Noise Equivalent Power (NEP) around 5×10^{-17} V/W.

DESIGN OF THE INSTRUMENT

The main advantage of the filled array in comparison with the classical 2Fλ feedhorn arrays is that there is no need to jiggle the focal plane. In order to fully sample the focal plane with the CEA bolometer array at $\lambda = 865$ μm, the final f-number (F), has to be equal to 1.7 (the pixel size is 750 μm and corresponds to a distance of 0.5Fλ at the focal plane). The optical layout of the camera consists of two high-density polyethylene (HDPE) lenses with parallel rays between them. Lens 2 is placed at the exit pupil of the system and defines the final f-number.

Infrared and far infrared lights are blocked by a black polyethylene filter on the liquid nitrogen stage and a pyrex filter on the 2K helium stage. The overall transmission of these filters and lenses has been measured at 865 µm to be equal to 40%, which is very similar to theoretical calculations. The production of the bandpass mesh filters is under discussion with the manufacturer.

The cryo-cooler is provided by L. Duband s team at CEA / SBT in Grenoble [4]. It is a double stage ^{3}He-^{4}He sorption cooler, which has no moving parts and can provide temperature down to 260 mK with a cold heat sink at less than 5 K.

Performances have been estimated at λ = 865 µm in the case of a future implementation of the instrument on the 3m diameter KOSMA telescope [5]. The pixel angular resolution is (30 x 30) arcsec and the field of view is (7.7 x 7.7) arcmin for a (16 x 16) array. The incident background flux is 5 pW per pixel, and for an atmospheric opacity of 0.3, the Noise Equivalent Flux Density is NEFD = 500 mJy/Hz$^{1/2}$ per pixel, corresponding to 1 sigma in 1s of integration.

FUTURE DEVELOPMENT

Simulations of observations with a ground based telescope have started in order to test the different possible modes of observations with the CEA bolometer array. A simulation program modeling the behaviour of one pixel already exists and is well representative of experimental measurements. Tests with a 256 pixels array are planned for the 2001/2002 winter on the KOSMA telescope. Different working modes will be tested (On The Fly , scanning), as well as the 5 different possible electronic readout modes.

Study of a focal plane for the 1.3 mm wavelength will start in the next months. A major problem is the fact that indium bumps cannot be extended to the height required for millimeter detection. Different technological solutions will be tested.

REFERENCES

1. Cesarsky, C. J., et al., Astron. Astrophys., 315, L32-L37, (1996).
2. Agnese, P., Buzzi, C., Rey, P, Proc. XXVth Conf. On Infrared Technology and Application, Orlando, SPIE 3698, 1999.
3. Efros, A., J. Phys. C : *Solid State Physics*, **9**, 2021, (1976).
4. Duband, L., Hui, L., Lange, A., *Cryogenics*, **30**, 263 (1990).
5. Degiacomi, C., Schieder, R, Stutzki, J., Winnewisser, G., *Optical Engineering*, **34**, 9, 2701 (1995).

Mesh Filters for the Mm/Submm Atmospheric Windows

Soglasnova V.A., Maslov I.A.

Space Research Institut of the Rissian Academy of Science, Profsoujznaya 84/32, 117810 Moscow, Russia

Abstract. The set of the band-pass freestanding metal mesh filters was developed for the spectral selection of the windows of the atmospheric transparency. The filter transmissions are more then 80%, the bandwidths are close to the widths of the atmospheric windows. The meshes are very elastic and can be also used as the reflective surfaces in the Fabry-Perot and in the multi-layes band-pass filters.

The freestanding metal meshes with a pattern of the cross-holes have resonant band-pass transmission at the wavelength near to the grid constant (G). The bandwidth of the transmission can be from 10 up to 50% and depends from the ratio the size of the holes to the grid constant [1]. The transmission characteristics of the resonant meshes and the spectra of the atmospheric transparency (including the 200-μm window [2]) are shown in Fig. 1. The position of the maximum transmission and the bandwidth of the filters are conformed with atmospheric windows.

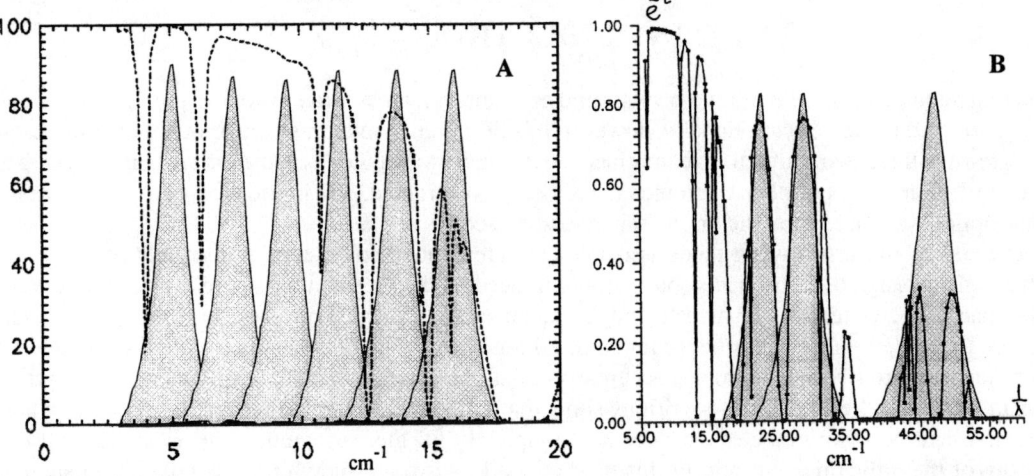

FIGURE 1. The atmospheric transparency and the set of the band-pass mesh filters (shadowed) for the on-ground astronomical observations in the mm-windows (A) and in the sub-mm windows (B) at the extra-dry sites (Antarctica)

The developed technology allows to produce high quality freestanding metal meshes with the grid constant G up to 50 μm with any ratio of the hole size to the G. The thickness of the meshes is 15–20 μm. The lack of any supporing substrate in the meshes provides them great optical and physical stability in the working in the vacuum and at the cryogenic temperatures.

REFERENCES

1. Soglasnova V.A., Gemünd P.-H., Kreysa E., Maslov I.A. "Freestanding Resonant Meshes as the Bandpass Filters and the Beamsplitters of the Far-Infrared Radiation" in *Proceedings ESA SP-388*. The 30th ESLAB Symposium "Submillimeter and Far-Infrared Space Instrumentation". 24-26 September 1996. ESTEC, Noordwijk, The Netherlands, 1996.
2. Maslov V.A., Soglasnova V.A. "The Spectral Filters for the Astronomical Observations through the Atmospheric 200 μm Window" in *Proceedings JENAM-2000*. Joint European and National Astronomical Meeting, May 29 - June 3, 2000. Moscow, Russia, 2000.

Partially-Coherent Long-Wavelength Optical Simulation Techniques for Microwave Background Astronomy

S. Withington,[*], C.Y.Tham,[*] and G.Yassin,[*]

[*]*Cavendish Laboratory, Madingley Road, Cambridge, UK*

Abstract. We outline a procedure for modelling the behaviour of partially-coherent long-wavelength optical systems. The procedure is of considerable importance to microwave background astronomy, where sequences of high-throughput optical components are often used. In the paper, we give a symbolic description of the basic method, and illustrate its use by calculating the radiation pattern of an overmoded waveguide bolometer. To confront a number of fundamental issues, we consider the case where the absorbing disc does not completely fill the waveguide. We also demonstrate the technique by showing how the optical behaviour of an imaging array can be determined by propagating all of the beams in the array simultaneously.

INTRODUCTION

When constructing instruments for microwave-background astronomy, it is essential to control precisely the amount of stray light that is coupled to the detector. Equally, however, it is desirable to maximise the throughput of the telescope. In an attempt to resolve these two conflicting requirements, few-mode waveguide bolometers are being developed. For wide-field mapping, large arrays of planar bolometers are being constructed, and in these cases a cold aperture must be placed in the optical system to limit the throughput of the telescope.

The millimetre and submillimetre-wave fields associated with few-mode bolometers are partially coherent, and this complicates the optical design of instruments considerably. Indeed, at the present time, there is no accurate technique for propagating partially-coherent vector fields through sequences of reflecting surfaces. In this paper, we outline a general procedure for modelling the behaviour of few-mode, long-wavelength optical systems. In its most fundamental form, the technique is rigorous, and can provide accurate simulations of many different experimental configurations.

To illustrate the power of the formalism, we briefly show that it is possible to simulate the behaviour of waveguide bolometers, even when the absorbing element does not completely fill the waveguide. This example, will force us to confront many of the difficulties inherent in this kind of work. Also, we show that it is possible to simulate the behaviour of large-format imaging arrays by propagating all of the beams in the array, through the optical system, simultaneously. The basic idea is to work in terms of the second-order statistical properties of the total field, rather than the individual beams themselves. This approach allows the array, and the individual pixels in the array, to be in any state of coherence.

DYADIC ANALYSIS OF PARTIALLY-COHERENT VECTOR FIELDS

For any given physical arrangement, the details of an analysis tend to be complicated, and so here we shall simply give a symbolic description of the basic concepts used.

We define the *electric* space-domain cross spectral dyadic [1], $\overline{\overline{E}}(\bar{r}_1,\bar{r}_2)$, as the ensemble average of the dyadic product of the field at one point $\bar{r}_2$ and the complex conjugate of the field at a second point $\bar{r}_1$:

$$\overline{\overline{E}} \equiv \overline{\overline{E}}(\bar{r}_1,\bar{r}_2) = \langle \bar{E}(\bar{r}_2)\bar{E}^*(\bar{r}_1) \rangle, \qquad (1)$$

which for a Cartesian system becomes

$$\overline{\overline{E}}(\bar{r}_1,\bar{r}_2) = E_{xx}\hat{x}\hat{x} + E_{xy}\hat{x}\hat{y} + E_{yx}\hat{y}\hat{x} + E_{yy}\hat{y}\hat{y}. \qquad (2)$$

The cross spectral dyadic contains complete information about the second-order statistical properties of the electric field, and from it we can extract all commonly used, classical, measures of behaviour: Stokes parameters, etc. For the purposes of this paper, we will assume that it is sufficient to work in terms of transverse field components alone, and that $\overline{\overline{E}}(\bar{r}_1,\bar{r}_2)$ has only 4 components. In the case of free space optics, 9 components are formally present, but the condition that $\nabla \cdot \bar{E} = 0$ imposes linear dependence. We shall not consider such topics here.

We can also define the *magnetic* space-domain cross spectral dyadic

$$\overline{\overline{H}} \equiv \overline{\overline{H}}(\bar{r}_1,\bar{r}_2) = \langle \bar{H}(\bar{r}_2) \bar{H}^*(\bar{r}_1) \rangle, \tag{3}$$

and the *mixed* space-domain cross spectral dyadic

$$\overline{\overline{N}} \equiv \overline{\overline{N}}(\bar{r}_1,\bar{r}_2) = \langle \bar{E}(\bar{r}_2) \bar{H}^*(\bar{r}_1) \rangle. \tag{4}$$

It is straightforward to show that $\overline{\overline{N}}(\bar{r}_1,\bar{r}_2)$ is derivable from $\overline{\overline{E}}(\bar{r}_1,\bar{r}_2)$ through:

$$\overline{\overline{N}}(\bar{r}_1,\bar{r}_2) = \frac{1}{jk\eta_o} \overline{\overline{E}}(\bar{r}_1,\bar{r}_2) \times \nabla_1, \tag{5}$$

where the subscript on ∇ emphasises the position variable that is used in the space derivatives. η_o is the impedance, and k the wave number, of free space. Here, we are assuming that by positioning the curl on the right of the dyadic we are operating on the right-hand vector in the dyadic product.

Similarly, $\overline{\overline{H}}(\bar{r}_1,\bar{r}_2)$ can be derived from $\overline{\overline{E}}(\bar{r}_1,\bar{r}_2)$ through

$$\overline{\overline{H}}(\bar{r}_1,\bar{r}_2) = \frac{-1}{(k\eta_o)^2} \nabla_2 \times \overline{\overline{E}}(\bar{r}_1,\bar{r}_2) \times \nabla_1. \tag{6}$$

(5) and (6) suggest strongly that it is sufficient to work in terms of $\overline{\overline{E}}$ alone. Often, however, it is most convenient to work through an analysis using all three quantities, and then use the appropriate dyadic, depending on whether one wants to know the average power flow, which would come from $\overline{\overline{N}}$, or the correlations in the electric field, which would come from $\overline{\overline{E}}$. We shall not discuss this subtleties of this point here, but simply work in terms of $\overline{\overline{E}}$.

The space-domain dyadics are operators, which describe the coherence of a field in an abstract way. In other words, we do not have to specify a coordinate system in order to be able to talk about the spatial correlations of a vector field.

We can, however, generate a particular representation by projecting the dyadic onto a basis set. Because all optical systems have finite throughput [2, 3], suitable basis sets are countable. The basis sets can be plane waves, waveguide modes, paraxial modes, etc. Symbolically, we can write

$$A_{i,m;j,n} = \left\langle \bar{\Psi}_{i,m} | \overline{\overline{E}} | \bar{\Psi}_{j,n} \right\rangle, \tag{7}$$

where we have used the Dirac notation to represent the projection of the operator $\overline{\overline{E}}$ onto the basis set $\{|\bar{\psi}\rangle\}$. It is assumed the basis sets are complete and orthonormal. In (7), i, j denote two modes having different spatial forms, and m, n denote the polarisation states of i and j respectively. For example, i, j could label the transverse wavevectors of two plane waves, and m, n are their individual states of polarisation. In practice, the matrix elements defined by (7) are evaluated by taking *right* and *left* scalar products with the mode functions, and evaluating a double integral over the surface at which the expansion needs to be performed. In this way, we have represented the statistical properties of the total field by specifying the correlations between orthogonal modes.

If the *coherence matrix* **A**, which comprises the elements A_{mn}, is known at some surface, then it is often simple to propagate the matrix to another surface through an expression of the form

$$\mathbf{A}' = \mathbf{S}\mathbf{A}\mathbf{S}^\dagger, \tag{8}$$

where S is the scattering matrix that propagates the modes of the basis set, and $\dagger$ denotes the conjugate transpose. For waveguide, plane-wave, and paraxial modes, the scattering matrix is simply a diagonal matrix of phase factors, which describe how the modes slip in phase with respect to each other as they propagate.

After propagation, the cross spectral dyadic can be re-assembled according to

$$\overline{\overline{E}}(\bar{r}_1,\bar{r}_2) = \sum_{m,n,i,j} A'_{i,m;j,n} |\bar{\Psi}_{i,m}\rangle \langle \bar{\Psi}_{j,n}|, \tag{9}$$

where $|\bar{\Psi}_{i,m}\rangle \langle \bar{\Psi}_{j,n}|$, which is an operator, is a dyadic also.

NATURAL MODES

In many cases, we may wish to propagate the statistical properties of a field from one plane to another by using techniques other than scattering matrices. For example, we may wish to trace a partially-coherent vector field through a large antenna using the principles of physical optics and the geometrical theory of diffraction. The solution is straightforward.

We can diagonalise the coherence matrix to find a representation of the form

$$A'_{i,m;j,n} = \delta_{m,n}\delta_{i,j}\lambda_{i,m}. \tag{10}$$

The associated eigenmodes of **A** are individually fully coherent, but mutually incoherent. In other words, we can diagonalise the coherence matrix to find the natural modes of the field. Note that the natural modes of the field are not the same as the natural modes of the optical system, which would be found by diagonalising the scattering matrix **S**. In summary, the eigenvectors of **A** comprise the mode coefficients of the natural modes in the original basis set, and the eigenvalues give, essentially, the power in each of the natural modes. Once the eigenmodes and eigenvalues are known they can be propagated independently using rigorous electromagnetic theory, say through an antenna, and then the statistical properties of the total field re-constructed at the output surface.

A considerable amount of structure emerges when fields are described in this way. For example, in the case of paraxial free-space modes, if we have an incoming field described by the electric coherence matrix **A**, and we wish to know the power coupled to a detector having the electric coherence matrix **B**, then the coupled power is proportional to Trace**AB** [4]. In the case of waveguides, a similar expression is found, but now **A** and **B** must be interpreted as the mixed coherence matrices [5].

SOURCES

To implement the above procedure there must be some surface in the system where the space-domain cross spectral dyadic is known. The best approach is often to identify a surface where a blackbody field can be assumed to exist. So, for example, a disc of absorber, such as a bolometer, can be modelled as a blackbody source of finite size. In this case, the space-domain cross spectral dyadic has the form

$$\overline{\overline{E}}(\bar{r}_1, \bar{r}_2) = \overline{\overline{I}} I(\bar{r}_1) \delta(\bar{r}_2 - \bar{r}_1), \tag{11}$$

where $\overline{\overline{I}}$ is the idem factor, which ensures random polarisation, and $I(\bar{r}_1)$ the intensity over the surface of the source. In reality, (11) can only be approximately true for a variety of reasons. It is, however, formally, correct for paraxial analysis. In the case of non-paraxial optics, the correlations between the field components of a blackbody source can, and must, be taken into account.

SUMMARY OF GENERAL ANALYSIS PROCEDURE

We can summarise the procedure for modelling the behaviour of partially-coherent optical systems as follows:

- Assume a space-domain cross spectral dyadic for the source.
- Find the matrix elements in a convenient basis set, usually the eigenmodes of the optical system: plane waves, waveguide modes, Gaussian modes for paraxial optics, etc.
- Coherence matrices can be added to produce a composite matrix that contains all of the information about a number of mutually incoherent sources, whatever the states of coherence of the individual sources. This procedure allows imaging arrays to be constructed.
- Propagate the coherence matrix using scattering matrices.
- Alternatively, diagonlise the coherence matrix to get the natural modes of the field, and then propagate each natural mode using classical techniques: physical optics, GTD, etc.
- Add in intermediate noise sources as required.
- Re-assemble the cross spectral dyadic at the output surface.
- Calculate the power flow, coupling efficiencies, Stokes parameters, etc.

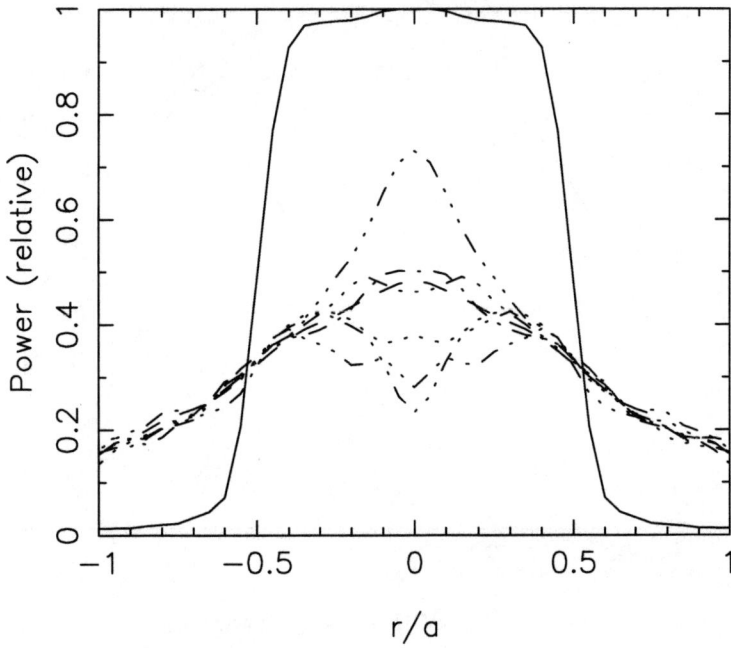

FIGURE 1. The intensity of the field in a waveguide having diameter 2a, when an absorbing disc, which only half fills the waveguide, is allowed to radiate. 239 waveguide modes where used, and the different curves correspond to distances from the source of 0,2.5,5,10,15,20,30 and 50mm.

WAVEGUIDE BOLOMETER

In this section, we illustrate the technique by showing a few results from the analysis of a waveguide bolometer. The bolometer simply comprises a disc of absorbing material in a circular waveguide. The waveguide flares into a corrugated horn at one end, and, to make the problem more demanding, we have assumed that the disc does not completely fill the waveguide. If the disc did fill the waveguide all of the modes would, to a good first order, be excited equally, and the far-field radiation pattern would simply be the sum of the intensity patterns produced by the individual waveguide modes. Because the disc does not completely fill the waveguide, correlations are induced between the waveguide modes, and the waveguide modes are no longer the eigenmodes of the field.

The analysis is carried out by projecting the space-domain cross spectral dyadic of the absorbing disc onto the basis set comprising propagating waveguide modes. These modes are then traced down the waveguide, and horn, by using a diagonal scattering matrix whose elements are differential phase factors. We have assumed that the flare of the horn is sufficiently gentle that no intermodal scattering takes place, but we have taken into account the change in the phase velocity that occurs when a mode travels down a waveguide of increasing size. We have also included a parabolic phase factor to provide a first-order description of the spherical phase cap that exists at the aperture of the horn. We could have easily used a complete scattering matrix, based on mode matching, to describe the horn, and in this way we could have incorporated any number of discontinuities [6]. In fact, an overmoded partially-coherent corrugated horn could be analysed in this way.

On reaching the aperture, we must find some way of propagating the statistical properties of the radiation field through free space. Propagation is achieved by diagonalising the coherence matrix to find the natural modes, taking the Fourier transform of each natural mode, and then recombining the modes on a far-field surface. In contrast to simply adding intensities, the full analysis described here allows the spatial form of the coherence length and polarisation state of the outgoing field to be determined accurately.

In Fig. 1, we show the intensity of the field across a waveguide at different distances from an absorbing disc that only half fills the waveguide. The results shown are based on a full vector calculation. The waveguide had a diameter of 3.33 mm, and the wavelength was chosen so that 239 modes could propagate. At the disc the modes recombine to give the correct intensity distribution, and on propagation the field spreads. The field only comprises those modes that propagate in the waveguide, not the evanescent modes, and therefore even at the disc the intensity distribution is

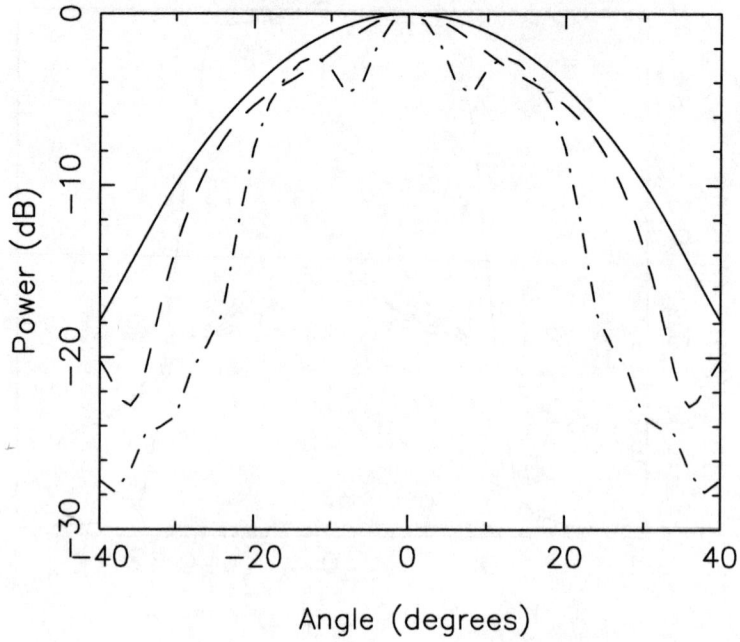

FIGURE 2. Far field radiation patterns of the horn described in the text. The solid line corresponds to the case when 2 modes propagate ($ka=2$), the dashed line to when 8 modes propagate ($ka=4$), and the dotted line to when 51 modes propagate ($ka=10$).

smoothed.

In Fig. 2, we show far-field radiation patterns of a horn having a length of 200mm and an aperture diameter of 10mm. For comparison purposes, the dimensions were chosen to be the same as those used in Murphy [7], which is turn were chosen to produce an essentially diffraction-limited horn. The feeding waveguide was 66mm long and 3.33mm diameter, and again the source comprised an absorbing disc that only half filled the waveguide.

When the disc completely filled the waveguide, the same radiation patterns as Murphy [7] were produced, verifying the integrity of the software and technique in this limit. When the radius of the disc was only half that of the waveguide, the beams patterns shown in Fig. 2 were produced. Beam patterns were calculated for wavelengths of 5.2mm (2 modes), 2.6mm (8modes), and 1.0mm (51 modes), in order to investigate different degrees of over-moding. The detail structure in the beams was not present when the disc filled the waveguide, and is due to the correlations between the modes causing interference.

IMAGING ARRAYS

Appropriately scaled Gaussian modes are close to the eigenmodes of paraxial optical systems, and therefore they form an excellent basis set for modelling, to first order, the behaviour of many millimetre and submillimetre-wave astronomical instruments. Also, for paraxial systems the full vector solution is simply a sum of two scalar solutions, and the three-dimensional spatial analysis of each scalar problem is simply the product of two two-dimensional problems. Hence, here, we will illustrate the method through the scalar analysis of a two-dimensional optical system.

The system we wish to study is a small-format imaging array. Because Gaussian-Hermite modes form a complete orthonormal set, it is possible to propagate a full imaging array using a single set of on axis Gaussian-Hermite modes. The key point is that we propagate the second order statistical properties of the total field rather than propagating the individual beams themselves. Our approach allows the array to be in any state of coherence.

For example, planar bolometers are often used for mapping broadband thermal continuum emission. In this case, considering the detectors to be radiating, the field across the surface of any single pixel is incoherent, and the fields associated with different pixels uncorrelated. Planar bolometers are, however, sensitive to radiation approaching from large angles, and therefore, to minimise stray-light coupling, the heat sensitive element is often placed in a length of few-moded waveguide. In this case, the fields associated different pixels are uncorrelated, but now the fields associated

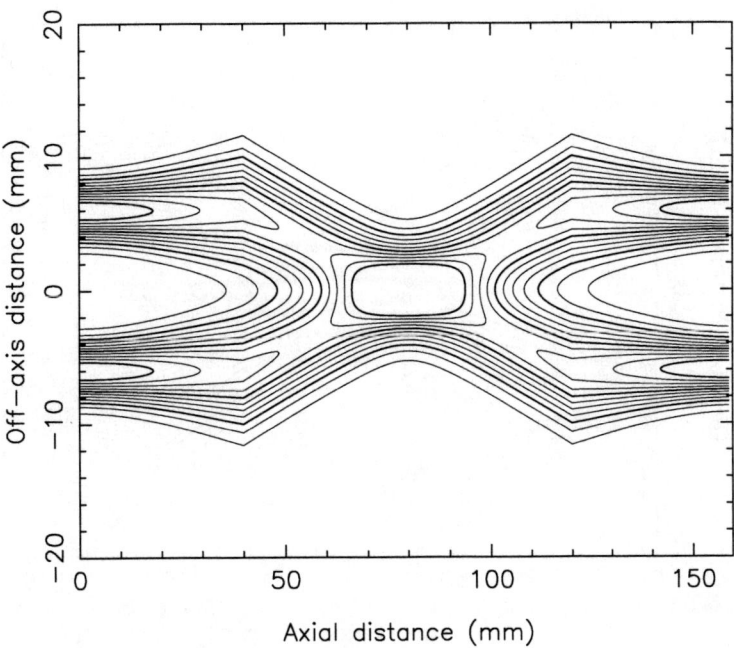

FIGURE 3. Two individually coherent but mutually incoherent incoherent Gaussian beams propagating through a Gaussian Beam Telescope. Thin lenses having focal lengths of 40mm are placed at distances of 40mm and 120mm from the source plane.

individual pixels are highly, but not completely, coherent. For narrow-band spectral-line observations, submillimetre-wave heterodyne mixing is used, and here, the signal beams are individually fully coherent. Finally, in a heterodyne imaging receiver, it is necessary to generate an array of local-oscillator beams. These beams are usually generated from a single source, either by a phase grating or a sequence of power splitters. Whatever the precise physical arrangement, the local oscillator beams are both individually and mutually fully coherent.

Because we are only interested in a scalar analysis, we shall use Gaussian-Hermite modes as the basis set. Each mode has the form

$$\psi_m(x,z) = \left(\frac{\sqrt{2}}{w(z)}\right)^{1/2} h_m\left(\frac{\sqrt{2}x}{w(z)}\right) \exp[\pm j\theta(z)] \exp\left[\mp \frac{j\pi x^2}{\lambda R(z)}\right] \exp[\mp jkz] \qquad (12)$$

where

$$h_m(u) = \frac{H_m(u)\exp\left[-\frac{u^2}{2}\right]}{\left(\sqrt{\pi}2^m m!\right)^{1/2}}, \qquad (13)$$

and

$$\theta(z) = (m+1/2)\frac{z}{z_c}, \qquad (14)$$

also $H_m(u)$ is the Hermite polynomial of order m in u.

It is important to appreciate that the functions $h_m(u)$ are orthonormal in the sense that

$$\int_{-\infty}^{+\infty} h_m(u) h_n(u) \, du = \delta_{mn} . \qquad (15)$$

In these equations the symbols have their usual meanings [8]. $w(z)$ characterises the scale size of the beam at a plane, $R(z)$ characterises the large-scale radius of curvature of the phase front, and $\theta(z)$, the phase slippage between modes, characterises the form of the field as the beam propagates and diffracts. All of these parameters are simple functions of z; z_c is the confocal distance.

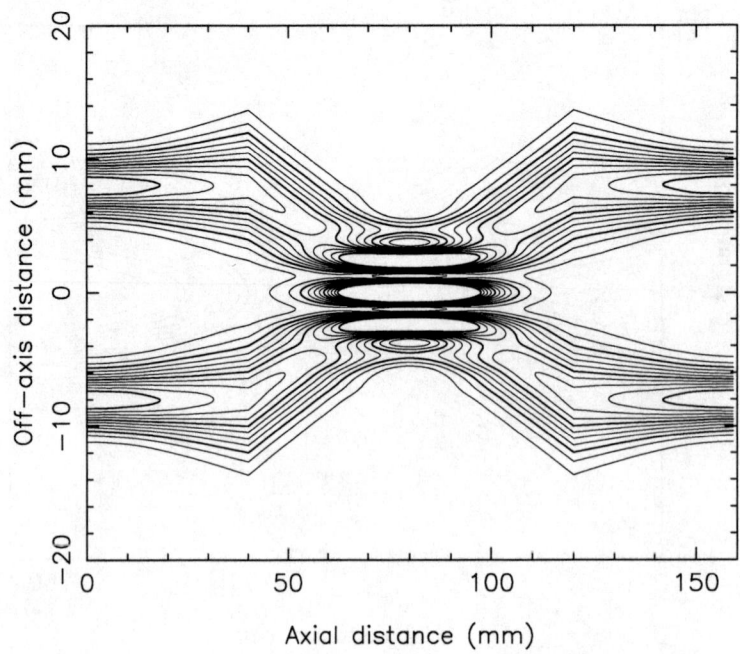

FIGURE 4. Two individually coherent, fully correlated Gaussian beams propagating through a Gaussian Beam Telescope. The geometry used was the same as Fig. 3

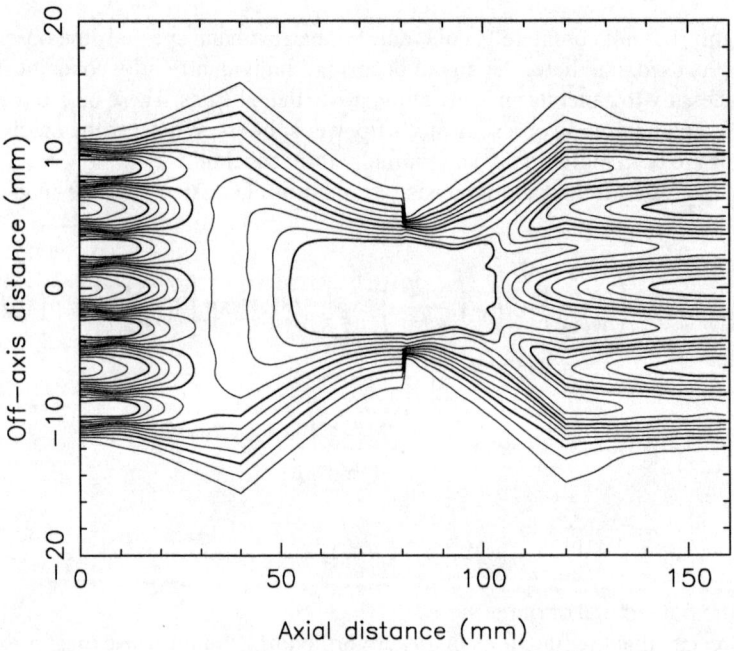

FIGURE 5. Four mutually incoherent cosine beams propagating through a Gaussian Beam Telescope with an aperture at the secondary focus. The aperture has a total width of 8mm, and the rest of the system is the same as Fig. 3

On the basis of these modes, we have constructed a model of a Gaussian Beam Telescope, where two thin lenses are separated by the sum of their focal lengths. We choose a system where thin lenses, having focal lengths of 40mm, are placed at 40mm and 120mm from the source plane. The wavelength is 1mm.

First, we constructed a source coherence matrix corresponding to two off-axis individually fully coherent but mutually uncorrelated Gaussian beams. Here, the Gaussians are placed at -6mm and +6mm, and the 1/e width of each is 2mm. The coherence matrix is traced through the Gaussian Beam Telescope using scattering matrices, and the intensity determined at each plane [9]. The resulting beam pattern is shown in Fig. 3. Clearly, the beams propagate through the system as expected, and are re-imaged at the final plane. Because the two beams are incoherent, they simply add in intensity when they overlap. All of this behaviour is contained completely within the single coherence matrix.

We now construct a source coherence matrix comprising two off-axis individually coherent and fully correlated Gaussian beams. In this case, the source beams are positioned at -8mm and +8mm, and the 1/e width of each beam is 2mm. The result is shown in Fig. 4. The most notable feature is that fringes are now formed at the second focal plane where the beams overlap. The form of these fringes is precisely what would be expected from Fourier analysis.

Finally, we analyse a more demanding problem. Here we place 4 individually coherent, but mutually incoherent, cosine fields in the source plane. The source beams were placed at -9mm,-3mm,+3mm, and +9mm, and they all have a width of 6mm. We also place an aperture, having a total width of 8mm, at the second focal plane. The aperture was modelled through a scattering matrix based on recursion calculations [10]. The result is shown in Fig. 5. Clearly, the beams simply add in intensity as expected, and the aperture truncates the field. At the image plane, the fields reappear, but now reduced in intensity and degraded in resolution. Again, we emphasise that the beam were not traced individually.

CONCLUSION

We have described a procedure for modelling the behaviour of long-wavelength partially coherent optical systems. The technique has many potential applications: the analysis of planar bolometers with cold apertures, the analysis of overmoded waveguide bolometers, the modelling of complete imaging systems, and the simulation of large reflector antennas.

REFERENCES

1. Withington, S., Yassin, G. and Murphy, J.A., *Dyadic analysis of partially coherent submillimetre-wave antenna systems,"* To be published in IEEE Trans. Antennas Propagat., 49, (2001).
2. Toraldo Di Francia, G., *Degrees of Freedom of an Image*, J. Opt. Soc. Am., 59, 799-804, (1969).
3. Bedinelli, M., Consortini, A., Ronchi, L., and Frieden, B.R., *Degrees of Freedom, and Eigenfunctions for the noisy Image*, J. Opt. Soc. Am., 64, 1498-1502, (1974).
4. Withington, S., and Yassin, G., *Power Coupled Between Partially-Coherent Vector Fields in Different States of Coherence*, Accepted for publication in J. Opt. Soc. Am. A.
5. Withington, S., and Yassin, G., *Power Coupled Between Waveguide Fields in Different States of Coherence*, in preparation.
6. Wexler, A., *Solution of Waveguide Discontinuities by Modal Analysis*, IEEE Trans. Microwave Theory Tech., 15, (1967).
7. Murphy, J.A., *Radiation Patterns of few-moded horns and condensing lightpipes*, Infrared Physics, 31, 291-299, (1991).
8. Goldsmith, P.F., *Quasioptical Systems*, IEEE Press, New York, 1998.
9. Withington, S., and Yassin, G., *Modal Analysis of Partially Coherent Submillimetre-wave Quasi-Optical Systems*, IEEE Trans. Antennas Propagat., 46, 1651-1659, (1998).
10. Murphy, J.A., Withington, S., and Egan, A., *Mode Conversion at diffracting apertures in millimetre and submillimerte-wave optical systems*, IEEE Trans. Microwave Theory Tech., 41, 1700-1702, (1993).

Corrugated Horn Design for HFI on PLANCK

J. A. Murphy[1], R. Colgan[1], E. Gleeson[1], B. Maffei[2], C. O'Sullivan[1] and P.A.R. Ade[2]

[1] National University of Ireland, Maynooth, Co. Kildare, Ireland
[2] University of Wales, Cardiff, CF24 3YB, UK.

Abstract. In this paper we report on the back-to-back corrugated feed horn design for the High Frequency Instrument on the PLANCK Surveyor. Special single moded Gaussian profiled horns have been developed for the 4 lowest frequency channels (100 GHz, 150 GHz, 217 GHz & 353 GHz). These feed structures produce very pure Gaussian radiation patterns with sidelobe levels reduced well below −30dB. Similar few moded horn antennas are being proposed for the highest frequency channels (545 GHz and 850 GHz) to give non-diffraction limited performance with high throughput. The modelling of the horns uses a rigorous electromagentic mode matching technique that is found to give good agreement with measurement.

INTRODUCTION

On the PLANCK Surveyor the telescope is directly fed by two arrays of horn antenna feeds with no front-end optics to condition the radiation patterns of the individual horns. The sidelobe levels of the illuminating horn feeds on the telescope must therefore be sufficiently low so that strong background sources do not cause contamination of the faint signal. Simple single moded and multi-moded horn structures such as conical horns and Winston cones are inadequate in this application to reach the level of rejection required without compromising the resolution. Therefore, more complex corrugated horn designs that produce much lower level sidelobes have to be considered [1,2,3]. It is important to have reliable modelling tools for predicting with a high degree of confidence the radiation patterns of such horns antennas, especially since we wish to profile the horn shape in order to optimize performance. We consider two alternate electromagnetic models as will be discussed in the next section of this paper.

Primarily we are concerned with the feed horn design of the HFI (High Frequency Instrument) on PLANCK [4]. Shown is Figure 1 is the basic quasi-optical concept for a single detector channel [5]. The corrugated front horn controls the beam pattern on the telescope, while the waveguide section controls the modes that can propagate. The back horn feeds the filter stack, which defines the frequency band and rejects any stray IR radiation. The 100mK horn is over-moded and so can accept all incident radiation. From a modelling point of view this arrangement can be idealized as a black body cavity illuminating the back horn (see Figure 2(a)) [6]. The high frequency channels on the HFI employ multi-moded horns with the number of modes being determined by the wave-guide filter properties (guide radius and corrugation depths). We can apply the same electromagnetic models both to the lower frequency single mode and the higher frequency multi-mode horns [7]. This procedure and related issues such as bandwidth and radiation characteristics are considered in detail in later sections of the paper.

MODELLING OF HORN ANTENNAS

Two electromagnetic methods are available for modelling horn antennas. The first based on based on mode-matching, and involving complex scattering matrix calculations, is accurate with exact horn profile definition needed for the analysis. The second approach based on hybrid modes is more approximate in that the corrugated waveguide wall is regarded as a non-uniform impedance surface.

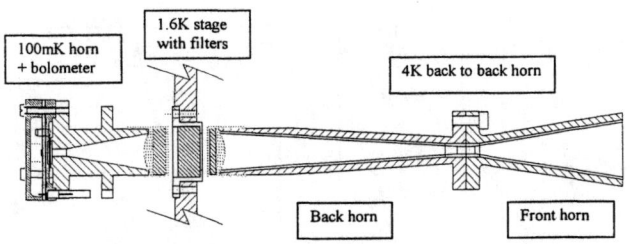

Figure 1: Back to back horn design

The mode matching approach assumes that the waveguide can be regarded as a sequence of very short smooth walled segments making up the slots and ridges of the corrugations [8]. Since the structures are cylindrically symmetric pure cylindrical waveguide modes are used for describing the propagation of the electromagnetic radiation. At each corrugation step account is taken of all of the scattering that occurs at the sharp transition in the guide width (see Figure 2(b)). One of the main advantages of the approach is that horns with an arbitrary shaped profile can be handled readily and the fields in the actual horn flare can be modelled right up to the horn aperture. The radiation pattern of the horn can then be written in terms of either a fully coherent or a partially coherent sum of the far field patterns of the modes in the horn aperture (depending on whether the horn is single moded or multi-moded). These features have been incorporated into a software program called SCATTER.

In the case of the surface impedance model (appropriate when there are many corrugations per wavelength) an average surface impedance can be assumed [1]. The true of modes of propagation in that case turn out to be hybrid mode combinations of TE and TM modes (true TE or TM modes do not satisfy the boundary conditions). Solving the waveguide equation for these hybrid modes is rather complex compared to smooth walled guide (see Figure 3, & [1]). As the modes approach the cut-off condition that the waveguide number β approaches zero, the so called HE modes become pure TM, while the corresponding EH modes become TE-like. The modes also have a high frequency cut-off which is approached as the corrugation slots get deeper for a given guide radius, as can be seen in Figure 3. Being modes of propagation these hybrid modes are not scattered in a corrugated waveguide. Furthermore, for a simple conical corrugated horn we can make the approximation that essentially the mode propagates into the horn without scattering and attains a spherical phase error with radius of curvature equal to the horn axial length.

In corrugated conical horns with resonant grooves ($\lambda/4$ deep) balanced hybrid modes are formed with tapered fields at the aperture, as illustrated in Figure 4. Thus, in a perfect non scattering horn the modal radiation patterns are not coherently related to each other, and so to obtain the overall beam pattern for a multi-moded horn the individual modal patterns must be added in quadrature. However, if scattering is expected to occur (as in a shaped overmoded corrugated horn) we have to apply the mode matching software. The application of hybrid modes is thus limited to conical horns and the simple analysis of waveguide filters.

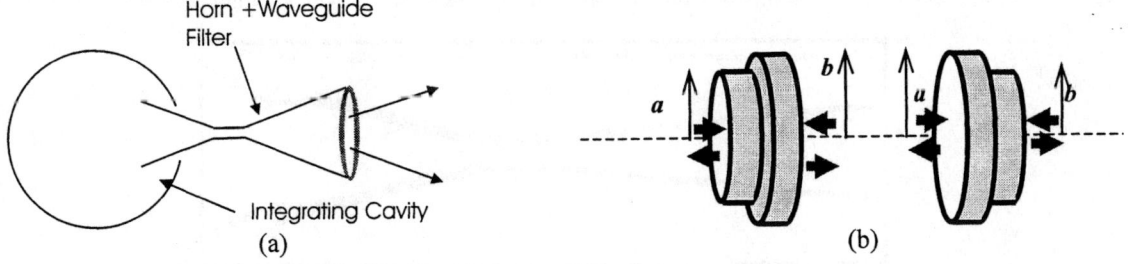

FIGURE 2. (a) . Idealisation of Back-to-back Feedhorn Design & (b) waveguide step details

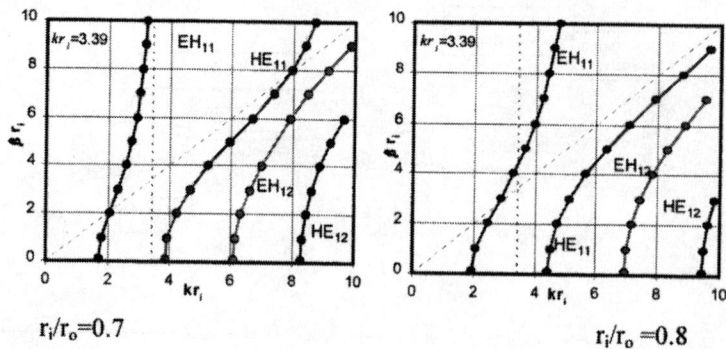

Figure 3: Dispersion curves for modes of azimuthal order 1 in a waveguide filter showing $kr_i=3.39$.

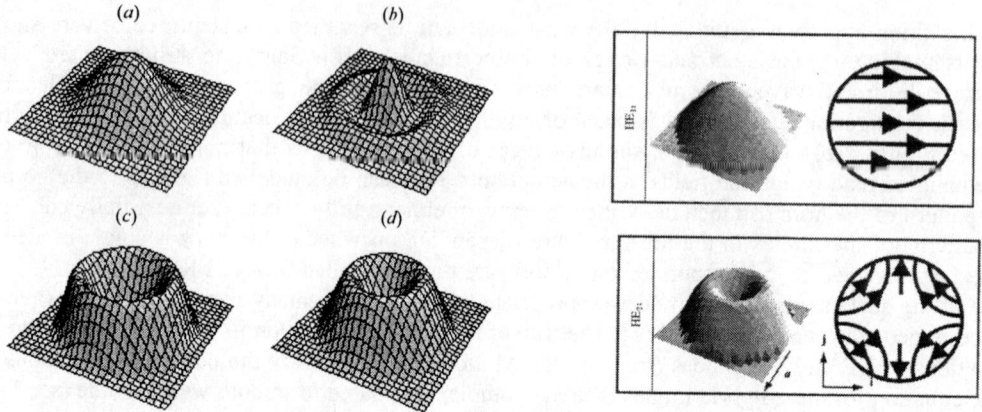

Figure 4: Aperture field intensities for (a) HE_{11}, (b) HE_{12}, (c) EH_{12} and (d) HE_{21} balanced hybrid modes.

SINGLE MODE CORRUGATED HORN DESIGN

Corrugated feed horns have several advantages for application to microwave background experiments. Such horns produce linearly polarised symmetric beam patterns with low sidelobe levels. However, the PLANCK demands on edge taper make it hard to achieve acceptable resolution with conical corrugated horns as the sidelobe levels are higher than the required −30dB at the primary reflector. This has led to the development of so called profiled Gaussian horns which produce a beam pattern with an extremely high purity Gaussian beam (see Figure 5 & [9,2]). Sidelobe levels well below −30dB can be achieved, and there is a high level of agreement between the mode matching model predictions and the measurements as is clear in Figure 6. However, the horns tend to have a narrower bandwidth (in terms of beam pattern quality) than linearly tapered conical corrugated horns.

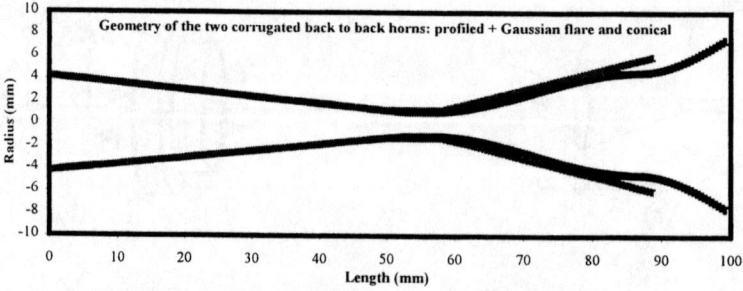

Figure 5: Superposition of the geometry of the two corrugated prototype horns.

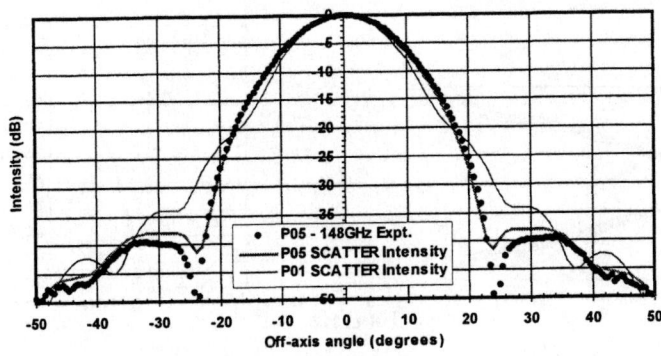

Figure 6: Far-field beam pattern (experimental and modal matching).

The phase centre is estimated by fitting a spherical phase correction to the far-field phase error, which arises because the far field calculations are referenced to the horn aperture. Empirically, by running several examples, it was determined that the horn shape influences how far down the horn the phase centre is located, with this position being adjustable to some degree by varying the opening angle of the Gaussian flare on the horn (see Table 1 and Figure 9, which summarises the horn design achieved by B. Maffei).

Experimental measurement of the transmission profile over a wide frequency range using an FTS arrangement at QMW shows that the transmission pass bands are limited for both conical and flared corrugated horns (see Figure 7). This is not experienced with smooth walled horns, and implies in fact, for multi-mode corrugated horn operation, one needs wave-guide filter slots that are not too deep otherwise the fundamental HE_{11} is cut-off. However, if the slots are too shallow the EH_{11} mode can propagate (there is not a sharp cut-off for the band). Evidence for this is illustrated in Figure 9 in which the edge of the pass-band for the 143 GHz horn was examined as a function of slot depth. Careful modelling using the mode matching approach was therefore required to optimize the wave-guide filter operation for either single mode operation (where the wave-guide acts as the high pass filter for the system) or multi-mode operation (in which the pass-band is defined by a quasi-optical filter behind the back to back horn, as shown in Figure 1).

TABLE 1. List of Single Mode Horn Parameters

Frequency	Aperture Radius	Horn Length	Phase Centre Position
100 GHz	7.81 mm	88.9 mm	4.4 mm
143 GHz	5.49 mm	68.4 mm	3.6 mm
217 GHz	4.16 mm	82.5 mm	3.2 mm
353 GHz	3.69 mm	75.2 mm	1.8 mm

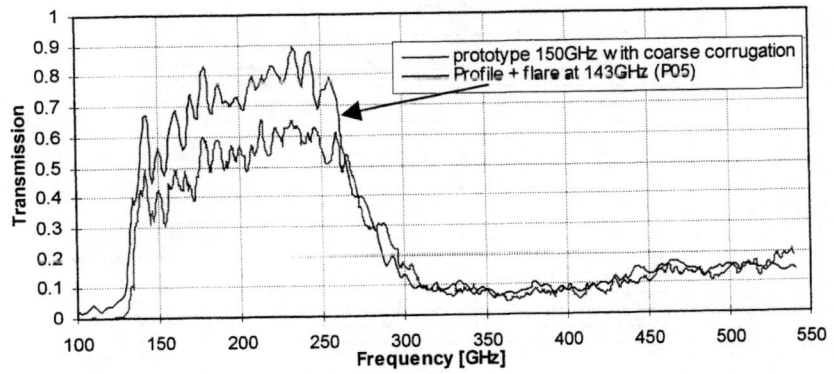

Figure 8: Transmission of corrugated waveguide filter shows band for fundamental mode is limited.

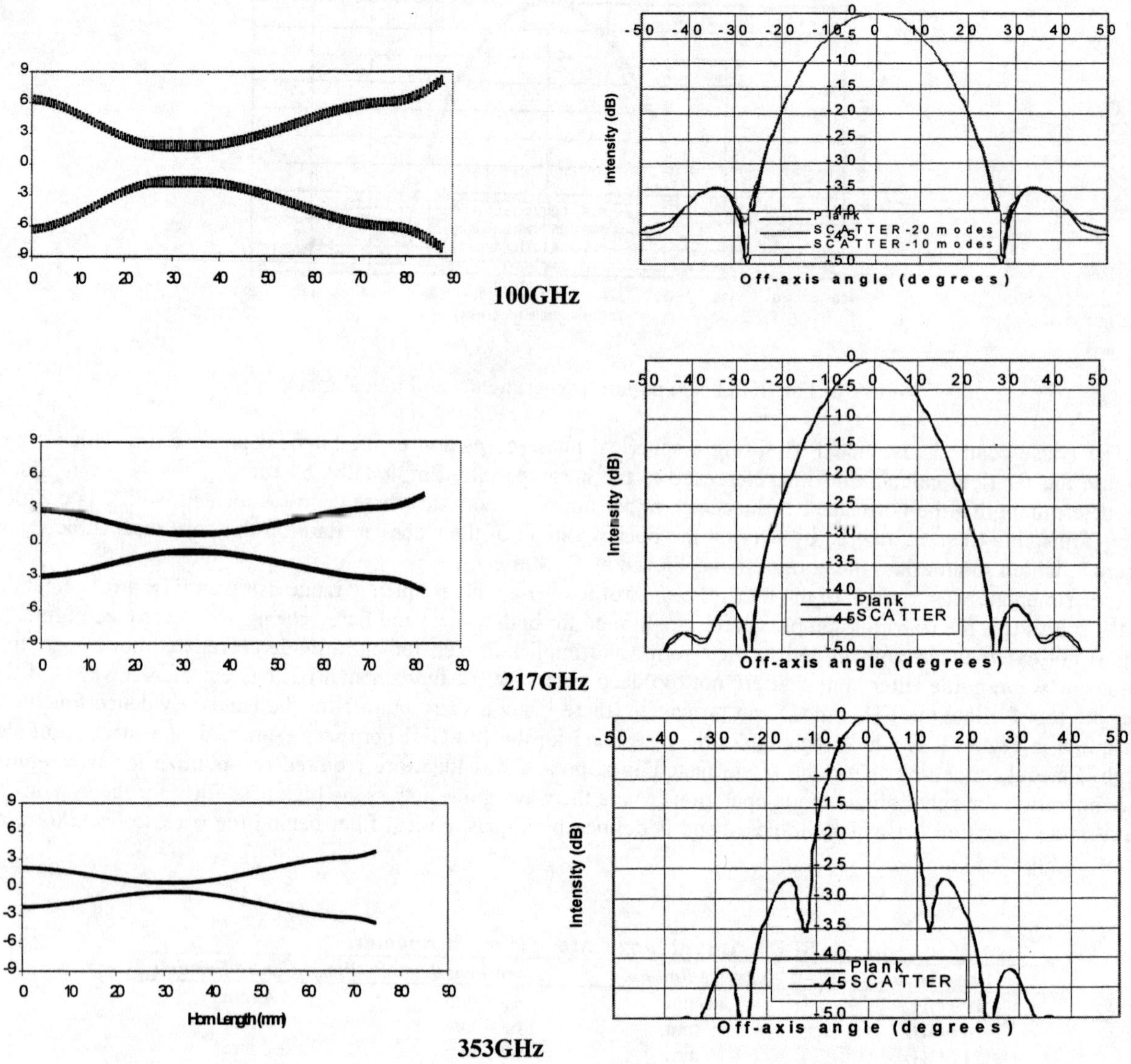

Figure 7: Horn Profiles (exaggerated, distances in mm) & Beam Patterns with two model predictions

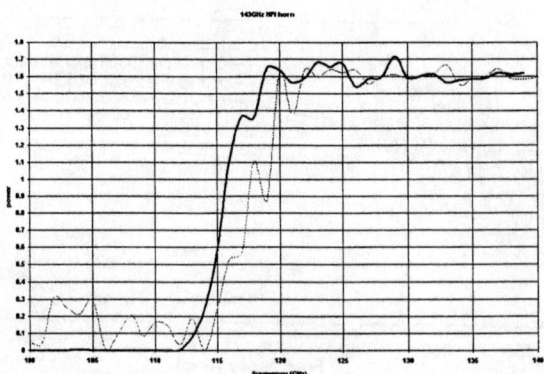

Figure 9. Model predictions for transmission at band edge (deep slots - thick line, shallow slots – thin line)

MULTI-MODE HORN DESIGN

Multi-moded horns provide high throughput when diffraction limited resolution is not required on a telescope. This is the case for the two high frequency channels on PLANCK (at 545GHz and 857 GHz). Such overmoded horns produce somewhat broader beams on the sky than would be the case if the system fed by the same horns were single moded. Because of extreme sensitivity to background contamination we still of course require very low edge taper and spill over levels here, as was the case for the single mode channels. In the case of the multi-moded horns it is effectively the waveguide filter that determines which modes can propagate over a certain frequency band. Figure 10 shows the low frequency and high frequency cut-offs for the hybrid modes that propagate in corrugated waveguides.

In order to experimentally verify the proposed multi-mode horn designs for PLANCK a 150 GHz scale mode of the proposed 545GHz design was manufactured and tested at QMW. A conical taper was chosen and by performing measurements at 150GHz higher accuracy could be achieved, which is important to test the predictive reliability of the modelling software and the assumptions made about the number of modes that can propagate (see Table 2 & Fig. 11). By also testing the same horn at 200GHz we mimic the operation of the PLANCK 850 GHz channel, in which a relatively large number of modes should be allowed to propagate (see Table 2 & Fig. 12).

In general, as illustrated by Figs. 11 & 12, for a given horn shape and aperture radius the measured far field radiation pattern of the horn (which illuminates the telescope) becomes broader also and tends to become top-hat like for long horns and many modes. Also shown on the plot are the model predictions for the horn. By comparng Figs. 11 & 12 one notes that as more modes are added for a given horn geometry, by simply operating such a horn at progressively higher frequencies, the beam pattern tends to have about the same width (this is, of course, expected from consideration of conservation of throughput for such a system). There is relatively good agreement between model and measurement, but with some indication of transmission loss for the highest order modes that can propagate.

TABLE 2. List of propagating pure TE/TM modes (for the SCATTER software) for prototype muilti-moded horn at 150GHz, cut-off wavenumbers for r_i=1.62mm.

Frequency/GHz	Kr_1	Propagating Modes
100	3.39	TE11, TM01, TE21
150	5.09	TE11, TM01, TE21, TM11, TE01, TE31
200	6.78	TE11,TM01,TE21,TM11,TE01,TE31,TM02,TM21,TE12,TM31,TE51,TE22

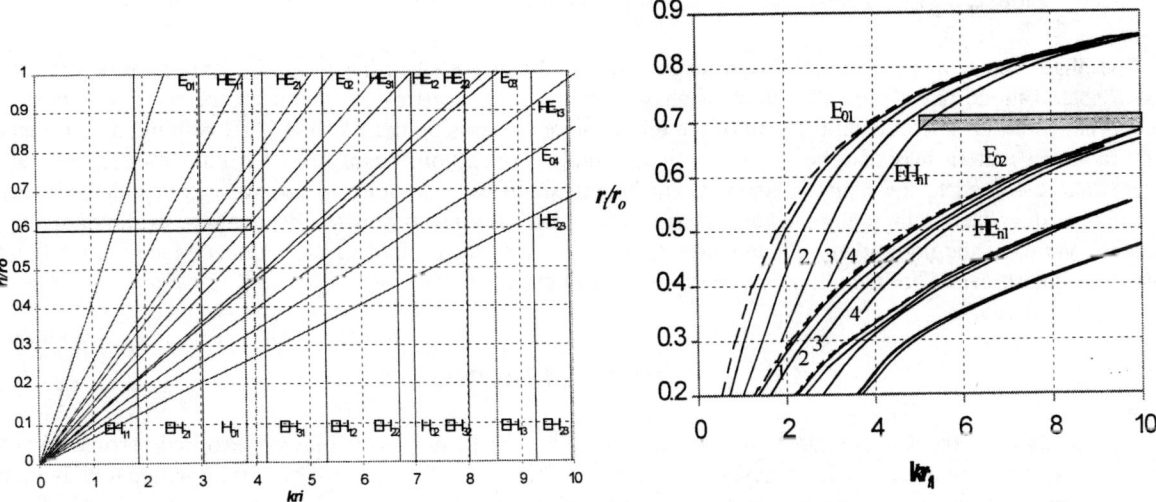

Figure 10: Low frequency & High frequency cut-off charts for lower order hybrid modes.
High frequency cut-off chart. For r_i/r_o=0.7, HE_{11} cutoff at $kr_i \approx 10$

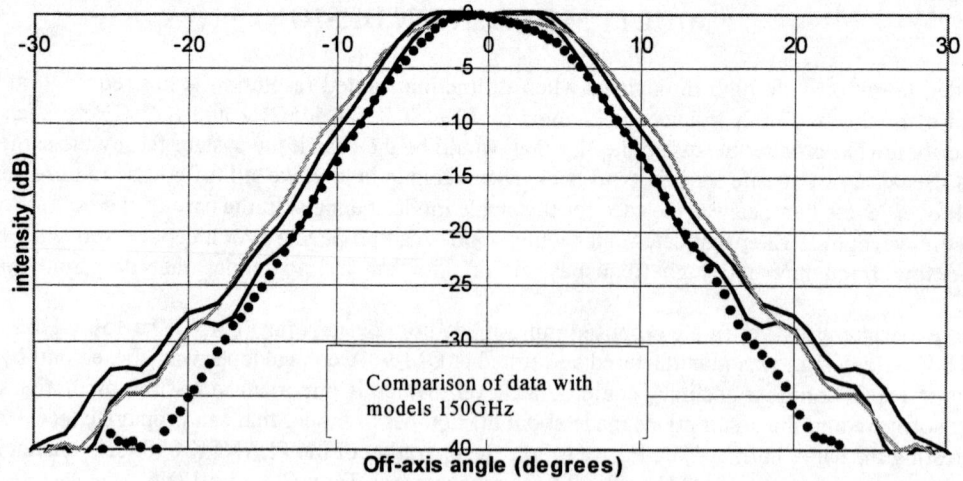

Figure 11. Beam pattern of prototype scale mode horn for 150GHz (test data - dots, hybrid model - light grey, mode matching - black (n=2 & n=3)).

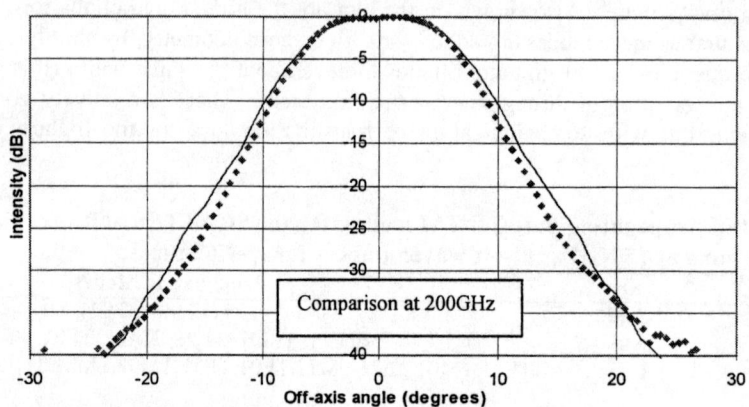

Figure 12: Beam pattern at 200 GHz for multi-moded scale horn (test data - dots, model – line)

The phase centre of a multimoded horn is an ill-defined concept since the fields are only partially coherent and each spatial mode has a separate centre of phase curvature. In terms of operation on the telescope what is most important is either maximising the gain of the telescope or minimising the FWHM (3dB resolution). The quality of the radiation pattern may also be important since a multi-moded horn beam results from summing several modal fields and can lead to the main telescope beam having significant structure. Thus, we can calculate for a perfect telescope (oversized with negligible edge taper and aberrational effects) how the beam shape or the on-axis gain on the sky varies as the horn aperture is moved relative to the true focal plane of the telescope (see Figure 13). In this way the "phase-centre" of a multi-moded horn can be calculated.

CONCLUSIONS

We summarize the design strategy for the HFI horn array for PLANCK as follows. Profiled corrugated horns are to be used for single moded horn design. Such horns offer very low edge taper levels compared with conical corrugated horns leading to optmised resolution. Futhermore, the position of phase centre can be adjusted by varying opening angle of the flare.

The definition of the multi-moded horns is now being finalized. Here again profiled horns are to be used to control position of the "phase centre" although the effect on the beam pattern edge taper is less pronounced. Aberrations appear to lead so some distortion and have to be included in the determination of resolution [10].

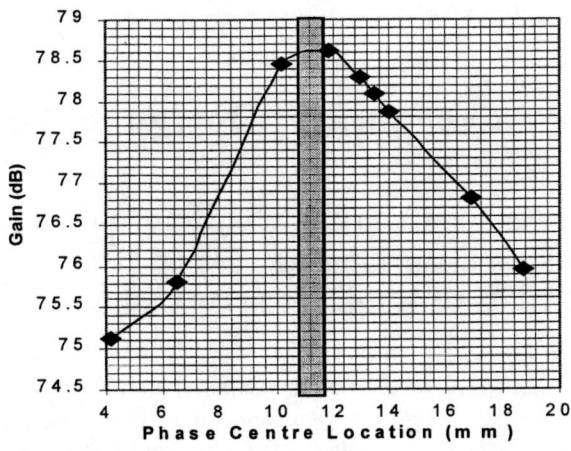

Figure 13. Gain of 545GHz multi-mode horn as function of aperture position with respect to telescope focal plane.

ACKNOWLEDGMENTS

The authors wish to acknowledge the financial assistance of Enterprise Ireland.

REFERENCES

1. Clarricoats P. J. B. and Olver A. D., Corrugated Horns for Microwave Antennas, IEE Electromagnetic Waves Series 18, Peter Perigrinus, London,1984, pp 20-57.
2. Maffei B. et al, International Journal of Infrared and Millimeter Waves 21, 2023-2034 (2000).
3. Padman R. and Murphy J.A., Infrared Physics 31, 441-446 (1991).
4. Lamarre J-M. et al, Astro. Letters and Communications, 37, 161-170 (2000).
5. Church S. E. et al, "A compact high-efficiency feed structure for Cosmic Microwave Background Astronomy at Millimetre Wavelengths", Proceedings 30th ESLAB Symposium Submillimetre and Far-Infrared Space Instrumentation, ESA SP-388, edited by E.J. Rolfe. 1996, pp 77-80.
6. Murphy J. A. and Padman R., Infrared Physics 31, 291-299 (1991).
7. Murphy J. A. et al, Infrared Physics and Technology, 41, 515-528 (2001).
8. Olver A. D. et al, Microwave Horns and Feeds, IEE Electromagnetic Waves Series 39, IEEE Press, New York, 1994, pp 100-122.
9. del Rio C. et al, IEEE Transactions Antennas & Propagation AP-47, 1440-1448 (1999).
10. Yurchenko V. B., Murphy J.A. and Lamarre J-M, International Journal of Infrared and Millimeter Waves 22, 173-184 (2001).

Millimetre-Wave Optics Design & Verification

Créidhe O'Sullivan*, J. Anthony Murphy*, Stafford Withington[†], Ghassan Yassin[†], Eli Atad-Ettedgui[¶], William Duncan[¶], David Henry[¶], Willem Jellema[§] and Herman van de Stadt[§]

*National University of Ireland Maynooth, Co.Kildare, Ireland
[†] Cavendish Laboratory, Madingley Road, Cambridge, UK
[¶] UK Astronomy Technology Centre, Royal Observatory, Edinburgh, Scotland
[§] SRON, Groningen, The Netherlands

Abstract. Microwave background astronomy requires very high performance millimetre-wave optical systems. However, compact quasi-optics are difficult to design with any confidence using techniques developed for visible wavelengths. In this paper we investigate the performance of existing software design tools (ASAP, CODE V, GLAD) as well as a Gaussian beam mode analysis technique not yet available as commercial software. We have devised a set of test cases and used these to study the underlying methodologies and physics of these packages and we look at their ability to analyse millimetre systems and components. We have used GRASP as our benchmark software.

INTRODUCTION

In this paper we investigate the performance of a range of commercial optical design software packages (GLAD, ASAP and CODE V) in analysing the behaviour of millimetre-wave optical systems. These packages are not specifically intended for use at millimetre wavelengths but they represent the only types of optical design tools available. We investigate approximations that are inherent in the theoretical method on which the software analysis is based, and how these impact on long-wavelength predictions.

Several test examples have been chosen that highlight some of the discrepancies that can arise between field predictions from the different software packages when applied to the millimetre-wave regime. We have taken the physical optics package, GRASP, a software tool for reflector antenna design and analysis, as our benchmark software against which the results of the other packages are compared. It should be noted that because of its complexity and computational intensity, GRASP is more suited to design verification rather than initial instrument design. Future work will involve an experimental verification of some GRASP results.

ANALYSIS TECHNIQUES

At optical wavelengths, away from any abrupt changes in intensity distribution, energy can be considered to be transported along light rays obeying certain geometrical laws [1]. Ray-tracing packages, based on this assumption, have proved to be very successful. In the millimetre regime, however, the wavelength may be an appreciable fraction of component sizes and so cannot be neglected. Diffraction effects become important and the approach of geometrical optics is inadequate. Diffraction problems tend to be difficult and rigorous solutions are rare. Other approximations, suitable for this wavelength regime, must be used for speed and for analysing complex systems.

In optical design a source field must be propagated from one optical component to the next. Techniques such as the Method of Moments attempt to calculate source fields in a rigorous manner, others, such as those used by the software packages we considered, make simplifications. When a field is incident upon an aperture, for example, it is often assumed that the field over the opaque region is zero, whereas over the transparent region it is the same as it was in the absence of the aperture. The field over the input surface is generally a vector field but the assumption of paraxial propagation (made in ASAP, CODE V and GLAD) reduces it to a product of scalar solutions that propagate independently.

Propagating a field accurately onto the next optical component (solving the wave equation) requires diffraction integrals to be evaluated for each field point calculated. Rather than evaluating the integrals directly, it is possible to decompose the assumed source field into modes and then propagate the modes as required. This often simply consists of slipping the mode phases with respect to each other. A plane wave analysis [2] has the significant advantage that it is not limited to paraxial fields. Gaussian modes, on the other hand, are solutions of the paraxial wave equation [3]. Appropriately scaled, they allow an efficient representation of paraxial systems. In the Gabor approach [4], a field is decomposed into a discrete set of Gaussian beams shifted both laterally and in phase slope. In this paper we compare some results obtained using the commercial packages ASAP, GLAD, and CODE V (beam propagation algorithm). These are scalar diffraction packages based on a modal analysis of fields. GLAD and CODE V decompose the fields into plane waves, ASAP uses Gabor modes. In addition we have used results from the 'in-house' software package, PROFILE, which is based on a Gaussian Beam Mode Analysis (GBM) technique specifically applied to submillimetre-wave optical systems (see *e.g.* [5]).

The term physical optics, as we use it here, refers to the calculation of the field radiated by a reflector using an approximate surface current distribution determined from the incident magnetic field. The field on that part of the reflector not directly illuminated by the incoming field is assumed to be zero. We use GRASP, which combines Physical Optics and the Physical Theory of Diffraction (PTD, developed to correct for edge effects), as our benchmark.

TEST CASES

Our aim is to study the essential differences between packages using a relatively small number of components. As well as examples chosen from regimes where the approximations made by all packages are valid and good agreement would be expected, we have probed more extreme examples, though still typical of quasi-optical systems in the far infra-red. The cases we describe in this paper, aperture stops and off-axis reflectors, are the basic fundamental components of many quasi-optical systems. They illustrate many of the essential features of modelling techniques and the results we present allow the estimation of the accuracy with which more complex multi-element systems can be analysed. Apertures and off-axis mirrors are readily modelled by GRASP.

We have investigated components with diameters down to a few wavelengths and quasi-collimated beams with F-number between 3 (typical of a horn antenna) and 30 (typical of quasi-collimated beams in interferometers and diplexers). We have used uniform illumination by an infinite plane wave in certain examples. The off-axis mirrors we investigate have a large angle-of-throw typical of many current optical designs (e.g. HIFI [6]). In the following section we describe some of the example test cases chosen.

TEST CASE RESULTS

Apertures

The first set of test cases modelled the diffraction effects of beam truncation at an aperture stop in a screen. The near and far field intensity patterns were calculated for uniform plane wave illumination of apertures of radius between 3λ and 30λ. Some results are shown in Figures 1 and 2. Figure 1 shows an example of the near-field results for the paraxial packages (GBM analysis results closely matched those of GLAD). For the paraxial packages the results simply scale with aperture diameter as expected. These examples were chosen so that an on axis minimum is predicted using a simple Fresnel diffraction calculation. The paraxial packages do produce this on-axis minimum. However, when compared with GRASP (Figure 2), some interesting differences arise particularly in the case of the smallest aperture (radius = 3λ). In the GRASP data there is no on-axis minimum and the overall pattern is smoother.

In terms of paraxial Fresnel diffraction, the on-axis minimum in these examples can be predicted by summing the contribution to the overall fields of neighbouring Fresnel zones [1]. The examples were chosen so that an even number of zones (four in our case) are seen to fill the aperture, and when viewed from the ouput plane their contributions approximately cancel. However, the non-constant obliquity factor from zone to zone is not included in such calculations. The GRASP plots suggest such an omission has a significant effect on the beam patterns calculated for the smallest aperture, where the angle associated with the obliquity factor is largest, destroying the perfect cancellations of neighbouring Fresnel zones. For the largest (30λ) aperture, where the on-axis minimum is present in the GRASP data, the obliquity angle of the outermost zone is relatively small (7.6°). In that case the off-axis structure predicted by the paraxial packages is in very good agreement with those predicted GRASP. The similarity of the GRASP near-field results when calculated with and without PTD show that it is the obliquity factor rather than edge effects that is the dominant source error in these cases. This indicates that propagation over short distances ($z < 2.5\,a$) can result in errors in the fine structure of the beam pattern for collimated beams.

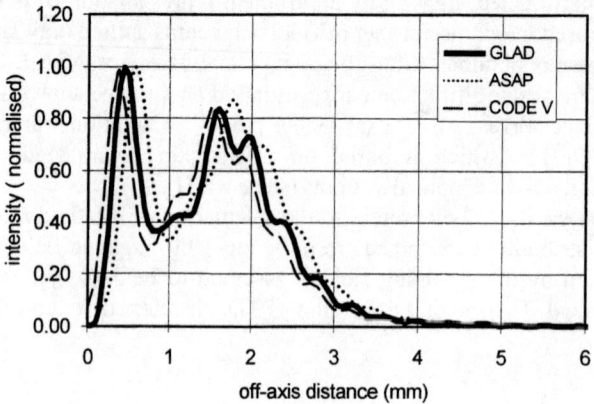

FIGURE 1. Intensity distribution in the near field ($z_{out} = 25$mm) of an aperture of radius 10mm calculated using GLAD, ASAP and CODE V. $\lambda = 1$mm.

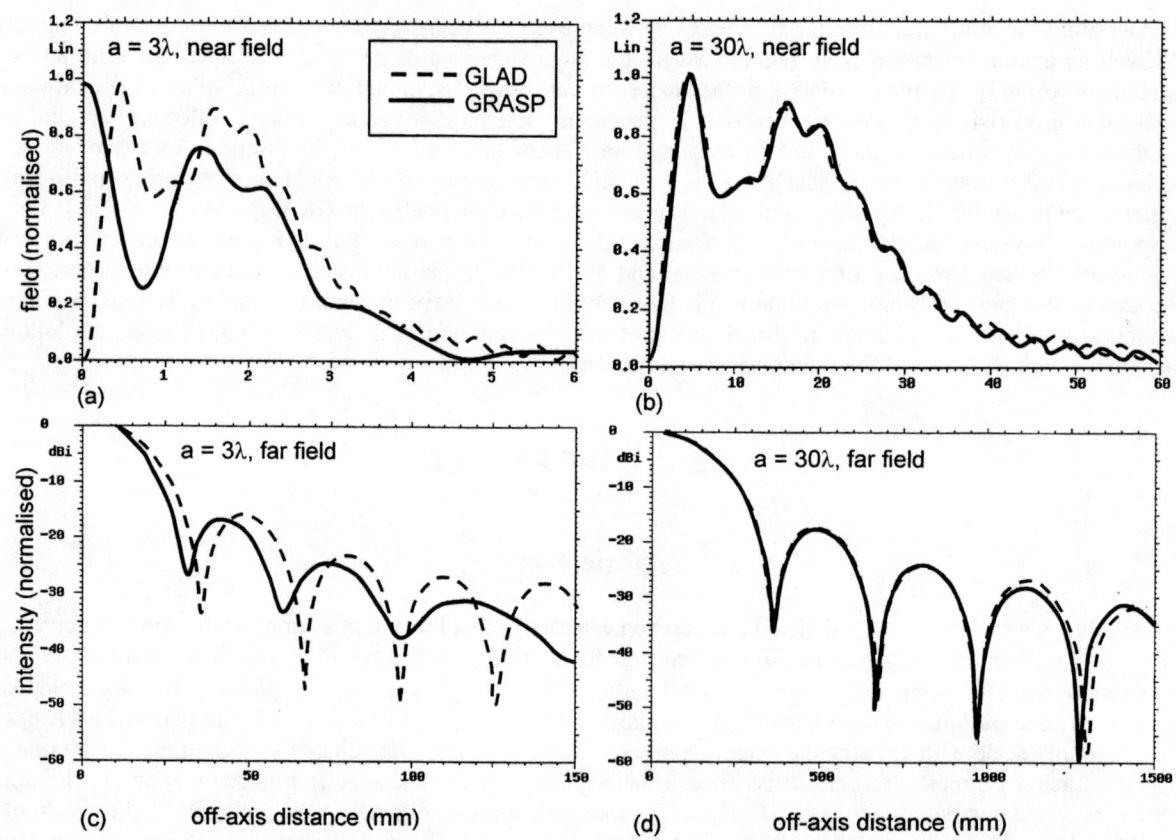

FIGURE 2. Beam amplitude in the near ($z_{out} = a^2/4\lambda$) and far ($z_{out} = 20a^2/\lambda$) field of an aperture of radius $a = 3\lambda$ and $a = 30\lambda$, calculated using GLAD and GRASP. GLAD is taken to be representative of the paraxial packages. $\lambda = 1$mm in all cases.

In the far field, by contrast, the angles associated with the obliquity factor are always relatively small and one would expect much closer agreement in the form of the beam between GRASP and the paraxial packages. This is clearly seen in the plots of Figure 2 (c) and (d) which show the far field patterns of the two apertures. Our data show that the paraxial packages are in broad agreement with each other, predicting the familiar Fraunhofer radiation pattern.

The discrepancy that appears in Figure 2(c) in the smallest aperture result is not due to an obliquity term, but rather due to the fact that the beam spreads out into a large (non-paraxial) angle. For the modal approach used in some packages, for example, the highest order modes may not be propagating paraxially. There is also the issue that, in far field calculations involving the paraxial approximation, it is assumed that $\theta \approx \sin \theta \approx \tan \theta$. An increasing lateral discrepancy between GRASP and the other paraxial programs will occur at off-axis distances corresponding to large angles as viewed from the beam waist.

Off-Axis Mirrors

This second set of test cases involved modelling diffraction effects associated with re-imaging a coherent beam at off-axis paraboloidal or ellipsoidal mirrors of finite size (see Figure 3).

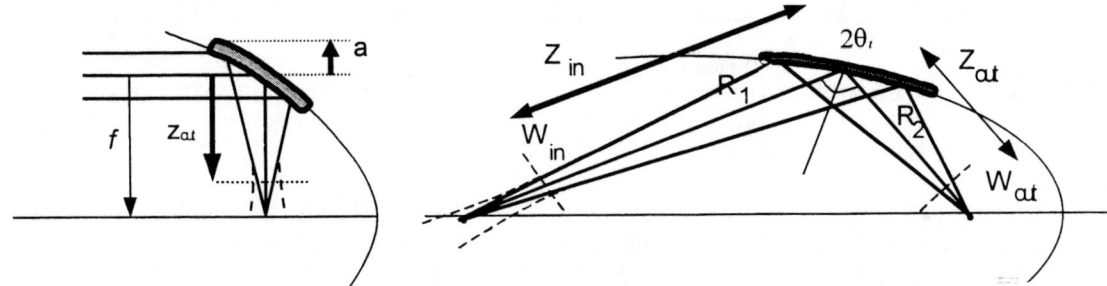

FIGURE 3. Off-axis mirror test cases. (a) Paraboloidal mirrors of focal length f and projected aperture a are used to reflect parallel beams forming a waist at z_{out}. (b) Ellipsoidal mirrors reflect wavefronts with a finite radius of curvature (R_1) producing an output beam waist at z_{out}.

In the short wavelength limit ellipsoids and hyperboloids are perfect phase-transforming reflectors of wavefronts with a finite radius of curvature. Paraboloidal mirrors are used as reflectors of parallel wavefronts. For strongly diffracting beams, however, phase aberrations occur beacause the radius of curvature of the phase front is not simply linearly related to distance along the beam. In the submillimetre regime, projection effects give rise to cross-polar scattering and spatial aberrations even at the wavelength for which the mirror was designed. We chose the test cases in this section to investigate the ability of the selected software packages to handle off-axis reflection and the resulting phase and amplitude distortions. All our test cases involved a 90° angle of throw. The sources investigated were Gaussian beams, with plane wave illumination chosen for one of the parabolic examples to mimic the operation of a telescope. There were software bugs in the first release of the CODE V beam propagation algorithm and so we do not include the results here

Figure 4 shows poor agreement between the packages, especially in the plane of where we would expect some asymmetry. Both ASAP and GLAD fail to predict the correct sidelobe structure or level. GLAD underestimates the sidelobe level by up to 10dB and predicts an almost symmetric beam. The sidelobe level and asymmetry calculated by ASAP are closer to those of GRASP, but the main beams differ by several dBs. There is better agreement in the other plane where the output beam is expected to be symmetric. GLAD and GRASP match down to -25dB, with GLAD's sidelobe level again lower, but in this case by about 5dB. ASAP, on the other hand overestimates the sidelobes by up to 10dB. The main beams show good agreement out to ~15°.

The approximations inherent in the packages must be considered in order to understand this level of disagreement. In the Gabor approach an electromagnetic field is decomposed into individual Gaussian beamlets, which can then be propagated through optical components using ray tracing methods. While ASAP is based on such a Gaussiam beam decomposition, the full Gabor representation is not implemented. One elementary Gaussian beam, rather than a fan of beams, is used to represent the field at each point on a spatial grid. This causes problems when attempting to model structure in beams on the scale of a few wavelengths. GLAD makes several simplifications which we might expect to affect the results of these test cases. The first is that it is restricted to apertures placed normal (or with small tilts) to the beam. Mirrors are infinite and their edges are defined by placing a suitable aperture in front. The level of trucation by the mirror is therefore an approximation. The second source of possible errors is the level of amplitude distortion caused by the mirror. The mirror can be designed so that it acts as an almost perfect phase transformer producing very little phase aberration, but projection effects will always introduce amplitude distortion. GLAD calculates aberrations by ray-tracing through the volume of conic sections before switching back to the ususal diffraction propagation. When doing this it considers the optical path length difference introduced, and uses these to calculate the phase aberration imposed on the beam. Amplitude distortions do not appear to be adequately modelled in this analysis.

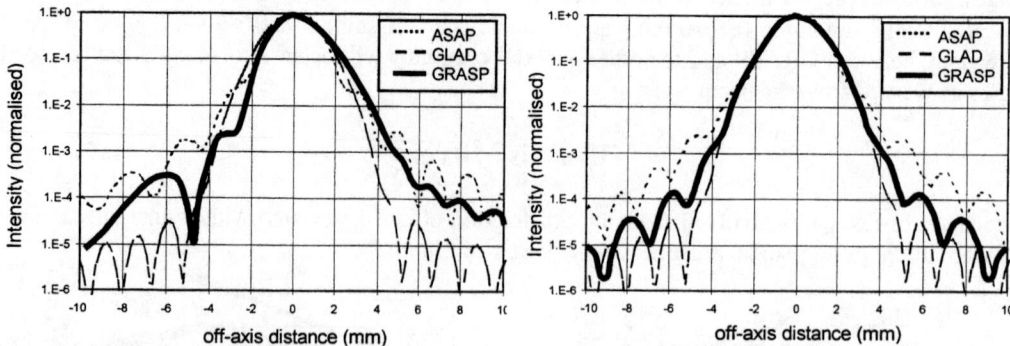

FIGURE 4. Intensity pattern at the output beam waist ($z_{in} = z_{out} = f = 12.57$mm) for an ellipsoidal mirror of projected aperture $a = 1.5 \times W$. (a) shows a cut in the plane of asymmetry, (b) in the plane of symmetry. $\lambda = 1$mm and $W_{in} = 2$mm in both cases. The beams were calculated using ASAP, GLAD and GRASP.

In some test cases the Gaussian input beam was replaced with a scalar horn aperture field. Both ASAP and GLAD can take an arbitrary complex field as input. Modelling the sharp cut-off in the horn field posed some particular problems for ASAP, however. Main beam predictions could be improved by using an alternative decomposition of the input field into a single fan of Gaussian beams centred on the origin. A full Gabor representation of the field was not possible. ASAP was the only paraxial package to predicted differences due to the polarisation direction of the input field, though they were significantly smaller than those calculated by GRASP.

CONCLUSIONS

In conclusion, it is clear that none of the commercially available software investigated is ideally suited to model submillimetre optical systems. Under certain conditions discussed they do give good results but it is important to bear their limitations in mind particularly when interested in sidelobe structure. All our results have been compared with those of GRASP, which we take to be correct, and our test cases have been restricted to those that can be easily modelled by it. We have tested some GRASP predictions against both Method-of Moments calculations and experimental results at microwave frequencies, and found good agreement. Future work will look at an experimental verification of some GRASP results.

ACKNOWLEDGMENTS

This work has been carried out as part of an ESA research contract (13043/98/NL/NB) 'Far IR Optics Design and Verification Tools'. The authors would like to acknowledge Peter de Maagt, Gert Ulbrich and Erico Armandillo of ESA for their useful comments. We thank TICRA for their help with GRASP.

REFERENCES

1. Born, M., and Wolf, E., Principles of Optics, Cambridge University Press, 1999.
2. Lawrence, G., "Optical system analysis with Physical Optics codes", SPIE no 766-18, O-E/Lase, 1987.
3. Goldsmith, P. F., Quasioptical Systems: Gaussian Beam Quasioptical Propagation and Applications, IEEE Press, 1998.
4. Einziger, P. D., Raz, S., and Shapira, M., J. Opt. Soc. Am. A, 3, 508-522 (1986).
5. Murphy, J. A., and Withington, S., Infrared physics & Technology, 37, 205-219, (1996).
6. de Graauw, Th., and Helmich, F.P., "Herschel-HIFI: The Heterodyne Instrument for the Far-Infrared" in The Promise of the Herschel Space Observatory, edited by G. L. Pilbratt et al., ESA SP-460, 45-51, (2001)

Electromagnetic Modelling of Few-Moded Winston Cones in the Far-Infrared

Emily Gleeson[1], J. Anthony Murphy[1], Sarah E. Church[2], Ruth Colgan[1], Créidhe O'Sullivan[1]

[1] *National University of Ireland, Maynooth, Co. Kildare, Ireland.*
[2] *Stanford University, Stanford, CA 94305, USA.*

Abstract. Winston cones have traditionally been used as detector feeds in far-infrared cosmological experiments, such as SuZIe, the Sunyaev-Zel'dovich Infra-red Experiment [1] on the CSO. They are usually designed using ray tracing, which becomes a very poor approximation when the number of spatial modes propagated by the horn is small in number, often the case at the longest wavelengths. We describe a more accurate approach involving electromagnetic modelling of Winston cones using a rigorous electromagnetic mode matching technique. It is straightforward to also consider the case of few-moded corrugated Winston cones, which offer lower sidelobe levels than smooth walled cones which is important for high sensitivity experiments. Furthermore, the mode matching technique allows more complex structures such as back-to-back Winston cones and the detector cavities to also be analysed.

INTRODUCTION

A Winston cone is an off-axis parabolic profiled horn designed to maximise the collection of incoming rays within a field of view determined by its dimensions (see Figure 1). Equation 1 describes the Winston cone geometry where z is the distance from the horn throat, ρ is the horn radius and d is the diameter of the horn throat. Equation 2 defines the maximum angle, α, in the geometrical limit at which the incoming rays reach the exit aperture of the cone where D is the diameter of the horn aperture.

$$z^2 = \left(\frac{2\rho+d}{2d}\right)^2 \left[(2\rho)^2 - d^2\right] \tag{1}$$

$$\sin(\alpha) = \frac{d}{D} \tag{2}$$

Winston cones are usually modelled using ray tracing but this clearly becomes a poor approximation when few modes propagate. A mode matching technique can be used to achieve high accuracy and also enable complex structures like back-to-back Winston cones and detectors with cavities to be modelled. We also wish to consider the effect of corrugating the horn antenna to reduce sidelobe levels in an analogous way to a conical corrugated horn.

In the mode matching technique [2] the corrugated horn [3], [4] structure is regarded as a sequence of cylindrical waveguide segments with the radius stepping between the top and bottom of the corrugation slots. The smooth walled profile is approximated with a series of cylindrical monotonically increasing radii giving a stair like profile. The natural modes of propagation for each segment are the TE and TM modes of a uniform cylindrical waveguide. There is a sudden change in the guide radius at the interface between two segments and the power carried by the individual modes is scattered between the backward propagating modes in the first guide segment and the forward propagating modes in the second guide segment. The mode matching technique is based on matching the total transverse field in the two guides at the junction so that the total power is conserved and that the usual boundary conditions apply to the fields at the conducting walls. Track is also kept of the evanescent modes in the guide as these can propagate as far as the next corrugation.

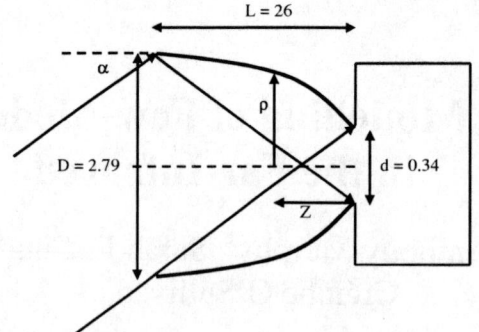

FIGURE 1. Schematic diagram of a Winston cone with a cavity.

The modal field at the aperture propagates to the farfield without scattering. If the waveguide filter is well matched to an integrating cavity, all the waveguide modes at the cavity entrance to the waveguide horn structure are assumed to be equally excited (carry an equal amount of power).

EXAMPLES

The farfield of a Winston cone is 'top-hat like' as a result of the extreme profile in its geometry. The design frequency used in the following examples is 545GHz. In figure 2 the farfield patterns of different Winston cones (smooth walled and corrugated back-to-back cones and smooth walled and corrugated cones with a 4mm cavity) are compared (dimensions as in Figure 1). The corrugated horns have a higher level of sidelobe rejection than the smooth walled horns. This is necessary for CMB experiments so that strong sources do not contaminate the results. The power pattern is in decibels to emphasise the sidelobe structure.

In figure 3 the frequency dependence of the farfield radiation patterns of some Winston cone geometries are compared. As the frequency decreases (260GHz) the horn becomes single-moded and as the frequency increases (1000GHz), more modes propagate and the beam becomes more top-hat like. At very high frequencies the farfield pattern resembles the geometrical beam for a smooth walled horn.

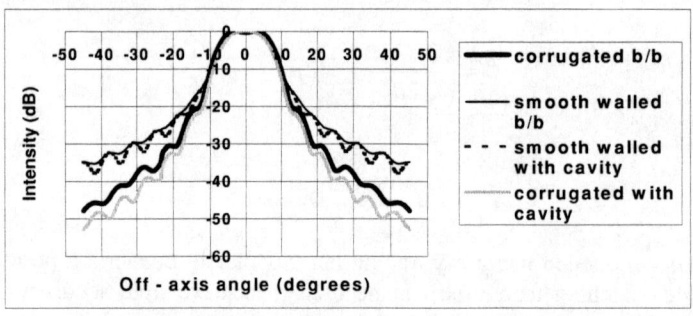

FIGURE 2. Farfield Radiation Patterns.

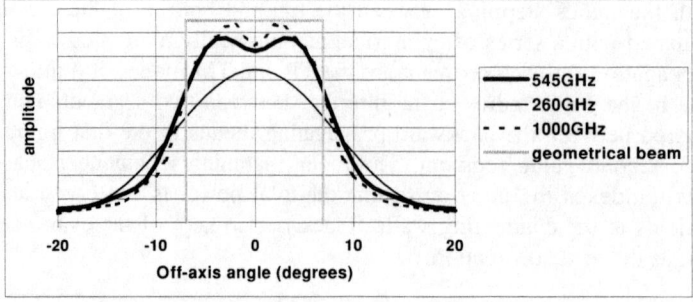

FIGURE 3. Frequency dependence of the farfield of Winston cones.

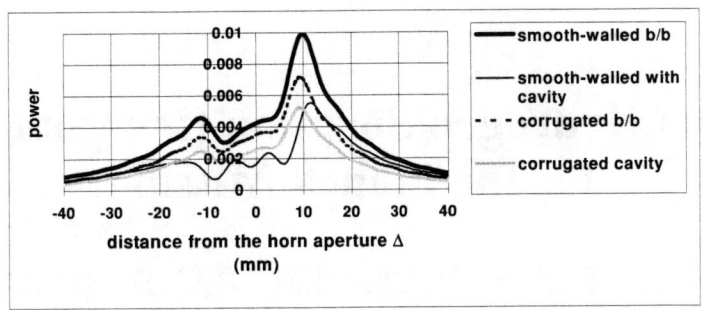

FIGURE 4. Phase-centre of Winston cones (negative Δ implies behind the aperture).

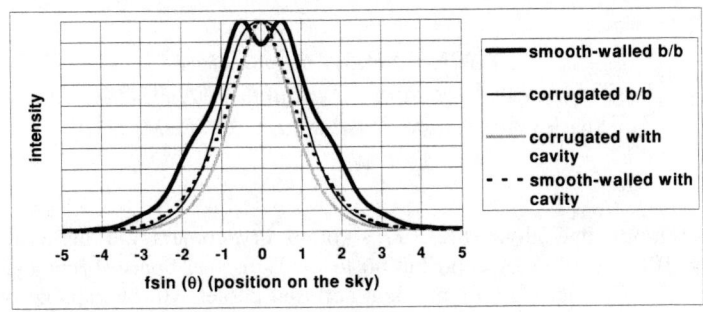

FIGURE 5. Radiation Patterns on the sky.

The phase-centre is determined by coupling the farfield radiation pattern of the horn to a telescope, for different axial positions, Δ, of the horn, and then maximising the on-axis power (gain). This is illustrated in figure 4 with phase-centre values ranging from 9-12mm for the Winston cones in question (defined in Figure 1).

In an imaging telescope configuration the radiation pattern on the sky is obtained by fourier transformation of the partially spatially coherent farfield of the horn truncated at the telescope aperture, where the horn is positioned for maximum on-axis gain. Figure 5 illustrates these patterns for various Winston cones.

ACKNOWLEDGEMENTS

The authors would like to acknowledge the support of Enterprise Ireland who fund this research.

REFERENCES

1. http://www.astro.caltech.edu/~lgg/suzie/sins.html
2. Murphy, J.A., Colgan, R., O'Sullivan, C., Maffei, B., Ade, P., *Infrared Physics and Technology*, (in press, 2001).
3. Clarricoats P.J.B., and Olver A.D., *Corrugated Horns for Microwave Antennas,* Peter Peregrinus Ltd, London, 1984, (chapter 3).
4. Olver A.D., Clarricoats P.J.B., Kishk A.A. and Shafai L., *Microwave Horns and Feeds,* IEEE Press, 1994, (chapter 9).

Two Hydrogen Sorption Cryocoolers for the Planck Mission

G. Morgante [1,2], D. Barber[2], P. Bhandari[2], R.C. Bowman[2], P. Cowgill[2],
D. Crumb[3], T. Loc[2], A. Nash[2], D. Pearson[2], M. Prina[2], A. Sirbi[2],
M. Schemlzel[2], R. Sugimura[2], L.A. Wade[2]

[1] *TESRE-CNR – Bologna, Italy*
[2] *Jet Propulsion Laboratory, Pasadena, CA, 91109, USA*
[3] *Swales Aerospace, Pasadena, CA, 91107, USA*

Abstract. Two continuous operation 18K/20K sorption cryocoolers are being developed by the Jet Propulsion Laboratory (JPL) as a NASA contribution to the European Space Agency (ESA) Planck mission, currently planned for a 2007 launch. Each individual sorption cooler will be capable of providing a total of about 200 mW of cooling power at 18 K and 1.2 W at 20 K, given a passive radiative precooling at 50 K. These coolers work by thermally cycling a metal-hydride to absorb and desorb hydrogen gas, used as the working fluid in a Joule-Thomson (J-T) refrigerator. The major advantage of the sorption coolers is their truly vibration-free operation capability together with the fact that they can be readily scaled to perform over a wide range of cooling powers. The hydrogen sorption coolers will directly cool the Planck Low Frequency Instrument (LFI) HEMT amplifiers to approximately 20 K and will provide precooling at 18 K to the RAL 4 K closed-cycle Helium J-T cooler for the High Frequency Instrument (HFI). The concept design, the cooler operations and the predicted performances of the flight models are here presented.

INTRODUCTION

The ESA Planck mission [1] will use ultra-high sensitivity, cryogenically-cooled detectors: receivers based on HEMT amplifiers for the Low Frequency Instrument (LFI) [2,3] and bolometric detectors for the High Frequency Instrument (HFI) [4,5]. Three distinct cryocoolers are used in the cryogenic chain [6]: a Hydrogen Sorption Cooler to reach 20 K (from Jet Propulsion Laboratory USA); a Helium Joule-Thomson (J-T) Mechanical Cooler to provide a 4K pre-cooling stage (from Rutherford Appleton Laboratory UK); and an Open Cycle Dilution Refrigerator (OCDR) to reach 0.1 K (from Centre de Recherche sur les Tres Basses Temperatures FR). Pre-cooling, in the temperature range 50K-60K, will be achieved passively in space.

The hydrogen sorption cooler will operate at a nominal temperature of 20 K with a cooling power of more than 1 W. The LFI will directly receive its cooling from the sorption cooler [6,7]. For the HFI, operating at 0.1 K, the sorption cooler will serve as a pre-cooling stage at 18 K for the RAL mechanical cryocooler and an open-cycle helium dilution refrigerator [6,7]. Continuous cycle hydrogen sorption cryocoolers have been previously discussed [8], but the requirements of the Planck mission necessitate extending the state-of-the-art for absorption cryocoolers in almost every aspect of performance [9,10,11,12,13].

PLANCK SORPTION COOLER DESCRIPTION AND OPERATION

The sorption cooler performs a simple thermodynamic cycle based on hydrogen compression up to 50 bar, hydrogen gas pre-cooling by the three radiators at about 150 K, 100 K and 50 K, further cooling due to the heat recovery by the cold low pressure gas stream, expansion through a J-T expansion valve and evaporation at the cold stage. A schematic of the flight Planck Sorption Cooler (PSC) is shown in FIG 2. The cooler is normally described as consisting of three assemblies: (1) Cold End Assembly is defined as the hardware colder than the 50 K pre-cooler; (2) Hydride Compressor Assembly includes the sorption beds, stabilization tanks and the warm filter; and (3) Connecting Piping that includes heat exchangers, a cold charcoal filter and tubing between the compressor assembly and the 50 K precooler.

The Compressor Assembly

The key element of the 20 K sorption cooler is the compressor, an absorption machine that pumps hydrogen by thermally cycling several sorbent compressor units.

The principle of operation of the sorption compressor is based on the properties of a unique sorption material ($La_{1.0}Ni_{4.78}Sn_{0.22}$) which can absorb large amounts of hydrogen at relatively low pressures and low temperature, and which will desorb to produce high-pressure hydrogen when heated in a limited volume. Heating of the sorbent is accomplished by electrical resistance heaters, while the cooling is achieved by thermally connecting the compressor element to a radiator. In order not to lose excessive amounts of heat during the heating cycle, a heat switch is provided to alternately isolate the sorbent bed from the radiator during the heating cycle, and to connect it to the radiator thermally during the cooling cycle.

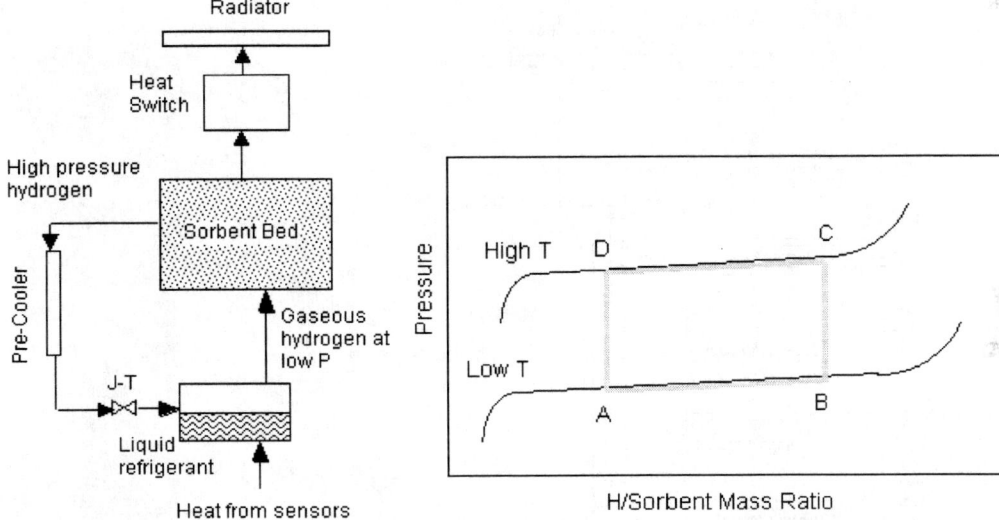

FIGURE 1. On the left side the concept of a hydrogen sorption cooler is shown; the right side shows the basic operation of the metal hydrides.

The mass of hydrogen stored in a unit mass of sorbent at equilibrium is plotted on the horizontal axis vs. the pressure on the vertical axis in the right side of Figure 1. Each curve is an "isotherm" at constant temperature. The sorbent in contact with hydrogen gas can exist at equilibrium anywhere on one of the isotherms. If hydrogen is added to or removed from the sorbent at constant temperature, the system will move along an isotherm. Absorption of hydrogen at constant low temperature is represented by the line A → B. Heating from "low T" to "high T" in a confined space is represented by B → C. Continuously removing hydrogen from the sorbent at constant temperature is represented by C → D. Cooling depleted sorbent from high to low temperature in a confined space is represented by D → A.

As a sorption compressor element (i.e. sorbent bed, see Figure 2, right side) is taken through these four steps in a cycle, it will intake low pressure hydrogen and output high-pressure hydrogen on an intermittent basis. In order to produce a continuous stream of liquid refrigerant, we need to employ several such sorption beds and stagger their phases so that at any given time, one is desorbing while the others are either heating, cooling, or re-absorbing low pressure gas. In such a system, there is a basic clock time period over which each step of the process is conducted. The Planck Sorption Cooler Compressor Assembly, shown in Figure 2, is composed of six identical sorption compressor elements, each filled with metal hydride and provided with independent heating and cooling. Each compressor element is connected to both the high pressure and low-pressure sides of the plumbing system through check valves, which allow gas flow in a single direction only. In addition to the compressors, there are five one-liter high-pressure stabilization tanks connected to the high pressure side of the system to damp out oscillations of the high pressure gas, and a low pressure stabilization sorbent bed to damp out pressure fluctuations of the low pressure gas.

If the high-pressure hydrogen travels from the compressors through a series of heat exchangers and radiators, which provide pre-cooling to below the inversion temperature, and then is expanded through a Joule-Thomson expansion orifice (J-T), the gas will partially liquefy, producing liquid refrigerant at low pressure for sensor systems. Heat from the sensors evaporates liquid hydrogen, and the low-pressure gaseous hydrogen is re-circulated back to the sorbent for compression.

THE COLD END

The principal Cold End components, shown in Figure 2, include the Joule-Thomson (J-T) expander, the first liquid reservoir (LR1) for cooling the High Frequency Instrument (HFI), a second liquid reservoir (LR2) for cooling the Low Frequency Instrument (LFI) and a system of heat exchangers (HXCG4 in Figure 2) used to boil off excess liquid hydrogen.

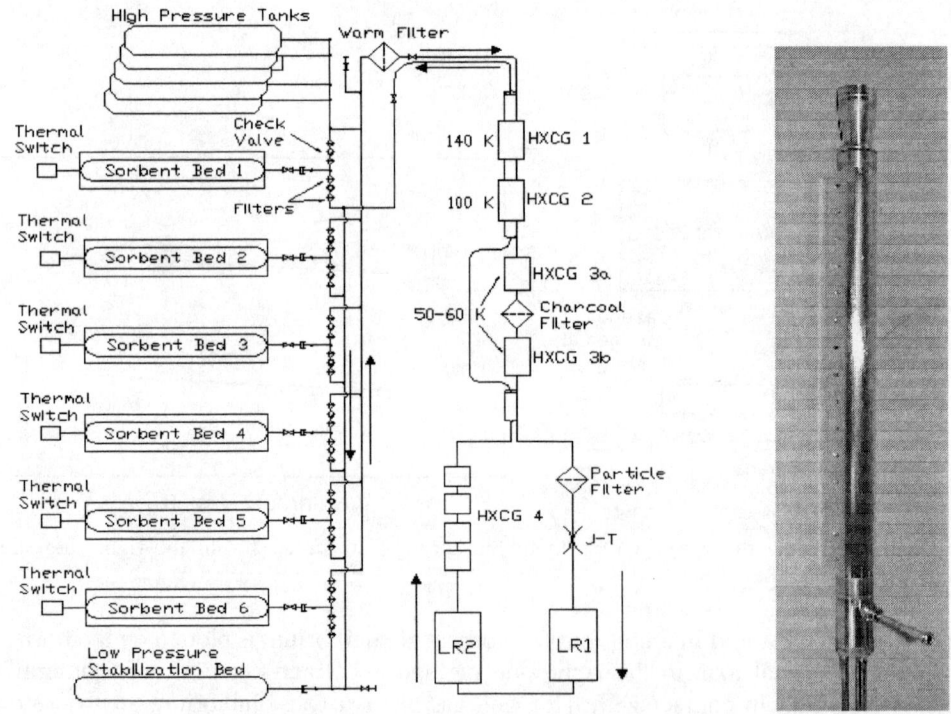

FIGURE 2. *Left*: General schematic of the Planck Sorption Cooler. *Right*: picture of a single sorbent bed.

A porous material was selected to create the J-T impedance needed for refrigerant expansion. The 316L porous material is contained inside a housing made of 316L stainless steel. The porous plug is sintered inside the housing to create a stronger contact with the housing wall. A plug of sintered powder was chosen because its large diameter is inherently contamination resistant: the presence in the line of particle filters and a charcoal trap helps to reduce the risk of possible contamination and clogging.

Joule-Thomson expansion of the Planck cooler's hydrogen refrigerant produces a mixture of liquid and gas in the range between 50% to 90% liquid as a function of inlet gas temperature. For the Planck Cooler the hydrogen liquid-gas mixture will flow through two separate reservoirs where the inlet and outlet of the reservoirs are on opposite ends: LR1 providing a maximum of 135 mW at 18 K as a pre-cooling stage for HFI and LR2 supplying ~1W of cooling power at 20 K for the LFI. The reservoir wall is SS 316L of thickness 0.5mm (0.020"). A copper clamp braised to the outside of each reservoir serves as the interface between the cooling power produced and the instrument heat load. The liquid is retained inside these reservoirs by copper foam to avoid any temperature gradients inside the reservoir.

It is expected that some the Planck cooler will provide somewhat more refrigeration than the instruments require. In zero gravity, the extra liquid produced could eventually flow back up in the low-pressure line, creating temperature fluctuations within the Cold End. The aim of the heat exchanger system (HXCG 4) shown in Figure 2, is to evaporate all the extra-liquid without over-heating the exit gas: this is achieved by controlling the temperature of the exchanger.

PLANCK SORPTION COOLER SPECIFICATIONS SUMMARY

Two sorption coolers will be flying on the Planck mission: one nominal and a redundant unit. They represent the first of a new generation of high cooling power, relatively low power consumption, long-life and continuous cryocoolers for space applications.

The principal characteristic of the Planck Sorption Cooler can be summarized as follows:

- Temperature stability: over one cycle, <0.1 K peak-to-peak; over life: within ranges 20K +2/-2.5K for LFI 18K +1/-0.5K for HFI. Smoothness of oscillations is more important than size.
- Life: 18 months flight operations + 6 months ground operations. Storage life: at least 4 years.
- Redundancy: two independent coolers will be combined to provide 100% redundancy
- Mass and volume: compressor assembly limited to 50 kg in 0.8m x 0.8m x 0.25m volume. Cryostat limited to 2.5 kg.
- Input power: 520 W rejected at 270K +10K/-20K radiator, on warm spacecraft bus.
- Cycle time: nominal 667s switch time, 4000s nominal cycle time.
- J-T expansion of real gas provides refrigeration:
  - No Moving Parts in the Cold End
  - Compressor can be remotely located, enabling passive cooling
  - 50 K passive precooling substantially enhances performance
- Compression accomplished thermally using sorption compressors:
  - Thermal Compressors Produce No Vibration
  - Low pressure refrigerant absorbed at ~270 K
  - High pressure refrigerant desorbed at ~465 K
- No moving parts in the cooler: the system is virtually vibration-free
- Heating and cooling beds in sequence allows continuous refrigeration

SUMMARY

Two hydrogen sorption coolers are being developed for the ESA Planck Mission. They will likely prove pathfinders for many future astrophysics missions, which will benefit from their technological inheritance.

ACKNOWLEDGEMENTS

The research described in this paper was carried out at the Jet Propulsion Laboratory, California Institute of Technology, under a contract with the National Aeronautics and Space Administration.

REFERENCES

1. Bersanelli, M., Bouchet, F.R., Efstathiou, G., Griffin, M., Lamarre, J.M., Mandolesi, N., Norgaard-Nielsen, H.U., Pace, O., Polny, J., Puget, J-L., Tauber, J., Vittorio, N., and Volonté, S., "ESA, COBRAS/SAMBA Report on the Phase A Study", **D/SCI(96)3** (1996)
2. Mandolesi, N. et al., "The Low Frequency Instrument, a Proposal Submitted to the ESA in response to the Announcement of Opportunity for the FIRST/Planck mission" (1998)
3. Mandolesi, N., Bersanelli, M., Butler, R.C., Burigana, C., Maino, D., Mennella, A., Morgante, G., Valenziano, L., and Villa, F., this Conference proceedings
4. Puget, J-L., et al., "The High Frequency Instrument, a Proposal Submitted to the ESA in response to the Announcement of Opportunity for the FIRST/Planck mission" (1998)
5. Lamarre, J.M., et al., this Conference proceedings
6. Collaudin, B. and Passvogel, T., *Cryogenics* **39**, 157 (1999).
7. Wade, L. A., et al., "Hydrogen Sorption Cryocoolers for the Planck Mission," in Advances in Cryogenic Engineering 45A, edited by Q-S. Shu, et al., Kluwer Academic/Plenum, New York, pp. 499-506 (2000)
8. Freeman, B.D., Ryba, E.L., Bowman, R.C., and Phillips, J.R., *Int. J. Hydrogen Energy*, **22**, 1125 (1997).
9. Bowman, R. C., et al., "Performance, Reliability and Life Issues for Components of the Planck Sorption Cooler" presented at CEC, Madison, WI, July, 2001, in Advances in Cryogenic Engineering, Vol.47.
10. Bhandari, P., Prina, M., Ahart, M., Bowman, R. C., and Wade, L. A., "Sizing and Dynamic Performance Prediction Tools for 20 K Hydrogen Sorption Cryocoolers," in *Cryocoolers 11*, edited by R. G. Ross, Jr., Kluwer Academic/Plenum, New York, pp. 541-549 (2001)
11. Bhandari, P., Bowman, R. C., Chave, R. G., Lindensmith, C. A., Morgante, G., Paine, C., Prina, M., and Wade, L. A., *Astro. Lett. And Communications*, **37**, pp. 227-237 (2000).
12. Paine, C. G., Bowman, R. C., Pearson, D., Schmelzel, M. E., Bhandari, P., and Wade, L. A., " Planck Sorption Cooler Initial Compressor Element Performance Tests," in *Cryocoolers 11*, edited by R. G. Ross, Jr., Kluwer Academic/Plenum, New York, pp. 531-540 (2001)
13. Sirbi, A., Bowman, Jr., R.C., Wade, L.A., and Barber, D.S., *Advances in Cryogenic Engineering*, **47** (2001)

Excess Noise in Cryogenic Detectors due to Vibrating 1 K Pots

A. Raccanelli, L. A. Reichertz, E. Kreysa

Max-Planck-Institut für Radioastronomie, auf dem Hügel 69, D-53121Bonn, Germany

Abstract. Temperatures between 4 K and 1 K are obtained by pumping on a ^{4}He bath. In order to reduce the consumption of liquid helium, it is common to pump only on a small volume of He inside a small chamber (1 K pot), continuously replenished from the main bath through an impedance. 1 K pots are a necessary cooling stage in dilution refrigerators, which are used to cool many different kind of sensitive detectors to temperatures in the mK range, e.g. gravitational antennas, bolometers for far-infrared radiation or high resolution particle detection. It is known that continuously filled 1K pots are affected by vibrations that are a source of excess noise, and thus can decrease the performance of the detectors. We present our study on the origin of these vibrations and a solution we have found to eliminate this excess noise from the detectors.

INTRODUCTION

Our group is working with an array of 19 composite bolometers [1] with NTD-Ge thermistors for 2 mm wavelength continuum observations. Since the atmospheric emission is relatively low at this wavelength, the array is cooled to a temperature below 100 mK by means of a ^{3}He/^{4}He-dilution refrigerator to avoid being limited by the system noise.

A necessary cooling stage in our refrigerator -as in the most of all continuously operating dilution refrigerators- is the 1 K pot, a small pumped helium chamber continuously replenished from the main bath at 4.2 K through a flow impedance. It is known that continuously filled 1 K pots are the source of vibrations that can result in electrical, thermal and mechanical noise [2].

Suggested solutions to the problem of vibrational noise in the 1 K pot include: either mechanical decoupling [3] of the experiment, or regulating the helium flow from the main bath and adjusting the helium level inside the 1 K pot itself [4]. These methods provide only a partial attenuation of the noise and do not eliminate of its origin; in some experiments this may not be a sufficient solution.

INVESTIGATION

In our experiments, the noise due to the 1 K pot was revealed as an excess of low frequency noise of almost two orders of magnitude, varying in amplitude from time to time (Fig. 3).

We installed a piezoelectric microphone so that it was in contact with the top flange of the 1 K pot, in order to understand whether the noise corresponded to mechanical vibrations of the pot. This allowed us to study the problem without having to run the dilution refrigerator and to restrict our attention to the 1 K pot.

We also installed a thermometer and a resistive heater on the capillary that acts as flow impedance from the main bath. The thermometer was mounted next to the pot, while the heater was soldered near the main bath (Fig. 1).

The details of our investigation can be found elsewhere [5]. What we have found is that there are three ways to stop the vibrations: a) stop pumping on the 1 K pot and allowing the pressure to rise above about 50 mbar, corresponding to a temperature higher than the lambda point; b) interrupting the helium flow from the main bath by activating the heater with the application of electrical power; c) pumping on the main bath in order to reduce the temperature of the helium flowing into the 1 K pot below the lambda point, as measured by the thermometer on the capillary. This indicated a clear dependence on the superfluidity of helium inside the pot and on the temperature of the helium flowing from the main bath, more precisely on its phase.

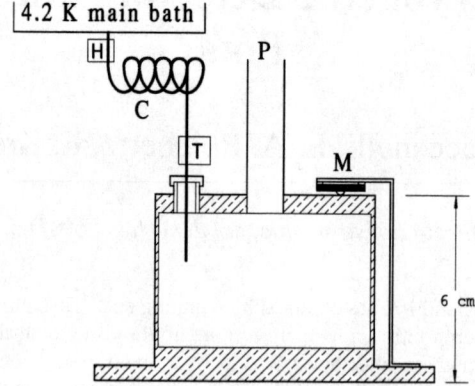

FIGURE 1. Sketch of our 1 K pot. H) Resistive heater; T) Thermometer; C) Capillary acting as a flow impedance from the main bath at 4.2 K to the 1 K pot; M) Piezoelectric microphone; P) 1 K pot pumping line.

The solution was therefore to have the helium already in the superfluid state before introducing it into the 1 K pot. We realized a simple heat exchanger internal to the pot by inserting some tens of cm of thin walled copper-nickel capillary inside the 1 K pot so that it was immersed in the superfluid helium bath. The helium coming from the main bath through the impedance flows inside this capillary and thermalizes to the bath temperature.

In this way the problem was completely solved (Fig. 2), as it was then confirmed by measurement of the noise spectra of the bolometric detectors (Fig. 3).

A full theoretical treatment of the phenomenon would be complex and is beyond the aim of this paper. We offer the following possible qualitative explanation of the nature of the problem: due to the peak in the heat capacity at T_λ, the injection of HeI into the HeII bath inside the 1 K pot leads to a high heat load, distributed on a relatively small surface. The superfluidity is locally broken by the He I injection, then generating quantum turbulence in the adjacent region.

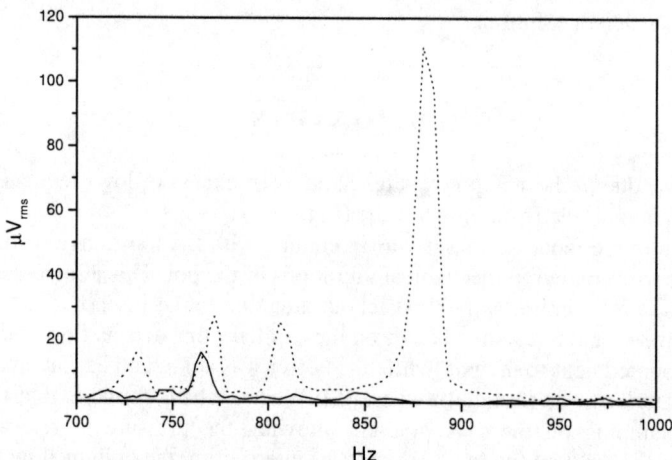

FIGURE 2. Noise spectrum of the microphone signal, before (dotted) and after (solid) the solution. The line at about 880 Hz is the mechanical resonance of the pot.

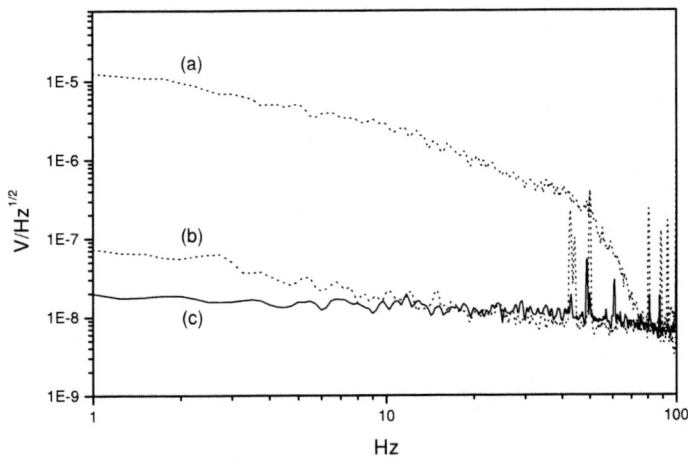

FIGURE 3. Noise spectra of the bolometer signal before (dotted) and after (solid) the solution. The noise varied in amplitude from time to time from a maximum (a) to a minimum (b). In the noise-off state (c) the 1/f roll off is shifted to a frequency lower than the frequencies of interest for our experiment.

CONCLUSIONS

The source of vibrational noise originating in continuously filled 1 K pots has been studied, revealing that the turbulence arises from the He I injection into the He II bath. Lowering the temperature of the helium below T_λ before mixing it with the superfluid cryogen will eliminate the vibrations.

REFERENCES

1. Reichertz LA, Esch W, Gemünd HP, Gromke J, Kreysa E, *Nuclear Instruments and Methods in Physics Research* **A 444,** 423-6 (2000).
2. Bhatia RS, Bock JJ, Ade PAR, Benoît A, Bradshaw TW, Crill BP, Griffin MJ, Hepburn ID, Hristov VV, Lange AE, Mason PV, Murray AG, Orlowska AH, and Turner AD. *Cryogenics* **39,** 701-15 (1999).
3. Pirro S, Alessandrello A, Brofferio C, Bucci C, Cremonesi O, Coccia E, Fiorini E, Fafone V, Giuliani A, Nucciotti A, Pavan M, Pessina G, Previtali E, Vanzini M, and Zanotti L, *Nuclear Instruments and Methods in Physics Research* **A 444,** 331-35 (2000).
4. Lawes G, Zassenhaus GM, Koch S, Smith EN, Reppy JD, and Parpia JM, *Review of Scientific Instruments* **69** (12), 4176-78 (1998).
5. Raccanelli A, Reichertz LA, Kreysa E, *submitted to Cryogenics*, (2001).

2K1BC Workshop
Experimental Cosmology at millimetre wavelengths

SZ EFFECT, MM SOURCES AND SIMULATIONS

The Sunyaev-Zeldovich Effect: Recent Progress and Future Prospects

Yoel Rephaeli

School of Physics & Astronomy, Tel Aviv University - Tel Aviv 69978, Israel
Center for Astrophysics and Space Sciences - University of California, San Diego
La Jolla, CA, 92093-0424

Abstract. Intense theoretical and experimental work on the small scale structure of the cosmic microwave background (CMB) radiation, and the increasingly more realistic description of the astrophysics of clusters of galaxies and their cosmological evolution, have motivated extensive recent work on the Sunyaev-Zeldovich (S-Z) effect. This review begins with a brief summary of the effect and the very significant observational progress in using interferometric arrays to obtain detailed images of the effect in a large number of moderately distant clusters. These measurements yielded important information on cluster masses and the value of the Hubble constant; I review these results and discuss the prospects for improved determination of these quantities from future observations of a large sample of nearby clusters with bolometric array telescopes. The high sensitivity and angular resolution of planned ground, stratospheric, and space telescopes will result in measurements of a large number of distant clusters over most of the sky, and the associated small angular scale CMB anisotropy they induce; I discuss this anisotropy and its expected cosmological yield.

INTRODUCTION

Intracluster (IC) Comptonization of the CMB results in a transfer of photons from the Rayleigh-Jeans (R-J) to the Wien side of the (Planck) spectrum. A detailed quantitative description of this process was given by Zeldovich & Sunyaev (1969) and Sunyaev & Zeldovich (1972), whose deep insight motivated extensive discussions of the great value of the effect as a diagnostic tool of clusters, and its use as a very valuable cosmological probe. The effect and its significance were reviewed by Sunyaev & Zeldovich (1981), a few years before the effect was convincingly measured for the first time with a single-dish radio telescope. Growing realization of the cosmological importance of the effect has led to major improvements in observational techniques, and to extensive modeling and theoretical work. The use of interferometric arrays, and the substantial progress in the development of sensitive radio receivers, have led to first images of the effect (Jones et al. 1993, Carlstrom et al. 1996). Over 40 clusters have already been imaged by the OVRO and BIMA arrays (Carlstrom et al., 2001). Theoretical treatment of the S-Z effect has also improved, starting with the work of Rephaeli (1995a), who performed an exact relativistic calculation and demonstrated the need for such a more accurate description. For general reviews, see Rephaeli 1995b, Birkinshaw 1999, and Carlstrom et al. (2001) for an update on the observational status.

The (essentially) redshift independence of the S-Z effect is, of course, an extremely valuable property. Measurements of the effect yield directly the integrated pressure of the hot IC gas, and the dynamical mass of the cluster, as well as indirect information on the evolution of clusters. Of general interest is the ability to determine the Hubble (H_0) constant and the density parameter, Ω, from S-Z and X-ray measurements. This method to determine H_0, which has clear advantages over the traditional galactic distance ladder method based on optical observations of galaxies in the nearby universe, has been yielding increasingly more precise results. However, substantial systematic uncertainties – due largely to modeling of the thermal and spatial distributions of IC gas – have to be significantly reduced before the full potential of this method yields competitive results for these cosmological parameters. Sensitive spectral and spatial mapping of the effect, and the minimization of systematic uncertainties in its measurements, are still the main challenges of current and near future S-Z work.

This is a brief review of some of the recent theoretical and observational work on the S-Z effect.

COMPTONIZATION IN CLUSTERS

To describe the interaction of the CMB with a hot electron gas the exact frequency re-distribution function has to be calculated in the context of a relativistic formulation. The original Sunyaev & Zeldovich (1972) treatment is based on a solution to the Kompaneets (1957) equation, a *nonrelativistic* diffusion approximation to the exact kinetic (Boltzmann) equation describing the scattering. Their calculation yields a simple expression for the CMB (temperature T) intensity change resulting from scattering of the CMB by electrons with *thermal* velocity distribution (temperature T_e), $\Delta I_t = i_o y g(x)$, where $i_o = 2(kT)^3/(hc)^2$, and $y = \int (kT_e/mc^2) n \sigma_T dl$, is the Comptonization parameter, an integral over the electron density (n) and temperature; σ_T is the Thomson cross section. The spectral function $g(x) = [x^4 e^x/(e^x-1)^2][x(e^x+1)/(e^x-1)-4]$, where $x \equiv h\nu/kT$ is the non-dimensional frequency, is negative in the R-J region and positive at frequencies above a critical value, $x = 3.83$, corresponding to ~ 217 GHz. Typically in a rich cluster $y \sim 10^{-4}$ along a line of sight through the center, and the magnitude of the relative temperature change due to the thermal effect is $\Delta T_t/T = -2y$ in the R-J region.

The effect has a second component when the cluster moves at a finite (peculiar) velocity in the CMB frame. This *kinematic* (Doppler) component is $\Delta I_k = [x^4 e^x/(e^x-1)^2](v_r/c)\tau$ where v_r is the line of sight component of the cluster peculiar velocity, τ is the Thomson optical depth of the cluster. The related temperature change is $\Delta T_k/T = -(v_r/c)\tau$ (Sunyaev & Zeldovich 1980).

The nonrelativistic description of the two components of the S-Z effect by Sunyaev & Zeldovich (1972) is sufficiently accurate only at low gas temperatures and at low frequencies. This approximation is inadequate for use of the effect as a precise cosmological probe. Electron velocities in the IC gas are high, and the relative photon energy change in the scattering is large enough to require a relativistic calculation. Using the exact probability distribution in Compton scattering, and the relativistically correct form of the electron Maxwellian velocity distribution, I calculated ΔI_t in the limit of small τ, keeping terms linear in τ (Rephaeli 1995a). Results of this semi-analytic calculation demonstrate that the relativistic spectral distribution of the intensity change is quite different from that derived by Sunyaev & Zeldovich (1972). Deviations from their expression increase with T_e and can be substantial, and are particularly large near the crossover frequency, which shifts to higher values with increasing gas temperature.

The semi-analytic calculations (Rephaeli 1995a) led to various generalizations and extensions of the relativistic treatment. Challinor & Lasenby (1998) generalized the nonrelativistic Kompaneets equation and obtained analytic approximations to its solution for the change of the photon occupation number by means of a power series in $\theta_e = kT_e/mc^2$. Itoh et al. (1998) adopted this approach and improved the accuracy of the analytic approximation by expanding to fifth order in θ_e. Sazonov & Sunyaev (1998) and Nozawa et al. (1998) have extended the relativistic treatment also to the kinematic component obtaining – for the first time – the leading cross terms in the expression for the total intensity change ($\Delta I_t + \Delta I_k$) which depends on both T_e and v_r. An improved analytic fit to the numerical solution, valid for $0.02 \leq \theta_e \leq 0.05$, and $x \leq 20$ ($\nu \leq 1130$ GHz), was given by Nozawa et al. (2000). Since in some rich clusters $\tau \sim 0.02 - 0.03$, the approximate analytic expansion to fifth order in θ_e necessitates also the inclusion of multiple scatterings, of order τ^2. This has been accomplished by Itoh et al. (2000), and Shimon & Rephaeli (2001).

Analysis of S-Z measurements necessitates use of the relativistically exact expressions for ΔI_t and ΔI_k, especially when the effect is measured at high frequencies. This is particularly essential when the effect is used for the purpose of determining precise values of the cosmological parameters. Moreover, since the ability to determine peculiar velocities of clusters depends very much on measurements very close to the crossover frequency, its exact value has to be known. This necessitates knowledge of T_e since in the exact relativistic treatment the crossover frequency is no longer independent of T_e, but is approximately given by $\simeq 217[1 + 1.167kT_e/mc^2 - 0.853(kT_e/mc^2)^2]$ GHz (Nozawa et al., 1998a). Also necessary is the use of a relativistically correct expression for the (spectral) bremsstrahlung emissivity when determining the gas temperature from X-ray measurements (Rephaeli & Yankovitch 1997). In the latter paper first order relativistic corrections to the velocity distribution and electron-electron bremsstrahlung were taken into account in correcting values of H_0 that were previously derived using the nonrelativistic expression for the emissivity (see also Hughes & Birkinshaw 1998). Nozawa et al. (1998b) have performed a more exact calculation of the relativistic bremsstrahlung Gaunt factor.

Compton scattering polarizes the CMB due to the (angular dependence of the cross section and) likely motion of

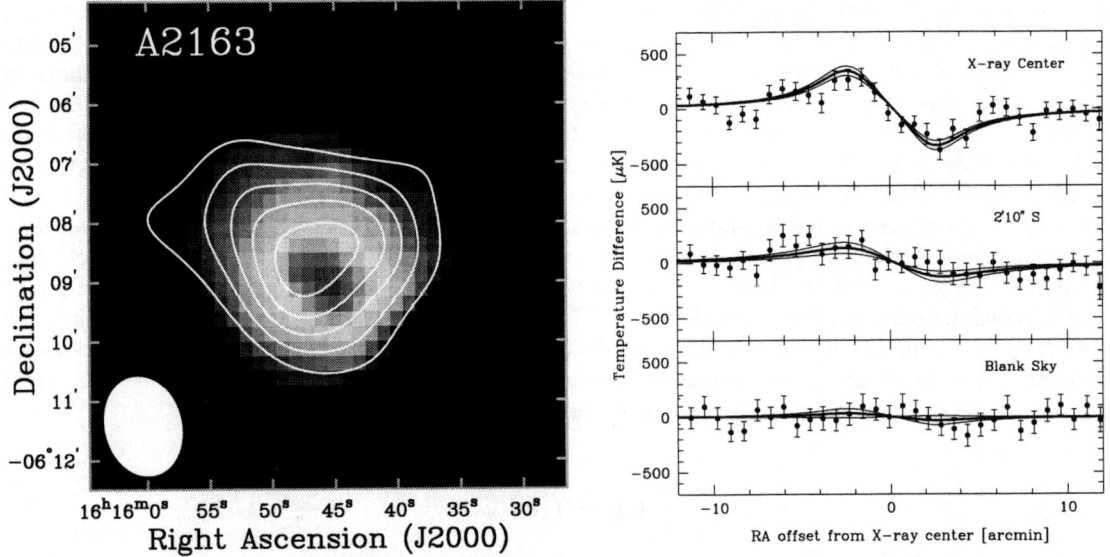

FIGURE 1. S-Z and X-ray views of the cluster A2163 (from Carlstrom et al., 2001). The left frame shows the contour image (with the ellipse indicating the FWHM beam size) obtained from interferometric BIMA measurements (at 28.5 GHz), superposed on a (false color) ROSAT X-ray image. The S-Z profile from co-added drift-scan measurements (at a central frequency of 142 GHz) with the SuZIE array (Holzapfel et al. 1997a) across the center, $\sim 2.2'$ to the South, and across a blank sky region, are shown in the right frame; the lines show predicted S-Z profiles.

the cluster (in the CMB frame), and also possibly aspherical electron distribution (Sunyaev & Zeldovich 1981). The leading term in the degree of net (integrated over the cluster) linear polarization is proportional to $(v_\perp/c)^2\tau$, where $v_\perp$ is the cluster velocity component transverse to the line of sight. An additional contribution to the degree of polarization is $\propto (v_\perp/c)\tau^2$, although no net polarization of this order is produced, unless the gas distribution is aspherical (Sazonov & Sunyaev, 1999). Itoh et al. (2000) have included relativistic corrections in the expression they derived for the kinematically induced polarization. The level of the polarized signal in direction to a fast moving cluster is generally expected to be below 1 μK; its detection by next generation CMB experiments will therefore be very challenging.

MEASUREMENTS

Measurements with interferometric arrays have greatly advanced work on the S-Z effect. Interferometric arrays have several major advantages over a single dish, including insensitivity to changes in atmospheric emission, sensitivity to specific angular scales and to signals which are correlated between array elements, and high angular resolution that enables nearly optimal subtraction of signals from point sources. With the improved sensitivity of radio receivers it became feasible to use existing arrays – beginning with the Ryle telescope (Jones et al. 1993) – to image the effect in moderately distant clusters. Extensive program of S-Z observations then followed with the BIMA and OVRO arrays; images of more than 40 moderately distant clusters (in the redshift range $0.17 < z < 0.89$) have already been obtained at ~ 30 GHz (Carlstrom et al. 1996, 2001). As an illustration of the higher sensitivity and resolution of interferometric images over the traditional one dimensional drift scans, the BIMA image of A2163 (Carlstrom et al., 2001) is shown in Figure 1 above a profile of the effect in the same cluster from measurements with the small SuZIE array (Holzapfel et al., 1997a). The BIMA contour plot is superposed on a ROSAT X-ray image (false color) of the cluster; note the relative smallness of the X-ray size in comparison with the S-Z size of the cluster.

The array configurations of BIMA, and (the specific) OVRO telescope system used by Carlstrom for the above measurements, are not suitable for S-Z observations of nearby clusters. A more optimal system for imaging the effect over larger angular scales is the CBI, a new interferometric array of small (0.9m) dishes, with spatial resolution in the $3' - 10'$, operating in the 26-36 GHz spectral range. Work with the CBI began at the Atacama desert (Chile), and has already resulted (Udomprasert et al., 2000) in measurements of the effect in 9 clusters.

Important advantages of high frequency measurements of effect are use of the spectral dependence for enhanced diagnostic capability, and a more advantageous measurement of the kinematic component. High frequency observations were made with the SuZIE array, the PRONAOS and MITO telescopes, and the Diabolo bolometer. Three moderately distant clusters were measured with the SuZIE 2 × 3 element array: A1689 & A2163 (Holzapfel et al. 1997a, 1997b; the S-Z profile across A2163 is shown in Figure 1), and A1835, which was observed at three spectral bands centered on 145, 221, 279 GHz (Mauskopf et al. , 2000). PRONAOS, an atmospheric 2m telescope, measured the effect in A2163 at four broad spectral bands in the combined range of 285-1765 GHz (Lamarre et al. 1998). This seems to have been the first detection of the effect by a balloon-borne experiment. The MITO 2.6m telescope (located in the Italian Alps), which operates at four high-frequency bands but currently has a large $\sim 17'$ beam, was used to observe the effect in the Coma cluster (D'Alba et al. , 2001). The sample of observed clusters now includes the distant (z=0.45) cluster RXJ 1347 which was measured to have the largest determined Comptonization parameter, $y = 1.2 \times 10^{-3}$ (Pointecouteau et al. 1999). The observations were made with the Diabolo bolometer operating at the IRAM 30m radio telescope. The Diabolo has a $0.5'$ beam, and a dual channel bolometer (centered on 2.1 and 1.2 mm). Four other clusters were also observed with the Diabolo bolometer (Desert et al. 1998).

RESULTS

The highly desirable properties of the S-Z effect – its well understood nature and redshift independence – make it an ideal cosmological probe. These major advantages, and the expectation that the considerable systematic uncertainties associated with the modeling of IC gas can be reduced, have led enhanced interest in using the effect to determine cluster properties and global cosmological parameters, as well as general aspects of cluster evolution. Basic methodologies and comprehensive discussions can be found in the reviews by Sunyaev & Zeldovich (1981), Rephaeli (1995b), and Birkinshaw (1999).

IC gas density and temperature profiles have so far been largely determined from X-ray measurements. High spatial resolution S-Z measurements can, in principle, yield these distributions out to larger radii due to the linear dependence of ΔI_t on n, as compared to the n^2 dependence of the (thermal bremsstrahlung) X-ray brightness profile. The profile of the cluster (total) mass, $M(r)$, can be derived directly from the gas density and temperature distributions by solving the equation of hydrostatic equilibrium. The method (which has already been employed in many analyses using X-ray deduced gas parameters, e.g. Fabricant et al. 1980), was used by Grego et al. (2001) to determine total masses and gas mass fractions of 18 clusters based largely on the results of their interferometric S-Z measurements. Isothermal gas with the familiar density profile, $(1 + r^2/r_c^2)^{-3\beta/2}$, was assumed. The core radius, r_c, and β were determined from analysis of the interferometric S-Z data, whereas the X-ray deduced value of the temperature was used. From these, the gas mass fraction was determined at a (fiducial) radius where the cluster mass density is presumed to be 500 times the critical density and the cluster baryon fraction is close to its universal value. In the currently popular open and flat, Λ-dominated CDM models, mean values in the range $(0.06 - 0.09)h^{-1}$ (where h is the value of H_0 in units of 100 km s^{-1} Mpc^{-1}) were obtained for the gas mass fraction. Use of more realistic temperature profiles that can now be measured by XMM and *Chandra* will reduce the substantial modeling uncertainties in these mass estimates.

Measurement of cluster radial velocities from the kinematic component of the S-Z effect is feasible only when observations are made in a narrow spectral band near the critical frequency, where the thermal effect vanishes while the kinematic effect – which is usually swamped by the much larger thermal component – is maximal (Rephaeli & Lahav 1991). This necessitates knowledge of the exact spectral shape of the thermal component near the critical frequency (which depends on the gas temperature). SuZIE is the first experiment with a spectral band centered on the crossover frequency. Measurements of the clusters A1689 and A2163 (Holzapfel et al. 1997b) and A1835 (Mauskopf et al. 2000) yielded substantially uncertain results for v_r (170^{+815}_{-630}, 490^{+1370}_{-880}, and 500 ± 1000 km s^{-1}, respectively). Balloon-borne measurements of the effect with PRONAOS have also resulted in a statistically insignificant value for the peculiar velocity of A2163 (Lamarre et al. 1998). The hope is that the more sensitive balloon-borne telescopes with bolometer arrays that are currently under development will be better optimized for higher quality measurements near the crossover frequency.

As is well known, the ability to measure H_0 and the cosmological density parameter, Ω, from S-Z and X-ray observations is essentially based on the different density dependences of Comptonization and thermal bremsstrahlung. Measuring ΔI_t, the X-ray surface brightness, and their spatial profiles yields the angular diameter distance, d_A. Due to large observational and systematic uncertainties (which are discussed in the reviews by Rephaeli 1995b, and Birkinshaw 1999) it is unrealistic to obtain a precise value of H_0 from measurements of any single cluster. Some of the modeling and other systematic errors can be significantly reduced by averaging over values of H_0 from measurements of a large number of clusters. First such averaging over only eight available values of H_0 yielded $H_0 \simeq 58 \pm 6$ km s^{-1} Mpc^{-1} (Rephaeli 1995b); however, the database was very non-uniform. A similar mean value (60 km s^{-1} Mpc^{-1}) was deduced by Birkinshaw (1999) based on a somewhat updated data set. The interferometric BIMA and OVRO S-Z survey provides the first relatively uniform dataset for the determination of H_0. From the full set of 33 available values (from single dish as well as interferometric measurements) for d_A, Carlstrom et al. (2001) deduce $H_0 = 60$ km s^{-1} Mpc^{-1} in an open cosmological model with (matter density parameter) $\Omega_M = 0.3$, and $H_0 = 58 \pm 3$ km s^{-1} Mpc^{-1} for a flat model with $\Omega_M = 1$. Observational errors are $\sim 5\%$, and the overall systematic uncertainty is estimated to be $\sim 30\%$ (Carlstrom et al. , 2001).

The value of Ω can be deduced from a plot of d_A vs. redshift (Hubble diagram). With the effect already measured in clusters with $z \leq 0.9$, it is possible to begin using this method to obtain limits on Ω. However, current limits are not yet very interesting (see figure 11 of Carlstrom et al. , 2001) due to the very large uncertainties in the values of d_A.

S-Z Induced Anisotropy

CMB anisotropy induced by Compton scattering in clusters – the main source of secondary anisotropy on angular scales of few arcminutes – was first modeled (Rephaeli 1981) in the context of a simple model for IC gas evolution. The magnitude of the temperature anisotropy, $\Delta T/T$, can be as high as $few \times 10^{-6}$, if the gas evolution in clusters is not too strong (Colafrancesco et al. 1994). Because of this, and the intense interest in CMB anisotropy on arcminute scales – multipoles (in the representation of the CMB temperature structure in terms of spherical harmonics) $\ell \geq 1000$ – the S-Z anisotropy has been studied extensively in recent years. The basic goal continues to be the calculation of this anisotropy (which is commonly characterized by the ℓ dependence of its power spectrum) in various cosmological, large scale structure, and IC gas models. The primary goal is precise global parameter determinations from the analysis of large stratospheric and satellite databases. This anisotropy is also a major tool in the study of clusters and their evolution.

Planned long duration balloon-borne experiments and satellites are expected to result in detections of thousands of clusters, and in detailed mappings of the small angular scale anisotropy. Cluster counts and the induced CMB anisotropy have been investigated in many recent works. For example, Colafrancesco et al. (1997), and Kitayma et al. (1998), have calculated the S-Z anisotropy and cluster number counts in an array of open and flat cosmological and dark matter models. S-Z maps and power spectra were generated by da Silva et al. (1999) from hydrodynamical simulations. It is expected that planned multi-frequency surveys will be able to determine the S-Z part of the anisotropy, perhaps with sufficient precision to determine its power spectrum and higher order correlations (Cooray et al. 2000). Many thousands of clusters are expected to be detected during the planned Planck survey.

The dependence of the power spectrum on the cosmological and dark matter models, and on detailed modeling of IC gas and its evolution, (with the added complexity of describing dynamical as well as hydrodynamical evolution) obviously makes its calculation parameter intensive. It is therefore not surprising that results for the predicted level and shape of the power spectrum vary considerably among the various calculations. Moreover, because the power spectrum depends very steeply on some of the poorly known parameters (involving, $e.g.$ details of the collapse and virialization of clusters and their mass spectrum), it is difficult to assess how realistic are some of the published results, which are indeed sometimes inconsistent. Less parameter intensive method is the more direct characterization of the anisotropy by simulations of the S-Z sky, based largely on results from cluster X-ray surveys and the use of simple scaling relations (as was first implemented by Markevitch et al. 1992).

To illustrate the main features of the S-Z power spectrum we show in Figure 2 results from the work of Sadeh & Rephaeli (2001) who have calculated power spectra of the S-Z and primary anisotropies in an array of cosmological and dark matter models. The S-Z power spectrum, using a Press & Schechter cluster mass function, was normalized by the observed X-ray luminosity function. The primary power spectrum (solid line) was calculated using the CMBFAST code of Seljak & Zaldarriaga (1996). The plots are of the predicted power spectrum, $C_\ell(\ell+1)/2\pi$ vs. multipole ℓ, in a flat cosmological model with $\Omega_\Lambda = 0.7$ (where Λ is the cosmological constant) and CDM density parameter $\Omega_M = 0.3$. IC gas was assumed to evolve in a simple manner consistent with the results of the EMSS survey (which was carried out by the Einstein satellite), as parameterized by Colafrancesco et al. (1994). In this model the S-Z anisotropy (which is dominated by the thermal effect) is appreciable already at $\ell \sim 1500$; for example, it comprises $\sim 10\%$ of the total power at $\ell \simeq 1840$, and becomes dominant for $\ell > 3000$. It is clear therefore that the S-Z anisotropy has to be included in the detailed modeling of the small scale structure of the CMB.

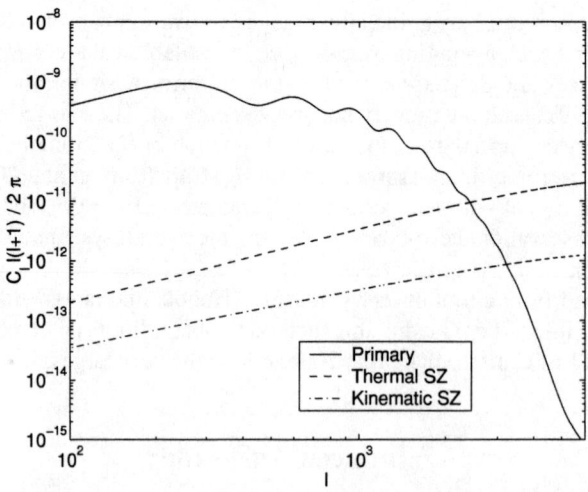

FIGURE 2. Primary and S-Z power spectra in the flat CDM model (Sadeh & Rephaeli 2001). The solid line shows the primary anisotropy as calculated using the CMBFAST computer code of Seljak & Zaldarriaga (1996). The dashed line shows the thermal S-Z power spectrum, and the dotted-dashed line is the contribution of the kinematic component.

FUTURE PROSPECTS

We have just begun exploiting the great potential of the S-Z effect as a cosmological probe. Very useful information on the gas mass fraction and total masses of clusters has already been obtained, and an important independent way to determine H_0 has produced significant results. Many dedicated S-Z projects – ground-based and balloon-borne telescopes equipped with bolometric multi-frequency arrays – will become operational in the near future. These will advance the field very significantly through improved sensitivity, higher spatial resolution, and expanded spectral coverage; use of the spectral characteristics of the S-Z effect will enhance its diagnostic power. Better understanding and control of systematics will continue to be the prime consideration in selecting observational strategies. Highest quality results will likely be obtained from measurements of the effect in nearby ($z \leq 0.1$) clusters. S-Z measurements of a large number of clusters with the many new experiments, and the high quality spectral and spatial X-ray data that are now available from observations with the XMM and *Chandra* satellites, will greatly improve the precision of the derived values of cluster gas mass fraction and total masses of clusters. The higher quality S-Z and X-ray measurements, and better control of systematic uncertainties, will result in a much reduced overall uncertainty of just $\sim 5\%$ in the value of H_0. The much larger number of clusters with more precise values of d_A, and expanded redshift coverage, will enable also the measurement of Ω. Towards the end of this decade the Planck satellite is expected to produce a comprehensive S-Z survey of clusters and detailed maps of the CMB anisotropy they induce.

REFERENCES

1. Birkinshaw, M., *Phys.Rep.*, **310**, 97 (1999).

2. Carlstrom, J.E., Joy, M. and Grego, L., *ApJ*, **456**, L75 (1996).

3. Carlstrom, J.E. et al. , astro-ph/0103480 (2001).

4. Colafrancesco, S., Mazzotta, P., Rephaeli, Y. and Vittorio, N., *ApJ*, *433*, 454 (1994).

5. Colafrancesco, S., Mazzotta, P., Rephaeli, Y. and Vittorio, N., *ApJ*, *479*, 1 (1997).

6. Challinor, A. and Lasenby, A., *ApJ*, *510*, 930 (1998).

7. Cooray, L., Hu, W. and Tegmark, M., astro-ph/0002238 (2000).

8. D'Alba, L. et al. , astro-ph/0010084 (2001).

9. da Silva, A.C. et al. , astro-ph/9907224 (1999).

10. Desert, F.X. et al. , *New Astron.*, **3**, 655 (1998).
11. Fabricant, D.M., Lecar, M. and Gorenstein, P., *ApJ*, *241*, 552 (1980).
12. Grego, L. et al. , *ApJ*, in press (2001).
13. Holzapfel, W.L. et al. , *ApJ*, **480**, 449 (1997a).
14. Holzapfel, W.L. et al. , *ApJ*, **481**, 35 (1997b).
15. Hughes, J.P. and Birkinshaw, M., *ApJ*, **501**, 1 (1998).
16. Itoh, N., Kohyama, Y. and Nozawa, S., *ApJ*, **502**, 7 (1998).
17. Itoh, N., Nozawa, S. and Kohyama, Y., astro-ph/0005390 (2000).
18. Jones, M. et al. , *Nature*, **365**, 320 (1993).
19. Kitayama, T et al. , *PASJ*, **50**, 1 (1998).
20. Kompaneets, A.S., *Soviet Phys.-JETP*, **4**, 730 (1957).
21. Lamarre, J.M. et al. , *ApJ*, **507**, L5 (1998).
22. Markevitch, M. et al. , *ApJ*, **395**, 326 (1992).
23. Mauskopf, P.D. et al. , *ApJ*, **538**, 505 (2000).
24. Nozawa, S., Itoh, N. and Kohyama, Y., *ApJ*, **507**, 530 (1998a).
25. Nozawa, S., Itoh, N. and Kohyama, Y., *ApJ*, **508**, 17 (1998b).
26. Nozawa, S. et al. , *ApJ*, **536**, 31 (2000).
27. Pointecouteau, E. et al. , *ApJ*, **519**, L115 (1999).
28. Rephaeli, Y., *ApJ*, **351**, 245 (1981).
29. Rephaeli, Y., *ApJ*, **445**, 33 (1995a).
30. Rephaeli, Y., *ARAA*, **33**, 541 (1995b).
31. Rephaeli, Y. and Lahav, O., *ApJ*, **372**, 21 (1991).
32. Rephaeli, Y. and Yankovitch, D., *ApJ*, **481**, L55 (1997).
33. Sadeh, S. and Rephaeli, Y., preprint (2001).
34. Sazonov, S.Y. and Sunyaev, S.Y., *ApJ*, **508**, 1 (1998).
35. Sazonov, S.Y. and Sunyaev, S.Y., *MN*, **310**, 765 (1999).
36. Seljak, U. and Zaldarriaga, M., *ApJ*, **469**, 437 (1996).
37. Shimon, M. and Rephaeli, Y., preprint (2001).
38. Sunyaev, R.A., *Comm.Ap.Sp.Phys.*, **7**, 1 (1977).
39. Sunyaev, R.A. and Zeldovich, Y.B., *Comm.Ap.Sp.Phys.*, **4**, 173 (1972).
40. Sunyaev, R.A. and Zeldovich, Y.B., *MN*, **190**, 413 (1980).
41. Sunyaev, R.A. and Zeldovich, Y.B., *Astrophys. Sp. Phys. Rev.*, **1**, 1 (1981).
42. Udomprasert, P.S., Mason, B.S. and Readhead, A.C.S., astro-ph/0012248 (2000).
43. Zeldovich, Y.B. and Sunyaev, R.A., *Comm.Ap.Sp.Phys.*, **4**, 301 (1969).

Non-thermal vs. Thermal SZ Effect in Galaxy Clusters

S. Colafrancesco[*], P. Marchegiani[†] and E. Palladino[†]

[*]*Osservatorio Astronomico di Roma, Via Frascati 33, I-00040 Monteporzio, Italy*
[†]*Dip. di Fisica, Università La Sapienza, Roma, Italy*

Abstract. Many of the clusters in which SZ effects have been detected there is also evidence for radio halo sources and EUV or hard X-ray excesses. So it is of interest to assess whether the detected effects are in facts from the thermal or non-thermal electron populations. We present here the results of an exact derivation of the non-thermal SZ effect in galaxy clusters. Such a derivation is made using the full relativistic formalism and overcoming the limitations of the Kompaneets and of the single scattering approximations. We show that the spectral shape of the non-thermal SZ effect depends not only from the frequency but also from the cluster physical parameters, like the electron pressure, optical depth and temperature. We also evaluate the total SZ effect for a combination of thermal and non-thermal electron population residing in the same volume like is the case in radio-halo clusters. We discuss both the spectral and the spatial features of the non-thermal SZ effect considering two radio-halo clusters: A2163 and Coma. We finally discuss how the combined observations of the non-thermal SZ effect and of the non-thermal phenomena occurring in galaxy clusters (radio-halos, EUV and hard X-ray excesses) provide crucial constrains of the spectrum of the relativistic electron population and, in turn, set constrains on the nature of non-thermal phenomena in galaxy clusters.

THE SZ EFFECT

Compton scattering of the Cosmic Microwave Background (CMB) by hot Intra Cluster (IC) gas is a relatively simple process whose spectral and spatial imprints on the radiation serve as indispensable cosmological and astrophysical probes. Such a scattering produces a systematic shift of the CMB photons from the Rayleigh-Jeans (RJ) to the Wien side of the CMB spectrum [1, 2, 3]. And since the CMB photons are democratic particles in the universe, they can be Compton scattered by any electron population along their path to the observer. Thus, three main processes can be considered as the main sources on Compton scattering of the CMB: *i*) the thermal SZ effect produced by the electrons of the hot IC gas, *ii*) the kinematic SZ effect produced by the bulk motion of the IC gas and *iii*) the up-scattering of CMB photons by relativistic electrons with a non-thermal energy distribution which are responsible of the radio halo emission and other non-thermal phenomena in galaxy clusters.

An approximate description of the scattering of an isotropic Planckian radiation field by a non-relativistic Maxwellian electron population can be obtained by means of the solution of the Kompaneets [3] equation. The resulting change in the spectral intensity, ΔI_{th}, due to the scattering of CMB photons by a thermal electron distribution is

$$\Delta I_{th} = 2\frac{(k_B T_0)^3}{(hc)^2} y_{th} g(x) , \qquad (1)$$

where $x = h\nu/k_B T_0$ is the a-dimensional frequency and the spectral shape of the non-relativistic effect is contained in the function

$$g(x) = \frac{x^4 e^x}{(e^x-1)^2}\left[x\frac{e^x+1}{e^x-1} - 4\right] \qquad (2)$$

which is zero at $x_0 = 3.83$ (or $\nu = 217$ GHz for a value of the CMB temperature $T_0 = 2.726$), negative at $x < x_0$ (in the RJ side) and positive at $x > x_0$ (in the Wien side). The Comptonization parameter, y_{th}, due to the thermal SZ effect is given by

$$y_{th} = \frac{\sigma_T}{m_e c^2}\int d\ell\, n_e k_B T_e , \qquad (3)$$

where n_e and T_e are the electron density and temperature of the IC gas, respectively, σ_T is the Thomson cross section, valid in the limit $T_e \gg T_0$, k_B is the Boltzmann constant and $m_e c^2$ is the rest mass energy of the electron. The Comptonization parameter in eq.(3) can be written as $y_{th} = \frac{\sigma_T}{m_e c^2} \int d\ell \, P_{th}$, where the relevant dependence from the total kinetic pressure, $P_{th} = n_e k_B T_e$, of the IC gas along the line of sight ℓ appears.

The previous description of the thermal SZ effect is obtained under the Kompaneets approximation and in the single scattering regime of the true photon redistribution function (see, e.g., [4] for details). As such, it only provides a reasonable approximation of the SZ effect in galaxy clusters for low temperatures ($T_e \lesssim 5$ keV) and low optical depth ($\tau = \sigma_T \int d\ell n_e \lesssim 10^{-3}$). However, recent X-ray observations have revealed the existence of many high-temperature clusters [5] with $k_B T_e$ up to ~ 17 keV [6]. The calculation of the thermal SZ effect from these hot clusters requires to take into account relativistic corrections [7, 8, 9, 4].

Another general assumption which is made in the calculation of the SZ effect is the use of a single population of thermal electrons which constitute the hot ($T \sim 10^7 - 10^8$ K), optically thin ($n_e \sim 10^{-3} - 10^{-2}$ cm^{-3}, $R \sim$ a few Mpc) Intra Cluster Medium (ICM). This assumption is based on the evidence that the IC gas is mainly constituted by thermal electrons (and protons) which are responsible for the X-ray emission observed in many clusters through thermal bremsstrahlung radiation (see [10] for a review). Nonetheless, in addition to the thermal IC gas, many galaxy clusters contain a population of relativistic electrons which produce a diffuse radio emission (radio halos and/or relics) via synchrotron radiation in a magnetized ICM (see, e.g., [29] for a recent observational review). The electrons which are responsible for the radio halo emission must have energies $E_e \gtrsim$ a few GeV to radiate at frequencies $\nu \gtrsim 30$ MHz and reproduce the main properties of the observed radio halos (see, e.g., [12, 13] and references therein). Some of the nearby clusters also show the presence of an EUV/soft X-ray excess [14, 15, 16] and of an hard X-ray excess [17, 18, 19, 15] over the thermal bremsstrahlung radiation. These emission excesses over the thermal X-ray emission may be produced either by Inverse Compton Scattering (hereafter ICS) emission of CMB photons off the relativistic electrons or by a combination of thermal (reproducing the EUV excess [20]) and suprathermal (reproducing the hard X-ray excess by non-thermal bremsstrahlung [21, 22, 23]) populations of distinct origins. Using a phenomenological complex spectrum (double power-law) with slopes $\alpha_r \sim -2.5$ at $E \gtrsim E_*$ and $\alpha_x \sim -0.5$ at $E \lesssim E_*$, with the break set at $E_* \sim 200 - 400$ MeV, one is able to reproduce the whole set of non-thermal phenomena which are present in several galaxy clusters [24].

Many of the clusters in which SZ effects have been detected there is also evidence for radio halo sources and EUV or hard X-ray excesses [25]. So it is of interest to assess whether the detected effects are in facts from the thermal or non-thermal electron populations. Here we discuss the spectral and spatial features of the SZ effect which have been calculated [26] using an exact derivation of the spectral distortion induced by a combination of a thermal and a non-thermal electron populations which are present at the same time in the ICM. We use $H_0 = 50$ km s^{-1} Mpc^1 and $\Omega_0 = 1$ throughout the paper unless otherwise specified.

THE NON-THERMAL SZ EFFECT

Several limits to the non-thermal SZ effect are available in the literature [4] from observations of galaxy clusters which contain powerful radio halo sources (such as A2163) or radio galaxies (such as A426), but a few detailed analysis of the results in terms of putting limits to the non-thermal SZ effect have been possible. Also the problem of detecting the non-thermal SZ effect in radio-halo clusters is likely to be severe because of the associated synchrotron radio emission. In fact, at low radio frequencies, such a synchrotron emission could easily dominate over the small negative signal of the SZ effect. At higher frequencies there is in principle more chance to detect the non-thermal SZ effect, but even here there are likely to be difficulties in separating the SZ effect from the flat-spectrum component of the synchrotron emission.

From the theoretical point of view, preliminary calculations [4, 27, 21] of the non-thermal SZ effect have been carried out in the diffusion approximation ($\tau \ll 1$), in the limit of single scattering and for a single non-thermal population of electrons. Specifically [27] and [21] considered the SZ effect produced by the supra-thermal tail of the Maxwellian electron distribution in the Coma cluster and concluded that the effect, even though of small amplitude, could be measurable in the sub-mm region by the next coming PLANCK experiment. However, it has been shown [24] that the suprathermal electron distribution faces with several crucial problems, the main being the large heating that these electrons would induce through Coulomb collisions in the ICM. The large energy input of the suprathermal distribution in the ICM of Coma would heat the IC gas up to unreasonably high temperatures, $k_B T_e \sim 10^{16}$ K, which are not observed.

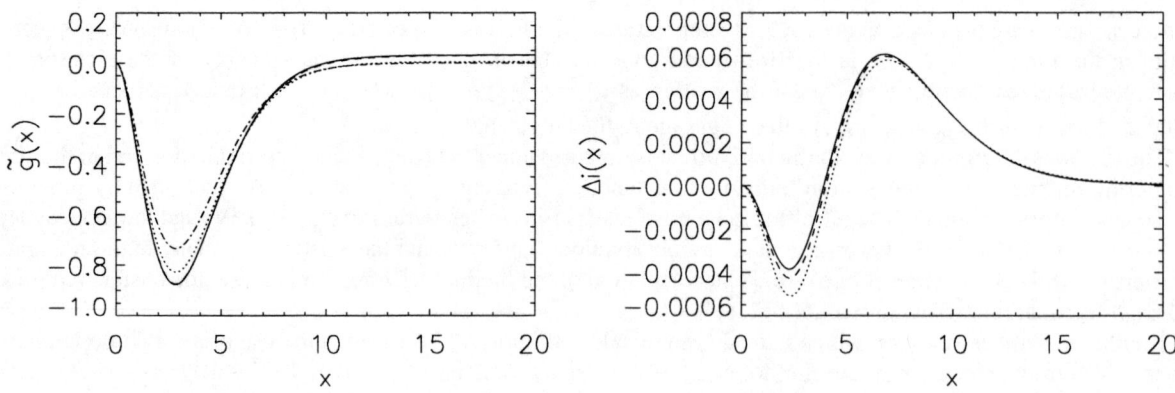

FIGURE 1. The function $\tilde{g}(x)$ (left panel) and the total spectral distortion (right panel) in units of $2(k_B T_0)^3/(hc)^2$ produced by the combination of a thermal electron distribution with $kT = 8.5$ keV and a non-thermal electron distribution with a double power-law spectrum (see Sect.1 for details) with $p_1 = 0.5$ (dotted), 1 (dashes) and 10 (dot-dashes).

Matters are significantly more complicated if the full relativistic formalism is used. However, this is necessary, since many galaxy clusters show extended radio halos and the electrons which produce the diffuse synchrotron radio emission are certainly highly relativistic so that the use of the Kompaneets approximation is invalid [4]. Moreover, the presence of thermal and non-thermal electrons in the same location of the ICM renders the single scattering approximation unreasonable, so that the treatment of multiple scattering among different electronic populations coexisting in the same cluster is necessary to describe correctly the overall SZ effect.

Our theoretical approach

The spectrum of the Comptonized radiation is given by

$$I(x) = \int_{-\infty}^{+\infty} ds \, I_0(x e^{-s}) P(s) , \qquad (4)$$

where $s \equiv ln(\nu'/\nu)$ and $I_0(x) = 2\frac{(k_B T_0)^3}{(hc)^2}\frac{x^3}{e^x-1}$ is the incident CMB spectrum. Following the results of Colafrancesco, Marchegiani & Palladino [26], the redistribution function of the CMB photons, $P(s)$, can be obtained in an exact form as

$$P(s) = \frac{1}{2\pi}\int_{-\infty}^{+\infty} \tilde{P}(k) e^{iks} dk \qquad (5)$$

in terms the anti Fourier transform of $\tilde{P}(k)$ given by

$$\tilde{P}(k) = e^{-\tau}\left[1 + \tau \tilde{P}_1(k) + \frac{1}{2}\tau^2 \tilde{P}_1^2(k) + \ldots\right] = e^{-\tau} e^{\tau \tilde{P}_1(k)} = e^{-\tau[1-\tilde{P}_1(k)]} , \qquad (6)$$

where

$$\tilde{P}_1(k) = \int_{-\infty}^{+\infty} P_1(s) e^{-iks} ds . \qquad (7)$$

The function $\tilde{P}(k)$ in eq.(6) takes into account the effect of multiple scattering and the function

$$P_1(s) = \int_0^{\infty} dp f_e(p) P_s(s;p) , \qquad (8)$$

takes into account the effect of the specific electron distribution $f_e(p)$ in the full relativistic case.

To compare the exact calculations of the SZ spectral distortion with those obtained in the non-relativistic limit it is useful to write the distorted spectrum in the general form

$$\Delta I(x) \equiv I(x) - I_0(x) = 2\frac{(k_B T_0)^3}{(hc)^2} y \tilde{g}(x) . \qquad (9)$$

The spectral shape of the SZ effect is contained in the function

$$\tilde{g}(x) = \left(\frac{\Delta I}{I_0}\right) \frac{1}{y} \frac{x^3}{e^x - 1} = \frac{\Delta i(x)}{y} \qquad (10)$$

(see Fig.1) where $\Delta i \equiv \Delta I \frac{(hc)^2}{2(k_B T_0)^3}$. The Comptonization parameter y is defined, in our general approach, in terms of the pressure P of the considered electron population:

$$y = \frac{\sigma_T}{m_e c^2} \int P d\ell . \qquad (11)$$

For a thermal electron population in the non-relativistic limit one has $P_{th} = n_e k_B T_e$, and we re-obtain the Compton parameter in eq.(3).
Also in the case of a non-thermal electron population we can write the spectral distortion $\Delta I_{non-th}(x)$ in the general form

$$\Delta I_{non-th}(x) = 2\frac{(k_B T_0)^3}{(hc)^2} y_{non-th} \tilde{g}(x) \qquad (12)$$

which is analogous to the standard case of eq.(1), where the non-thermal Compton parameter is given by:

$$y_{non-th} = \frac{\sigma_T}{m_e c^2} \int P_{rel} d\ell , \qquad (13)$$

where P_{rel} is the pressure of the non-thermal, relativistic electron distribution here considered. Thus, for a non-thermal population we can write, at first order in τ:

$$\tilde{g}(x) = \frac{\Delta i}{y} = \frac{\tau[j_1 - j_0]}{\frac{\sigma_T}{m_e c^2} \int P_{rel} d\ell} \equiv \frac{m_e c^2}{\langle k_B T_e \rangle} [j_1 - j_0] \qquad (14)$$

where we introduced the quantity

$$\langle k_B T_e \rangle = \frac{\sigma_T}{\tau} \int P_{rel} d\ell = \frac{\int P_{rel} d\ell}{\int n_e d\ell} , \qquad (15)$$

which is the analogous of the average temperature for a thermal electron populations (in this last case in fact, $\langle k_B T_e \rangle \equiv k_B T_e$ obtains).
For galaxy clusters which contain two different electronic populations, like radio-halo clusters, one has to evaluate the spectral distortion produced by both populations on the CMB radiation. Such a derivation has been provided by Colafrancesco et al. [26]) and we refer to this paper for details. We assume here that the two populations are independent and that no change in the thermal population is induced by the non-thermal electrons. This condition is reasonable for electrons with energies $\gtrsim 150$ MeV which do not suffer Coulomb collisions in the ICM [12, 13] and hence do not heat the IC gas. Electrons with $E_e \gtrsim 150$ MeV loose energy mainly through ICS and synchrotron losses. Radio halo emission in galaxy clusters is produced by electrons with energy $E_e \approx 16.4$ GeV $B^{-1/2}(\nu_r/GHz)^{1/2}$ which yield $E_e \sim 1.6 - 52$ GeV for a typical $B = 1\mu G$ IC magnetic field. For such energies the non-thermal electrons do not interact strongly with the thermal IC gas. Under this hypothesis, the probability that a CMB photon is scattered by an electron of population A (say thermal) is not affected by the fact that the same photon has been scattered by an electron of population B (say non-thermal). Thus, at first order in τ we expect that the total SZ effect is given by the sum of the first order SZ effects due to each single population. At higher order approximation in τ, however, one has to consider the effect of repeated scattering [26]. It is intuitive to think that in this last case the total SZ effect is not only given by the sum of the separate SZ effects at higher orders in τ and the terms describing the cross–scattering between CMB photons and the electrons of populations A and B have to be taken into account. The spectral shape of the total SZ effect produced by a specific combination of a thermal plus a non-thermal electron distribution is shown in Fig.1.

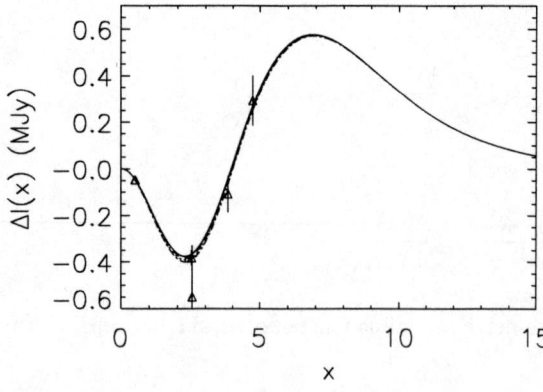
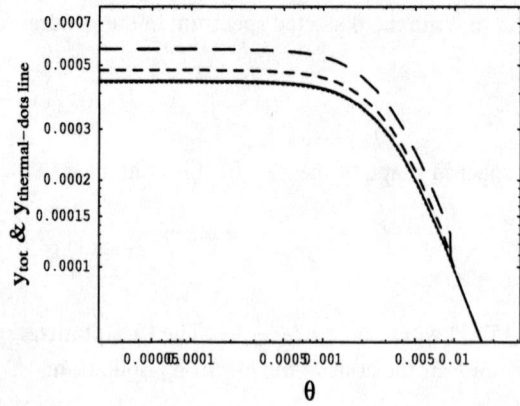

FIGURE 2. Left: the spectral distortion observed in A2163 and the fit obtained with a thermal SZ effect (solid) and that calculated with the addition of another non-thermal population of electrons with $\bar{P} = 0.12$ (dashes), 0.23 (dots) and 0.35 (dot-dashes). The best fit is obtained for a value $\bar{P} = 0.23$. **Right**: the spatial dependence of the total SZ effect produced by the combination of the thermal and relativistic electrons in Coma. We show the thermal SZ effect (dots) and the toral SZ effect with $\bar{P} = 0.05$ (solid), 0.49 (short dashes) and 1.48 (long dashes). Calculations are done for an a-dimensional frequency of $x = 2.5$ where the total SZ effect has its minimum value.

Specific predictions

We apply the previous considerations to the case of A2163, an X-ray bright galaxy cluster which host a giant radio halo. We evaluate the total SZ effect produced by a population of thermal electrons with parameters derived from X-ray observations [28] and a population of relativistic electrons with a non-thermal spectrum which fits the radio halo spectrum of the cluster and which is spatially distributed like the thermal one, as suggested by the observations [29]. The case of a single power-law spectrum for the non-thermal electron population is not acceptable since the best fit requires a value $\bar{P} = 7.8$ which is physically unreasonable. The case of a broken power-law spectrum fits better the SZ spectrum of A2163 (see Fig.2) and yields a more reasonable value $\bar{P} = 0.23$. The best fit improves sensitively when this non-thermal component is added in the evaluation of the total SZ effect.

The presence of a non-thermal SZ effect also produces spatial variations of the SZ radial profile of the cluster. Colafrancesco et al. (2001) found that the spatial dependence of the total Comptonization parameter is

$$y_{tot} = \frac{\sigma_T}{m_e c^2} P_{th}^0 \left\{ r_{c,X} Y_{th}(\theta) \cdot g(x) + r_{c,rad} \bar{P}_0 Y_{non-th}(\theta) \cdot \tilde{g}(x) \right\}. \tag{16}$$

which is a function of the angular distance θ from the cluster center and of the a-dimensional frequency x and of the pressure ratio $\bar{P}_0 = P_{rel}(\theta = 0)/P_{th}(\theta = 0)$ at the cluster center. Here, $r_{c,X}$ and $r_{,rad}$ are the core radii of the spatial distribution of the thermal and relativistic electron distributions, respectively. The functions $Y_{th}(\theta)$ and $Y_{non-th}(\theta)$ describe the spatial behaviour of the thermal and non-thermal contributions to the SZ effect, respectively (see Palladino, Colafrancesco and Marchegiani, these Proceedings, for more details). In Fig.2 we show the spatial behaviour of the total SZ effect in the case of Coma (see also Palladino, Colafrancesco and Marchegiani 2001, these Proceedings for more details).

DISCUSSION

The spectra of the thermal and non-thermal SZ effects are distinctly different and the overall spectrum of the SZ effects measures the energy densities in the thermal and in the radio halo source separately. Since many galaxy clusters where SZ measurements are available show also the presence of an extended radio halo, it is relevant to understand quantitatively the relevance of the non-thermal SZ effect in these clusters and to disentangle it from the true thermal SZ effect which is used for cosmological studies.

We have presented here some of the results of a new set of calculations [26] for the spectral shape of the non-thermal SZ effect and for its spatial behaviour adopting an exact approach which uses the full relativistic formalism and goes beyond the Kompaneets and the single scattering approximation. Moreover, we evaluated the total SZ effect from the combination of a thermal and a non-thermal electron populations residing in the same galaxy cluster.

Beyond the relevance of the study of the non-thermal SZ effect as a bias for the cosmologically relevant thermal SZ effect, the non-thermal SZ effect has also a crucial astrophysical relevance as a diagnostic for the presence of any relativistic population of electrons with a non-thermal energy spectrum. In fact, the non-thermal SZ effect actually measures the total pressure of the non-thermal electron population and hence yield constrains to its energy spectrum, as discussed here for the case of A2163.

The presence of a non-thermal SZ effect also influence the spatial profile of the total SZ effect of a radio-halo cluster. In fact, the region occupied by the relativistic electrons which produce the radio halo emission shows an increment (decrement) of the signal at frequencies near the minimum (maximum) of the SZ effect. This happens as a consequence of the difference in the spectral shapes of the non-thermal SZ effect with respect to the thermal one.

The specific spectral and spatial features of the non-thermal SZ effect allow to detect it through a multi-frequency observation with high sensitivity and narrow-band detectors. The best observational strategy is to observe in the frequency range $x \sim 2 \div 8$, where the spectral features allow clearly to disentangle the non-thermal SZ effect from the thermal one. The PLANCK surveyor experiment has the capabilities to detect and map the non-thermal SZ effect in a large number of nearby radio-halo clusters. However, dedicated experiment with high sensitivity and narrow band spectral coverage are also adequate to detect the non-thermal SZ effect in radio-halo galaxy clusters.

REFERENCES

1. Zel'dovich, Ya.B. and Sunyaev, R.A. 1969, Astrophysics Space Sci., 4, 173
2. Sunyaev, R.A. and Zel'dovich, Ya.B. 1972, Comments Astrophys. Space Sci., 4, 173
3. Sunyaev, R.A. and Zel'dovich, Ya.B. 1980, ARA&A, 18, 537
4. Birkinshaw, M. 1999, Physics Report, 310, 97
5. Mushotzky, R.F. and Scharf, C.A. 1997, ApJ, 482, L13
6. Tucker, W. et al. 1998, ApJ, 496, L5
7. Rephaeli, Y. 1995, ApJ, 445, 33
8. Itoh, N., Kohyama, Y. and Nozawa, S. 1998, ApJ, 502, 7
9. Challinor, A. and Lasenby, A. 1998, ApJ, 499, 1
10. Sarazin, C.L. 1988, 'X-ray emission from clusters of galaxies', Cambridge University Press
29. Feretti, L. 2001, in 'Constructing the Universe with Clusters of Galaxies', Eds. D. Gerbal and F. Durret
12. Blasi, P. and Colafrancesco, S. 1999, Astroparticle Physics, 122, 169
13. Colafrancesco, S. and Mele, B. 2001, ApJ, 562, 1
14. Lieu, R. et al. 1999, ApJ, 510, L25
15. Kaastra, J. et al. 1999, ApJ, 519, 1119
16. Bowyer, S. 2000, AAS, 32, 1707
17. Fusco-Femiano, R. et al. 1999, ApJ, 513, L21
18. Fusco-Femiano, R. et al. 2000, ApJ, 534, L7
19. Rephaeli, Y., Gruber, W. and Blanco, P. 1999, ApJ, 511, L21
20. Lieu, R. et al. 2000, A&A, 364, 497
21. Blasi, P. Stebbins, A. and Olinto, A. 2000, ApJ, 535, L71
22. Dogiel, V.A. 2000, A&A, 357, 66
23. Sarazin, C.L. and Kempner, J.C. 2000, ApJ, 533, 73
24. Petrosian, V. 2001, ApJ, in press (preprint astro-ph/0101145)
25. Colafrancesco, S. 2001, in preparation
26. Colafrancesco, S., Marchegiani, P. and Palladino, E. 2001, A&A, submitted
27. Ensslin, T. and Kaiser, C. 2000, A&A, 360, 417
28. Markevitch, M. et al. 1996, ApJ, 465, L1
29. Feretti, L. et al. 2001, A&A, 373, 106 (preprint astro-ph/0104451)

Balloon-borne and Ground-based Sub-millimetre Cosmological Surveys: Breaking the "Redshift Deadlock"

David H. Hughes[*], Itziar Aretxaga[*], Edward Chapin[*] and Enrique Gaztañaga[*]

[*]*Instituto Nacional de Astrofísica, Óptica y Electrónica, Luis Enrique Erro 1, Tonantzintla, Puebla, Pue., Mexico*

Abstract. In recent years sensitive submillimetre (sub-mm) and millimetre (mm) wavelength surveys have provided the opportunity to study star-formation in the high-z Universe. Identifying the formation epoch of clusters, massive galaxies and the first generations of stars, and understanding their subsequent evolution is now a realistic possibility. In this paper we describe how the combination of ambitious balloon-borne (BLAST) [1] and ground-based sub-mm (or mm) surveys can provide the essential redshift information, with sufficient precision, to break the current *deadlock* that is preventing an accurate description of the star formation history of the sub-mm starburst galaxy population.

INTRODUCTION

It is now generally accepted that the density of star formation activity in galaxies is roughly constant between $z \sim 1-4$. A series of ground-based sub-mm 850μm surveys [37, 18, 3, 13, 29, 36, 16], undertaken with the SCUBA camera [20] on the 15-m James Clerk Maxwell Telescope (JCMT), have contributed significantly (initially controversially) to this understanding which is in marked contrast to the situation 4 years ago, prior to the first SCUBA results, when the optical surveys suggested the density of star-formation declined by a factor of ~ 5 over the same redshift range [40, 31]. However one important caveat to this opening statement is the fact that we actually have little accurate information on the redshift distribution of the sub-mm galaxy population which currently consists of ~ 100 galaxies identified in 850μm surveys at a level > 2 mJy. The lack of confident optical and radio identifications, and thus precise redshifts for the sub-mm galaxies, was a justifiable cause for the early scepticism of the claims, based only on sub-mm data, that there existed no evidence for a decline in density of star formation at $z > 2$ [33].

There are a few cases of spectroscopically-determined redshifts for sub-mm galaxies associated with AGN [17, 27]. In almost all cases these galaxies have been identified in sub-mm surveys of lensing clusters [37, 28]. Their optical spectroscopic redshifts are consistent with the overall redshift distribution, which places the population of sub-mm galaxies between $z = 1-4$. The evidence for this redshift distribution of sub-mm galaxies is based on simple comparisons of their IR-radio spectral energy distributions (SEDs), usually one or two detections plus a few limits, with some template, often the ULIRG galaxy Arp220.

This raises an obvious question:*What classes of local galaxies offer the most useful analogues to the high-z sub-mm population?* In order to address this question we must measure the full rest-frame X-ray to radio SEDs of individual high-z sub-mm galaxies, measure the dispersion in the shapes of their SEDs and their range of luminosities (which depend on redshift), and thus determine the accuracy with which they can be characterised by a limited number of local template SEDs.

Optical and IR follow-up observations of those few SCUBA sources for which unambiguous identifications exist have revealed that the sub-mm counterparts are often extremely red objects (EROs with I-K > 6 [30, 39, 25]), although in a few cases the sub-mm counterparts are blue galaxies, with weak AGN (*e.g.* SMM02399-0136 [26]). However the similar surface densities of SCUBA sources and EROs suggest that the sub-mm sources may be related to a population of optically-obscured galaxies [39]. Also the evidence indicating that SCUBA galaxies are forming stars at a very

---

[1] Balloon-borne Large-Aperture Sub-millimeter Telescope - P.I. Mark Devlin (University of Pennsylvania, Philadelphia, USA)

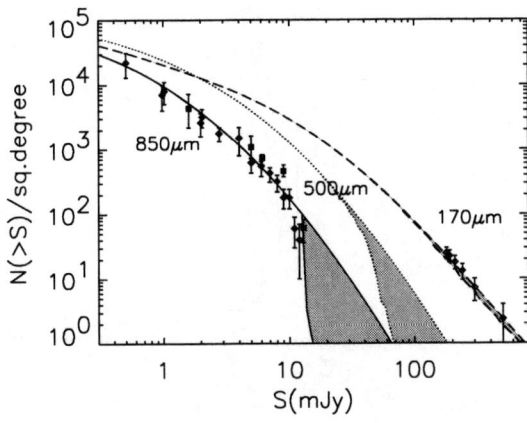 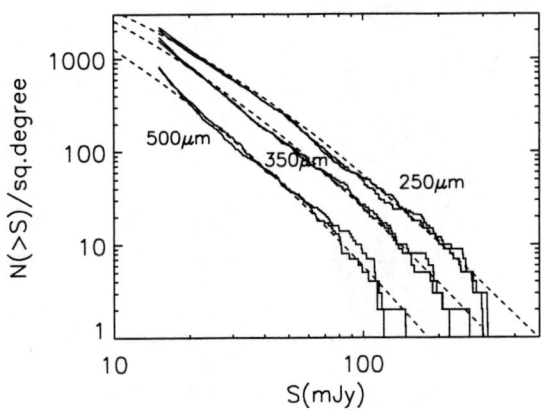

FIGURE 1. Integrated source-counts at FIR to sub-mm wavelengths for a model in which the IRAS 60μm luminosity function is evolved as $(1+z)^3$ upto $z = 1.5$, and maintained constant (with no further evolution) for $1.5 < z < 6$. The SED of Arp220 is adopted to represent the whole galaxy population. *Left panel:* The curves represent the source-counts at 850μm (solid line), 500μm (dotted) and 170μm (long-dashed). The measured source-counts from SCUBA surveys at 850μm [36] and the ISOPHOT FIRBACK survey at 170μm [11] are shown as diamonds. The flux-densities of the MAMBO 1.3 mm source-counts are scaled upwards by a factor of 2.25 to represent the equivalent measurements at 850μm, assuming the MAMBO sources lie at redshifts $1 < z < 6$, and are shown as filled squares [5]. The models also show the effect of a high-luminosity cut-off (at $L_{FIR} > 10^{13} L_\odot$) in the sub-mm population. The shaded areas illustrate the region of parameter space that needs to be searched with future experiments to improve our understanding of the evolution of the luminous high-z sub-mm galaxies. *Right panel:* Integrated source-counts for extragalactic *BLAST* surveys at 250μm, 350μm and 500μm. The dashed-lines show the same evolutionary model described for the left-panel. The histograms illustrate the extracted source-counts at each wavelength from two different Monte-Carlo simulations of 1 sq. degree surveys. These simulations explore evolutionary models with small differences in the strength of evolution and the redshift distribution of sources. The details of the Monte-Carlo simulations are summarised here and elsewhere [1].

high-rate ($> 100 M_\odot$/yr) implies that local dusty ULIRGs may also be suitable analogues of high-z sub-mm sources. Unfortunately the SEDs of this local class of luminous star-forming galaxies vary significantly with increasing FIR luminosity [35]. Furthermore, there are few ULIRG galaxies at $z > 1$ for which complete SEDs exist.

Despite the diversity in the properties of sub-mm galaxies [27], efforts have been made to place them in the context of an evolutionary model of star formation. Thus, in order to make some progress on describing the redshift distribution of the high-z sub-mm galaxy population, it has been necessary to compare their SEDs with those of local and distant galaxies.

For example, in a SCUBA 850μm survey of the northern Hubble Deep Field [18], the SEDs of Arp220 and Mrk231 were adopted as templates from which redshifted colours could be calculated under different cosmologies (given that their SEDs, when normalised at 850μm, bound those of the majority of other ULIRGs, radio-quiet quasars, Seyfert galaxies with circum-nuclear starforming rings, and starburst galaxies). It was on the strength of their colours between 15, 170, 450, 850μm and 1.4 GHz that the HDF-N galaxies were placed at redshifts $0.9 < z < 3.8$. Taking into account the sub-mm luminosities of these galaxies, the star formation density was estimated to be $\sim 0.2\, h M_\odot \text{yr}^{-1} \text{Mpc}^{-1}$ in the redshift interval $2 < z < 4$, and thus it was argued that a significant amount of star-formation is obscured by dust at high-z, and had been *missed* in earlier optical studies.

Similar investigations have also used the 850μm – 1.4 GHz colour to measure photometric redshifts [7, 8, 38] and have concluded that galaxies selected from MAMBO (1.2 mm) and SCUBA (850μm) surveys have median redshifts of $z = 2.5$ and $z = 1.9$, respectively.

In summary, the consensus is that the population of sub-mm galaxies is distributed between redshifts $z = 1 - 4$, yet the details of this distribution are unknown. It is common to describe the redshift of individual galaxies as "*in the range of ...., with a best guess of ....*". Without unambiguous optical, IR or radio identifications there is currently no direct means to measure the redshifts of individual sub-mm galaxies, or the overall distribution with any greater precision.

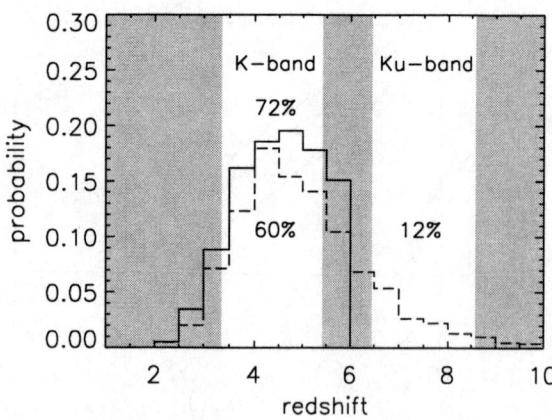
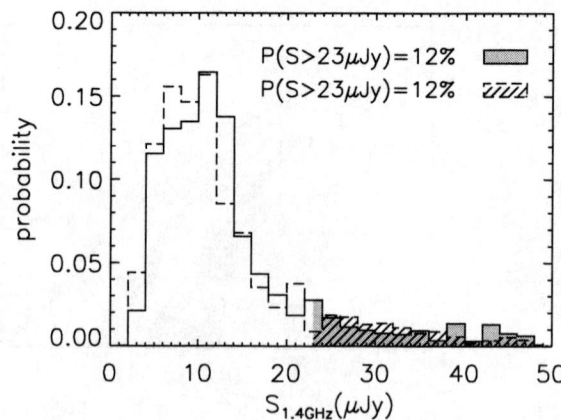

FIGURE 2. *Left panel:* Redshift probability distribution for the sub-mm galaxy HDF850.1 [18] assuming an evolutionary model with a cut-off at $z = 6$ (solid-line) and a model with constant evolution between $2 < z < 10$ (dashed-line). The open (unshaded) windows show the redshift ranges within which the 100-m GBT can detect the J=1-0 CO-line with the K-band (18.5 – 26.0 GHz) and Ku-band (12 – 15.4 GHz) receivers. The confidence (given as a percentage) of detecting the J=1-0 CO-line from HDF850.1 in each band is calculated for both evolutionary models. *Right panel:* Probability distribution of the 1.4 GHz flux density of HDF850.1 derived for the same two evolutionary models described in the left panel. In both models the probability of measuring a flux above the published 3σ limit of 23μJy [32] is only 12% and thus is consistent with the non-detection.

STATUS OF SUB-MILLIMETRE SURVEYS

The existing SCUBA and MAMBO surveys are limited in their ability to constrain the evolutionary models of the sub-mm galaxy population. The practical reasons for these limitations have been described elsewhere [23] and can be summarised as follows: restricted wavelength coverage (enforced by the limited number of atmospheric windows for ground-based observatories); low spatial resolution (resulting in both a high extragalactic confusion limit and poor positional accuracy); and low system sensitivity (a combination of detector and electronic noise, size of telescope aperture and telescope surface accuracy, sky transmission and sky noise) which restricts even the widest and shallowest sub-mm surveys to areas < 0.3 sq. degrees [36, 16].

Partial remedies to this situation on the short-timescale (within the next 3 years) can be found in two new large-aperture, single-dish, ground-based, millimetre-wavelength telescopes: the 100-m Green Bank Telescope (GBT) which has recently been commissioned, and the 50-m Large Millimetre Telescope (LMT) which is under construction and will *see* first-light in late 2003. Both telescopes will provide significantly greater survey sensitivities than the current confusion-limited data. Also the development of new large-format filled-array cameras, SCUBA-II (employing 6000–10000 TES detectors and a multi-plexed SQUID readout array) for the JCMT, and SHARC-II (using a 12×32 array of silicon pop-up bolometer detectors) for the 10-m Caltech Submillimetre Observatory, will dramatically enhance the efficiency of conducting ground-based 850μm and 350μm surveys respectively. In addition large sub-mm interferometers, which are also under construction (*e.g.* SMA) will soon provide accurate positions for a limited number of sources.

Despite these new ground-based facilities, there will still remain a large uncertainty in the redshifts for the majority of sub-mm galaxies. Furthermore the lack of data at short sub-mm wavelengths ($< 450\mu m$) means there can be little constraint on the rest-frame FIR luminosities and SFRs of individual high-z galaxies and thus the star formation history of the entire sub-mm population.

Beyond 2008, ALMA, on a high-altitude, dry Chilean site, will undoubtedly solve many of these difficulties with its powerful combination of observing wavelengths between 350μm and 3 mm, and high spatial resolution. The expected launch of the Herschel satellite in 2008 will also bring the chance to map large areas of the extragalactic sky at 250–500μm with high sensitivity. However, before 2008, long-duration (~ 15 day) balloon-borne experiments (*e.g. BLAST*, a Balloon-borne Large-Aperture Sub-millimetre Telescope [10]) will conduct sub-mm surveys with signficantly greater sensitivity than is available from even the largest ground-based sub-mm observatories.

In this paper we describe simulations of these future *BLAST* surveys which demonstrate how the FIR–sub-mm

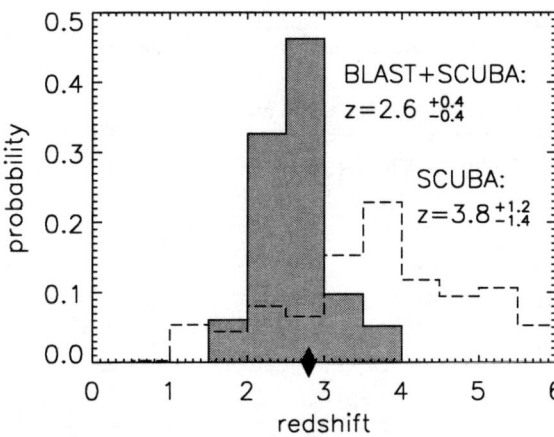
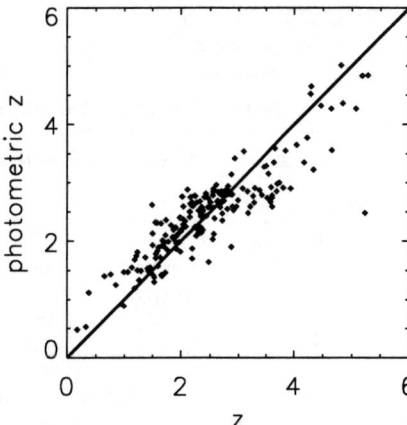

FIGURE 3. *Left panel:* The redshift distribution for a typical galaxy, $6 \times 10^{12} L_\odot$, detected at 850μm ($4\sigma = 10$ mJy), with a 450μm non-detection limit ($3\sigma < 75$ mJy) in the UK 8mJy SCUBA survey [36] (shown as the dashed histogram). The addition of data at 250, 350 and 500μm from a follow-up deep *BLAST* survey ($S_{250\mu m} = 31 \pm 5$ mJy, $S_{350\mu m} = 38 \pm 5$ mJy, $S_{500\mu m} = 29 \pm 5$ mJy) significantly improves the redshift constraints. The redshift probability distribution for the identical SCUBA galaxy with follow-up *BLAST* observations is shown as the shaded histogram. The black diamond marks the true redshift, $z = 2.8$, of the mock galaxy. Photometric redshift determinations within a 68% confidence level are indicated in the panel labels. *Right panel:* Photometric redshift vs. true redshift relationship for 200 galaxies simultaneously detected in a 1 sq. degree survey at 250, 350 and 500μm with *BLAST* and also at 850μm in a future wide-area ground-based SCUBA survey.

SEDs of galaxies can be used to measure their photometric redshifts with an accuracy of $\Delta z \sim \pm 0.6$ over the range $0 < z < 6$ and break the current *redshift deadlock* that limits our understanding of the evolution of the sub-mm galaxy population [1]. From such precision in the individual galaxy redshifts, we can determine the star formation history for sub-mm galaxies brighter than $3 \times 10^{12} L_\odot$ with an accuracy of $\pm 35\%$. The following cosmological model is adopted: $\Omega_\Lambda = 0.7, \Omega_M = 0.3, H_0 = 67 \text{kms}^{-1} \text{Mpc}^{-1}$.

MONTE-CARLO SIMULATIONS OF SUB-MM PHOTOMETRIC REDSHIFTS

Measurements of photometric redshifts with FIR – sub-mm – radio data [18, 6, 7, 8, 12, 14, 4] do not usually include a rigorous analysis of the errors. The need to improve this situation, and understand the accuracy with which sub-mm observations can provide redshifts for individual galaxies, motivated the development of Monte-Carlo simulations that take into account the imprecise knowledge of the luminosity function of sub-mm galaxies, the dispersion in the luminosity-dependent SEDs, absolute calibration and observational photometric errors.

The goal of the Monte-Carlo simulations, which are described in detail elsewhere [1, 2, 22], is to assign a confidence (probability) to the photometric redshift of any individual galaxy, and hence provide a statistical measure of the redshift distribution for the sub-mm population.

One example of the method, applied to known sub-mm survey sources, is shown in Fig. 2. The redshift probability distribution for the brightest sub-mm source in the *Hubble Deep Field*, HDF850.1 [18], indicates that $z_{\text{phot}} = 4.5 \pm 0.9$ with 68% confidence, and $z_{\text{phot}} = 4.5^{+1.5}_{-1.2}$ with 90% confidence (for the model with a cut-off in the luminosity evolution at $z = 6$). This sub-mm galaxy has a marginal (2.4σ) 8.4 GHz radio detection (VLA 3651+1226 [32]), no detection at 1.4 GHz ($3\sigma < 23\mu$Jy, [33]) and no detected X-ray, optical, IR or FIR counterpart, despite an accurate interferometric position at 236 GHz and the unprecedented depth of the HDF surveys [21, 41, 42].

The Monte Carlo simulations demonstrate that, given the depth of the 1.4 GHz observations, it is no surprise that HDF850.1 was not detected at this frequency, and that a 3σ sensitivity of 6μJy at 1.4 GHz is necessary in future radio observations (with the upgraded VLA) before there is 85% probability of detecting HDF850.1.

The clear advantage of having a constrained redshift probability distribution for individual sub-mm galaxies, without optical counterparts, is that we can now determine the likelihood that a redshifted rotational CO transition-line falls

TABLE 1. Examples of the number of detected galaxies and their redshift distributions in alternative 50 hour *BLAST* surveys to be undertaken during a series of future long duration balloon flights in Antarctica.

50 hour 250 μm *BLAST* survey strategies: D=2.0 m, NEFD=236 mJy s$^{1/2}$

survey area (sq. degrees)	1σ depth	no. of pixels	no. of detected galaxies		no. of > 5σ galaxies	
			> 5σ	> 10σ	$z > 1$	$z > 3$
1.0	5 mJy	18334	835	265	765	147
2.0	7 mJy	36668	1012	291	927	151
4.0	10 mJy	73336	1100	294	988	147
9.0	15 mJy	165006	1111	247	1023	129
36.0	30 mJy	660024	990	246	895	105

into the frequency range of any particular spectral-line receiver on the next generation of large mm-cm telescopes. For example, Fig.2 demonstrates that, depending on the adopted evolutionary model, the J=1–0 CO line of HDF850.1 has a 60–72% probability of being detected in the K-band receiver (18.0–26.5 GHz) of the 100-m GBT. There is only a 12% chance that the redshift of HDF850.1 is high enough for J=1–0 CO line to be detected in the Ku-band.

BALLOON-BORNE SUB-MILLIMETRE EXTRAGALACTIC SURVEYS

We have presented an example of the redshift probability distributions for one of the best-studied sub-mm galaxies, HDF850.1 [18]. However a similar amount of sensitive follow-up data is not available for most sub-mm sources. In general the information on the SEDs of the sub-mm population consists of a low-S/N detection at 850μm, an upper-limit at 450μm, a 1.4 GHz radio observation (frequently a non-detection) and a very shallow 170μm ISOPHOT flux limit. Armed with only these data, the typical redshift distribution for an individual galaxy that can be derived from the SCUBA colours is shown in Fig.3

This inability to accurately determine the redshift distribution of sub-mm galaxies motivated a successful proposal to build and fly a Balloon-borne Large-Aperture Sub-millimetre Telescope (*BLAST* [10], see also Devlin - these proceedings, and http://www.hep.upenn.edu/blast). *BLAST* has a 2-m primary aperture and is equipped with large-format bolometer cameras operating at 250, 350 and 500μm which are copies of the SPIRE focal-plane cameras for the Herschel satellite. *BLAST* is scheduled for a test-flight in November 2002. During a series of long-duration (15 day) balloon flights in Antarctica from 2003 onwards, *BLAST* will conduct large-area sub-mm surveys (Table 1), and will provide sensitive rest-frame FIR data that can constrain the redshifts, and hence the luminosities and SFRs of the sub-mm galaxy population.

An application of our Monte-Carlo simulations to *BLAST* observations shows that the majority of scatter in the sub-mm colours will be due, in approximately equal parts, to the dispersion in the SEDs of galaxies in our library of templates, and the observational and calibration errors. Since we can control to some extent the quality of the experimental data, it is the uncertainty in the observed (or model) SEDs of high-z starburst galaxies that will ultimately limit the accuracy of these photometric redshift predictions. Theoretical SEDs, based on radiative transfer models for high-z starburst galaxies [15], can help in principle. However the low S/N, and restricted wavelength coverage of the available data are consistent with such a broad range of theoretical models that they currently provide little discriminatory power in determining the most appropriate SEDs to use as templates. Consequently one of the most straight-forward achievements of *BLAST*, but no less important, will be simply to provide an accurate empirical model of the SED for 1000's of high-z galaxies at these critical short sub-mm wavelengths (250–500μm).

We have developed realistic galaxy survey simulations, which incorporate evolutionary models that are consistent with the observed sub-mm and mm source-counts, to assist in the design of these future balloon-borne sub-mm experiments [19, 24]. These simulations, which produce artifical multi-wavelength images at different spatial resolutions, address real issues confronting existing current and forthcoming surveys. For example they describe how the measured source-counts can be affected by telescope resolution, extragalactic and galactic source-confusion, survey sensitivity and noise and sampling variances due to clustering and shot-noise.

The discrepancy between the bright-end source-counts in the 850μm SCUBA surveys and 1.3 mm MAMBO surveys [36, 14, 5], and a visual inspection of their reconstructed maps, suggests that the clustering of sub-mm galaxies may

TABLE 2. Comparison of the 1σ accuracy, Δz, of optical [34] and sub-mm photometric redshifts (this paper) in redshift bins between $0.5 < z < 5.5$. At redshifts > 1.5, the accuracy of the photometric redshifts derived from the combined *BLAST* (250, 350, 500μm) and SCUBA 850μm data is equivalent to, or better than, the optical estimates.

redshift bin	optical	*BLAST*	*BLAST* + 850μm
z	Δz	Δz	Δz
0.5 − 1.5	0.15	0.34	0.47
1.5 − 2.5	0.33	0.31	0.34
2.5 − 3.5	0.6	0.71	0.34
3.5 − 4.5	0.7	1.14	0.69
4.5 − 5.5	1.0	2.14	0.60

be influencing the statistics (Fig.1.). Taken at face value, the steepening of the 850μm counts for sources > 10 mJy can be interpreted as evidence for an under-density of galaxies in the survey-field providing those particular data, or perhaps a high-luminosity cut-off in the luminosity function for galaxies $> 10^{13} L_\odot$. Only wider-area (≥ 0.5 sq. deg.), bright sub-mm surveys can distinguish between these possible alternatives.

Simulations of future *BLAST* surveys have guided the choice of survey areas, with the intention of avoiding any artifacts in the final data due to galaxy clustering. The same simulations have also guided the choice of survey depths to ensure that the deepest *BLAST* surveys reach below the extragalactic confusion-limit in the presence of foreground cirrus contamination. A minimum survey area of 1.0 sq. deg is sufficient to reduce the clustering noise to the level of $\sim 3\%$ in the counts. The extraction of sources from these simulated images, smoothed to the appropriate resolution, indicates that the extragalactic 3σ confusion limit of *BLAST* at 250-500μm will be $\sim 20 - 30$ mJy. For comparison the traditional estimate of extragalactic confusion (1 source/30 beams [9]) suggest that confusion will begin to dominate below 30 mJy. Hence, during a single long-duration balloon flight from Antarctica, *BLAST* will conduct a series of simultaneous large-area surveys at 250, 350 and 500μm, down to a 1σ sensitivity of 5 mJy, detecting in each survey ~ 1000 luminous sub-mm galaxies ($> 10^{12} L_\odot$) between $0.5 < z < 6$ (Table 1).

COMPARISON OF OPTICAL AND SUB-MM PHOTOMETRIC REDSHIFTS

The comparison of photometric redshifts of galaxies derived from sub-mm *BLAST* data (250, 350 and 500μm), using the method described above [1], with the true redshifts in our mock catalogues demonstrates that an accuracy of $\Delta z \sim \pm 0.6$ can be achieved over the range $0.5 < z < 6$, although, as Table 2 shows, the scatter increases significantly for $z > 3.5$. The accuracy of photometric redshifts measured from the optical-IR SEDs of galaxies [34] varies between $\pm 0.15 - 1.0$ over the same redshift range. The inclusion of longer wavelength data (*e.g.* 850μm) to the *BLAST* data increases the photometric redshift accuracy to $\Delta z \sim \pm 0.4$, which means that, despite the greater wealth of data describing the optical SEDs of galaxies, sub-mm photometric redshifts can be determined with greater accuracy than those at optical wavelengths. Furthermore the combination of *BLAST* and 850μm data extends the redshift range over which we can measure reliable redshifts to $0.5 < z < 6$ (Fig.3 and Table 2).

Further examples of the probability redshift distributions for galaxies detected in *BLAST* surveys, and a demonstration of the improved photometric accuracy that can be gained with the increased sensitivity of future SPIRE/Herschel observations are presented by Aretxaga *et al.* (these proceedings).

MEASURING THE STAR FORMATION HISTORY FROM SUB-MM SURVEYS

Taking all of the above observational errors and uncertainties in the luminosity function and SEDs into account, a deep *BLAST* survey, combined with a wide-area shallow SCUBA survey ($S_{850\mu m} > 6$ mJy), can recover the SFR density at $0.5 \leq z \leq 5.5$ with an accuracy of 35% due mainly to $L_{\rm FIR} \geq 10^{12} L_\odot$ galaxies (Fig.4). The error in the SFR density is dominated by the choice of the appropriate SED used to derive the 60μm luminosities of the mock galaxies (which have intrinsically different SEDs), and not by the accuracy of the redshift determinations. The completeness of the

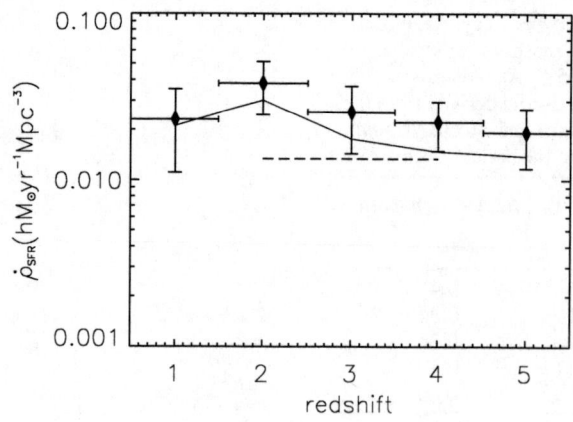
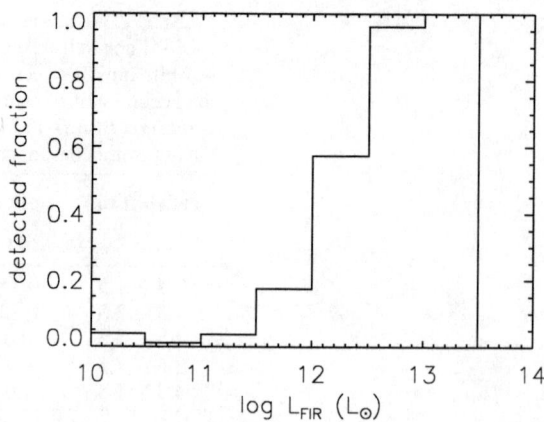

FIGURE 4. *Left panel*: Global star formation rate history of galaxies detected in at least 2 filters from a combined *BLAST* ($3\sigma = 15$ mJy) and wide-area SCUBA 850μm survey ($3\sigma = 6$ mJy). Filled diamonds represent the recovered values of the SFRs using the photometric redshift distribution. We compare these values with the input SFR in the catalogue (shown as a solid line) before errors are introduced. The error bars are dominated by the uncertainty of the SED that should be used to derive the 60μm luminosities of the mock galaxies (which have intrinsically different SEDs), and not by the accuracy of the redshift determinations. If a single SED is adopted (Arp220 for example), the derived SFR is recovered with an accuracy better than the size of the diamonds. The dashed horizontal line shows an estimate of the minimum SFR density for sub-mm galaxies that have been detected in the UK 8mJy SCUBA 850μm survey [36], assuming that the SCUBA galaxies lie between $2 < z < 4$. *Right Panel*: Fraction of galaxies, per luminosity interval, detected at all redshifts in the mock combined *BLAST* and wide-area SCUBA survey. The completeness of the survey is ~98% for galaxies $> 3 \times 10^{12} L_\odot$.

survey is $\sim 98\%$ above $3 \times 10^{12} L_\odot$. The SFR density derived from the mock combined *BLAST* and SCUBA survey is consistent with the preliminary results from the UK 8 mJy SCUBA survey [36], and provides a good illustration of the advantage of being able to derive photometric redshifts for the entire survey sample over a large redshift range (Fig.4).

CONCLUSIONS

The product of the simulations of sub-mm surveys described in this paper (see also Aretxaga *et al.*- these proceedings) is the ability to calculate the probability distribution for the redshift of any individual galaxy, taking into account observational errors and the uncertainty in the appropriate template SED, without the requirement to first identify the optical, IR or radio counterpart. These simulations demonstrate that the combination of balloon-borne and ground-based sub-mm data between 250–850μm for a statistical sample of galaxies can provide the rest-frame FIR luminosity distribution, and hence the star formation history of the entire sub-mm population with unprecedented accuracy.

ACKNOWLEDGMENTS

This work has been partly supported by CONACYT grants 32180-E and 32143-E.

REFERENCES

1. Aretxaga, I., Hughes, D.H, Chapin, E., Gaztañaga, E., 2001, these proceedings
2. Aretxaga, I., Hughes, D.H, Chapin, E., Gaztañaga, E.,Dunlop, J.S., Devlin, M., 2001, in prep.
3. Barger, A.J., *et al.*, *Nature* **394**, 248 (1998)
4. Barger, A.J., Cowie, L.L, Richards, E.A., *A.J.* **119**, 2092 (2000)
5. Bertoldi, F., *et al.*, "The Dawn of Galaxies: Deep MAMBO Imaging Surveys," in *Cold Gas and Dust at High Redshift*, edited by D.J.Wilner et al., Highlights of Astronomy 12, 2000, astro-ph/0010553,

6. Blain, A.W., *MNRAS* **309**, 955 (1999)
7. Carilli, C.L., and Yun, M.S., *Ap.J.* **513**, L13 (1999)
8. Carilli, C.L., and Yun, M.S., *Ap.J.* **530**, 618 (2000)
9. Condon, J.J., *Ap.J.* **188**, 279 (1974)
10. Devlin, M., 2001, "A Balloon-borne Large Aperture Submillimetre Telescope" in *Deep Millimetre Surveys: Implications for Galaxy Formation and Evolution*, edited by J.Lowenthal and D.H.Hughes, World Scientific, 2001, astro-ph/0012327
11. Dole, H., *et al.*, *A&A* **372**, 364 (2001)
12. Dunne,L., Clements, D.L., Eales, S.A., *MNRAS* **319**, 813 (2000)
13. Eales, S.A., *et al.*, *Ap.J.* **515**, 518 (1999)
14. Eales, S.A., *et al.*, *A.J.* **120**, 2244 (2000)
15. Efstahiou,A., Rowan-Robinson, M., Siebenmorgen, R., *MNRAS* **313**, 734 (2000)
16. Fox, M.J., *et al.*, *MNRAS*, submitted, astro-ph/0107585 (2001)
17. Frayer, D.T., *et al.*, *Ap.J. Lett.* **506**, L7 (1998)
18. Hughes, D.H., *et al.*, *Nature* **394**, 241 (1998)
19. Hughes, D.H. and Gaztañaga, E., "Simulated Submillimetre Galaxy Surveys" in *Star formation from the small to the large scale*, edited by F.Favata *et al.*, ESA SP-445, 2000, astro-ph/0004002
20. Holland, W.S., *et al.*, *MNRAS* **303**, 659 (1999)
21. Hornschmeier, A., *et al.*, *Ap.J.* **541**, 49 (2000)
22. Hughes, D.H, Aretxaga, I., Chapin, E., Gaztañaga, E., 2001, in prep.
23. Hughes, D.H., "Cosmological Surveys at Submillimetre Wavelengths" in *Clustering at High Redshift*, edited by A.Mazure *et al.*, ASP Conf. Series 20, 2000, astro-ph/0003414
24. Gaztañaga, E. and Hughes, D.H., "Clustering in Deep (submillimetre) Surveys" in *Deep Millimetre Surveys: Implications for Galaxy Formation and Evolution*, edited by J.Lowenthal and D.H.Hughes, World Scientific, 2001, astro-ph/0103127
25. Gear, W.K., *et al.*, *MNRAS*, **316**, 51 (2000)
26. Ivison, R.I., *et al.*, *MNRAS* **298**, 583 (2001)
27. Ivison, R.J., *et al.*, *MNRAS* **315**, 209 (2000)
28. Knudsen, K.K., van der Werf, P.P., Jaffe, W., "A Submillimetre Selected Quasar in the Field of Abell 478", in *Deep Millimetre Surveys: Implications for Galaxy Formation and Evolution*, edited by J.Lowenthal and D.H.Hughes, World Scientific, astro-ph/0009024 (2000)
29. Lilly, S.J., *et al.*, *Ap.J.* **518**, 441 (1999)
30. Lutz, D., *et al.*, *A&A*, in press, astro-ph/0108131 (2001)
31. Madau, P., *et al.*, *MNRAS* **293**, 1388 (1996)
32. Richards, E.A., *et al.*, *Ap.J.* **116**, 1039 (1998)
33. Richards, E.A., *Ap.J.* **513**, L9 (1999)
34. Rowan-Robinson, M., submitted to *Ap.J.*, http://astro.ic.ac.uk/ mrr/photz/photzapj.ps (2001)
35. Sanders, D.B., and Mirabel, I.F., *ARA&A* **34**, 749 (1996)
36. Scott, S. *et al.*, *MNRAS*, submitted, astro-ph/0107446 (2001)
37. Smail, I., *et al.*, *MNRAS* **490**, L5 (1997)
38. Smail, I., *et al.*, *Ap.J.* **528**, 612 (2000)
39. Smail, I., *et al.*, *MNRAS* **308**, 1061 (1999)
40. Steidel, C.C., *et al.*, *Ap.J* **462**, L17 (1996)
41. Thompson, R., *et al.*, *A.J.* **117**, 17 (1999)
42. Williams, R.E. *et al.*, *A.J.* **112**, 1335 (1996)

The Microwave Spectra of Planets

Thérèse Encrenaz* and Raphaël Moreno**

*DESPA, Observatoire de Paris, F-92195 Meudon
**IRAM, Centre Universitaire, F-38406 Saint-Martin d'Hères

Abstract. This paper reviews our current knowledge of the spectra of planets, from the centimeter to the submillimeter range, both from an observational and a theoretical point of view. In the case of Venus, due to its very thick CO_2 atmosphere, the spectrum is dominated by the the pressure-induced CO_2 absorption, with some other contributions from minor constituents (SO_2, H_2O). In the case of Mars, which has a very tenuous atmosphere, the microwave spectrum is composed of discrete narrow molecular lines of CO, H_2O and their isotopes, superimposed over a continuum defined by the surface emissivity and temperature. The microwave spectra of the giant planets probe their tropospheres. They are dominated by collision-induced absorption, mostly due to H_2-H_2 and H_2-He collisions, with, in the case of Jupiter and Saturn, additional features due to NH_3 and PH_3. For all cases, observed and synthetic spectra are described, and the various sources of uncertainty are discussed.

INTRODUCTION

Planets are commonly used as calibration targets for infrared and radio observations. However, the analysis of their spectra shows that they do not radiate as simple blackbodies. Their spectra are the combination of a continuum (either due to the surface emission, in the case of Mars, or due to pressure-induced absorption and/or molecular absorption, in the other cases) and a series of discrete emission or absorption lines due to the presence of atmospheric minor contituents. Several sources of uncertainty may be associated to these components: in the case of Mars, surface heterogeneity and variability of the emissivity; in the case of Venus and the giant planets, at high temperature (Venus) or low temperature (giant planets), uncertainty in line shapes, thermal profiles and vertical distribution of the minor compounds.

VENUS

Venus is characterized by a thick CO_2-dominated atmosphere, with a surface pressure of 92 bars and a surface temperature of 730 K. Apart from CO_2, atmospheric constituents include N_2 (3%) and traces of H_2O, SO_2 and CO (in the upper atmosphere). A thick cloud, made of different layers of H_2SO_4 particles between 40 and 70 km, hides the surface at most wavelengths, including the visible range. There is no temperature inversion in the mesosphere, so that molecular signatures are always observed in absorption.

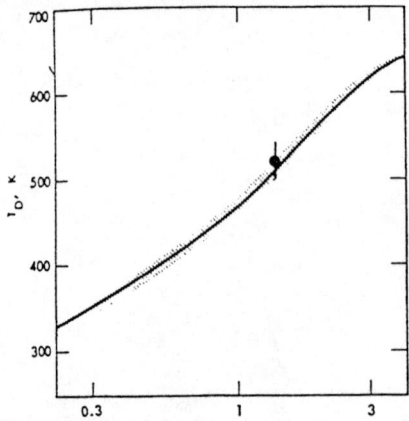

FIGURE 1. The Venus radio spectrum from 0.3 to 3 cm (measurement and synthetic spectrum). The figure is taken from Janssen and Klein (1981).

The continuum spectrum of Venus

At microwave wavelengths, the continuum radiation probes the lower troposphere of Venus, at altitudes of 10-20 km. The surface is never seen; the opacity is mostly due to the collision-induced absorption (CIA) of CO_2, due to CO_2-CO_2 and CO_2-N_2 collisions. In addition, H_2O and SO_2, which have mixing ratios of about 30 ppm and 180 ppm respectively at these altitudes, contribute to the absorption. Because of the high pressure, no individual line is detectable, and these contributions simply add to the observed continuum.

A synthetic spectrum of Venus between 0.3 and 3 cm (10-100 GHz) has been performed by Janssen and Klein (1981, [1]; Fig. 1) who compared it to radiometric measurements between 1.25 and 1.50 cm (20-24 GHz). They also estimated the expected effect of SO_2 and H_2O contributions (a few percent), assuming mixing ratios of 180 ppm and 0.13% respectively for these species. The H_2O abundance is now known to be much lower, as mentioned above (Pollack et al., 1993; [2]), which means that the effect of H_2O is negligible.

Mesospheric signatures

Superimposed over this continuum, discrete absorption features can be identified in the case of molecular constituents present in the mesosphere. This is the case of H_2O and CO, which have both been observed in the millimeter range by heterodyne spectroscopy. CO, present in the upper atmosphere only as a product of CO_2 photochemistry, was detected in the 1970s by Kakar et al. (1976; [3]) through its (1-0) transition at 115 GHz, and later monitored (Wilson et al, 1981) for diurnal variations (Wilson et al., 1981; [4]). The (2-1) transition was later used for measuring mesospheric winds (Lellouch et al., 1994; [5]). Water was detected above the clouds, at an altitude of about 90 km, through the 183 GHz H_2O transition and the 226 GHz HDO transition (Encrenaz et al., 1991, 1995; [6],[7]). These data provided information on the water vertical profile in the upper atmosphere.

Uncertainties and open problems

Several uncertainty factors limit our knowledge of the microwave spectrum of Venus.
-The radiometric data, limited to a few wavelengths (i.e. 1.35 cm) have an accuracy of about 2-3 percent (Janssen and Klein, 1981;[1]).
-The accuracy of the synthetic continuum spectrum (Fig. 1) is limited by the following uncertainties:
. the broadening of the molecular species by CO_2 at high temperatures (the uncertainty estimate is about a few percent at 500 K);
. the H_2SO_4 droplet absorption (a few percent, according to Janssen and Klein, 1981;[1]);
. the abundances and the possible variability of H_2O and SO_2 in the lower atmosphere of Venus. Indeed, there has been some controversy about these numbers; in addition, local variations cannot be a completely excluded; they would be expected if active volcanism were still present at the surface of Venus; which is still an open question.

MARS

The composition of the martian atmosphere (95% CO_2, 3% N_2, 2% Ar) is very similar to the one of Venus, but its physical parameters are very different: the surface pressure is only 6 mbars and the surface temperature is about 210 K, with very strong seasonal variations. The total pressure varies by as much as 30% over a full seasonal cycle (22 months); temperatures between 180 K and 270 K can be measured, depending upon the latitude and the season. Ice caps, made of H_2O and CO_2 ice are found at the poles. H_2O clouds and haze can appear occasionally, as well as H_2O frost at the surface.

As the atmosphere is very tenuous, the microwave spectrum of Mars directly measures the emission of the surface, apart from a few discrete, very narrow, gaseous molecular signatures. The radio flux thus gives access to the product Ts x e, where Ts is the surface temperature and e is the emissivity, and is independent of atmospheric properties.

Discrete spectral signatures have been identified in the radio (H_2O, 22 GHz; Clancy et al., 1992;[8]) and millimeter range (CO at 115 and 230 GHz; e.g. Clancy et al., 1996, [9]; H_2O and its isotopes, Encrenaz et al., 2001; [10]). These data have been used to monitor the thermal profile (from CO) and to study the vertical distribution of H_2O along the seasonal cycle.

Uncertainties

The main uncertainty in the Mars radio spectrum is the emissivity and the temperature of the surface, which result from the fact that the martian surface is highly heterogeneous. It is possible to calculate the expected surface temperature of Mars for a given location and time of the martian year, using global circular models (Forget et al., 1999;[11]) but an assumption has to be made upon its emissivity, which depends upon the nature of the soil. For the synthetic radio spectrum of Mars, the surface emissivity is often calculated from the Fresnel coefficient of reflection, using a surface dielectric constant of 2.5 (Lellouch et al., 1989;[12]). However, this calculation assumes an homogeneous surface composition over the disk.

Some attempts have been made to determine the mean surface emissivity of Mars in the far-infrared range, using the Infrared Space Observatory (ISO) data obtained with both the SWS and LWS spectrometers. The infrared spectrum of Mars is dominated by CO_2 and H_2O absorption lines throughout the whole infrared range. The H_2O distribution was first retrieved from SWS the near- and mid-IR water bands; then the surface emissivity in the far-infrared range was retrieved from a modelling of the H_2O lines at these wavelengths (Lellouch et al., 2000, [13]; Burgdorf et al., 2000;[14]). The retrieved emissivity seems to show significant variations with wavelengths (Fig. 2), which might be due to solid signatures. This result however requires further confirmation.

Another source of uncertainty is expected to come from seasonal effects. Dust storms are known to heat the atmosphere significantly (Clancy et al., 1996;[9]). As the surface reacts very quickly to the solar illumination, the surface temperature is expected to be modified as a consequence. The presence of aerosols, or clouds of H_2O/CO_2 ice, may also affect the measurement of the surface emissivity. One may hope to better constrain the surface emissivity when the martian surface is mapped from planetary orbiter missions. However the seasonal effects will remain difficult to predict accurately.

One could think that Mars is not a reliable calibration source in the present state of knowledge, due its heterogenous surface and its strong seasonal effects. Nevertheless, Mars has been used as a primary calibrator in the radio, millimeter and submillimeter domain with an absolute brightness temperature measurement of 207 +/- 7 K (Ulich, 1981; [15]). The main reason why Mars can still be used as a calibration source is because the heterogeneity and a least some ssseasonal effects are diluted in the beam size of current radiotelescopes, which is comparable to the planet's size.

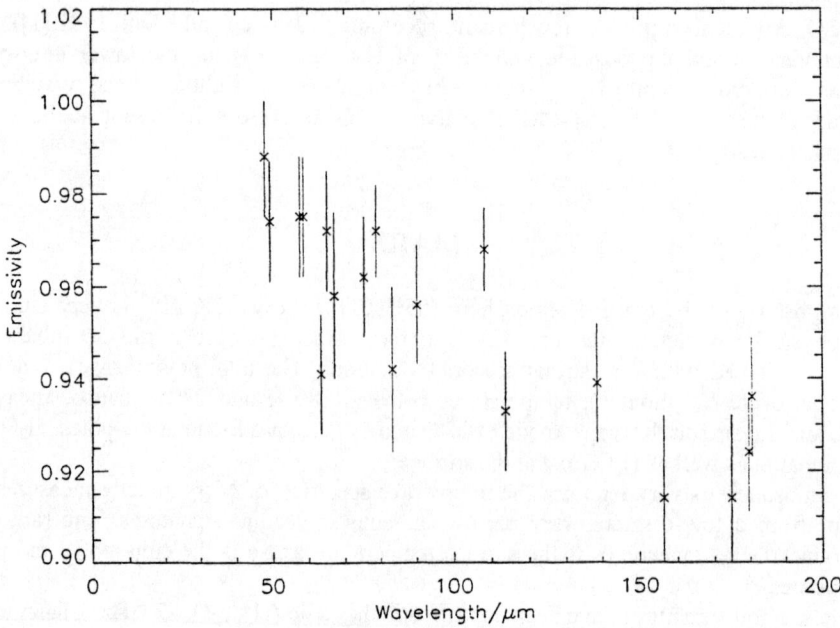

FIGURE 2. The far-infrared emissivity of Mars, as derived from ISO measurements. The figure is taken from Burgdorf et al. (2000).[36]

GIANT PLANETS

The observable atmosphere of the giant planets is mostly composed of hydrogen and helium, with traces of minor constituents (CH_4, NH_3, PH_3, H_2O). They are characterized by a troposphere, where the temperature decreases as the altitude increases, following the adiababtic gradient, and above it, a stratosphere where the temperature increases again with altitude. Between the troposphere and the stratosphere lies the tropopause, at a pressure level of about 100 mbars, where the temperature is minimum (110 K for Jupiter, 80 K for Saturn, 50 K for Uranus and Neptune). As a result of temperature inversion, in the thermal regime (infrared and radio ranges), molecular features can be seen in absorption or in emission, depending whether the radiation probes the deep layers (troposphere) or the higher layers (stratosphere).

The microwave spectrum of the giant planets probes the troposphere, at regions ranging from 0.1 to a few bars.It is, in all cases, dominated by the collision-induced spectrum of hydrogen, which has two strong components (H_2-H_2 and H_2-He collisions) and a weaker one (H_2-CH_4 collisions). If N_2 is present in Neptune's atmosphere at a level of a few percent, which has been envisaged after the detection of HCN in this planet (Rosenqvist et al., 1992, [16]; Marten et al., 1993;[17]), the H_2-N_2 collisions might have to be considered also in the case of Neptune.

Superimposed over this continuum, molecular signatures are observed, either in the troposphere or in the stratosphere. In the case of Jupiter and Saturn, NH_3 and PH_3 are abundant enough to be observed as broad tropospheric absorption lines. As the lorentz broadening coefficient of molecules is typically 0.1 cm^{-1}/atm (or 3 GHz/ atm), these features are difficult to detect by heterodyne spectroscopy, for which the typical bandwidth is currently 0.5 GHz. However, these species have been observed in the far-infrared range with ISO-SWS (Davis et al., 1996, [18]; Burgdorf et al., 2001;[19]) and more recently, in the case of Jupiter, with the CIRS instrument at the time of the Cassini flyby (December 2000). Uranus and Neptune are too cold for NH_3 and PH_3 to be directly detectable; these species are expected to condense below the atmospheric levels where they would be detectable. However, their continuum around 1.35 cm is expected to be influenced by the strong inversion band of NH_3. The same remark is true for H_2S, which has been detected by Galileo in the deep atmosphere of Jupiter, and which is expected to be present also in the other giant planets, on the basis of cosmic abundances.

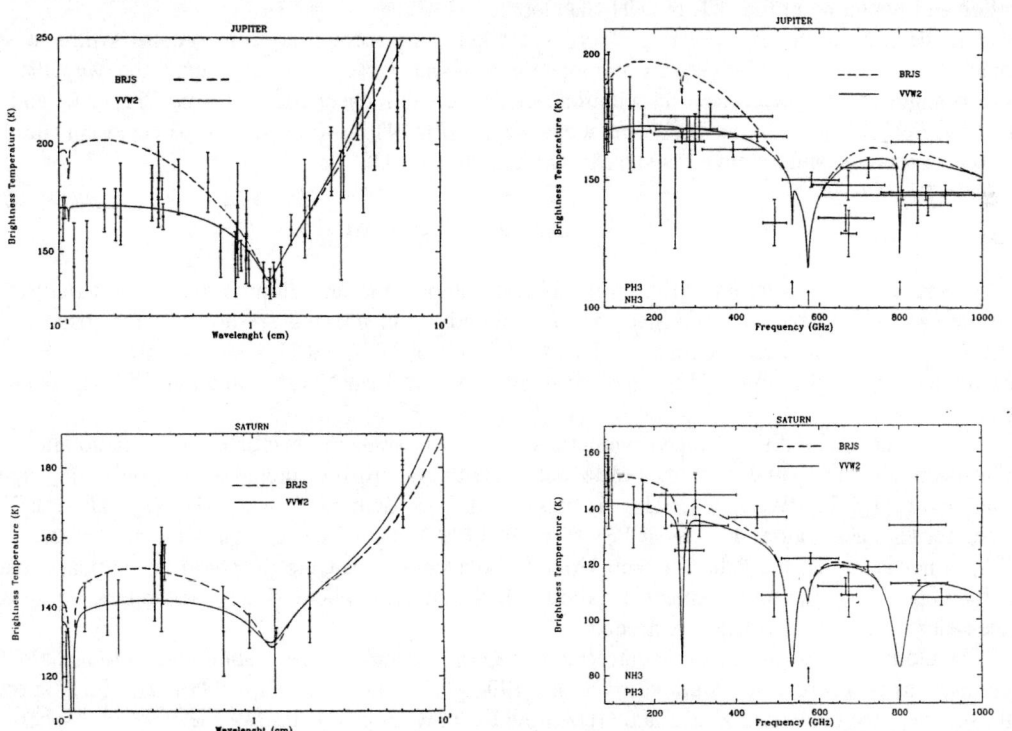

FIGURE 3. Left: The microwave spectra of Jupiter (top) and Saturn (bottom) between 1 mm and 10 cm. Right: The millimeter/submillimeter spectra of Jupiter (top) and Saturn (bottom). The figure is taken from Moreno (1998).

CO has been found to be present in Neptune in detectable amounts (at the 1 ppm level), in the stratosphere and probably also in the troposphere. HCN has been also detected in Neptune's stratosphere but is probably of external origin. H_2O, detected by ISO-SWS in the stratospheres of all giant planets, has been detected at 557 GHz (538 microns) on Jupiter and Saturn with heterodyne spectroscopy from the SWAS satellite (Bergin et al., 2000;[20]). There is no stratospheric emission feature observed so far in the microwave spectrum of Uranus, nor (apart from H_2O) in the one of Saturn; CH_4, however, is expected to show emission features at high spectral resolution in the far-infrared/submillimeter spectrum of Saturn. In the case of Jupiter, stratospheric features were detected following the collision of comet Shoemaker-Levy 9 with the planet, in July 1994, due to the formation of new species by shock chemistry (CO, CS, HCN, OCS); some of them remained observable several years after the collision (Moreno et al., 2001;[21]).

Jupiter and Saturn

The microwave spectra of Jupiter and Saturn strongly depend upon the far wings of the inversion band of NH_3, entered at 1.35 cm. A major uncertainty comes from the choice of the line profile, which is crucial. Several profiles have been proposed, based on a combination of laboratory measurements and theoretical calculations (Moreno, 1998; [22]), in particular the so-called:
-BRJS profile (theoretical profile by Ben-Reveun (1996; [23]), adjusted to fit laboratory data by Joiner and Steffes (1991;[24];
-VVW2 profile (a modified Van Vleck-Weisskopf profile; de Pater and Massie, 1985, [25]; Lellouch and Destombes, 1985, [26]; Moreno, 1998; [22]).

Fig. 3 shows a comparison between the radiometric measurements and the disk-averaged synthetic spectra of Jupiter and Saturn, in the radio range and in the submillimeter range respectively, assuming standard atmospheric models (Moreno, 1998;[22]). It can be seen that the uncertainty in the NH_3 line profile induces large differences in the calculated spectra, but the VVW2 lineshape fits the radiometric measurements of Jupiter and Saturn better than the BRJS lineshape. Note that some authors (Joiner and Steffes, 1991; [24], de Pater and Mitchell, 1993; [27]) have suggested that the mssing opacity using the Ben Reuven profile, could come from the absorption and scattering of the NH_3 and NH_4SH clouds.

In the case of Saturn, the rings were not taken into account to compute its synthetic spectra. The influence of the ring on the brightness temperature is about 10%, depending upon the wavelength, the ring inclination angle and the beam size. In addition, spectroscopic measurements of the PH_3 (1-0) and (3-2) lines, obtained by ground-based Fourier transform spectroscopy (Fig. 4), have been used to constrain the PH_3 vertical distribution (Weisstein and Serabyn, 1994, [28]; Orton et al., 2000;[29]).

Uranus and Neptune

There is more uncertainty in the tropospheric composition and cloud structure of Uranus and Neptune than in the case of Jupiter and Saturn, as many compounds condense and are not directly observable; however, thay can contribute to the radio continuum. This is the case of NH_3 and H_2S, but possibly also PH_3. If N_2 were present in Neptune's troposphere, H_2S might be more abundant than NH_3, a situation different from Jupiter and Saturn.

Fig. 5 shows the disk-averaged synthetic spectra of Uranus and Neptune, in the radio and submillimeter range respectively, compared to radiometric data. Synthetic profiles include CIA only (H_2-He-CH_4), NH_3 dominant over H_2S (VVW2 and BRJS profiles), and H_2S dominant over NH_3 (so-called BRDB profile, optimized for H_2S absorption; de Boer and Steffes, 1994;[30]). It can be seen that some absorption, in addition to the CIA, is needed to fit the data, for both Uranus and Neptune; but the proposed theoretical spectra are not constrained enough by the measurements to allow a better determination of the tropospheric composition. More accurate radiometric measurements are needed.

In addition to the stratospheric detection of CO on Neptune, there is some indication that CO is present in Neptune's troposphere also (Guilloteau et al., 1994,[31]; Naylor et al., 1994,[32]). FTS spectroscopy of Uranus and Neptune between 200 and 300 GHz provided only an upper limit of the PH_3 tropospheric abundance (Encrenaz et al., 1996;[33]).

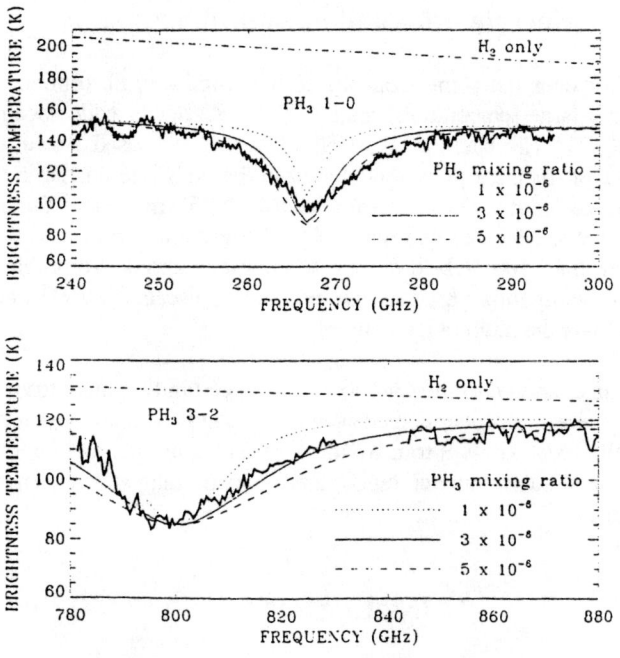

FIGURE 4. The millimeter and submillimeter transitions of PH_3 in Saturn, as observed with an FTS spectrometer at CSO. The figure is taken from Orton et al. (2000).

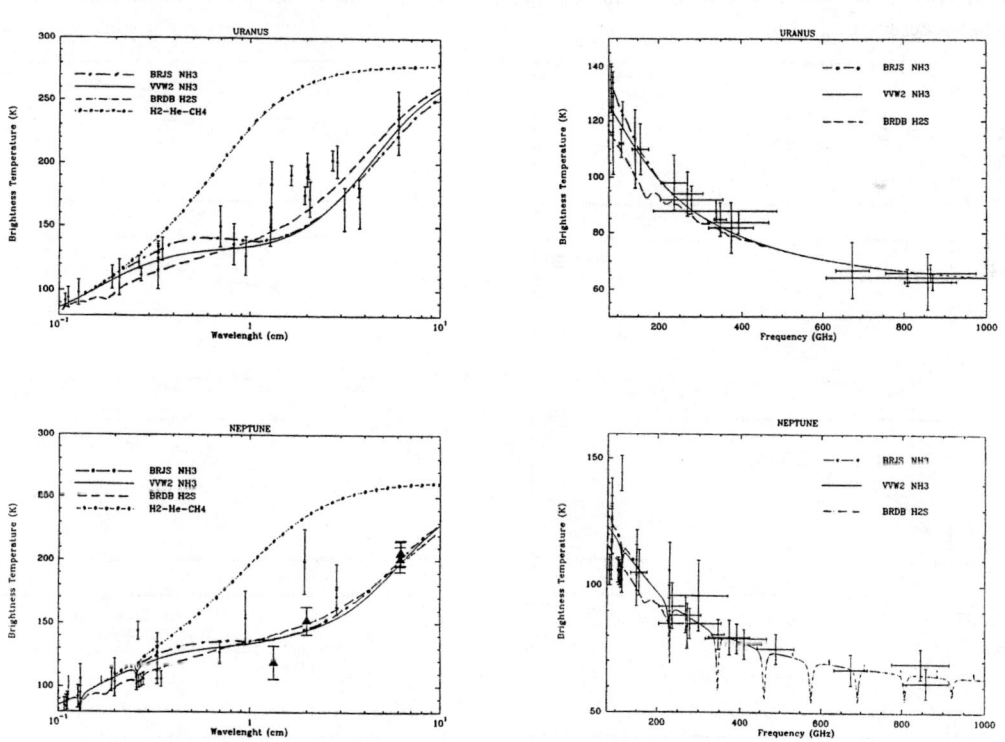

FIGURE 5. Left: The microwave spectra of Uranus (top) and Neptune (bottom) between 1mm and 10 cm. Right: The millimeter/submillimeter spectra of Uranus (top) and Neptune (bottom). The figure is taken from Moreno (1998).

Uncertainties and open problems

The main source of error does not come from the collision-induced absorption of hydrogen, which has been accurately modelled over a large temperature range (50-300 K) and a wide spectral range (0-6000 cm^{-1}; Borysow and Frommhold, 1986,[31]; Birnbaum et al., 1996;[32]). As discussed above, the main uncertainty in the modeling of the radio spectra of the giant planets comes from the NH_3 line profile in the far wings. The error bar is thus minimum at the center of the NH_3 inversion band at 1.35 cm, where the accuracy reaches a few percent for Jupiter and Saturn, but it becomes as high as 15-20% around 1 mm, and decreases again down to about 5 % at frequencies higher than 500 GHz (below 600 microns). Since the problem of the choice of the lineshape is not fully solved, for calibration purposes, we propose to use the VVW2 line profile fot NH_3, as it gives the best overall agreement over the radio range (Fig. 6).

In the case of Uranus and Neptune, another important source of uncertainty comes from the poor knowledge of the vertical distributions of the minor tropospheric species and the cloud structures. As a result, the uncertainty is maximum (as high as 25%) at the 168 GHz H_2S transition. A better accuracy (about 5%), in contrast, is reached in average over the millimeter and submillimeter range, atnd in particular at wavelengths lower than 1.3 mm where the accuracy is lower than 2%..

CONCLUSIONS

The present review shows that using planets for calibrating radio observations requires to take into account their spectral dependence, and bears intrinsic uncertainties. For all planets, one has to take into account the variable size of the disk, the phase effect, possible spatial variability over the disk due to surface morphology or climatology, and, in the case of Saturn, the ring contribution. The beamsize of the telescope is, of course, to be considered also. Under good observing conditions, calibrating radio-observations with planets permits to achieve an absolute accuracy of 5-10%.

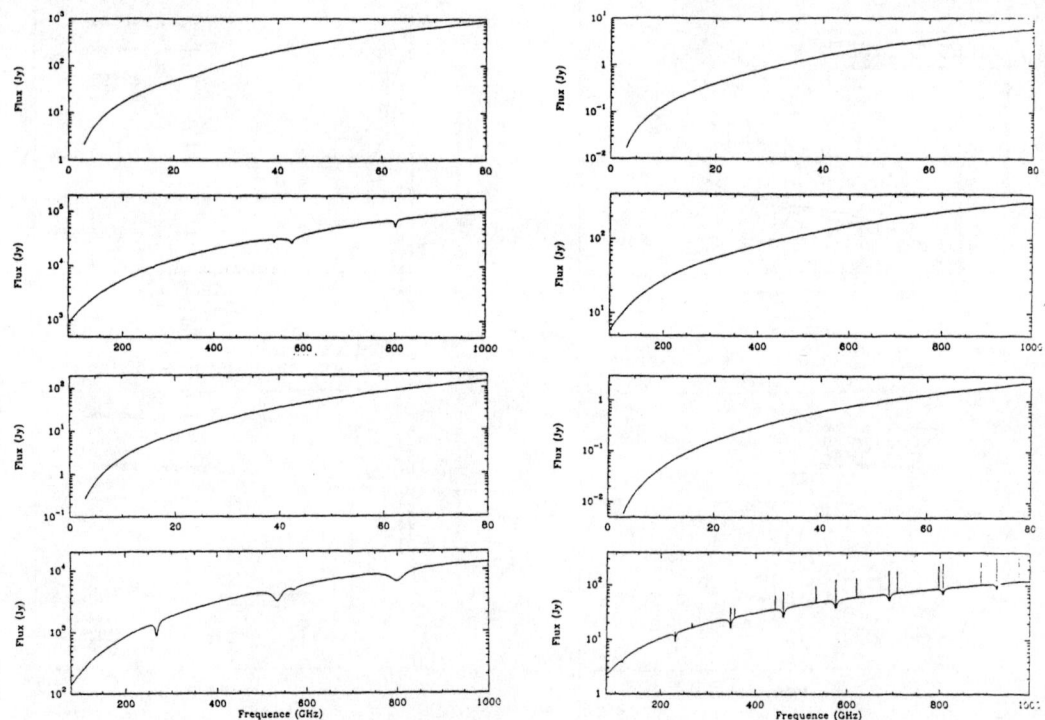

FIGURE 6. The 10-1000 cm^{-1} synthetic spectrum of the four giant planets. Top, left: Jupiter; bottom, left: Saturn; top, right: Uranus; bottom, right: Neptune. The figure is taken from Moreno (1998).

In the mm/submm radio domain, Uranus appears to be the best calibration source, in particular at wavelengths smaller than 1.3 mm, because it is apparently devoid of discrete features and because it is small compared to current radiotelescope beams. The planet is brighter than Neptune and exhibits no climatologic variations. Other possible calibration sources, apart from the planets, are Callisto or bright asteroids like Ceres. Both have been used, together with Uranus, for calibrating ISO-LWS data at far-infrared wavelengths. There is still an uncertainty, however, about the surface emissivity of these bodies at submmillimeter and millimeter wavelengths. If Callisto is used, one has to be aware of the possible contamination by Jupiter (about 9 arcmin away) in the case of astronomical observations using wide fields of view.

In the future, on can hope to get reliable spectra of planets in the submillimeter range with Herschel (apart from Venus) and in the millimeter/submillimeter range with ALMA. In addition, in the case of Saturn, information will be provided by the CIRS/Cassini spectrometer at frequencies higher than 300 GHz.

REFERENCES

1. Janssen M. P., and Klein, M. J., Icarus 46, 58-69 (1981)
2. Pollack, J. B. et al., Icarus 103, 1-42 (1993)
3. Kakar, R. K., Waters, J. W., and Wilson, W. J., Science 191, 379-380 (1976)
4. Wilson, W. J. et al., Icarus 45, 624-637 (1981)
5. Lellouch E., Goldstein, J. J., Rosenqvist, J., and Bougher, S. W., Icarus 108, 112-136 (1994)
6. Encrenaz, Th. Lellouch, E., Paubert, G., and Gulkis, S., Astron. Astrophys. 246, L63-L66 (1991)
7. Encrenaz, Th. et al., Icarus 117, 162-172 (1995)
8. Clancy, R. T., Grossman, A.W., and Muhleman, T. O., Icarus 100, 48-59 (1992)
9. Clancy, R. T. et al., Icarus 122, 36-62 (1996)
10. Encrenaz, Th. et al., Plan. Space Sci. 49, 731-741 (2001)
11. Forget, F. et al., J. Geophys. Res. 104, 24155-24176 (1999)
12. Lellouch, E., Gerin, M., Combes, F., and Encrenaz, Th., Icarus 77, 414-438 (1989)
13. Lellouch, E., et al., Plan. Space Sci. 48, 1393-1401 (2000)
14. Burgdorf, M. et al., Icarus 145, 79-90 (2000)
15. Ulich, B. L. Astron. J. 86, 1619-1626 (1981)
16. Rosenqvist, J. et al., Astrophys. J. 392, L99-L102 (1992)
17. Marten, A. et al., Astrophys. J. 406, 285-297 (1993)
18. Davis, G. R. et al., Astron. Astrophys. 315, L393-L396 (1996)
19. Burgdorf, M. et al., in 'The Promise of FIRST', G. L. Pilbratt et al., eds., ESA SP-460, in press (2001)
20. Bergin, e. et al., Astrophys. J. 539, L147-L150 (2000)
21. Moreno, R. et al., Plan. Space Sci. 49, 473-486 (2001)
22. Moreno, R., PhD Thesis, University Paris VI (1998)
23. Ben Reuven, A., Phys. Rev 145, 7-22 (1996)
24. Joiner, J. and Steffes, P. G., J. Geophys. Res. 96, 17463-17470 (1991)
25. De Pater, I. and Massie, S. T., Icarus 62, 143-171 (1985)
26. Lellouch, E. and Destombes, J.-L., Astron. Astrophys. 152, 405-412 (1985)
27. De Pater, I. and Mitchell, D. L., J. Geophys. Res. 98, 5471-5490 (1993)
28. Weisstein, E., and Serabyn, E., Icarus 109, 367-381 (1994)
29. Orton, G. S., Serabyn, E., and Lee, Y. T., Icarus 146, 48-59 (2000)
30. De Boer, D. R. and Steffes, P. G., Icarus 109, 324-335 (1994)
31. Guilloteau, S., Dutrey, A., Marten, A., and Gautier, D., Astron. Astrophys. 279, 661-667 (1994)
32. Naylor, D.A. et al., Astron. Astrophys. 291, L51-L53 (1994)
33 Encrenaz, Th., Serabyn, E., and Weisstein, E. W., Icarus 124, 616-624 (1996)
34. Borisow, A , and Frommhold, L., Astrophys. J. 304, 849-865 (1986)
35. Birnbaum, G., Borisow, A., and Orton, G. S., Icarus 123, 4-22 (1996)
36. Burgdorf, M. et al., in 'ISO beyond the Peaks', A. Salama et al. eds., ESA SP-456, pp.9-12 (2000)

Big Bang Nucleosynthesis, Cosmic Microwave Background Anisotropies and Dark Energy

Monique Signore* and Denis Puy[†]

*Observatoire de Paris-DEMIRM, Paris (France), email: monique.signore@obspm.fr
[†]Institute of Theoretical Physics, Zürich and PSI-Villigen (Switzerland), email: puy@physik.unizh.ch

Abstract. [1] Over the last decade, cosmological observations have attained a level of precision which allows for very detailed comparison with theoretical predictions. We are beginning to learn the answers to some fundamental questions, using information contained in Cosmic Microwave Background Anisotropy (CMBA) data.
In this talk, we briefly review some studies of the current and prospected constraints imposed by CMBA measurements on the neutrino physics and on the dark energy. As it was already announced by Scott [1], we present some possible *new physics* from the Cosmic Microwave Background (CMB).

INTRODUCTION

Since the 80's, cosmologists introduce for the baryonic density (ρ_b or Ω_b) of the Universe, a concordance interval where predicted and measured abundances of light elements (^7Li, ^4He, D) were consistent, within their uncertainties. At the end of 90's, the determination of primeval D is supposed to be accurate enough to pin down ρ_b or Ω_b; and the *concordance intervals* for D, ^7Li, ^4He and Ω_b were predicted by D measurements, see Burles et al. [2] and Signore-Puy [3]. Fig. (1) shows this problem of concordance from the works of Burles-Nollett-Turner [2].
In the year 2000, observations of CMBA have become a competitive means for estimating Ω_b. Then CMBA results can be a cross check of Big Bang nucleosynthesis (BBN) results and can lead also to new constraints on BBN theory, ie neutrino physics.
On the other hand from observations of type Ia-supernova at high z carried out by two major teams -*Supernovae Cosmology Project* [4] and *High-z Supernovae Team* [5]- there is some direct evidence that the present Universe is accelerating. For a pedagogical and recent review on this subject see [6] and references therein. Moreover, recent measurements of CMBA and of baryon fraction in galaxy clusters indicate that the Universe is flat and that the matter contributes about one third of the critical density $\Omega_M \sim 1/3$; and about two thirds of the critical density constitutes the *dark energy*: $\Omega_\Lambda \sim 2/3$. The nature of this dark energy is the new challenge for cosmology and fundamental physics. Now, an important question is: besides the *SNIa experiments* (see SNAP [7]), can CMBA measurements provide constraints on the nature of the dark energy ?

CONSTRAINTS ON NEUTRINO PHYSICS

As already said above, until recently, BBN -from observations of primordial D abundances- provided the only precision estimates of Ω_b. Considering the most recent primordial deuterium data, Burles, Nollett and Turner [8] give:

$$\Omega_b h^2 = 0.017 - 0.024 \; (95\% \; CL) \qquad (1)$$

where h is the Hubble parameter in units of 100 km sec^{-1} Mpc^{-1}. In the past year, the first results which may rightly be called precision CMBA measurements have been obtained from BOOMERANG [9] and MAXIMA [10]. A higher

---

[1] Talk presented by M. Signore

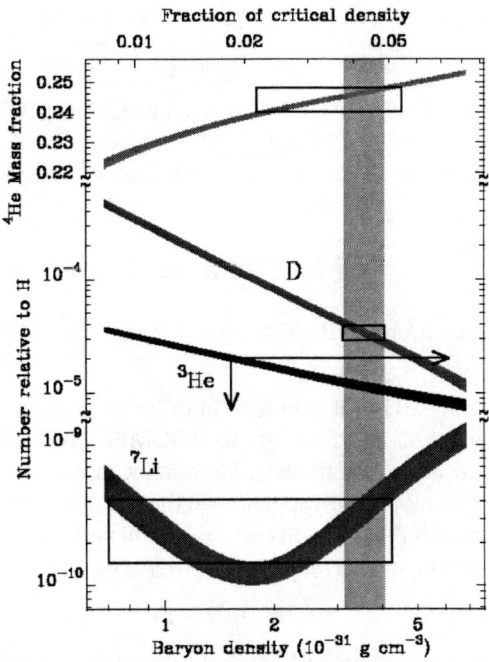

FIGURE 1. The situation of BBN in 1999. It shows the concordance intervals for each element (2σ uncertainties) and the baryon density predicted by primordial deuterium measurements, from Burles-Nollett-Turner [2].

baryon density than that predicted from BBN has been claimed:

$$\Omega_b h^2 \sim 0.03. \tag{2}$$

The discrepancy between BBN and CMBA estimates for Ω_b led to the suggestions that one must consider some *new physics* which appeared between the BBN epoch ($T \sim 1$ MeV) and the CMB epoch ($T \sim 1$ eV) in order to understand these different values of Ω_b ([11], [12], [13], [14], [15]).
But the more recent data from BOOMERANG [16], MAXIMA [17] and DASI [18] show that there is no more difference between BBN and CMBA estimates for Ω_b:

$$\Omega_b h^2 = 0.02, \tag{3}$$

and therefore no need of *new physics* to reconcile BBN and CMBA data. However, some cosmologists, in particular Kneller et al. [19], Hannestad [20], Hansen et al. [21] consider these CMBA data sets of high precision to constrain, independently of BBN, this *new physics*, that is to say the *neutrino physics*.

BBN limit on N_ν

First, let us introduce N_ν - the equivalent number of standard model neutrino species- through the energy density ρ:

$$N_\nu \equiv \frac{\rho}{\rho_{\nu_o}} \tag{4}$$

where ρ_{ν_o} is the energy density of a standard neutrino species. This is a way of expressing the energy density in light non-interacting species. As noted in [20], the standard model predicts:

$$N_\nu \sim 3.04 \tag{5}$$

due to the fact that the neutrinos are not completely decoupled during the $e^+ - e^-$ annihilation; see Steigman [22] for a detailed neutrino counting and a discussion on the above value, Eq. (5). The abundances of primordial ^4He, D, ^7Li

can be used to determine BBN limit on N_v. For example, Lisi et al. [23] give the following bound adopted also by [20]:

$$3 \leq N_{v,BBN} \leq 4 \quad (95\% \text{ CL}). \tag{6}$$

Let us also mention the work done by Kneller et al. [19] who consider for BBN predictions the three parameters η -the baryon to photon ratio ($\eta_{10} = 10^{10} \times \eta = 274\Omega_b h^2$ - ΔN_v the asymmetrical part (or degenerate part) of N_v and $\xi_v \equiv \mu_v/T_v$ where μ_v and T_v are respectively the chemical potential and the temperature of the v-species.

CMBA limit on N_v

A bound on N_v has also been derived from CMBA data by many authors [19] [20] [21]. Let us only summarize the main points of these studies:

- *i)* Kneller et al. [19] used the CMFAST software [24] in order to calculate the cosmic background fluctuation spectrum as a function of η and ΔN_v and compare to BOOMERANG [16], MAXIMA [17] and DASI [18] observations for four different cosmological models. They show, in the ($\eta - \Delta N_v$) plane, the four very different shapes of the confidence interval contours corresponding to the four cosmological models. Their results point out the sensitivity of the *new physics* (ΔN_v) to the other cosmological parameters.
- *ii)* The analysis of the CMBA data by three groups [19], [20], [21] lead to robust upper bounds on N_v

$$N_v < 7-17 \tag{7}$$

which are much weaker than that given from BBN data, the right hand side of equation (6) !
- *iii)* Adding large scale structure data to CMBA data Hannestad [20] gives a non trivial lower bound:

$$N_v > 1.5 \quad (95\% \text{ CL}) \tag{8}$$

which is the first independent indication of the presence of a cosmological neutrino background, predicted by the standard model, and already seen in BBN data, the left hand side of equation (6).
- *iV)* It seems that there is no significant indication of non standard physics -i.e. no *new physics*- contributing to N_v at the recombination epoch [19] [20].

CONSTRAINTS ON DARK ENERGY

Recent observations [4] [5] of type Ia-supernovae indicate that the Universe may be presently dominated by an additional *dark energy* with a negative pressure such that the Universe is presently accelerating (see also [6]). Combined observations of type Ia-supernovae [4], CMBA [25] and cluster evolution [26] for which the results have been done in the form of likelihood contours in the Ω_M and Ω_Λ plane are reported in Fig. (2).
Ω_M and Ω_Λ are defined by

$$\Omega_M = \frac{8\pi G \rho_o}{3H_o}, \quad \Omega_\Lambda = \frac{\Lambda}{H_o} \tag{9}$$

where the index o refers to the present epoch, ρ, H and Λ being respectively, the energy density, the Hubble parameter and the cosmological constant -see [6] for instance.

Let us recall that the Friedman-Lemaitre equation can be written as

$$\Omega_M + \Omega_\Lambda + \Omega_k = 1 \tag{10}$$

with a term of matter plus radiation Ω_M, a term of *dark energy* Ω_Λ and a curvature term such that

$$\Omega_k = \frac{k}{R_o^2 H_o^2} \tag{11}$$

where R is the cosmic scale factor and k is the curvature constant.
What is the nature of this dark energy ? This is the present challenge for cosmology and particle physics. The simplest interpretation of this dark energy is the *cosmological constant* Λ (vacuum energy) for which the equation of state:

$$w \equiv \frac{P}{\rho} \tag{12}$$

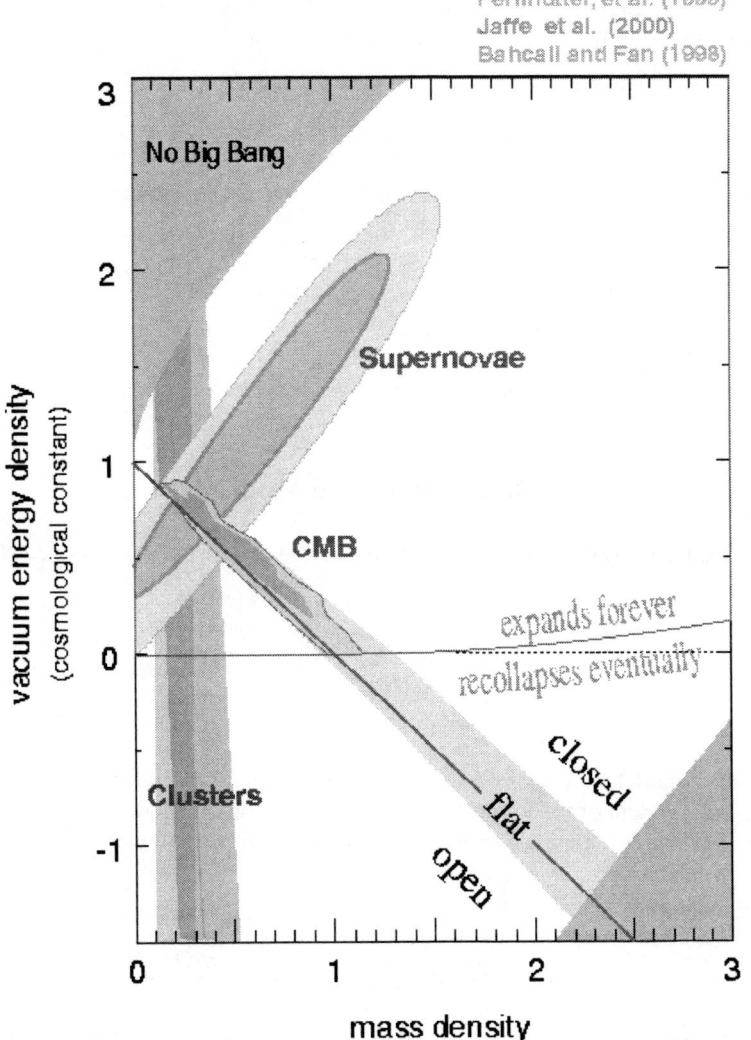

FIGURE 2. From [7], Confidence regions in the ($\Omega_M - \Omega_\Lambda$) plane for high z-supernovae (see Perlmutter et al. [4]), CMBA [25]) and cluster evolution (Bahcall & Fan [26]) measurements. The consistent *overlap* is a strong indicator for the existence of Ω_Λ, i.e. of a cosmology constant or dark energy.

is equal to -1.

It is important to know if this cosmological *constant*, as inferred by observations, is truly constant or if the observations point out some form of cosmic evolution often called *quintessence Q*, for which the equation of state w_Q is such that:

$$-1 \leq w_Q \leq 0. \tag{13}$$

In this case, the vacuum energy is the result of a scalar field Q slowly evolving along an effective potential or getting trapped in a local minimum and which only interacts with the other fields via gravity- In any case -cosmological constant or quintessence- one is faced with two problems:

- *i*) a fine tuning problem: why the vacuum energy, is so small ? From particle physics, one might expect: $\Lambda/8\pi G \sim m_{planck}^4$, and it is off by about 120 orders of magnitude.
- *ii*) a cosmic coincidence problem: why Ω_M and Ω_Λ are nearly equivalent now ?

Since w is, in general, time vrying, the first step toward solving the dark energy problem is to determine $w(t)$ or $w(z)$. Before considering some models of quintessence, let us only recall that:

- • for the case where the dark energy is the cosmological constant Λ:

$$w_\Lambda = -1 \tag{14}$$

- • some authors -for instance, Huey et al. [27]- introduce an effective (constant) equation of state w_{eff} defined by:

$$w_{eff} \sim \frac{\int \Omega_Q(z)\,\omega(z)\,dz}{\int \Omega_Q(z)\,dz} \tag{15}$$

- • for topological defects:

$$w_{string} \sim -\frac{1}{3} \quad \text{and} \quad w_{wall} \sim -\frac{2}{3} \tag{16}$$

On Quintessence Models

Many quintessence effective potentials exist in the litterature -see, for instance Weller & Albrecht [28]. Here, let us only mention:

- • The cosmological tracker solutions [29] [30]
  with, in particular, the inverse tracker potential of Ratra & Peebles [31]

$$V(Q) = M^{(4+\alpha)} Q^{-\alpha} \tag{17}$$

where M and α are parameters such as $\Omega_Q \sim 2/3$ at present. The tracker solutions evolve on a common evolutionary track independent of the initial conditions. All the tracker models have in common that the density in the dark energy at late times dominates over all the other density contributions and therefore the expansion of the Universe starts accelerating.

- • The Supergravity Potential:

$$V_{SUGRA}(Q) = M^{(4+\alpha)} Q^{-\alpha} \exp\left[\frac{1}{2}\left(\frac{Q}{M_{pl}}\right)^2\right], \tag{18}$$

which is related to the supersymmetry breaking -see in particular Binetruy [32], Brax & Martin [33]. M and α are chosen such that the supersymmetry breaking occurs above the electroweak scale. A discussion on this potential is found in Kolda & Lyth [34].

In all of these models, the energy density of the field Q is given by the kinetic and potential components:

$$\rho_Q = \frac{1}{2}\dot{Q}^2 + V(Q) \tag{19}$$

while the pressure is given by the difference

$$P_Q = \frac{1}{2}\dot{Q}^2 - V(Q). \tag{20}$$

Moreover, we assume that the field Q is homogeneous on large scales. Therefore the equation of state of the quintessence is given by:

$$w_Q = \frac{P_Q}{\rho_Q}. \tag{21}$$

Fig. (5) in Weller & Albrecht [28] shows the evolution -in the range of redshift $[0-2]$- of the equation of state of the dark energy component:

$$w_Q = w_Q(z) \tag{22}$$

for all of these models they discuss in [28] and, in particular, for the two potentials considered here, Eqs (17) and (18).

Constraints on the equation of state of dark energy

We have seen that searches for SNIa at high z have already provided a strong evidence for an accelerating present Universe [4], [5], [6]. By analyzing a simulated data set as might be obtained by the proposed SNAP satellite [7], Weller & Albrecht [35] claim that it will be possible to discriminate among different dark energy solutions.

Fig. (3) shows the separation of three dark energy models in the $(\Omega_M - w_o)$ plane where w_o is such that $w = w_o + w_1 z$, although Maor et al. [36] show that this method is indeed very limited.

However, as already seen through the Fig. (2), the result can be better by combining SNIa constraints with other complementary measurements. A low-z measurement such as a cluster survey, an intermediate-z measurement such as a SNIa survey and a high-z measurement such as CMBA measurements can provide complementary constraints.

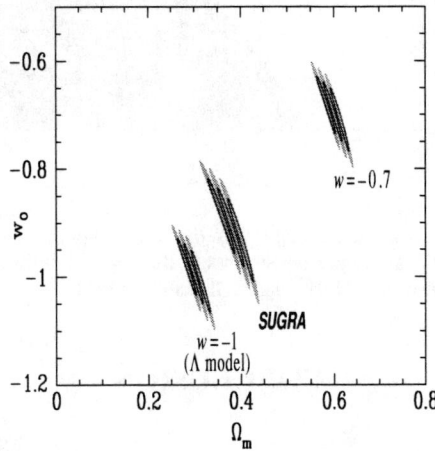

FIGURE 3. From Weller & Albrecht [35], separation of three dark energy models in the $(\Omega_M - w_o)$ plane.

Fig. (4) from Hu et al. [37] and Fig. (5) from Huterer & Turner [40] indicate how the constraints from several measurements would constrain a model when the Universe is assumed flat ($\Omega_k = 0$) and ω is supposed to be constant

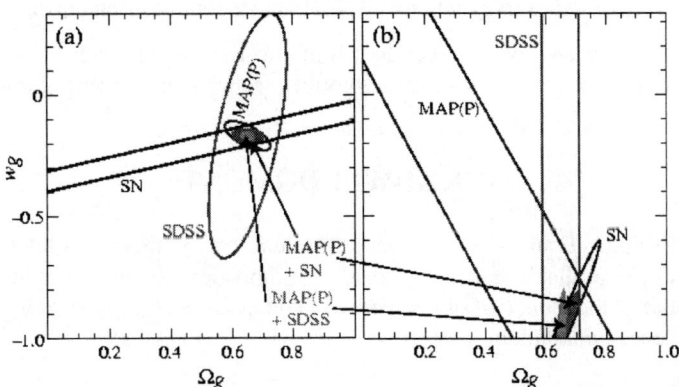

FIGURE 4. From Hu et al. [37], here $\Omega_g = \Omega_Q = 1 - \Omega_M$, confidence regions in the $(\Omega_g - w_g)$ plane from CMB, SN and large scale structure survey (68 % CL). Here SN means constraints of a supernova program such as SNAP [7], SDSS means constraints of Sloan Digital Sky Survey [38], MAP means constraints of the MAP satellite [39]; (P) means polarization information. (a) : Left curves, $w_g = w_Q = -1/6$, $\Omega_M \sim 1/3$. (b) : Right curves, $w_g = w_Q = -1$, $\Omega_M \sim 1/3$. Note the complementarity nature of the data sets.

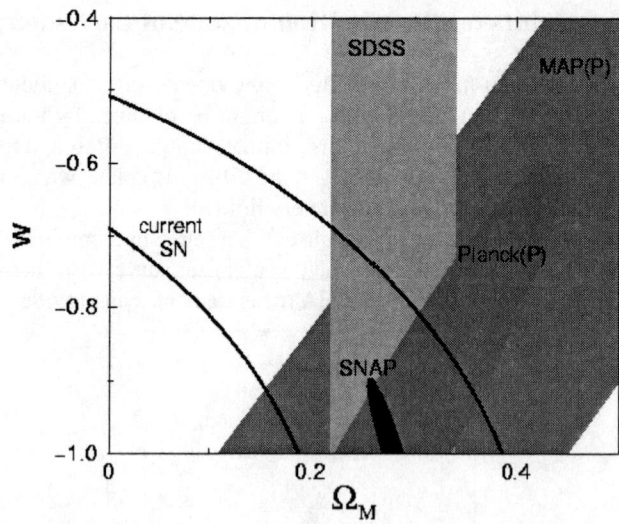

FIGURE 5. From Huterer & Turner [40], Confidence regions in the (Ω_M, w) plane for the case b) of Fig. (4). Here, <u>SDSS</u>, <u>MAP</u>, <u>P</u> have the same meaning as Fig. in (4b). <u>SNAP</u> means constraints of the SNAP satellite [7], <u>current SN</u> means present constraints using about 50 SNIa, <u>PLANCK</u> means constraints of the PLANCK satellite [41].

CONCLUSION

We have seen that:

- *i*) The accurate determination of the primeval deuterium abundance pins down the baryon density of the Universe: $\Omega_B h^2 \sim 0.02$. New CMBA data (BOOMERANG, MAXIMA, DASI) lead also to $\Omega_B h^2 \sim 0.02$ and can significantly constrain neutrino physics if an additional cosmological constrain is imposed.

- *ii*) While luminosity-distance measurements of type Ia SN calibrated candles have recently shown that our Universe is accelerating now, the resent question is: what is the dark energy ? Particle physics theory proposes a number of alternatives to a non-zero vacuum energy/cosmological constant: quintessence in particular.
With future CMBA measurements (MAP, PLANCK) it should be possible to measure a constant equation of state of dark energy within a 10-30 % accuracy. To determine ω to 5 % and to begin to probe a time-varying equation of state requires also a large SNIa survey (such as SNAP) and a count of rich clusters of galaxies.

The research on the nature of dark energy, however, is still in a nascent stage. Over the next decade a variety of new strategies and more precise applications of old strategies could very well answer this question once and for all.

ACKNOWLEDGMENTS

The authors would like to thank Francesco and Bianca Melchiorri for their helpful discussions and continuous encouragements. We would like to thank Marco De Petris, Massimo Gervasi and Fernanda Luppinacci for organizing a superb meeting at a wonderful location. Part of the work of D. Puy has been supported by the Dr *Tomalla* Foundation and the Swiss National Science Foundation.

REFERENCES

1. Scott D., `astro-ph/9911325`.
2. Burles S., Nollett K., Turner M., `astro-ph/9903300`.
3. Signore, M., Puy, D., *New Ast. Rev.*, **43**, 185 (1999).
4. Perlmutter, S. et al., *ApJ*, **517**, 565 (1999).
5. Riess, A., et al., *Astron J.*, **116**, 1009 (1998).
6. Signore, M., Puy, D., *New Ast. Rev.*, **45**, 409 (2001).

7. SNAP, http://snap.lbl.gov.
8. Burles S., Nollett K., Turner M., *ApJ*, **552**, L1 (2001); astro-ph/0010171.
   O'Meara J., Tytler D., Kirman D. et al., astro-ph/0011179.
9. De Bernardis P., Ade P., Bock J. et al., *Nature*, **404**, 955 (2000).
10. Hanany S., Ade P., Balbi A. et al., *ApJ Lett.*, **545**, L5 (2000).
11. Lesgourgues J., Peloso M., *Phys. Rev. D*, **62**, 1301 (2000); astro-ph/0004412.
12. Hannestad S., *Phys. Rev. Lett.*, **85**, 4203 (2000); astro-ph/0005018.
13. Orito N., Kajino T., Mathews G., Boyd R., astro-ph/0005446.
14. Esposito S., Mangano G., Melchiorri A. et al., *Phys. Rev. D*, **63**, 3004 (2000); astro-ph/0007419.
15. Mangano G., Melchiorri A., Pisanti O., *Nucl. Phys. Proc. Supp.*, **100**, 369 (2001); astro-ph/0012291.
16. Netterfield C., Ade P., Bock J. et al., astro-ph/0104460.
17. Lee A., Ade P., Balbi A. et al., astro-ph/0104459.
18. Halverson N., Leitch E., Pryke C. et al., astro-ph/0104489.
19. Kneller J., Sherrer R., Steigman G., Walker T., astro-ph/0101386.
20. Hannestad S., astro-ph/0105220.
21. Hansen S., Mangano G., Melchiorri A. et al. astro-ph/0105385.
22. Steigman G., astro-ph/0108148.
23. Lisi E., Sarkar S., Villante F., *Phys. Rev. D* **59**, 3520 (1999).
24. Seljak U., Zaldarriaga M., *ApJ* **469**, 437 (1996).
25. Jaffe A., Ade P., Balbi A. et al., astro-ph/0007333.
26. Bahcall N., Fan X., *Proc. Nat. Acad. Sc.* **95**, 5956 (1998); *ApJ* **504**, 1 (1998).
27. Huey G., Wang L., Dave R. et al., *Phys. Rev. D* **59**, 3005 (1999); astro-ph/9804285.
28. Weller J., Albrecht A., astro-ph/0106079.
29. Steinhardt P., Wang I., Zlatev I., *Phys. Rev. D* **59**, 3504 (1999).
30. Zlatev I., Wang I., Steinhardt P., *Phys. Rev. Lett.* **82**, 896 (1999).
31. Ratra B., Peebles P., *Phys. Rev D* **37**, 3406 (1988).
32. Binetruy P., *Phys. Rev D* **60**, 3502 (1999).
33. Brax Ph., Martin J., *Phys. Lett B* **468**, 40 (1999).
34. Kolda C., Lyth D., *Phys. Lett. B* **458** 197 (1999).
35. Weller J., Albrecht A., *Phys. Rev. Lett.* **86**, 1939 (2001); astro-ph/0008314.
36. Maor I., Brustein R., Steinhardt P., *Phys. Rev. Lett.* **86**, 6 (2001).
37. Hu W., Eisenstein D., Tegmark M., White M., *Phys. Rev. D* **59**, 3512 (1999); astro-ph/9806362.
38. SDSS (2001), http://www.astro.princeton.edu/BBook
39. MAP (2001), http://map.gsfc.nasa.gov
40. Huterer D., Turner M., *Phys. Rev. D* **60**, 1301 (1999); astro-ph/0012510.
41. MAP (2001), http://astro.estec.esa.nl/Planck/

Primordial molecules at millimeter wavelengths

Denis Puy* and Monique Signore†

*Institute of Theoretical Physics, Zürich and PSI-Villigen (Switzerland), email: puy@physik.unizh.ch
†Observatoire de Paris-DEMIRM, Paris (France), email: monique.signore@obspm.fr

Abstract. [1] Chemistry plays a particular role in astrophysics. After atomic hydrogen, helium and their ions, the Universe probably contains more mass in molecules than in any other species. Molecule formation in the early, pre-galactic Universe may have had much to do with the formation of galaxies themselves. In this context the possible interaction between primordial molecules and photons of the Cosmic Microwave Background (CMB) is very important through the theoretical perspectives and constraints which could give some information on the theory of the large scale structure formation.

In this paper we recall the more recent progresses on the chemistry of the early Universe, and describe the importance of molecules in the formation phase of proto objects. A special attention is done concerning the *case of LiH*.

INTRODUCTION

Molecules are found in a large variety of astronomical environments. They are now widely used as diagnostic probes of the physical conditions in which they occur. The diversity of molecular environments has helped to stimulate interest in a variety of different chemical processes.

According[2] to the standard Big Bang cosmology, the space of the Universe expands adiabatically, cooling from an extreme initial temperature and density. Thus at about one second after the Big Bang, the temperature of the Universe remains hot and around 10^{10} K. At this stage, the collisions between neutrons and protons can form deuterium, and open the way of the primordial nucleosynthesis through other fusion reactions (see Signore & Puy [1] and references therein). At about 100 sec the nucleosynthesis epoch is over, thus most neutrons are in ^4He nuclei, and most protons remain free while smaller amounts of D, ^3He and ^7Li are synthetized. The low densities, the Coulomb barriers and stability gaps at masses 5 and 8 worked against the formation of heavier elements.

After the nucleosynthesis period atoms form by recombination of these primordial nucleus with free electrons, leading to the thermal decoupling between the matter and the radiation (see Puy-Signore [2]). The radiative recombination processes were then not reversed by photoionization and electron impact ionization, as they had been earlier because the supply of energetic photons and electrons had diminished. The Universe was transformed into a neutral state, apart from the few relict ions and electrons left behind in the expansion.

The chemistry of the early Universe is the chemistry of the elements H, its isotope D, helium and their isotopic forms. The ongoing physical reactions are immense after the recombination of hydrogen, the main processes are collisional (ionization, radiative recombination, attachment...) and radiative due to the presence of the CMB (photoionization, photodetachment...). During the last decade, a large litterature has been developed on the chemical networks of the primordial chemistry (Lepp & Shull [3], Puy et al. [4], Stancil et al. [5], Galli & Palla [6], Puy & Signore [7] and [2] for historical description). Thus primordial molecules such as H_2, HD and LiH formed

The existence of a significant abundance of molecules can be crucial on the dynamical evolution of collapsing objects. Because the cloud temperature increases with contraction, a cooling mechanism can be important for the structure formation, by lowering pressure opposing gravity. This is particularly true for the first generations of objects.

---

[1] Talk presented by D. Puy.
[2] see the talk of Signore-Puy in this proceedings and Signore & Puy [1].

Thus, in the first part of this communication, we recall the main and recent results of the molecular influence on the formation of the proto-objects, particularly that of H_2 and HD molecules. Interactions between primordial molecules and CMB could be important. In the second part, we will recall the potential importance of LiH molecules on the CMB anisotropies. In particular, the scattering process of CMB photons on LiH molecules could play an important role and lead to produce secondary anisotropies on the spectrum of CMB. We will conclude this communication on the possible outlooks.

IMPORTANCE OF H_2 AND HD FOR COSMOLOGY

As we have seen, early Universe chemistry has been previously investigated by many authors. H_2 and HD molecules are the most abundant molecules, and could play a non-negligible role on the formation of the first objects of the Universe. These molecules could contribute to cooling function and lead to dynamical influence on the collapse mechanism. Moreover although the abundance of H_2 molecules is rather insensitive to the choice of cosmological model, the abundance of HD molecules shows large variations.

H_2 MOLECULE

Eddington [8], then Strömgren [9] were the firsts to suggest that H_2 might exist in the interstellar space. Herzberg [10] described the quantum mechanics of homonuclear molecules in some details, which opened important theoretical works on H_2 in astrophysics. Although the medium is free of grains after the cosmological recombination, the formation of H_2 comes into play through the ions H^- and H_2^+. In the post-recombination medium the radiation is hotter than the matter. In this context the radiative excitation of the rotational levels, which is here more efficient than the collisional excitation produces a heating -see Puy et al. [4].
Lepp & Shull [3] were the firsts to point out this important characteristics, which was confirmed by Puy et al. [4] with a better estimation of the thermal function. In the gravitational collapse the situation is very different [11]; the temperature of CMB is below than the matter inside of the collapse. Thus the thermal balance between matter and radiation leads to produce a molecular cooling function, H_2 becoming a good coolant agent of the collapse.
The possibility to observe high redshift systems is mainly associated with the presence of quasars, i.e. strong background sources. Most of our knowledge of the Universe between $z = 1$ and 5 comes from the study of the Lyman-α absorbers in the optical range. In a mini-survey for molecular hydrogen in eight high-redshift damped Lyman-α systems, Petitjean et al. [12] confirmed the presence of H_2 in a system toward PKS 1232+082 ($z = 2.3377$). They show that there is no evidence for any correlation between H_2 abundance [3] and relatively heavy element depletion into dust grains.

HD MOLECULE

The role of deuterium was analyzed by Palla et al. [13], then completed by Stancil et al. [5], in the context of the chemical and thermal evolution of the gas component in the post-recombination Universe and more recently by Flower [14]. Puy & Signore [15] revealed that the HD molecule is the main cooling agent, a result which was confirmed later by many authors such as Okumurai [16], Uehara & Inutsuka [17] and Flower et al. [18]. Thus HD molecules could have important consequence on the problem of fragmentation of primordial clouds in order to form first structures like massive stars. Searches for a primordial signature of HD is crucial.
One must pointed out that very recently Varshalovich et al. [19] have analyzed the spectrum of the quasar PKS 1232+082 obtained by Petitjean et al. [12]. HD molecular lines have been identified in an absorption system at the redshift $z = 2.3377$, this is the first detection of HD molecules at high redshift.

---

[3] In this case the upper limits on the molecular fraction derived in nine of the systems are in the range $1.2 \times 10^{-7} - 1.6 \times 10^{-5}$.

THE CASE OF LIH

From an initial idea of Zel'dovich, Dubrovich [20] showed that resonant elastic scattering must be considered as the most efficient process in coupling matter and radiation at high redshift. He noted that the cross section for resonant scattering between cosmic microwave background (CMB) photons and molecules is several orders of magnitude larger than that between CMB and electrons; even a modest abundance of primordial molecules would produce significant Thomson scattering.

During an elastic scattering between CMB photons and primordial molecules, a photon is absorbed and reemitted at the same frequency but not in the same direction. This process could have negligible effect because of low abundances of primordial molecules. Dubrovich [21], Maoli et al. [22], Signore et al. [23] showed that this effect could alter the primary spatial distribution of the CMB anisotropies. More precisely resonant scattering of CMB photons on *LiH* molecules can be particularly efficient for smoothing the primary anisotropies. Maoli et al. [22] pointed out that primordial molecules such as *LiH* may play significant role in altering the amplitude and power spectrum of CMB anisotropies; the effect depends essentially on the *Li* abundance and the lithium chemistry. They found that primary CBR anisotropies may be erased or attenuated for angular scales below 10° and frequencies below 50 GHz, if *LiH* primordial abundance, relative to H, exceed 10^{-10}. In 1996 Stancil, Lepp and Dalgarno [24] implemented the first complete post-recombination lithium chemistry, and concluded that the final abundance of primordial *LiH* is below 10^{-18}, which ruled out this possibility of erasing the primary anisotropies.

Different studies focused on the chemical evolution in primordial clouds by solving a chemical reaction network within idealized collapse models (see Puy & Signore [11], Anninos & Norman [25], Abel et al. [26]). Puy & Signore [15] examined the evolution of primordial molecules in a context of gravitational collapse and showed that primordial molecules could coexist in a collapsing proto-cloud, particularly during the first phase of gravitational collapse. Maoli et al. [27] emphasized the role that elastic resonant scattering through *LiH* molecules can be produced in this collapsing structure. If the scattering source has a non-zero component of the peculiar velocity along the line of sight, they showed that the elastic scattering is no more isotropic in the observer frame and molecular secondary anisotropies are produced in the CMB spectrum. The angular scale of these secondary anisotropies are therefore directly related to the size of the primordial clouds. Bougleux & Galli [28], then Puy & Signore [29] studied the chemistry of primordial *LiH* in a collapsing protocloud, from the chemical network of Stancil et al. [24] and the fully quantum mechanical treatment, of the radiative association of the excited *Li* states, developed by Gianturco & Gori-Giorgi [30]. We concluded that with this chemical network the *LiH* abundance is closed to 3×10^{-18}, leading to very low secondary anisotropies in the CMB.

Nevertheless a precise analysis of the chemical network shows that most of reaction rates are quite uncertain. For example the reaction rate of the main reaction which dissociates the *LiH* molecules:

$$LiH + H \to Li + H_2 \quad (1)$$

is constant and independent of the temperature and of the density !

OUTLOOK

Although astrochemical observations started in the visible, they are dominated by the radio and above all by millimitre and sub-millimetre observations. Tentative of direct searches of primordial molecules were developed this last decade, but the results were not at the level of efforts of teams of observers (see for example De Bernardis et al. [31], Signore et al. [32], Combes & Wiklind [33]). Recently Papadopoulos et al. [34] revealed the discovery of large amounts of low-excitation molecular gas at redshift $z \sim 3.91$. Shibai et al. [35] investigated the observability of hydrogen molecules in absorption. They argued that the absorption efficiency of the hydrogen molecules become comparable with or larger than that of the dust grains in the metal-poor condition expected in the early Universe. Thus the absorption measurement of the hydrogen molecules could be an important technique to explore the primordial gas clouds that are contracting into first-generation objects.

The HERSCHEL satellite [36] could prove the origins of structure and the chemistry at the early interprotostructure medium and furnish a spectral atlas for molecules (see Encrenaz et al. [37]). A submillimetre spectra of protoclouds of gas could offer important constraints on the critical chemistry, on dynamics, on the heating and the cooling processes that occur in the primordial gas before and during gravitational collapse of protostructure. The research of primordial lines with HERSCHEL [36] could open an important new field of cosmology: the cosmochemistry.

Theory is essential for many aspects of astrochemistry (or cosmochemistry). Chemical models require chemical rates and these are not always available from experimentalists. The calculations of the minimum energy pathway and dynamical calculations are crucial. This last point is particularly important for lithium chemistry. Recently Zaldarriaga & Loeb [38] explored the imprint of the resonant 6708 Åline opacity of neutral lithium on the temperature anisotropies of the CMB at observed wavelengths of 250-350 μm. They showed that the standard CMB anisotropies would be significantly modified in this wavelength band. The primordial chemistry and particularly *LiH* could give same conclusions and important consequences on the temperature and polarization anisotropies.

Very recently LoSecco et al. [39] argued that extragalatic cold molecular clouds could lead to a significant absorption on the CMB. They speculate that the use of very high resolution spectrometers on large aperture telescopes might facilitate a 1-2 order of magnitude improvements in the CMB temperature measurements at high redshifts. Such accurate observations would enable us to constrain the anisotropy, inhomeogeneity of the Universe and the protochemistry.

We have seen the possibility of fragmentation of proto-clouds by the cooling due to H_2 and *HD* molecules. This process could lead to the formation of primordial massive stars, which could be a possible source of contamination in heavier elements at early epochs. Then, the gravitational collapse of following objects (galaxies...) could be strongly influenced by the existence of heavier elements such as *CO, CI* or *HCN*.

Chakrabarti & Chakrabarti [40] showed that a significant amount of adenine, a DNA base, may be produced during molecular collapse, through the *HCN* addition. Recently Sorrell [41] outlines a theoretical model for the chemical manufacture of interstellar amino acids and sugars. This chemistry model explains the existence of both the amino acid glycine and the sugar glycolaldehyde; this last component was recently detected in millimetre-wave rotational transition emission from the star-forming cloud Sagittarius B2 [42]. The formation of DNA bases could happen in the early history of the Universe. Pre-biotic molecules could have contamined the first objects and planets from the beginning [43]...

We are living in a golden age of astronomy, new observations with instruments such as NGST [44] PLANCK[45] and HERSCHEL [36], will push forward the frontiers of our ignorance. As Herbst wrote: *Astrochemistry may not tell us much about the first three minutes, but ultimately it should tell us our place in the Universe.*

ACKNOWLEDGMENTS

We are very grateful to Marco De Petris, Massimo Gervasi and Fernanda Luppinacci for the opportunity to attend this splendid workshop. The authors gratefully acknowledge Francesco Melchiorri, Roberto Maoli and Pierre Encrenaz for valuable discussions on this field. Part of the work of D. Puy has been supported by the Dr Tomalla Foundation and the Swiss National Science Foundation.

REFERENCES

1. Signore M., Puy D., *New Ast. Rev.* **43**, 185 (1999).
2. Puy D., Signore M., `astro-ph/0101157` (2001).
3. Lepp S., Shull M., *ApJ*, **280** 465 (1984).
4. Puy D., Alecian G. Le Bourlot J. et al., *AA* **267**, 337 (1993).
5. Stancil P., Lepp S., Dalgarno A., *ApJ* **509**, 1 (1998).
6. Galli D., Palla F., *AA* **335**, 403 (1998).
7. Puy D., Signore M., *New Ast. Rev.* **43**, 223 (1999).
8. Eddington A., *Observatory* **60**, 99 (1937).
9. Strömgren B., *ApJ* **89**, 526 (1939).
10. Herzberg G., *Spectra of Diatomic Molecules*, Van Nostrand Reinhold, New York, 658 pp (1950).
11. Puy D., Signore M., *AA* **305**, 371 (1996).
12. Petitjean P., Srianand R., Ledoux C., *AA* **364**, L26 (2000).
13. Palla F., Galli D., Silk J., *ApJ* **451**, 44 (1995).
14. Flower D., *MNRAS* **318**, 875 (2000).
15. Puy D., Signore M., *New Ast.* **2**, 181 (1997).
16. Okumurai K., *ApJ* **534**, 809 (2000)
17. Uehara H., Inutsuka S., *ApJ* **531**, L91 (2000).
18. Flower D., Pineau des Forêts G., *MNRAS* **3116**, 901 (2000).
19. Varshalovich D., Ivanchik A., Petitjean P. et al., *astro-ph/0107310* (2001).

20. Dubrovich V., *AA* **324**, 27 (1997).
21. Dubrovich V., *Astron. Lett.* **19**, 83 (1993).
22. Maoli R., Melchiorri F., Tosti D., *ApJ* **425**, 372 (1994).
23. Signore M., De Bernardis P., Encrenaz P. et al., *Astro. Lett. Comm.* **35**, 349 (1997).
24. Stancil P., Lepp S., Dalgarno A., *ApJ* **458**, 401 (1996).
25. Anninos P., Norman M., *ApJ*, **460** 556 (1996).
26. Abel T., Anninos P., Zhang Y., Norman M., *New Ast.* **3**, 181 (2000).
27. Maoli R., Ferrucci V., Melchiorri F. et al., *ApJ* **457**, 1 (1996).
28. Bougleux E., Galli D., *MNRAS* **288**, 638 (1997).
29. Puy D., Signore M., *New Ast.* **3**, 27 (1998).
30. Gianturco F., Gori-Giorgi P., *Phys. Rev. A* **54**, 1 (1996).
31. De Bernardis P., Dubrovich V., Encrenaz, P. et al., *AA* **269**, 1 (1993).
32. Signore M., Vedrenne G., De Bernardis P. et al., *ApJ Supp. Ser.* **92**, 535 (1994).
33. Combes F., Wiklind T., *AA* **334**, L81 (1998).
34. Papadopoulos P., Ivison R., Carilli C., Lewis G., *Nature* **409**, 58 (2001).
35. Shibai H., Takeuchi T., Rengarajan T., Hirashita H., astro-ph/0106102 (2001).
36. HERSCHEL mission, http://astro.estec.esa.nl/SA-general/Projects/First/ (2001).
37. Encrenaz P., Maoli R., Signore M., Proceedings of the ESA symposium, ESA SP-401, pp. 145–150 (1997).
38. Zaldarriaga M., Loeb A., astro-ph/0105345 (2001).
39. LoSecco J., Mathews G., Wang Y., astro-ph/0108260, (2001).
40. Chakrabarti S., Chakrabarti S.K., *AA* **354**, L6 (2000).
41. Sorrell W., *ApJ* **555**, L129 (2001).
42. Hollis J., Lovas F., Jewell P., *ApJ* **540**, L107 (2000).
43. Puy D., astro-ph/0011435 (2000).
44. NGST mission, http://ngst.gsfc.nasa.gov/ (2001).
45. PLANCK mission, http://astro.estec.esa.nl/Planck/ (2001).

Sunyaev-Zel'dovich Effect and Morphology of Galaxy Clusters

R. Piffaretti[*], Ph. Jetzer[†], D. Puy[*] and S. Schindler[**]

[*]Paul Scherrer Institute and Institute of Theoretical Physics, University of Zürich - CH
[†]Institute of Theoretical Physics, University of Zürich and Institute of Theoretical Physics, ETH Zürich - CH
[**]Astrophysics Research Institute, Liverpool John Moores University - UK

Abstract. We investigate the influence of the finite extension and the aspherical geometry of a galaxy cluster on the estimate of the Hubble constant through the Sunyaev-Zel'dovich (SZ) effect. An analysis of a recent *Chandra* image of the galaxy cluster RBS797 indicates a strong ellipticity and thus a pronounced aspherical geometry. We estimate the total mass of RBS797 assuming spherical or ellipsoidal geometry and show that in the latter case the mass is about 10-17% less than the one inferred for a spherical shape.

INTRODUCTION

The recent technical developments of millimetre receivers open new perspectives for more accurate measurements of the SZ effect and will thus trigger new developments on theoretical work (see the invited paper of Y. Rephaeli in this conference). Accurate measurements of the SZ effect are still difficult as well as their correct interpretation, indeed systematic errors can be significant. For example Cooray [1] showed the influence of projection effects and Sulkanen [2] pointed out that the shape of galaxy clusters could produce systematic errors. More recently, Puy et al. [3] investigated the SZ effect and the X-ray surface brightness for galaxy clusters with a non-spherical mass distribution. In the first part of this communication we review the "classical" systematic errors such as cluster extension and geometry, by discussing the recent millimetre measurements of Mauskopf et al. [4] of the SZ effect in Abell 1835. In the second part, we briefly comment on the possible geometrical influence on the determination of the total mass of galaxy clusters.

SZ EFFECT AND THE HUBBLE CONSTANT

Observations of galaxy clusters in the millimetre and X-ray wavebands give important information for cosmology. By combining the SZ intensity change and the X-ray emission observations, the angular diameter distance to galaxy clusters can be derived. Assuming a cosmological model, this leads to an estimate of the Hubble constant H_o.
Mauskopf et al. [4] determined H_o from X-ray measurements of A1835 obtained with *ROSAT* and from the corresponding millimetric observations of the SZ effect with the *Suzie* experiment. Assuming an infinitely extended, spherical gas distribution with an isothermal profile $\beta = 0.58 \pm 0.02$, $T_{eo} = 9.8^{+2.3}_{-1.3}$ keV, $n_{eo} = 5.64^{+1.61}_{-1.02} \times 10^{-2}$ cm^{-3}, they found $H_o = 59^{+36}_{-28}$ km s^{-1} Mpc^{-1}.
Since the hot gas in a real cluster has a finite extension, each of the observed quantities as the Compton parameter y and the X-ray surface brightness S_x will be smaller than those estimated based on the infinite extension assumption ($l \to \infty$). Since the Hubble constant is estimated from the ratio S_x/y^2, in Puy et al. [3] we showed that the relative error $\varepsilon_{H_0}^{fini}$ on the estimate of the Hubble constant, between a spherical distribution with and without finite extension, is given by:

$$\varepsilon_{H_0}^{fini} = \frac{H_0(\infty) - H_0(l)}{H_0(\infty)} = 1 - \frac{B(3\beta - \frac{1}{2}, \frac{1}{2}) \left[B(\frac{3}{2}\beta - \frac{1}{2}, \frac{1}{2}) - B_m(\frac{3}{2}\beta - \frac{1}{2}, \frac{1}{2}) \right]^2}{B^2(\frac{3}{2}\beta - \frac{1}{2}, \frac{1}{2}) \left[B(3\beta - \frac{1}{2}, \frac{1}{2}) - B_m(3\beta - \frac{1}{2}, \frac{1}{2}) \right]}. \tag{1}$$

The functions B and B_m are combinations of the classical Gamma-functions and the factor m is a cut-off relative to the finite extension (see Puy et al. [3]). In the same way we have analysed the relative error between spherical and aspherical geometries (without finite extension).

Figure 1 shows the influence of geometry and of the assumption of finite extension on the above result using the same input parameters of Mauskopf et al. [4]. The left panel shows that for a spherical geometry H_o displays a strong dependence on the cluster extension. The right panel gives the value of H_o assuming an infinite extended ellipsoid shaped cluster as a function of its axis ratio ζ_1/ζ_3.

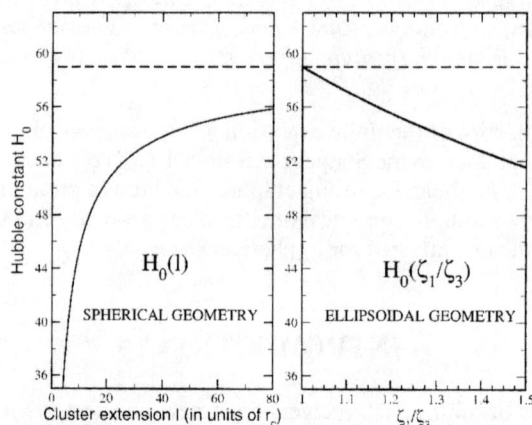

FIGURE 1. The Hubble constant derived from the data of Mauskopf et al. [4]. Fig. (a) shows the influence of finite extension, while Fig. (b) gives the value of H_o assuming an axisymmetric ellipsoidal geometry. In the latter case, oblate or prolate geometry give the same value of H_o when taking a line of sight through the cluster center, as is assumed here.

We see that it is crucial to know the shape of a cluster and its temperature profile. In this context the X-ray satellites *XMM* and *Chandra* have the necessary spatial and spectral resolution to address this problem on nearby cluster samples.

MORPHOLOGY OF GALAXY CLUSTERS

The β-model [5] is widely used in X-ray astronomy to parametrise the gas density profile in clusters of galaxies by fitting their surface brightness profile. In this fitting procedure spherical symmetry is usually assumed, also in cases where the ellipticity of the surface brightness isophotes is manifest. For example Fabricant et al. [6] showed a pronounced ellipticity of the surface brightness for the cluster Abell 2256, Allen et al. [7] obtained the same result for the profile of Abell 478 and Neumann & Böhringer [8] for CL0016+16.

The asphericity of the observed surface brightness let us also ponder on the possible asphericity of the intracluster medium, which can be modelled with an ellipsoidal β-model rather than with the less accurate spherical one.

Hughes & Birkinshaw [9] fitted the surface brightness of CL0016+16, which shows an axis ratio of major to minor axis of 1.176, with both circular and elliptical isothermal β-models obtaining for the best fit parameters $\beta^{circ} = 0.728^{+0.025}_{-0.022}$, $\sigma_c^{circ} = 0.679^{+0.045}_{-0.039}$ arcmin and $\beta^{ell} = 0.737^{+0.027}_{-0.022}$, $\sigma_c^{ell} = 0.746^{+0.044}_{-0.044}$ arcmin (σ_c^{ell} is the core radius along the major axis), respectively, with the latter model providing a considerably better fit.

More recently, a *Chandra* observation of the galaxy cluster RBS797 reveals a pronounced aspherical geometry [10]. The analysis of the image (see Figure 2) gives a strong ellipticity, where the axis ratio of major to minor axis varies slightly from 1.3 at a radius of 0.26 arcmin to 1.4 at a radius of 1.7 arcmin (as mentioned in Schindler et al. [10]). Our analysis of the surface brightness profile for RBS797 gives best fit parameters: $\beta^{circ} = 0.62^{+0.03}_{-0.03}$, $\sigma_c^{circ} = 7.32^{+0.7}_{-0.7}$ arcsec (see also [10]) and $\beta^{ell} = 0.59^{+0.02}_{-0.02}$, $\sigma_c^{ell} = 7.89^{+0.9}_{-0.9}$ arcsec (along the major axis), for the circular and elliptical models respectively [11].

FIGURE 2. *Chandra* image of the cluster RBS797. The cluster is rather regular with, however, an ellipticity of 1.3-1.4 in E-W direction. The center and the position angle ($\approx -70°$, N over E) of the various isophotes are almost the same over the entire radius range, from Schindler et al. [10].

Assuming hydrostatic equilibrium the total mass of a cluster can be estimated from the parameters provided by the surface brightness fit and clearly some care is needed if the galaxy cluster in question shows a pronounced ellipsoidal shape. Assuming hydrostatic equilibrium, the general expression for the total mass density ρ_{tot} is given by

$$\rho_{tot} = -\left(\frac{k_B}{4\pi G \mu m_p}\right)\vec{\nabla}\left[\frac{1}{\rho_g}\vec{\nabla}(T_g \rho_g)\right], \qquad (2)$$

where ρ_g is the gas mass density, T_g its temperature and μm_p is the mean particle mass of the gas.

In order to obtain the total mass of the cluster, one assumes a spherical geometry and integrates this equation over a sphere with radius R. For RBS797 we have additionally investigated the ellipsoidal geometry (for which we assume oblate or prolate shapes) and integrated the equation over an ellipsoid (concentric and similar to the gas core) with major semi-axis R [11]. We thus obtained mass estimates for the spherical shape: $M_{tot}^{sph}(R = 4\sigma_c^{circ} = 29.28\,\text{arcsec}) = 8.66^{+2.5}_{-2.3} \times 10^{13} M_\odot$ and $M_{tot}^{sph}(R = 30\sigma_c^{circ} = 219.6\,\text{arcsec}) = 6.89^{+2.0}_{-1.8} \times 10^{14} M_\odot$ and for ellipsoidal shapes, which are, compared at the same values of R, lower than those for the spherical symmetry by $\sim 10\%$ and $\sim 17\%$ for oblate or prolate shapes, respectively.

REFERENCES

1. Cooray, A. R., *AA*, **339**, 623 (1998).
2. Sulkanen, M., *ApJ*, **522**, 59 (1999).
3. Puy, D., Grenacher, L., Jetzer, Ph., Signore, M., *AA*, **363**, 415 (2000).
4. Mauskopf, P., Ade, P., Allen, W. et al. *ApJ*, **538**, 505 (2000).
5. Cavaliere, A., Fusco-Femiano, R. *AA*, **49**, 137 (1976).
6. Fabricant, D., Rybicki, G., Gorenstein P., *ApJ*, **286**, 186 (1984).
7. Allen, S., Fabian, A., Johnstone, D, White, D., Daines, S., Edge, A., Steward, G. *MNRAS*, **262**, 901 (1993).
8. Neumann, D., Böhringer, H. *MNRAS*, **289**, 123 (1997).
9. Hughes, J., Birkinshaw M. *ApJ*, **501**, 1 (1998).
10. Schindler, S., Castillo-Morales, A., De Filippis, E., Schwope, A., Wambsganss, J., *Astro-ph/0107504*, (2001)
11. Piffaretti, R., Jetzer, Ph., Puy, D., Schindler, S., in progress (2001).

Constraints on the Accuracy of Photometric Redshifts Derived from BLAST and Herschel/SPIRE Sub-mm Surveys

Itziar Aretxaga*, David. H. Hughes*, Edward Chapin* and Enrique Gaztañaga*

INAOE, Aptdo. Postal 51 y 216, 72000 Puebla, Mexico

Abstract. More than 150 galaxies have been detected in blank-field millimetre and sub-millimetre surveys. However the redshift distribution of sub-mm galaxies remains uncertain due to the difficulty in identifying their optical-IR counterparts, and subsequently obtaining their spectroscopic emission-line redshifts. In this paper we discuss results from a Monte-Carlo analysis of the accuracy with which one can determine redshifts from photometric measurements at sub-millimetre-FIR wavelengths. The analysis takes into account the dispersion in colours introduced by including galaxies with a distribution of SEDs, and by including photometric and absolute calibration errors associated with real observations. We present examples of the probability distribution of redshifts for individual galaxies detected in the future BLAST and Herschel/SPIRE surveys. We show that the combination of BLAST and $850\mu m$ observations constrain the photometric redshifts with sufficient accuracy to pursue a program of spectroscopic follow-up with the 100m GBT.

DESCRIPTION OF THE TECHNIQUE AND RESULTS

Determining the density of star formation as a function of redshift is the primary science objective of the Balloon-borne Large Aperture Submillimetre Telescope (BLAST, [1]) and other sub-mm/mm facilities. Using Monte-Carlo simulations that take into account realistic photometric and absolute calibration errors for future BLAST surveys, we show that it is possible to determine redshifts from BLAST data (at 250, 350 and $500\mu m$) with a 1σ average precision of $\Delta z \sim \pm 0.6$. A similar level of redshift accuracy is found for simulated observations with the Herschel/SPIRE camera which will operate at identical wavelengths to BLAST.

The power of this simple technique to derive redshifts arises from the unique ability of BLAST and SPIRE observations to bracket the ubiquitous rest-frame FIR peak (at $\sim 60-150\mu m$) in the spectral energy distribution (SED) of high-redshift ($1 \leq z \leq 4$) galaxies undergoing a significant amount of star formation.

To determine the accuracy of this method we have generated mock catalogues of galaxies between $z = 0$ and $z = 6$ using an evolving $60\mu m$ luminosity function [2] that reproduces the observed $850\mu m$ number counts. The sub-mm flux densities, and colours, of these mock galaxies are calculated from SEDs selected at random from a library of template starburst galaxies, ULIRGs and AGN (Fig.1). Observational noise is then added to the intrinsic fluxes: 1σ photometric errors of 5 and 2.5 mJy for the BLAST and Herschel/SPIRE observations respectively, and in both cases an absolute calibration error of 7%.

We are therefore able to determine the photometric redshift probability distribution for any galaxy detected in BLAST surveys by comparing its *measured* BLAST colours with the complete distribution of *simulated* colours and redshifts of galaxies in the mock catalogue (see Fig.2).

It is instructive to illustrate the discrepancy between the photometric redshifts determined from the colours of galaxies detected in our simulated BLAST (or SPIRE) surveys and their true mock catalogue redshifts. Figure 3 shows that over the entire redshift range of our simulations, $0 \leq z \leq 6$, the 1σ error in the photometric redshifts, Δz, derived from detections in 3 BLAST filters is ± 0.6. However beyond $z \sim 4$ the BLAST data alone systematically underestimate the redshifts, confusing $z \sim 4.5$ galaxies with $z \sim 2.5$ galaxies for example, as the longest wavelength BLAST filter ($500\mu m$) moves short-ward of the rest-frame FIR peak.

A significant improvement in the derived photometric redshifts occurs when the BLAST (or SPIRE) observations are complemented with data from a longer-wavelength survey (*e.g.* ground-based $850\mu m$ SCUBA data). This extension to the wavelength coverage (250–$850\mu m$) ensures that the photometric redshifts are uniformly distributed about the

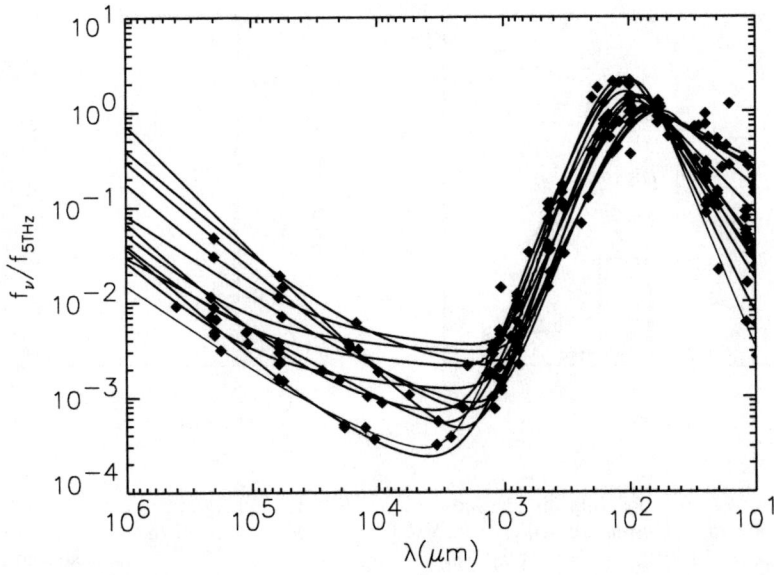

FIGURE 1. Rest-frame spectral energy distributions of 13 starburst galaxies, ULIRGs and AGN, normalized at 60μm. Lines represent the best fit models to the SEDs and include contributions from non-thermal synchrotron emission, free-free and grey-body thermal emission.

$z = z_{phot}$ regression line over the entire range $0 < z < 6$, with an improved average error of $\Delta z \sim \pm 0.4$ (Fig. 3). This method therefore continues to provide un-biased estimates of photometric redshifts for the most distant galaxies. Furthermore, an increased sensitivity in the observations (with SPIRE for instance) naturally translates into an increased accuracy of the redshift distributions.

OPPORTUNITIES FOR SPECTROSCOPIC FOLLOW-UP

Redshift accuracies can be further improved with heterodyne follow-up observations. We plan to use the recently commissioned 100m GBT to search for low-J CO molecular-line transitions. For example, for those BLAST and Herschel/SPIRE galaxies with photometric redshifts in the range $z \sim 4.0 - 4.8$, which have 1σ accuracy of $\Delta z \sim \pm 0.4$ when combined with 850μm observations, we can search for CO(1–0) in the K-band (18.0 – 26.5 GHz). The addition of a future GBT receiver operating at 26 – 40 GHz will extend the CO(1–0) search to galaxies at $2.3 \lesssim z \lesssim 4.8$. Similarly, a proposed 68 – 116 GHz receiver, will detect both the CO(1–0) and CO(2–1) transitions of the low-z population ($z \lesssim 2$).

To conclude, our simulations demonstrate that, within the next few years, large-area BLAST surveys will produce a catalogue of $\gtrsim 5000$ high z galaxies for which sub-mm photometric redshifts with an accuracy $\Delta z \sim \pm 0.4 - 0.6$ can be determined. Follow-up observations with GBT receivers will provide definitive molecular-line redshifts and dynamical-mass estimates. The combination of BLAST and GBT provide a powerful combination to break the redshift *deadlock* that hinders our ability to understand the evolution and nature of the sub-mm starburst galaxy population.

Acknowledgments: This work has been partly supported by CONACyT grants 32143-F, and 32180-E.

REFERENCES

1. Devlin, M. *et al.* 2001, in *Deep Millimetre Surveys: Implications for Galaxy Formation and Evolution*, eds. J.Lowenthal, D.H.Hughes, World Scientific
2. Saunders, W. *et al.* 1990, MNRAS, 424, 318

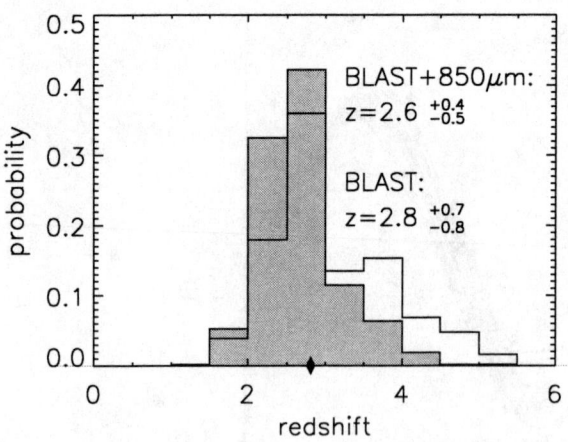
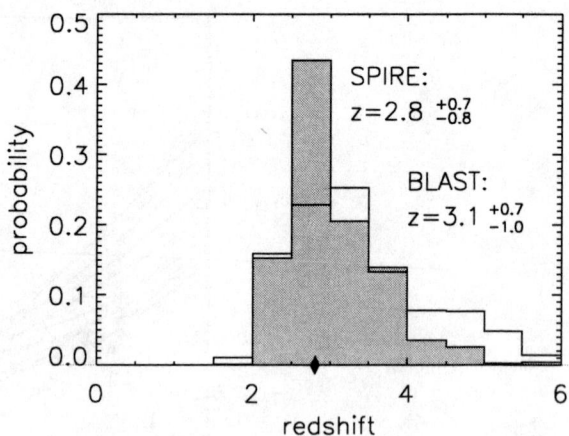

FIGURE 2. *Left panel:* Redshift probability distributions of a 4×10^{12} $L_\odot$ galaxy at $z = 2.8$, detected at 250, 350 and 500μm with BLAST, and also at 850μm in a ground-based deep survey. The intrinsic colours of this galaxy are determined from a scaled SED of NGC 1614: $S_{250\mu m} = 17$ mJy, $S_{350\mu m} = 23$ mJy, $S_{500\mu m} = 27$ mJy, $S_{850\mu m} = 5$ mJy. The unshaded and shaded distributions have been calculated from BLAST detections and from BLAST+850μm detections, respectively. The black diamond marks the true redshift of the mock galaxy. Photometric redshift determinations within a 68% confidence level are indicated within the panel. *Right panel:* A comparison of the redshift distributions for a 3×10^{12} $L_\odot$ galaxy at $z = 2.8$ determined from SPIRE and the less sensitive BLAST observations. The colours of this galaxy are similar to NGC 2992: $S_{250\mu m} = 18$ mJy, $S_{350\mu m} = 28$ mJy, $S_{500\mu m} = 23$ mJy, $S_{850\mu m} = 7$ mJy. The galaxy was detected simultaneously in the 3 SPIRE passbands, but only at 350 and 500μm with BLAST. The unshaded and shaded histograms correspond to the use of BLAST and SPIRE detections, respectively.

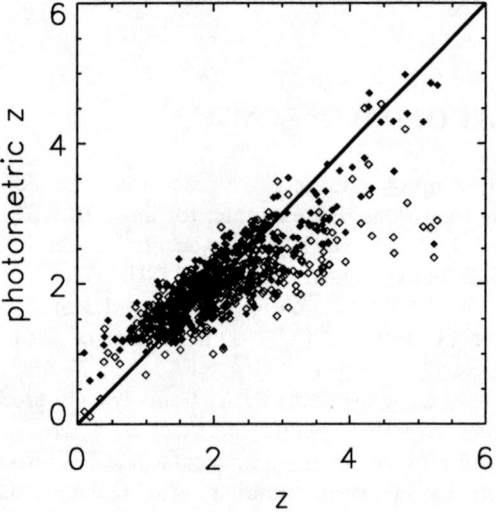
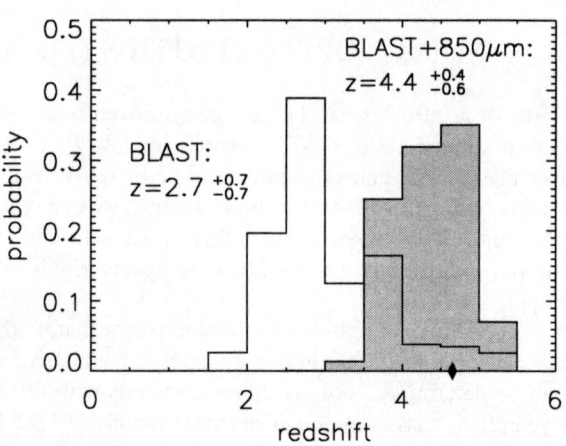

FIGURE 3. *Left panel:* Photometric redshift vs. true redshift relationship for 400 galaxies simultaneously detected at 250, 350 and 500μm in 1 sq. degree. Open symbols show the relationship inferred using only BLAST data to derive the redshifts. Filled symbols show the relationship when the redshifts are estimated using colours based on BLAST and complementary 850μm detections from a ground-based survey. The addition of 850μm measurements significantly increases the accuracy at $z \gtrsim 4$, since at these redshifts the BLAST filters sample the rest-frame mid-IR to FIR (~ 35–100μm). The longer wavelength data are required to bracket the rest-frame FIR peak, which provides the diagnostic power for the photometric redshift technique discussed in this paper. *Right panel:* Example of the correction attained at $z > 4$ when 850μm observations are included in the photometric redshift analysis. In particular, these are redshift distributions of a 1×10^{13} $L_\odot$ galaxy at $z = 4.65$, with observed fluxes $S_{250\mu m} = 18$ mJy, $S_{350\mu m} = 29$ mJy, $S_{500\mu m} = 27$ mJy, $S_{850\mu m} = 19.5$ mJy.

Simulating the Performance of Large-format Sub-mm Focal-plane Arrays

E. Chapin[*], D. H. Hughes[*], B. D. Kelly[†] and W. S. Holland[†]

[*]*Instituto Nacional de Astrofísica, Óptica y Electrónica, Apartado Postal 51 y 216, 72000, Puebla, Mexico*
[†]*UK Astronomy Technology Centre, Blackford Hill, EH9 3HJ, Edinburgh, UK*

Abstract. A robust measurement of the clustering amplitude of the sub-mm population of starburst galaxies requires large-area surveys ($\gg 1$ deg^2). The largest-format arrays subtend only 10 arcmin^2 on the sky and hence scan-mapping is a necessary observing mode. Providing realistic representations of the extragalactic sky and atmosphere, as the input to a detailed simulator of the telescope and instrument performance, allows important decisions to made about the design of large-area fully-sampled surveys and observing strategies. In this paper we present preliminary simulations that include detector noise, time-constants and array geometry, telescope pointing errors, scan speeds and scanning angles, sky noise and sky rotation.

GENERATING REALISTIC SYNTHETIC TIME-SERIES

A mapping simulator has been developed to generate synthetic bolometer time-series data and associated astrometric information which are then run through a realistic reduction pipeline (*http://www.inaoep.mx/~echapin/scansim.html*). This process allows one to develop and test the observing strategy and analysis software in advance of instrument delivery. Furthermore it is possible to assess the impact of instrumental design aspects on the science objectives.

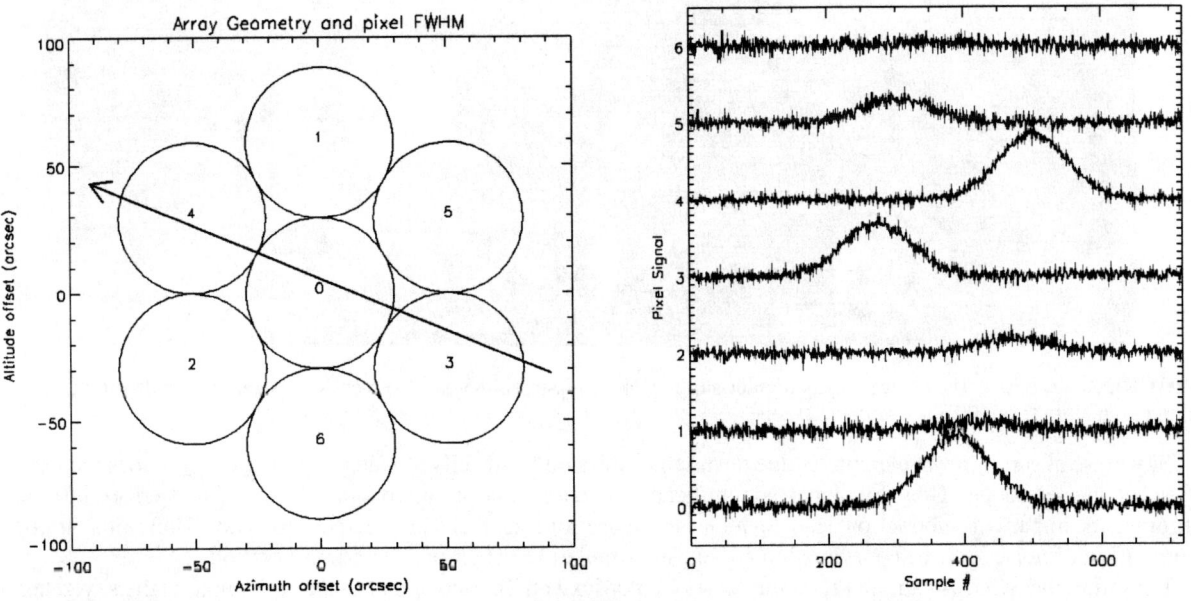

FIGURE 1. 7-element, 500μm 1-$F\lambda$ spaced hexagonal close-packed array geometry with 59″ FWHM beams, and bolometer signals for a scan across a point source (path of source indicated by arrow).

The simulated sky, typically generated by convolving a synthetic catalogue of extra-galactic sources with the telescope beam [1], is assigned equatorial coordinates on the celestial sphere. The geometry of the array is defined and the beam positions on the sky are determined relative to the telescope bore-sight. The telescope is then arbitrarily located and the beams are scanned across the simulated sky model (in some pattern on an Alt-Az coordinate system for a given local time) and, using a rigorous calculation of the astrometry, the positional information and *noiseless* flux density time-series is determined for each bolometer in the array.

More realistic time-series are produced by incorporating features characteristic of real instruments, the pointing performance of the telescope, and the non-negligible contribution of the atmosphere (sky noise and attenuation).

Sources of instrumental noise in the time-series include: 1) addition of an uncorrelated Gaussian system noise component; 2) multiplication by gain factors (or different responsivities) for each detector; 3) addition of independent low frequency $1/f$ noise for each bolometer time-series to model the slowly-changing detector baselines in long duration scans; 4) convolution with an $e^{-t/\tau}$ impulse function to mimic the instantaneous response of the detectors. The instrumental noise components for a linear detector are summarized by the following expression for the response:

$$I_{measured} = G \times F + I_{dark} \quad (1)$$

where G is the detector gain, F is the incident flux and I_{dark} is the *dark current* consisting of both broad-band and $1/f$ noise components. The *perfect* astrometry may also be *corrupted* by adding random and systematic pointing errors. Fig. 1 shows the time-series for a 7-pixel feed-horn coupled array scanned across a point source at 500μm.

SKY NOISE

FIGURE 2. A 360 × 180 degree^2 simulation of sub-mm sky emission at 850μm from cells of water vapour distributed in a flat plane at an altitude of 600m above the telescope.

Sky noise at sub-mm wavelengths is due to variable emission from cells of water vapour moving across the field-of-view of the telescope. This sky signal is significantly greater than the astronomical signal, and errors introduced through its imperfect subtraction lead to increased noise and artifacts in the reduced data. Thus, this important component of noise must be considered in all realistic simulations of ground-based observations.

Data from the SCUBA camera [2] on the James Clerk Maxwell Telescope on Mauna Kea suggests the sky generates $1/f$-like noise that is correlated across the array. An empirical model that reproduces the observed spectrum of signals measured with the SCUBA array at 850μm assumes the sky consists of cells of water vapour, generated by convolving Gaussian noise with a symmetrical $(1/f)^{11/6}$ function, where f is the 2-D spatial frequency, distributed in a flat-plane at some fiducial altitude, moving at the wind-speed above the telescope aperture (Fig. 2). This planar sky-model is passed across the simulated beams to produce a strong, variable sky signal.

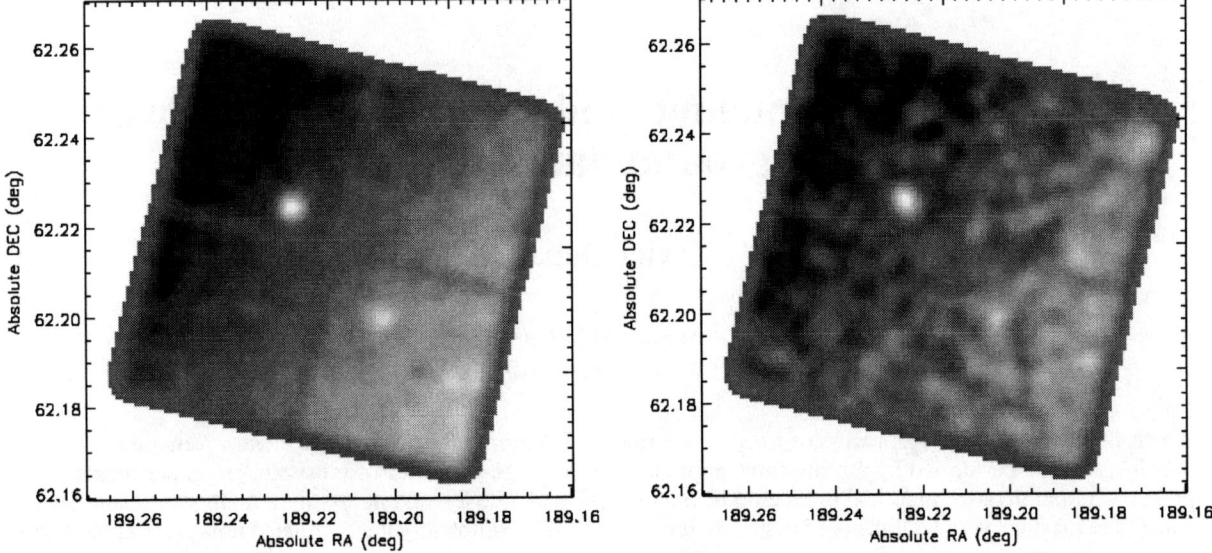

FIGURE 3. Simulations of one-second point-and-stare total power observations with a fully-sampled (0.5-$F\lambda$) 40×40 pixel monolithic array at 850μm with 14″ FWHM beams. In the case of perfect flat-fielding (left panel), two significant sources with a S/N of 25 and 12σ are clearly observed against a typical sky-background gradient. In contrast, the fainter of the sources blends into the background noise when the gains of the bolometers are only known to a precision of $\sim 1\%$ (right panel).

Each 850μm pixel in the SCUBA2 array under good weather conditions will be illuminated by a sky background power of 8pW with a photon noise in a 1s integration of 4.3×10^{-5}pW. While the total power of the sky may be removed by subtracting the median signal of the off-source sky bolometers, strong gradients still remain and can dominate the astronomical signals. DC sky measurements with SCUBA are used to estimate these gradients by transforming the time-series into a spatial signal via the recorded wind speed, assuming the emission is from a fixed altitude above the telescope, and scaling it accordingly for the different background loading of the SCUBA2 pixels. The estimated median change in the sky level across an 8 arcmin field of view is 7×10^{-4}pW (about an order-of-magnitude greater than the photon noise). These estimates of the expected gradients and DC power levels allow the proper scaling and bias of the planar-cloud model used to generate the sky component of bolometer signals in the simulation.

Effective subtraction of the sky in the case of total power measurements depends critically upon the knowledge of detector gains and drifts. In Equation (1), G and I_{dark} vary from pixel to pixel and need to be removed using separate operations. The gain G is slowing-changing, and characterized by performing flat-field observations with long integrations. The flat-fielding error must be smaller than the photon noise measured in the longest astronomical integration. In addition G is multiplied by the large sky signal ($\gg$ astronomical signal), small errors in its measurement will lead to residual artifacts upon dividing it out of the response. An initial estimate suggests that G will need to be measured to an accuracy of $\ll 1\%$. I_{dark} consists of broad-band (white) noise plus a slowly varying $1/f$ component; the drift may be subtracted by measuring "dark" frames (1s integrations) in the absence of sky signal every few minutes.

Fig. 3 demonstrates, for pointed total-power observations, two examples of the effects described above. In conclusion, the requirement to flat-field at sub-mm wavelengths with a precision of $\ll 1\%$ presents a major obstacle to the efficient use of large-format monolithic arrays on telescopes operated in total power mode.

REFERENCES

1. Hughes, D. H. and Gaztañaga, E., "Submillimetre Galaxy Surveys" in *Star Formation from the Small to the Large Scale*, edited by F. Favata et al., ESLAB Symposium 33, Noordwijk, 2000, pp. 29–36, astro-ph/0004002.
2. Holland, W. S. *et al.*, *MNRAS*, **303**, 659–672 (1999).

Iterative map-making methods for Cosmic Microwave Background data analysis

Xavier Dupac

Centre d'Étude Spatiale des Rayonnements, 9, avenue du Colonel Roche, BP 4346, F-31028 Toulouse cedex 4, France, dupac@cesr.fr

Abstract. The map-making process of Cosmic Microwave Background data involves linear inversion problems which cannot be performed by a brute force approach for the large timelines of most modern experiments. We present optimal iterative map-making methods, both COBE and Wiener, and apply these methods on simulated data. They produce very well restored maps, by removing nearly completely the correlated noise that appears as intense stripes on the simply pixel-averaged maps.

INTRODUCTION

The Cosmic Microwave Background (CMB hereafter) is being extensively studied nowadays, thanks to the improvement of instrument and detector performances. Since the COBE experiment (Smoot et al. 1992), which performed the first detection of CMB anisotropies, the size of time-ordered information (TOI hereafter) has been widely increased. The data processing and analysis is thus still a challenge for large timeline data, already in processing or still to come. The Cosmic Microwave Background data analysis is usually performed in three steps, the first being the map-making process, the second the C_l estimation from the maps and the noise covariance matrices, the third the cosmological parameters estimation from the C_l power spectrum. In this article, we focus on the map-making step, which, in the context of large timelines, cannot be performed by a brute-force approach, which would imply the manipulation and inversion of tera-element large matrices. Non optimal map-making methods have been developed, such as destriping using the scan intercepts (see Delabrouille 1998 and Dupac & Giard 2001). We aim in this paper to apply the optimal methods to large timelines, developing algorithms to avoid computation trouble. The map-making methods for CMB are known to be optimal when following desirable properties. These are linear map-making methods, well known as the COBE method (Janssen & Gulkis 1992) and the Wiener filter (Wiener 1949). The matrix expressions of these optimal methods can be found, e.g., in Tegmark (1997).

Applying these optimal methods to large timelines is not straightforward, because of computer limitations. The several tens or hundreds of mega-elements in a bolometer timeline have to be processed by vector-only methods, that we aim to present in this paper. We mean by "vector-only methods" computations in which one never needs to compute or even to store any large matrix.

OBSERVATION STRATEGY

We present here a large-coverage observation strategy, simulated from simple and usual technical requirements. We simulate balloon-borne CMB experiments, required to scan the sky making constant elevation circles at a rather constant rotation speed (as do Archeops: Benoît et al. 2001, and TopHat: http://topweb.gsfc.nasa.gov). This allows to observe a large area on the sky, thanks to the rotation of the Earth (and, eventually, to the moving of the balloon on the Earth). We have simulated a balloon-borne experiment timeline with a constant scanning elevation of 35 degrees above the horizon, a sampling frequency of 100 Hz and a rotation speed of the gondola of 2 rpm. The flight is 24 hours long, launched from a polar place (Kiruna for instance). The winter time in this region provides polar nights allowing to make 24 hours flight avoiding contamination from the sun. The coverage for this polar flight is 35 % of the sky.

SIMULATION PROCESS

We have made our simulated skies from three components: the CMB fluctuations, the dipole, and the Galaxy, at 2 mm wavelength. The simulated Universe we have simulated is Λ dominated, with $\Omega_\Lambda = 0.7$, $\Omega_{CDM} = 0.25$, $\Omega_{bar} = 0.05$, $H_0 = 50$ and a scalar spectral index of the fluctuations equal to 1. The C_l simulated spectrum (with no cosmic variance) is made thanks to the CMBFAST software (Seljak & Zaldarriaga 1996). The sky gaussian random field (i.e. a realization of a Universe with the cosmological parameters we chose) is then simulated with the SYNFAST tool of the HEALPix package (http://www.eso.org/science/healpix). The dipole is added thanks to its COBE/DMR determination (Lineweaver et al. 1996), and the Galaxy at 2 mm wavelength is extrapolated from the composite 100 μm IRAS-COBE/DIRBE all sky dataset (Schlegel et al. 1998). The noise we introduce in our simulated timelines can be characterized by its statistical power spectrum which follows a 1/f law: $l_{inf}.(1 + (f_c/f)^n)$, where l_{inf} is the level of white noise (i.e. the only noise at high frequency), f_c the cut frequency and n the power index. The noise we introduced is characterized by $n = 1$, $f_c = 0.1$ Hz and $l_{inf} = 100$ μK_{CMB} rms. This white noise rms is approximatively the level expected for Planck (Tauber 2000) bolometers with a 100 Hz sampling rate. Of course this quite low value is the level of the non-correlated noise in the timelines, but the total amount of noise introduced is much larger.

MAP-MAKING METHODS APPLIED ON LARGE TIMELINES

We use the HEALPix pixel scheme (http://www.eso.org/science/healpix) to make our maps. Reprojecting timelines on maps is not only a domain to domain transform, but the simplest way to estimate the true map of the sky, by averaging the samples of a same pixel on the sky. The noise is therefore reduced by a factor square root of the number of samples in the pixel (the weight). We will not investigate the beam deconvolution in this article, and therefore consider only 1-and-0 point-spread matrices.

The optimal map-making methods use the noise (N) and sky (S) covariance matrices: these are impossible to invert or even to store for such large timelines. However, if the noise is stationary in the time domain, as it is the case for 1/f noise and white noise (usual bolometer noises), the noise correlation matrix in the time domain is circulant, i.e. multiplying a vector by this matrix is a convolution, which is filtering in the Fourier domain. The sky covariance matrix is stationary in the map domain, as far as the Cosmic Microwave Background is a gaussian random field (for timelines without the Galaxy). In this case the sky covariance matrix in the map domain is circulant.

The COBE equation: $\tilde{x} = [A^t N^{-1} A]^{-1} A^t N^{-1} y$, cannot be directly applied with vector-only algorithms, because of the matrix inversions needed. (A is the point-spread matrix, N the noise covariance amtrix in the time domain, y the data timeline and $\tilde{x}$ the optimal reconstructed sky map.) Thus the trick is to solve rather:
$[A^t N^{-1} A] \tilde{x} = A^t N^{-1}$ y

This form prevents from the heavy inversion, but needs an iterative scheme. The general iterative scheme for this equation is:
$\alpha \tilde{x}_{n+1} = \alpha \tilde{x}_n + A^t N^{-1}$ y $- [A^t N^{-1} A] \tilde{x}_n$

where α is any linear operator on a vector, that is, any square matrix. We have tested this algorithm on simulations and real data from the Archeops experiment (Benoît et al. 2001), with α being a scalar. By testing the method with different α, we find that the identity is the best iterator.

Another scheme can be developed, by making the noise map converge instead of the sky map, as mentioned by Prunet (2001). This can be better, as the signal can be more tricky than the instrumental noise for the stability of the iterative scheme: hot galactic points for example may induce stripes on the maps. The noise-iterating scheme works with the following trick: we change the variable $\tilde{x}$ to $\check{x} = [A^t A]^{-1} A^t$ y $- \tilde{x}$. It is straightforward to show that this is the noise map plus the reconstruction error. It leads to:
$\alpha \check{x}_{n+1} = \alpha \check{x}_n + A^t N^{-1}$ z $- [A^t N^{-1} A] \check{x}_n$

where z $= A[A^t A]^{-1} A^t$ y $-$ y. If this algorithm converges, then the converging limit is exactly the optimal solution of the map-making problem. The case of $\alpha = I$ is actually the simplest iterator one can imagine, but works well on the simulations and real data that we have processed.

The iterative scheme that we have developed for the Wiener method is close to the COBE one:
$\alpha \check{x}_{n+1} = \alpha \check{x}_n +$ u $- [S^{-1} + A^t N^{-1} A] \check{x}_n$

where u $= A^t N^{-1}[A[A^t A]^{-1} A^t y - y] + S^{-1}[A^t A]^{-1} A^t y$. As we have shown for the COBE iterative method, here the converging limit is the exact solution of the Wiener map-making equation. This iterative scheme needs to handle both the N matrix, noise covariance matrix in the timeline, that we process as a filter in the Fourier domain like we do for the COBE iterative method, and the S matrix, sky covariance matrix in the map domain. Handling this as a matrix is not possible for a small scale pixelization that we need for CMB experiments of today, thus we have to process it as a filter in Fourier space, like we do for filtering timelines. The HEALPix RING scheme is stationary with respect to the sphere, because it pixelizes it making a ring around the sphere from the north pole to the south pole, with equal pixel surfaces. So filtering a HEALPix vector (i.e. a map) in Fourier space is optimal, to the condition that there must not be large holes in the map, that would harm the stationarity of the sky in the HEALPix scheme.

RESULTS AND CONCLUSION

We have applied these methods to the polar flight simulated data: the method reaches the residual noise at about 50 iterations, and this residual is about $21.6\,\mu K_{CMB}$ rms. We can check that we have reached the convergence by observing the evolution of the global residual noise rms, but also the evolution for some individual pixels, the map aspect and the C_l power spectrum. We have to compare this result to the white noise amount in the map: the rms level of white noise in the polar flight map is $21.06\,\mu K_{CMB}$ rms, which is very close to the residual noise amount. This shows how good the reconstruction is, as it is clear that the correlated noise is significantly removed from the map. The reconstructed map exhibits no visible difference with the true map. The noise spectrum is nearly the one of a white noise above about l=50, but exhibits some weak residual correlation at lower scales. Even if the residual noise amount is very low, this deviation from the flat spectrum could have to be taken into account for very precise measurements of the low l.

This kind of vector-only methods seems to us unavoidable to make optimal maps from CMB experiments of today, or still to come. The reduction of the information in CMB data is a heavy work, from gigabytes of rawdata to essentially 12 cosmological numbers with their error bars. Since the computer facilities are limited and unsufficient for brute force approaches (and it will be still the case for Planck data reduction), it is an interesting challenge to process each step of this reduction work without losing information. Using stationarity properties of a signal in a given domain (sphere, map, timeline...) to transform a matrix inversion problem into a vector-only solution, could be probably also developed for other CMB reduction steps, such as the component separation.

ACKNOWLEDGEMENTS

We would like to thank K.M. Górski and his collaborators for their so useful HEALPix package.

REFERENCES

Benoît, A., *et al.* : 2001, *Astrop. Physics, in press*
Delabrouille, J.: 1998, *Astron. Astrophys. Suppl.*, 127: 555
Dupac, X., Giard, M.: 2001, *Proc. of the "Mining the Sky" 2000 Colloq. in Garching*
Janssen, M.A., Gulkis, S.: 1992, *Proc. of the NATO adv. study inst., Les Houches*
Lineweaver, C.H., Tenorio, L., Smoot, G.F., Keegstra, P., Banday, A.J., Lubin, P.: 1996, *ApJ*, 470: 38
Prunet, S.: 2001, *Proc. of the Moriond 2000 Colloq.*
Schlegel, D.J., Finkbeiner, D.P., Davis, M.: 1998, *ApJ*, 500: 525
Seljak, U., Zaldarriaga, M.: 1996, *ApJ*, 469: 437
Smoot, G.F., *et al.* : 1992, *ApJ Lett.*, 396: L1
Tauber, J.: 2000, *IAU Symposium 204 in Manchester*
Tegmark, M.: 1997, *ApJ*, 480: L87
Wiener, N.: 1949, *Extrapolation and Smoothing of Stationary Time Series*, NY: Wiley

Spatial features of non-thermal SZ effect in galaxy clusters

E. Palladino*, S. Colafrancesco[†] and P. Marchegiani[†]

*Università "La Sapienza", P.le Aldo Moro 5, Rome, Italy
[†]Osservatorio Astronomico di Roma, Monteporzio, Italy

Abstract. We investigate the spatial behaviour of the total comptonization parameter y_{tot} evaluated for a galaxy cluster containing two population of electrons: the thermal population, with energy around some KeV and whose trace is evident in the X-ray emission of the ICM (Intra-Cluster Medium), and the relativistic population, which give rise to the radio halo emission found in several clusters of galaxies. We present the first results obtained from our analysis showing that there are remarkable features in such spatial trend, which might throw a new light in understanding the cluster internal processes.

Recently has been investigated the possibility that the thermal SZ effect, *i.e.* the shift in photon energy of the cosmic microwave background radiation (CMBR) due to its passage through the intra-cluster medium (ICM) ([1], [2], [3]), could be implemented by the so called non-thermal effect due to a relativistic electron population with energy of order tens of KeV, instead of the usual thermal amount of some KeV ([8], [5]).

Here we stress the fact that the presence of a non-thermal SZ effect also produces spatial variations of the total SZ radial profile of the cluster.

Let us consider a galaxy cluster in which are present two kind of electron populations: the first one is *thermal*, *i.e.* it follows a relativistic maxwellian velocity distribution, the second one is *non thermal*, *i.e.* its energy spectrum is relativistic in a certain range. In particular we have used a phenomenological complex spectrum (double-power law) with slopes $\alpha_r \sim 2.5$ at $E > E_*$ and $\alpha_x \sim -0.5$ at $E < E_*$ with the break set at $E_* \sim 200 \div 400$ MeV, suitable to reproduce the whole set of non thermal phenomena found in several galaxy clusters [6] . Consequently, the behaviour of this second population can not be treated with the Kompaneets approach usually used to describe the dynamic of the thermal one (for further details on a complete relativistic treatment of the non-thermal effect see [7]) . In a first approximation we consider these populations independently superposed, so it is possible to completely separate their contribution to the total SZ effect on the incoming photons of the CMBR (Colafrancesco, Marchegiani & Palladino, these Proceedings, for more details).

The inverse Compton scattering of the CMB photons by the electrons of the ICM is theoretically described by the so-called *comptonization parameter*. For a thermal radiation field and in the more general form such a contribution can be expressed as $y_{th} = y_{0,th} \cdot g(x)$, where $y_{0,th}$ physically represents the time spent by the radiation in the medium and it is proportional to the integral of the thermal electron kinetic pressure $P_{th} = n_e k_B T_e$, while the function $g(x)$ contains the dependence of the effect from the a-dimensional frequency $x = h\nu/k_B T_{CMB}$ of the background radiation.

According to the classical approach we can describe the kinetic pressure P_{th} using a theoretical model in which we assume known the spatial distribution and the temperature of the electron population. The model we have used is the *isothermal β-model* [9], according to which the temperature of the ICM is constant and equal to T_e and the electron density n_e has a spherical distribution parameterized by the relation

$$n_e(\mathbf{r}) = n_{e0} \left[1 + \left(\frac{r}{r_c} \right)^2 \right]^{-3\beta/2}, \tag{1}$$

where n_{e0} is the electron density at the center of the cluster, r_c is the core-radius and $\beta = \mu m_p v^2 / k_B T_e$ is the exponent observed to be in the range $0.6 - 1$ (see, e.g., [4]). The detailed values of such parameters are found from the analysis of X-ray emission maps produced by the thermal electrons of the cluster.

TABLE 1. Values of the parameters used for the COMA cluster study

z	$k_B T_e$ (keV)	$r_{c,X}$ ($h_{50}^{-1} Mpc$)	$r_{c,rad}$ ($h_{50}^{-1} Mpc$)	$r_{lim,X}$ ($h_{50}^{-1} Mpc$)	$r_{lim,rad}$ ($h_{50}^{-1} Mpc$)	β_X	β_{rad}	$n_{e0,X}$ ($h_{50}^2\ cm^{-3}$)
0.0232	8.5	0.42	0.4	4.2	1.25	0.75	0.8	$3.0 \cdot 10^{-3}$

Using the model (1), calling $P_{th}^0 = n_{e0} k_B T_e$ the central kinetic pressure and calculating the integral along the line of sight [4], we obtain the final expression for the thermal comptonization parameter:

$$y_{0,th} = \frac{\sigma_T}{m_e c^2} P_{th}^0 r_c Y_{th}(\theta, \theta_c). \tag{2}$$

In the function Y_{th}, whose explicit expression is given by

$$Y_{th}(\theta, \theta_c) = \sqrt{\pi} \frac{\Gamma(\frac{3}{2}\beta - \frac{1}{2})}{\Gamma(\frac{3}{2}\beta)} \left[1 + \left(\frac{\theta}{\theta_c}\right)^2 \right]^{\frac{1}{2} - \frac{3}{2}\beta}, \tag{3}$$

is contained the dependence of the effect from the spatial coordinates, *i.e.* from the angle θ between the center of the cluster and the direction of observation, while $\theta_c = r_c/D_A$ is the angular core radius as deduced from the X-ray data. The *angular diameter distance*, D_A, of the cluster, given in terms of the redshift, contains all the information relative to the assumed cosmological model; in all our work we have put null the cosmological constant and the curvature and the Hubble constant is in the form $H_0 = h_{50} 50$ km sec^{-1} Mpc^{-1}, with $h_{50} = 1$.

Taking into account that we use the same formalism to describe both the thermal and non thermal contributions to the total effect, we can soon write down all the expressions we have previously obtained also in the case of the relativistic population, in which we again parameterized the electron density with a spherical distribution, *i.e.* with a β-model. One fundamental aspect to underline is the difference between the relativistic distribution and the thermal one, which consists in the fact that the spherical radius of the first distribution must be truncated at some limiting radius r_{lim}, as it results from the measurements of the radio emission from clusters of galaxies, that in the case for example of COMA is found to be $r_{lim} = 1.25\, h_{50}^{-1}$ Mpc.

So for the non thermal comptonization parameter we have:

$$y_{0,non-th} = \frac{\sigma_T}{m_e c^2} P_{th}^0 r_{c,rad} \bar{P}_0 Y_{non-th}(\theta, \theta_{c,rad}), \tag{4}$$

where the subscript "*rad*" is used to remember that these quantities refer to the relativistic electrons that emits in the radio band of the electromagnetic spectrum (they in general belong to the *radio halo* of the cluster), while $\bar{P}_0 = P_{rel}^0/P_{th}^0$ is the ratio between the relativistic and the kinetic pressures of the electrons in the centre of the cluster. The function $Y_{non-th}(\theta, \theta_{c,rad})$, in which $\theta_{c,rad} = r_{c,rad}/D_A$ is the angular core radius of the radio halo, has the same expression as the one of eq. (3), in which we substitute the parameters proper of the non-thermal effect.

Now we have all the necessary information to put together the two previous contributions in a whole expression for the total comptonization parameter y_{tot} given by the sum $y_{th} + y_{non-th}$. Elaborating this expression we have the final relation used to obtain our results (here the subscript "*X*" refers to the thermal quantities given by the clusters X-ray maps):

$$y_{tot} = \frac{\sigma_T}{m_e c^2} P_{th}^0 \left\{ r_{c,X} Y_{th}(\theta, \theta_{c,X}) \cdot g(x) + r_{c,rad} \bar{P}_0 Y_{non-th}(\theta, \theta_{c,rad}) \cdot \tilde{g}(x) \right\}. \tag{5}$$

In Tab.1 we present the parameters used to make the analysis and to obtain the results we present for the COMA cluster.

In Fig.1 we show the behaviour of the comptonization parameter, where we have fixed four values of the adimensional frequency x (see Tab.2), to simulate the frequency channels in a common observational experiment devoted to measure clusters properties at millimeter and sub-millimeter wavelength and three values of the ratio $\bar{P}_0$ between the thermal and relativistic pressures of the electron populations, *i.e.* 0.05, 0.49, 1.48. In each panel we graph the spatial trend of the total SZ effect comptonization parameter y_{tot} together with the thermal one y_{th} founded for the same physical values, to make a comparison among the different curves, from the minimum of the effect ($x_{min} \sim 2.5$), the zero ($x_0 \sim 3.8$), to its maximum ($x_{max} \sim 6.2$).

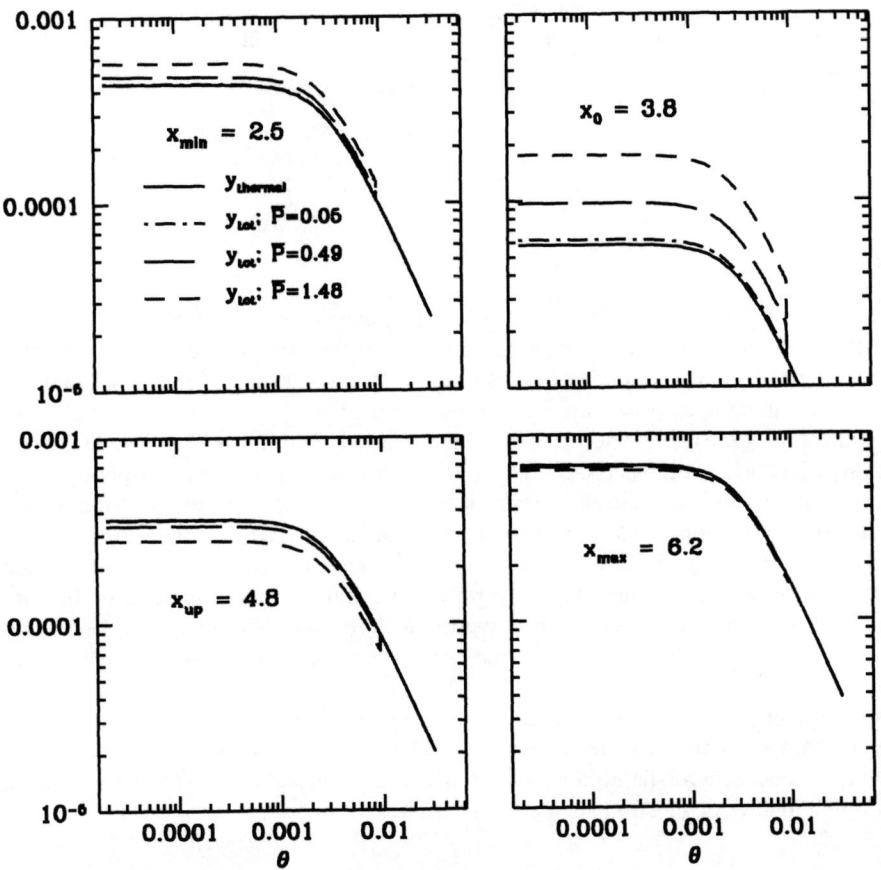

FIGURE 1. The spatial dependence of the total SZ effect produced by the combination of the thermal and relativistic electrons in COMA, with respect to the thermal one (*thin line*). The four panels refers to the four values of the frequency x and in each one there are plotted the case for $\bar{P}_0 = 0.05$ (*point-dash line*), for $\bar{P}_0 = 0.49$ (*long-dash line*) and $\bar{P}_0 = 1.48$ (*short-dash line*).

TABLE 2. The four values of x chosen to make the analysis (see text for details).

Channel	λ (mm)	ν (GHz)	x
1	2.1	142.857	2.51
2	1.4	214.286	3.77
3	1.1	272.727	4.80
4	0.85	352.941	6.21

As one can see, the detection of non-thermal effect is greater in the first two channels, even if the second channel presents a difficult in measuring the effect for the presence of the maximum of the kinetic one that can be in principle greater then the non-thermal one. Going towards higher frequencies ($x > 5$), we find that there is no observational evidence of the non-thermal effect with respect to the thermal one.

Another point is the jump one can see in both the upper panels: it is due to the fact that the relativistic population has a core radius smaller than the optical radius of the cluster (that is given by $r_{c,X}$). The possibility to detect such a jump in the spatial distribution is linked to the angular resolution of the experiment devoted to measure the effect: it is possible, in fact, that a window function of many arc-min, with respect to the angular dimension of the examined cluster, cannot see such a characteristic spatial feature, for the signal has to be convoluted with the instrumental response.

In conclusion we underline the importance of specific spectral and spatial features of the non-thermal SZ effect that can be detected through a multi-frequency observation with narrow-band detectors: the best observational strategy is to check the frequency range $x \sim 2 \div 8$ where the spectral features allow to distinguish the non-thermal effect to the thermal one.

This is very important since the SZ effect is a remarkable tool for cosmology (*i.e.* with the measurement of the Hubble constant H_0) and for better knowing the astrophysics of clusters. We stress that the PLANK surveyor experiment has the possibility to detect the total SZ effect (thermal and non-thermal one) in a a large number of nearby radio-halo (with a relativistic electron component) clusters.

REFERENCES

1. Sunyaev, R.A. and Zel'dovich, Ya.B. 1972, Comments Astrophys. Space Sci., 4, 173
2. Rephaeli, Y. 1995, ARA&A, 33, 541
3. Birkinshaw, M. 1999, Physics Report, 310, 97
4. Sarazin, C.L. 1988, 'X-ray emission from clusters of galaxies', Cambridge University Press
5. Blasi, P. Stebbins, A. and Olinto, A. 2000, ApJ, 535, L71
6. Petrosian, V. 2001, ApJ, in press (preprint astro-ph/0101145)
7. Colafrancesco, S., Marchegiani, P. and Palladino, E. 2001, A&A, submitted
8. Ensslin, T. and Kaiser, C. 2000, A&A, 360, 417
9. Cavaliere, A. and Fusco-Femiano, R. 1976, A&A, 49, 137

Utilization of a Center for Cosmic Structure To Stimulate Undergraduate Education-YCOOP

Martin S. Spergel

Department of Natural Sciences
York College of the City University of New York
Jamaica, New York, USA

Abstract. A program titled YCOOP* (York College Observatory Outreach Program) has joined with a consortium of universities with the goal of utilizing the research work product of the consortium's planned Center for Cosmic Structure to further undergraduate and secondary education.

YORK COLLEGE OBSERVATORY OUTREACH FOR EDUCATION

YCOOP in coordination with a consortium of institutions (see figure 1) has been developing a program (see Figure 2) to stimulate education in physics and earth science. This program will utilize both the Center's research work product and the work experiences provided by the proposed experimental resource, the planned Atacama telescope (see Figure 3).

The program will reach its goals by use of: web presentations; teacher training institutes; development and distribution of curricula materials; demonstration classes; and individual scientist mentoring students. The mentored students will be involved in systemic reduction of data as well as in the solution of problems associated with research projects in cosmology.

This NASA-OSS funded project* is providing support for a broad program to upgrade science education in Jamaica and other inner city areas. This program is developing educational infrastructure at York College of the City University of New York; at John Adams High School of the Board of Education of the City of New York; training in-service secondary school science teachers and establishing linkages with educational and NASA research institutions. Specifically, this program is developing at York College: minors in Astronomy, new courses in Astronomy, and is revising courses aimed at majors in Physics, physical and biological sciences, teacher education and general education majors. The NASA-OSS grant funding has strengthened the college's physics and astronomy program by the addition of a new faculty line. The grant has stimulated cooperative programs: with Princeton University that have offered research opportunities for York College faculty and undergraduates; with John Adams High School's program in Earth and Space Sciences, which is being upgraded through curriculum and high school faculty development. An in-service teacher-training program for secondary school science faculty is under preparation. Cooperative programs with another City University of New York-NASA/OSS grant programs have lead to strengthening of the university's program in space sciences. A collaboration has been established with the York College SEMAA (Science, Engineering, Mathematics and Aerospace Academy), a NASA sponsored, enrichment program for underrepresented children in 4-8th grades.

YCOOP is collaborating with Professors Bohdan Paczynski and David N. Spergel of Princeton University in developing student research projects. Professor Paczynski has been involved in shaping the output of the All Sky Automated Survey[2] is a project which final goal is photometric monitoring of approx. 10^7 stars brighter than 14 magnitude all over the sky for undergraduates. Professor David Spergel has been involved in introducing YCOOP undergraduates to cosmology and the MAP[3] satellite project.

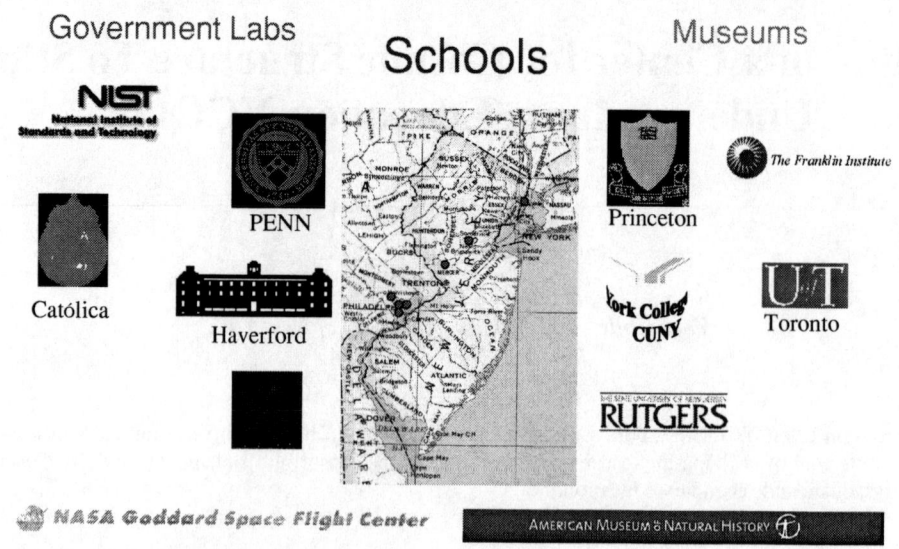

FIGURE 1. Consortium of institutions united through research, education and public outreach

Goals of the York College Observatory Outreach Progr.

1) Improve the Infrastructure of Science Education at York

   Revise Physics Major Curriculum introduce Cosmology and Astronomy as an incentive
   - Thermodynamics using CMB and stellar spectra
   - Mechanic using Satellite & Planetary motion and observation
   - Modern Physics using CMB and stellar spectral modeling

   Recruit Undergraduate research students from Physics, Science and Science Ed pools

2) Improve High School Physics Education in Inner City Minority Areas.
   - Visit HS and MS
   - Establish Observatory Hours for PS and Public Secondary Ed.
   - Retraining program for In-Service Science Teacher and those interested in obtaining their Secondary Science Ed Physics license

FIGURE 2. Education and Outreach goals of YCOOP.

ACT - Atacama Cosmology Telescope

- 6 Meter Aperture
- Low Cost - $3M
- Low Ground Pickup
- No Moving Optics
- Remote Controlled
- Legacy Telescope
- Flexible Focal Plane
- Near the ALMA Site

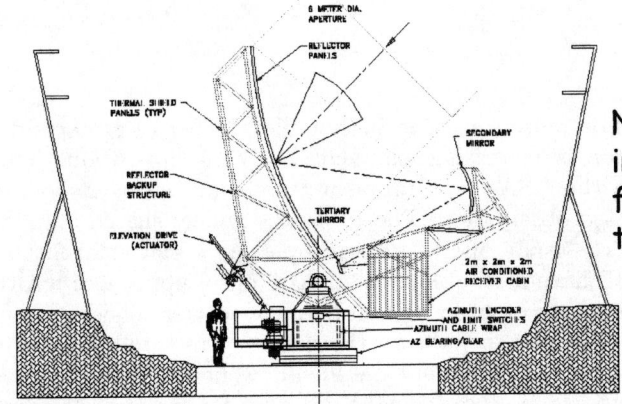

No existing telescope incorporates the features required for these measurements.

FIGURE 3. Planned principal experimental research resource of the Center for Cosmic Structure[1]

ACKNOWLEDGMENTS

*YCOOP is support in part by the Minority Institutions Division of the Office of Space Sciences of NASA, grant # NAG5 10152. We appreciate Professor Mark Devlin of the University of Pennsylvania for allowing for the use of Figure 3[1].

REFERENCES

1. Devlin, M.,"Millimeter and Submillimenter Observations from the Atacama Plateau and High Balloons", Proceeding of the 2K1BC Workshop - Experimental Cosmology @ mm-waves" 2001 this volume,
2. Paczynski,B., "The All Sky Automated Survey", Acta Astronomica 1997,47,467
3. Park,C.,Colley,W.N.,Gott,J.R.,Ratra,B.,Spergel,D.N.,&Sugiyama,N.,"Cosmic Microwave Background Anisotropy Correlation Function and Topology: Simulated Maps for MAP",Ap.J.,495,511 (1998)

CONCLUDING REMARKS

Francesco Melchiorri

Dipartimento di Fisica, Università "La Sapienza", P.le A. Moro,2 I-00185 Roma, Italy

The first impression derived from this very well organized meeting is that the number of projects devoted to the study of CMB anisotropies and polarization is increasing exponentially with time.About ten years ago, immediately after COBE,very few groups in the USA and Europe were still proposing balloon borne experiments; even less were involved in ground based observations.There was the widely shared opinion that those types of observations were not adequate for the study of CMB anisotropies.As a matter of fact several balloon borne experiments carried out before COBE have been able to estabilish (only) upper limits.Although these limits were stringent enough to force theoreticians to propose non-barionic matter models,the success of COBE in producing the first map of CMB sky seemed to indicate the necessity of satellite experiments in order to measure CMB anisotropies.Today we know that balloon experiments went very close to detecting anisotropies (it may be that some of them observed true cosmological signals,see for instance Melchiorri et al,92) while COBE was very close to failing (for instance,see Wilkinson 94).Both bolometers and radioreceivers were much noisier than today.In fact,detectors have shown a tremendous improvement in sensitivity during the past ten years, especially in the infrared and radio regions.The recently performed balloon-borne experiment BOOMERANG had a sensitivity of about 100 microKelvin per second of integration,about 200 times higher than that of the COBE satellite. As a consequence,the various systematics present in ground based and balloon experiments were immersed in the noise rendering the results unreliable. COBE did not improve the sensitivity of the detectors,but took advantage of the fact that satellite experiments are less prone to systematics. After COBE both detectors improved and long duration balloon flights have provided enough redundance of data for testing and removing systematics. BOOMERANG and MAXIMA have shown that it is possible to measure the power spectrum of CMB anisotropies with enough sensitivity to identify the first ,the second and (may be) the third acoustic peak.A new class of instruments (interferometers with a short baseline) has proved that the same job can be performed from ground.The beautiful DASI experiment has measured the power spectrum of CMB anisotropies from the South Pole. Another program called VSA is planning to provide similar results from a more comfortable place like Tenerife . This being the situation the choice of a good site for millimetric observations is again a central issue for CMB observations.Apart from the rare cases when atmospheric noise becomes lower than detector noise ,the best way of taking care of atmospheric fluctuations consists in subtracting them from the signals recorded at a "sensitive" frequency ν_1 by means of simultaneous observations at a higher "atmospheric"frequency ν_2. If we average the scans where the correlation among channels 1 and 2 was greater than a certain value C we get a residual noise

$$Noise^2 = \frac{DN^2 + 1/3 S_A^2(1-C)^3}{N_0(1-C)^\alpha}$$

Were DN is the noise of detector 1,S_A is the atmospheric noise in channel 1 and the number of scans with a correlation larger than C is fitted by empirical law $N = No(1-C)^\alpha$ .The best value for C is given by

$$1 - C = \left(\frac{DN}{S_A}\right)^{2/3}\left(\frac{\alpha}{3-\alpha}\right)^{1/3}$$

Therefore a site is fully characterized by the amplitude of S_A with respect to the noise of the detector and by the parameter α .The first quantity may be better specified if we assume that our detector is background limited i.e. its noise is equal to the quantum noise of a gray body with the same temperature and transmittance of the atmosphere:the quantity $W = DN/S_A$ is equal to 1 in the case of an ideal atmosphere ,while it is less than one if atmospheric fluctuations exceed the quantum noise.A plot in the plane W,α of the values characteristic of a site will allow compariing them quantitatively. I think that a comparison of this type will help all those planning ground based observations. A slightly more obscure situation is that relative to CMB polarization studies: while it appears possible to measure the degree of polarization by means of ground based experiments, many groups have proposed balloon-borne and space experiments.The first of these proposals which should provide results will be BOOMERANG-POL and/or MAP. High spatial resolution is needed to study the distribution of polarization around CMB bumps and this will be the goal of some ground -based experiments.

After this brief analysis of the new waves in Big Bang Cosmology let me add some less positive comments. It is possible to read in several articles the very exciting phrase "BOOMERANG and MAXIMA have opened the new era of Precision Cosmology".I think that this statement is only partially true: as the noise of detectors decreases it is not longer the limiting factor for accuracy.The problem of calibration now predominates. As shown by the analysis done by Therese Encrenaz,the brightness temperatures of planets are known in the millimetric region with an accuracy no better than 10with great accuracy,thanks to the precise measurements of COBE.Not all the experiments can use this source of calibration,however. BOOMERANG had problems due to the presence of the Sun ; even the MAXIMA calibration could have been contaminated by residual atmospheric emission (there is a well known North-South temperature gradient in the upper atmosphere,as it was observed in the Ulisse program,the first balloon experiment which employed the Dipole Anisotropy as calibrator).I think that experimental cosmologists should consider the possibility of planning a small satellite with several standard sources on board,in order to illuminate ground based and balloon borne experiments with well known powers.

I would like to conclude my remarks by noting that the choice of the name of Herschel for the ESA submillimeter satellite (FIRST was the previous name) appears not well deserved.Herschel was convinced the the radiant heat he observed towards the Sun was different from light and emitted only by the Sun (other celestial bodies being "cold"). He finally proposed that the Sun itself was cold and the observed radiant heat was produced by the fires caused by the inhabitants of our star! Almost at the same time as Herschel, Macedonio Melloni proved that radiant heat propagates with the speed of light, measured it quantitatively with the first infrared detector, the Thermopile, invented by himself. By using a thermopile and a NaCl lens Melloni measured the "ultrared" emitted by the Moon, thereby disproving Herschel's theory.He also studied the sky background around one micron by measuring it at various elevations and discovering the secant law for atmospheric emission. He concluded that the sky background at one micron should correspond to the integrated emission of about ten billions stars.In the French encyclopedia of Science , Melloni was quoted as the "Newton of Heat".The predominance of the anglo-saxon culture combined with the carelessness of the Italian delegation at ESA are the causes for the choice of the name of Herschel instead of Melloni.

I REFERENCES

1.B.Melchiorri,F.Melchiorri;"*Comparison between COBE Preliminary Results and Ulisse observations*", Current Topics in Astrofundamental Physics,Eds.N.Sanchez,A.Zichichi, World Scientific,1992,pg.148.

2.D.T.Wilkinson;"*A worning label for Cosmic Background Anisotropy Experiments*", Lake Louise Winter Institute Lecture ,February 1994

LIST OF PARTICIPANTS

Ali Shafinaz
University of Wales, Cardiff – 5 The Parade – CF24 3YB Cardiff – Wales – UNITED KINGDOM
Phone +44 02920875106 Fax +44 02920874806 - URL: www.astro.cf.ac.uk/groups/instrumentation - alis@cf.ac.uk

Antoniucci Simone
Dipartimento di Fisica – Universita' La Sapienza – P.le A. Moro, 2 – 00185 Rome – ITALY
Phone +39 06 49914201 Fax +39 06 4957697 - s_antoniucci@hotmail.com

Antonucci Federica
Dipartimento di Fisica – Universita' La Sapienza – P.le A. Moro, 2 – 00185 Rome – ITALY
Phone +39 06 49914201 Fax +39 06 4957697 - fede.antonucci@tiscalinet.it

Aretxaga Itziar
Instituto Nacional de Astrofisica, Optica y Electronica - Luis Enrique Erro, #1, Tonantzintla - 72000 Puebla – MEXICO
Phone +52 266 3100 x.2316 Fax +52 2 247 2231 - itziar@inaoep.mx

Battistelli Elia
Dipartimento di Fisica G.Occhialini – Universita' di Milano Bicocca - Piazza della Scienza 3 – 20126 Milan - ITALY
Phone +39 02 64482827 Fax +39 02 64482888 – elia.battistelli@mib.infn.it

Benoit Alain
CNRS-CRTBT – 25 av. des martyrs – 38042 Grenoble - FRANCE
Phone +33 047 6889072 Fax +33 047 6875060 - benoit@polycnrs-gre.fr

Bocci Alessio
Dipartimento di Fisica – Universita' La Sapienza – P.le A. Moro, 2 – 00185 Rome – ITALY
Phone +39 06 49914201 Fax +39 06 4957697 - alessio.bocci@tin.it

Bock James
Jet Propulsion Laboratory – 91109 Pasadena, CA - USA
Phone (818)-354-0715 Fax (818)-354-8895 - jjb@astro.caltech.edu

Boscaleri Andrea
CNR-IROE - Via Panciatichi 64 – 50127 Florence - ITALY
Phone +39 055 4235 245 Fax +39 055 4235 245 - URL: www.iroe.fi.cnr.it - aboscale@iroe.fi.cnr.it

Bujakas Victor
Astro Space Center of P.N.Lebedev Physical Institute, Profsoiuznaya st.84/32, Moscow, 117810, RUSSIA
bujakas@ asc.rssi.ru

Candidi Maurizio
IFSI-CNR – Via Fosso del Cavaliere, 100 – 00133 Rome – ITALY
Phone +39 06 49934490 Fax +39 06 49934374 – URL: www.ifsi.cnr.it - candidi@ifsi.rm.cnr.it

Carretti Ettore
Istituto TeSRE – Via P. Gobetti, 101 – 40129 Bologna – ITALY
Phone +39 051 6398735 Fax +39 051 6398741 – carretti@tesre.bo.cnr.it

Cartwright John
California Institute of Technology MS 105-24, Caltech – 91125 Pasadena, CA – USA
Phone 626-395-4121 Fax 626-568-9352 – URL: www.astro.caltech.edu/~tjp/CBI - jkc@astro.caltech.edu

Cavaliere Francesco
Università degli Studi di Milano - Via Celoria 16 – 20133 Milan - ITALY
Phone +39 02 58357360 Fax +39 02 58357337 - Francesco.cavaliere@mi.infn.it

Chapin Edward
Instituto Nacional de Astrofisica, Optica y Electronica - Luis Enrique Erro, #1, Tonantzintla - 72000 Puebla – MEXICO
Phone +52 22472011 Fax +52 22472231 - echapin@inaoep.mx

Chouvaev Denis
Chalmers University of Technology - SE-412 96 Göteborg, SWEDEN
Phone +46 31 7725173 Fax +46 31 7723442 - chouvaev@fy.chalmers.se

Colafrancesco Sergio
Osservatorio Astronomico di Roma, Via Frascati, 33 – 00171 Monteporzio - ITALY
Phone +39 06 94286418 Fax +39 06 9447243 – URL: www.mporzio.astro.it/~cola/ - cola@coma.mporzio.astro.it

Cortiglioni Stefano
Istituto TeSRE – Via P. Gobetti, 101 – 40129 Bologna – ITALY
Phone +39 051 6398703 Fax +39 051 6398741 – cortiglioni@tesre.bo.cnr.it

D'Alba Livia
   Dipartimento di Fisica – Universita' La Sapienza – P.le A. Moro, 2 – 00185 Rome – ITALY
   Phone +39 06 49914201 Fax +39 06 4453397 – livia.dalba@roma1.infn.it

de Bernardis Paolo
   Dipartimento di Fisica – Universita' La Sapienza – P.le A. Moro, 2 – 00185 Rome – ITALY
   Phone +39 06 49914271 Fax +39 06 4453397 – paolo.debernardis@roma1.infn.it

De Grazia Moira
   Dipartimento di Fisica – Universita' La Sapienza – P.le A. Moro, 2 – 00185 Rome – ITALY
   Phone +39 06 49914201 Fax +39 06 4957697 - moira.degrazia@roma1.infn.it

De Petris Marco
   Dipartimento di Fisica – Universita' La Sapienza – P.le A. Moro, 2 – 00185 Rome – ITALY
   Phone +39 06 49914690 Fax +39 06 49914690 – marco.depetris@roma1.infn.it

Désert F.-Xavier
   Laboratoire d'Astrophysique, Observatoire de Grenoble - 414 rue de la piscine – 38041 Grenoble Cedex 9 – FRANCE
   Phone +33 4 76635512 Fax +33 4 76448821
   URL: www-laog.obs.ujf-grenoble.fr/~desert – Francois-Xavier.Desert@obs.ujf-grenoble.fr

De Troia Grazia
   Dipartimento di Fisica – Universita' La Sapienza – P.le A. Moro, 2 – 00185 Rome – ITALY
   Phone +39 06 49914201 Fax +39 06 4453397 – grazia.detroia@roma1.infn.it

Devlin Mark
   University of Pennsylvania - 209 South 33rd St. – 19104 Philadelphia PA - USA
   Phone 215-573-7521 Fax 215-898-2010 – URL: www.hep.upenn.edu/CBR - devlin@physics.upenn.edu

Di Stefano Giuseppe
   INGV - Via di Vigna Murata 605 – 00143 Rome - ITALY
   Phone +39 065180305 - DiStefano@ingv.it

Dupac Xavier
   CESR - 9, avenue du Colonel Roche, BP 4346 - 31028 Toulouse cedex 4 – FRANCE
   Phone +33 5 6155 8190 - dupac@cesr.fr

Encrenaz Pierre
   Observatoire de Paris - 61, avenue de l'Observatoire - 75014 Paris – FRANCE
   Phone +33 1 40512006 – pierre.encrenaz@obspm.fr

Encrenaz Therese
   Observatoire de Paris sec. de Meudon – Dèpartement de Recherche Spatiale – 5, place J. Janssen, 92190 Meudon - FRANCE
   Phone +33 1 45077691 Fax +33 1 45072806 - Therese.Encrenaz@obspm.fr

Gear Walter
   University of Wales, Cardiff – Box 913 - 5 The Parade – CF24 3YB Cardiff – Wales – UNITED KINGDOM
   Phone +44 02920875526 - gear@cf.ac.uk

Gervasi Massimo
   Dipartimento di Fisica G.Occhialini – Universita' di Milano Bicocca - Piazza della Scienza 3 – 20126 Milan – ITALY
   Phone +39 02 644828830 Fax +39 02 64482888 – massimo.gervasi@mib.infn.it

Giard Martin
   CESR-CNRS - 9 Ave du Colonel Roche, BP 4346 – 31028 Toulouse cedex 04 - FRANCE
   Phone 33 5 61 55 66 48 Fax 33 5 61 55 67 01- giard@cesr.fr

Gleeson Emily
   National University of Ireland Maynooth – Maynooth - IRELAND
   Phone +35350424669 - emily.m.gleeson@may.ie - emily.gleeson@ireland.com

Gromov Vladimir
   Astro Space Center of LPI RAN - Profsoyuznaya 84/32 – 117810 Moscow - RUSSIA
   Phone 7 095 3334088 Fax 7 095 3332378 – URL:http://fy.chalmers.se/~f4agro/Submillimetron/ - f4agro@chalmers.se

Hampai Dariush
   Dipartimento di Fisica – Universita' La Sapienza – P.le A. Moro, 2 – 00185 Rome – ITALY
   Phone +39 06 49914201 Fax +39 06 4957697 - belanner@tiscalinet.it

Hennessy Richard
   National University of Ireland Maynooth – Maynooth - IRELAND
   Phone +35350424669 - richard.g.hennessy@may.ie

Holler Christian
   Astrophysics Group, Cavendish Lab., Cambridge University -Madingley Road –CB3 0EG Cambridge –UNITED KINGDOM
   Phone +44-1223-337231 - cmh54@cam.ac.uk

Hristov Viktor
   California Institute of Technology - CalTech, Observational Cosmology, MS 59-33 – 91125 Pasadena – California - USA
   Phone 626 3953184 Fax 626 5849929 - URL:www.astro.caltech.edu/~lgg/ - vvh@astro.caltech.edu

Hughes David
 Instituto Nacional de Astrofisica, Optica y Electronica - Luis Enrique Erro, #1, Tonantzintla - 72000 Puebla – MEXICO
 Phone +52 22663100 (extension x.2305) Fax +52 22472231 - dhughes@xocoatl.inaoep.mx

Jones Michael
 Cavendish Laboratory, University of Cambridge - Madingley Road - CB3 0HE Cambridge – UNITED KINGDOM
 Phone +44 1223 337363 Fax +44 1223 354599 - mike@mrao.cam.ac.uk

Keating Brian
 California Institute of Technology, CALTECH - Mail Code 59-33 – 91125 Pasadena – California - USA
 Phone 626-395-8002 Fax 626-584-9929 - bgk@astro.caltech.edu

Kreysa Ernst
 Max-Planck-Institut fuer Radioastronomie - Auf dem Huegel 69 – 53121 Bonn - GERMANY
 Phone +49 228525269 Fax +49 228525229 - ekreysa@mpifr-bonn.mpg.de - kreysa@t-online.de

Lamagna Luca
 Dipartimento di Fisica – Universita' La Sapienza – P.le A. Moro, 2 – 00185 Rome – ITALY
 Phone +39 06 49914201 Fax +39 06 49914690 – URL: oberon.roma1.infn.it - luca.lamagna@roma1.infn.it

Lamarre Jean-Michel
 Istitut d'Astrophysique Spatiale - bat 121, Universite Paris-Sud - 91405, Orsay cedex - FRANCE
 Phone +33 1 69858577 Fax +33 1 69858675 - URL: www.ias.u-psud.fr - lamarre@ias.fr - jean-michel.lamarre@obspm.fr

Laurenza Monica
 Dipartimento di Fisica – Universita' La Sapienza – P.le A. Moro, 2 – 00185 Rome – ITALY
 Phone +39 06 49914201 Fax +39 06 4957697 - m.laurenza@tiscalinet.it

Leitch Erik
 University of Chicago - 5640 S. Ellis Ave. – 60637 Chicago – Illinois - USA
 Phone 773-702-7795 Fax 773-834-1891 - eml@polestar.uchicago.edu

Lucente Marco
 Dipartimento di Fisica – Universita' La Sapienza – P.le A. Moro, 2 – 00185 Rome – ITALY
 Phone +39 06 49914201 Fax +39 06 4957697 - marco.lucente@roma1.infn.it

Luzzi Gemma
 Dipartimento di Fisica – Universita' La Sapienza – P.le A. Moro, 2 – 00185 Rome – ITALY
 Phone +39 06 49914201 Fax +39 06 4957697 - gemma.luzzi@roma1.infn.it

Macculi Claudio
 Istituto TeSRE – Via P. Gobetti, 101 – 40129 Bologna – ITALY
 Phone +39 051 6398668 Fax +39 051 6398724 – macculi@tesre.bo.cnr.it

Mainella Giovanni
 THEMIS c/o Instituto de Astrofisica de Canarias , calle via Lactea s/n - La Laguna – 38200 Tenerife , Canarias - SPAIN
 Phone +34-619-472597 Fax +34-922-605194 - mainella@themis.iac.es

Mandolesi Nazzareno
 Istituto TeSRE – Via P. Gobetti, 101 – 40129 Bologna – ITALY
 Phone +39 051 6398668 Fax +39 051 6398724 – reno@tesre.bo.cnr.it

Maoli Roberto
 Dipartimento di Fisica – Universita' La Sapienza – P.le A. Moro, 2 – 00185 Rome – ITALY
 Phone +39 06 49914457 Fax +39 06 49914690 - roberto.maoli@roma1.infn.it

Masi Silvia
 Dipartimento di Fisica – Universita' La Sapienza – P.le A. Moro, 2 – 00185 Rome – ITALY
 Phone +39 06 49914690 Fax +39 06 49914690 - silvia.masi@roma1.infn.it

Mauskopf Philip
 University of Wales, Cardiff – Box 913 - 5 The Parade – CF24 3YB Cardiff – Wales – UNITED KINGDOM
 Phone +44 02920876170 Fax +44 02920874056 - URL: www.physics.cf.ac.uk - mauskopf@astro.cf.ac.uk

Melchiorri Bianca
 Dipartimento di Fisica – Universita' La Sapienza – P.le A. Moro, 2 – 00185 Rome – ITALY
 Phone +39 06 4454937 Fax +39 06 4453397

Melchiorri Francesco
 Dipartimento di Fisica – Universita' La Sapienza – P.le A. Moro, 2 – 00185 Rome – ITALY
 Phone +39 06 4454937 Fax +39 06 4453397 - francesco.melchiorri@roma1.infn.it

Melhuish Simon
 University of Wales, Cardiff – Box 913 - 5 The Parade – CF24 3YB Cardiff – Wales – UNITED KINGDOM
 Phone +44 02920875106 Fax +44 02920874056 - URL:www.physics.cf.ac.uk/groups/ - Simon.Melhuish@astro. cf.ac.uk

Mennella Aniello
 IFC-CNR - Via Bassini 15 – 20133 Milan - ITALY
 Phone+39-02-23699461 Fax+39-02-2666017 - URL: www.ifctr.mi.cnr.it - daniele@ifctr.mi.cnr.it

Miller Paul
 QMC Instruments Limited - Mile End Road, E1 4NS London - UNITED KINGDOM
 Phone +44 020 8980 1288 Fax +44 020 8981 8337 - p.miller@qmw.ac.uk

Morgante Gianluca
 Istituto TeSRE – Via P. Gobetti, 101 – 40129 Bologna – ITALY
 Phone +39 051 6398668 Fax +39 051 6398724 – morgante@tesre.bo.cnr.it

Murphy J. Anthony
 Dept. of Experimental Physics, National University of Ireland Maynooth – Maynooth - IRELAND
 Phone +353 1 7083771 Fax +353 1 7083313 – URL: www.may.ie - anthony.murphy@may.ie

Natale Vincenzo
 CAISMI-CNR - L.go E. Fermi 5 – 50125 Florence - ITALY
 Phone +39 0552752218 Fax +39 0552752287 - URL: arcetri.astro.it - natale@arcetri.astro.it

Nati Federico
 Dipartimento di Fisica – Universita' La Sapienza – P.le A. Moro, 2 – 00185 Rome – ITALY
 Phone +39 06 49914690 Fax +39 06 49914690 - federico.nati@roma1.infn.it

Orlando Angiola
 Dipartimento di Fisica – Universita' La Sapienza – P.le A. Moro, 2 – 00185 Rome – ITALY
 Phone +39 06 49914690 Fax +39 06 49914690 - angiola.orlando@roma1.infn.it

O'Sullivan Creidhe
 National University of Ireland Maynooth – Maynooth - IRELAND
 Phone +353 1 7083953 Fax +353 1 7083313 – URL: www.may.ie - creidhe.osullivan@may.ie

Palladino Emilia
 Dipartimento di Fisica – Universita' La Sapienza – P.le A. Moro, 2 – 00185 Rome – ITALY
 Phone +39 06 49914201 Fax +39 06 49914690 - emilia.palladino@roma1.infn.it

Pascale Enzo
 IROE / CNR - Via Panciatichi, 64 – 50127 Florence – ITALY
 Phone +39 0554235245 Fax +39 0554235204 - pascale@iroe.fi.cnr.it

Piacentini Francesco
 Dipartimento di Fisica – Universita' La Sapienza – P.le A. Moro, 2 – 00185 Rome – ITALY
 Phone +39 06 49914690 Fax +39 06 49914690 - francesco.piacentini@roma1.infn.it

Piccirillo Lucio
 University of Wales, Cardiff – Box 913 - 5 The Parade – CF24 3YB Cardiff – Wales – UNITED KINGDOM
 Phone +44 02920874056 Fax +44 02920875031 – lucio.piccirillo@astro.cf.ac.uk

Puy Denis
 Institute of Theoretical Physics, University of Zurich - Winterthurerstrasse, 190 - 8057 Zurich - SWITZERLAND
 Phone +41 1 6355820 Fax +41 1 6355704 - puy@physik.unizh.ch

Raccanelli Andrea
 Max-Planck-Institut fuer Radioastronomie - Auf dem Huegel 69 – 53121 Bonn - GERMANY
 Phone +49 228525270 Fax +49 228525229 - araccan@mpifr-bonn.mpg.de

Reichertz Lothar
 Max-Planck-Institut fuer Radioastronomie - Auf dem Huegel 69 – 53121 Bonn - GERMANY
 Phone +49 228525270 Fax +49 228525229 - reichertz@mpifr-bonn.mpg.de

Rephaeli Yoel
 School of Physics & Astronomy - Tel Aviv University – 69978 Tel Aviv - ISRAEL
 Phone +972 3 6407809 Fax +972 3 6405141 - yoelr@noga.tau.ac.il - yoelr@mamacass.ucsd.edu

Reveret Vincent
 CEA/Saclay, Service d'Astrophysique - Bat. 709, Orme des Merisiers – 91191 Gif Sur Yvette - FRANCE
 Phone 0033 169088056 Fax 0033 169086577 - reveret@discovery.saclay.cea.fr

Richards Paul L.
 Department of Physics, University of California – 94708 7300 Berkeley – California - USA
 Phone 510 6423027 Fax 510 6435204 - richards@physics.berkeley.edu

Romeo Giovanni
 INGV - Via di Vigna Murata 605 – 00143 Rome - ITALY
 Phone +39 065180305 - Romeo@ingv.it - Romeo.g@tiscalinet.it

Rossinot Philippe
 University of Wales, Cardiff – Box 913 - 5 The Parade – CF24 3YB Cardiff – Wales – UNITED KINGDOM
 Phone +44 02920875106 Fax +44 02920874056 – Philippe.Rossinot@astro.cf.ac.uk

Salimbeni Sara
 Dipartimento di Fisica – Universita' La Sapienza – P.le A. Moro, 2 – 00185 Rome – ITALY
 Phone +39 06 49914201 Fax +39 06 4957697 - bandasara@tiscalinet.it

Sandri Maura
    Istituto TeSRE – Via P. Gobetti, 101 – 40129 Bologna – ITALY
    Phone +39 051 6398705 Fax +39 051 6398724 – sandri@tesre.bo.cnr.it

Savini Giorgio
    Dipartimento di Fisica – Universita' La Sapienza – P.le A. Moro, 2 – 00185 Rome – ITALY
    Phone +39 06 49914201 Fax +39 06 49914690 - giorgio.savini@roma1.infn.it

Sbarra Carla
    Istituto TeSRE – Via P. Gobetti, 101 – 40129 Bologna – ITALY
    Phone +39 051 6398668 Fax +39 051 6398741 – sbarra@tesre.bo.cnr.it - Carla.Sbarra@cern.ch

Signore Monique
    DEMIRM, Observatoire de Paris – 61, Avenue de l'Observatoire – 75014 Paris - FRANCE
    Phone +33 1 40512112 Fax +33 1 40512002 - Monique.Signore@obspm.fr

Silverberg Robert
    NASA/Goddard Space Flight Center – Code 685 – 20771 Greenbelt – MD USA
    Phone +301 2867468 Fax +301 2861617 – Robert.Silverberg@gsfc.nasa.gov

Siringo Giorgio
    Max-Planck-Institut fuer Radioastronomie - Auf dem Huegel 69 – 53121 Bonn - GERMANY
    Phone +49 228525270 Fax +49 228525229 - gsiringo@mpifr-bonn.mpg.de

Sironi Giorgio
    Dipartimento di Fisica G.Occhialini – Universita' di Milano Bicocca - Piazza della Scienza 3 – 20126 Milan - ITALY
    Phone +39 02 644828830 Fax +39 02 64482888 - giorgio.sironi@mib.infn.it

Soglasnova Vera
    Space Research Institut RAN - Profsouznaya 84/32 - 117810 GSP-7 Moscow – RUSSIA
    Phone 333-23-67 Fax 333-12-48 - vera@mx.iki.rssi.ru

Spergel Martin
    York College of the City University of New York - Guy Brewer Bl'vd – Jamaica –11451 New York - USA
    Phone 718 262 2650 Fax 718 262 2652 – URL:natsci.york.cuny.edu/~cosmic/ -spergel@york.cuny.edu

Stark Antony
    Smithsonian Astrophysical Observatory - 60 Garden St. MS 78 – 02138 Cambridge – MA USA
    Phone 617-496-7648 Fax 617-496-7554 - URL:cfa-www.harvard.edu/~aas -aas@cfa.harvard.edu

Stringfellow Guy
    NASA Headquarters - Code SR, 300 E St SW – 20546 Washington DC - USA
    Phone 202 3580311 Fax 202 3583097 - Guy.Stringfellow@hq.nasa.gov

Subrahmanyan Ravi
    Australia Telescope National Facility (CSIRO) - Locked bag 194 – 2390 Narrabri - NSW 2390 - AUSTRALIA
    Phone +61 2 67904074 Fax +61 2 67904090 - rsubrahm@atnf.csiro.au

Taccetti Quintilio
    INGV - Via di Vigna Murata 605 – 00143 Rome - ITALY
    Phone +39 065180305 - Taccetti@ingv.it

Tartari Andrea
    Dipartimento di Fisica G.Occhialini – Universita' di Milano Bicocca - Piazza della Scienza 3 – 20126 Milan - ITALY
    Phone +39 02 64482826 Fax +39 02 64482888 - andrea.tartar@libero.it

Tascone Riccardo
    IRITI-CNR c/o Politecnico di Torino - Corso Duca degli Abruzzi, 24 – 10129 Turin - ITALY
    Phone +39 011 564 4061 Fax +39 011 564 4089 - tascone@polito.it

Taylor Angela
    Cavendish Astrophysics, University of Cambridge - Madingley Road – CB3 0HE Cambridge – UNITED KINGDOM
    Phone +44 1223 337304 Fax +44 1223 354599 - act21@cam.ac.uk

Terenzi Luca
    Istituto TeSRE – Via P. Gobetti, 101 – 40129 Bologna – ITALY
    Phone +39 051 6398733 Fax +39 051 6398724 – lteren@tesre.bo.cnr.it

Tofani Gianni
    CAISMI-CNR/Osservatorio Astrofisico di Arcetri - L.go E. Fermi 5 – 50125 Florence - ITALY
    Phone +39 0552752217 Fax +39 0552752287 - URL: arcetri.astro.it - tofani@arcetri.astro.it

Valenziano Luca
    Istituto TeSRE – Via P. Gobetti, 101 – 40129 Bologna – ITALY
    Phone +39 051 6398700 Fax +39 051 6398724 – valenziano@tesre.bo.cnr.it

Villa Fabrizio
    Istituto TeSRE – Via P. Gobetti, 101 – 40129 Bologna – ITALY
    Phone +39 051 6398733 Fax +39 051 6398724 – URL:tonno.tesre.bo.cnr.it/~villa/ - villa@tesre.bo.cnr.it

Wilson Grant
   Univ. of Massachussets - Dept. of Astronomy - 619E LGRT-B 710 North Pleasant St. – 01003-9305 Amherst – MA USA
   Phone 1 773 7022241 Fax 1 425 9628575 - wilson@oddjob.uchicago.edu

Withington Stafford
   Cavendish Laboratory,Dept. Physics, Univ. of Cambridge - Madingley Road - CB3 OHE Cambridge – UNITED KINGDOM
   Phone +44 1223 337393 Fax +44 1223 354599 - stafford@mrao.cam.ac.uk

Yurchenko Volodymyr
   Dept. of Experimental Physics, National University of Ireland Maynooth – Maynooth - IRELAND
   Phone +353 1 7083746 Fax +353 1 7083313 – v.yurchenko@may.ie

Zannoni Mario
   Dipartimento di Fisica G.Occhialini – Universita' di Milano Bicocca - Piazza della Scienza 3 – 20126 Milan - ITALY
   Phone +39 02 64482828 Fax +39 02 64482888 – mario.zannoni@mib.infn.it

Zeppilli Andrea
   Dipartimento di Fisica – Universita' La Sapienza – P.le A. Moro, 2 – 00185 Rome – ITALY
   Phone +39 06 49914201 Fax +39 06 4957697 - andrea.zeppilli@roma1.infn.it

AUTHOR INDEX

A

Abroe, M., 12
Ade, F., 39
Ade, P. A. R., 3, 12, 18, 107, 126, 213, 282
Aghanim, N., 116
Agnese, P., 270
Aguirre, J., 23
Ali, S., 126
Andre, P., 270
Andreone, D., 123
Aretxaga, I., 322, 354
Atad-Ettedgui, E., 290

B

Balbi, A., 12
Baralis, M., 140, 145, 150
Barber, D., 298
Battistelli, E. S., 92, 123, 164
Bava, E., 123
Benoit, A., 31, 116, 213
Bernard, J.-P., 116
Bernardi, G., 140, 145
Bersanelli, M., 193, 219, 224, 229, 242, 245
Bezaire, J., 23
Bhandari, P., 298
Bhatia, R., 213
Bock, J. J., 3, 12, 18, 39, 107, 213, 251
Boella, G., 123, 140, 145, 164
Bond, J. J., 39
Bond, J. R., 3
Bonometto, S., 140, 145
Borrill, J., 3, 12, 39
Boscaleri, A., 3, 12, 18, 39, 56, 140, 145
Bouchet, F. R., 213
Bowman, R. C., 298
Brunetti, L., 123
Bruscoli, M., 140, 145
Bujakas, V. I., 239
Burigana, C., 193, 224, 229, 242, 245
Butler, C., 229
Butler, R. C., 193, 219, 224, 242, 245

C

Camus, P., 116
Carlstrom, J. E., 65
Carretti, E., 140, 145, 150
Cartwright, J. K., 135
Cavaliere, F., 123, 164

Cecchini, S., 140, 145
Chapin, E., 322, 354, 357
Cheng, E. S., 23
Christensen, P. R., 23
Church, S. E., 295
Clarke, J., 259
Coble, K., 3, 39
Colafrancesco, S., 316, 363
Colgan, R., 282, 295
Collins, J., 12
Contaldi, C. R., 3, 39
Cordone, S., 23
Coron, N., 116
Cortiglioni, S., 140, 145, 150
Cottingham, D. A., 23
Cowgill, P., 298
Crawford, T., 23
Crill, B. P., 3, 18, 39
Crumb, D., 298
Cuttaia, F., 219

D

Dall'Oglio, G., 183
de Bernardis, P., 3, 12, 18, 39, 52, 59, 168, 213
De Gasperis, G., 3
De Grazia, M., 92
Delabrouille, J., 116
de Oliveira-Costa, A., 175
De Petris, M., 92, 123
Désert, F.-X., 116
De Troia, G., 3, 39, 168
Devlin, M., 44
De Zotti, G., 245
Di Stefano, G., 59
Doumayrou, E., 270
Dragovan, M., 65
Duncan, W., 290
Dupac, X., 360

E

Edgington, S., 107
Encrenaz, P., 202
Encrenaz, T., 330
Esch, W., 129

F

Fabbri, R., 140, 145
Farese, P., 3, 39, 183
Ferreira, P. G., 12

Ferretti, R., 229
Fixsen, D. J., 23

G

Ganga, K., 3, 39
Gastaud, R., 270
Gaztañaga, E., 322, 354
Gemünd, H.-P., 262
Gervasi, M., 123, 140, 145, 164
Giacometti, M., 3, 18, 39
Giard, M., 116, 213
Gildemeister, J. M., 259
Gleeson, E., 282, 295
Glenn, J., 107
Goldin, A., 107
Golwala, S., 107
Gromov, V. D., 205
Gundersen, J., 183

H

Haig, D., 107
Halverson, N. W., 65
Hanany, S., 12
Haynes, V., 126
Henry, D., 290
Hivon, E., 3, 18, 39
Holland, W. S., 357
Holler, C., 97
Holzapfel, W. L., 65
Hristov, V. V., 3, 12, 18, 39, 107
Hughes, D. H., 322, 354, 357

I

Iacoangeli, A., 3, 39

J

Jaffe, A. H., 3, 12, 39
Jellema, W., 290
Jetzer, P., 351
Johnson, B., 12
Jones, M. E., 79
Jones, W. C., 3, 39

K

Kardashev, N. S., 205
Keating, B. G., 175, 183

Kelly, B. D., 357
Klawikowski, S., 183
Knowles, B., 107
Knox, L., 23, 183
Kovac, J., 65
Kreysa, E., 129, 187, 262, 303
Kristensen, R., 23
Kuzmin, L. S., 205

L

Lacquaniti, V., 123
Lamagna, L., 92
Lamarre, J.-M., 116, 213
Lange, A., 107, 213
Lange, A. E., 3, 18, 39
Lanting, T., 259
Lee, A. T., 12, 259
Leitch, E. M., 65
Leonov, V. N., 239
Le Pennec, J., 270
Levy, A., 183
Loc, T., 298
Lubin, P., 183
Luzzi, G., 92

M

Macculi, C., 140, 145
Maffei, B., 213, 282
Maggi, S., 123
Maino, D., 193, 229
Malaspina, M., 242
Mandolesi, N., 193, 219, 224, 229, 242, 245
Marchegiani, P., 316, 363
Martinis, L., 3, 39
Marty, P., 116
Masi, S., 3, 18, 39, 52, 59, 168
Maslov, I. A., 273
Mason, P., 3, 39
Matsumu, T., 12
Mauskopf, A. E., 18
Mauskopf, P. D., 3, 12, 39, 107, 126
Melchiorri, A., 3, 39
Melchiorri, F., 92, 157, 370
Mennella, A., 193, 219, 224, 229, 242, 245
Meyer, S. S., 23
Monari, J., 140, 145
Montroy, T., 3, 39
Morelli, E., 140, 145
Moreno, R., 330
Morgante, G., 193, 219, 224, 229, 242, 245, 298
Morigi, G., 219

Murphy, A., 213
Murphy, J. A., 282, 290, 295
Myers, M. J., 259

N

Nash, A., 298
Natale, E., 123
Natale, V., 140, 145, 210
Nati, F., 52
Natoli, P., 3, 168
Netterfield, C. B., 3, 18, 39
Netterfield, P. D., 12
Nguyen, H., 107
Nicastro, L., 140, 145
Nogaard-Nelsen, H. U., 23

O

O'Dell, C. W., 175, 183
Olivieri, A., 150
Orlando, A., 92
O'Sullivan, C., 282, 290, 295

P

Padin, S., 135
Palladino, E., 316, 363
Pascale, E., 3, 12, 18, 39, 56, 140, 145
Passerini, A., 123, 164
Pearson, D., 298
Pearson, T. J., 135
Peverini, O. A., 140, 145, 150
Piacentini, F., 3, 18, 39, 59, 168
Piat, M., 213
Piccirillo, L., 126, 175, 183
Piffaretti, R., 351
Pisani, U., 123
Pisano, G., 168
Pogosyan, D., 3, 39
Pointecouteau, E., 116
Polenta, G., 3, 39
Pongetti, F., 3, 39, 59
Poppi, S., 140, 145
Prina, M., 229, 298
Prunet, S., 3, 18, 39
Pryke, C., 65
Puget, J. L., 213
Puy, D., 338, 346, 351

R

Rabii, B., 12
Raccanelli, A., 262, 303
Rao, S., 59
Readhead, A. C. S., 135
Reichertz, L. A., 129, 187, 262, 303
Rephaeli, Y., 309
Reveret, V., 270
Richards, P. L., 12, 259
Rodriguez, L., 270
Romeo, G., 3, 39, 59
Rossinot, P., 126
Rownd, B., 107
Ruhl, J. E., 3, 18, 39, 183

S

Sandri, M., 219, 224, 242, 245
Savini, G., 92, 157
Sbarra, C., 140, 145
Scaramuzzi, F., 3, 39
Schemlzel, M., 298
Schindler, S., 351
Schwan, D., 259
Shepherd, M. C., 135
Signore, M., 338, 346
Silverberg, R. F., 23
Sirbi, A., 298
Siringo, G., 187, 262
Sironi, G., 123, 140, 145, 164
Skidmore, J. T., 259
Smoot, G. F., 12
Soglasnova, V. A., 116, 273
Spergel, M. S., 367
Spieler, H. G., 259
Stark, A. A., 83
Stebor, N. C., 175
Steni, R., 123
Stompor, R., 12
Subrahmanyan, R., 102
Sudiwala, R., 213
Sugimura, R., 298

T

Tascone, R., 140, 145
Taylor, A. C., 72
Taylor, T. J., 135
Tegmark, M., 175
Terenzi, L., 219, 224, 242, 245
Tham, C. Y., 274
Thorpe, J. R., 123
Timbie, P. T., 23, 126, 175, 183

Tofani, G., 123
Torre, J. P., 213
Trinchero, D., 150
Troitsky, V. F., 239
Tucci, M., 140, 145

V

Valenziano, L., 193, 219, 224, 229, 242, 245
van de Stadt, H., 290
Ventura, G., 140, 145
Villa, F., 193, 219, 224, 242, 245
Vittorio, N., 3

W

Wade, L. A., 298
Weferling, B., 129
Wilson, G. W., 23

Winant, C. D., 12
Withington, S., 274, 290
Wu, J. H. P.,, 12

Y

Yassin, G., 274, 290
Yoon, J., 259
Yurchenko, V., 213, 234
Yvon, D., 52

Z

Zannoni, M., 123, 140, 145, 164